19

TESI

THESES

tesi di perfezionamento in Matematica sostenuta il 8 novembre 2013

COMMISSIONE GIUDICATRICE
Luigi Ambrosio, Presidente
Dorin Bucur
Giuseppe Buttazzo
Gianni Dal Maso
Lorenzo Mazzieri
Andrea Carlo Giuseppe Mennucci
Edouard Oudet
Michel Pierre

Bozhidar Velichkov
Laboratoire Jean Kuntzmann (LJK)
Universite Joseph Fourier
Tour IRMA, BP 53
51 rue des Mathematiques
38041 Grenoble Cedex 9 France

Existence and Regularity Results for Some Shape Optimization Problems

Bozhidar Velichkov

Existence and Regularity Results for Some Shape Optimization Problems

EDIZIONI
DELLA
NORMALE

ISBN 978-88-7642-526-4 ISBN 978-88-7642-527-1 (eBook)
DOI 10.1007/978-88-7642-527-1

Contents

Preface

The shape optimization problems naturally appear in engineering and biology. They aim to answer questions as:

- What a perfect wing may look like?
- How to minimize the resistance of a moving object in a gas or a fluid?
- How to build a rod of maximal rigidity?
- What is the behaviour of a system of cells?

The shape optimization appears also in physics, mainly in electrodynamics and in the systems presenting both classical and quantum mechanics behaviour. For explicit examples and further account on the applications of the shape optimization we refer to the books [21] and [72].

Here we deal with the theoretical mathematical aspects of the shape optimization, concerning existence of optimal sets and their regularity. In all the practical situations above, the shape of the object in study is determined by a functional depending on the solution of a given partial differential equation (shortly, PDE). We will sometimes refer to this function as a *state function*. The simplest state functions are provided by solutions of the equations

$$-\Delta w = 1 \qquad \text{and} \qquad -\Delta u = \lambda u,$$

which usually represent the torsional rigidity and the oscillation modes of a given object. Thus our study will be concentrated mainly on the situations, in which these state functions appear, *i.e.* when the optimality is intended with respect to *energy* and *spectral* functionals.

In Chapter 1 we provide some simple examples of shape optimization problems together with some elementary techniques, which can be used to obtain existence results in some cases and motivate the introduction of the *quasi-open* sets as natural objects of the shape optimization. We also discuss some of the usual assumptions on the functionals, with respect to which the optimization is performed. In conclusion, we give some justification for the expected regularity of the state functions on the optimal sets.

In Chapter 2 we deal with the case when the family of shapes consists of the subsets of a given ambient space, satisfying some compactness assumptions. A typical example of such a space is a bounded open set in the Euclidean space $\mathbb{R}^d$ or, following the original terminology of Buttazzo and Dal Maso, a *box*. The first general result in this setting was obtained by Buttazzo and Dal Maso in [33] and the proof was based on relaxation results by Dal Maso and Mosco (see [52] and [53]). The complete proof was considerably simplified in [21] (see also [30] for a brief introduction to this technique), where only some simple analytic tools were used. This Chapter is based on the results from [37], where we followed the main steps from [21], using only variational arguments. This approach allowed us to reproduce the general result from [33] in non-linear and non-smooth settings as metric measure spaces, Finsler manifolds and Gaussian spaces. Some of the proofs in this chapter are considerably simplified with respect to the original paper [37] and some new results were added.

Chapter 3 is dedicated to the study of the *capacitary measures*, *i.e.* the measures with respect to which the Sobolev functions can be integrated. The aim of this chapter is to gather some results and techniques, basic for the theory of shape optimization and general enough to be used in the optimization of domains, potentials and measures. Our approach is based on the study of the *energy state functions* instead of functionals associated to capacitary measures. The main ideas and results in this chapter are based on the work of Bucur [19], Bucur-Buttazzo [22] and Dal Maso-Garroni [51]. The exact framework, in which the modern shape optimization techniques can be applied, is provided by the following space of capacitary measures

$$\mathcal{M}_{\text{cap}}^{T}(\mathbb{R}^d) = \Big\{\mu \text{ capacitary measure}: \ w_\mu \in L^1(\mathbb{R}^d)\Big\},$$ [1]

and was originally suggested by Dorin Bucur.

Chapter 4 is dedicated to the study of shape subsolutions, *i.e.* the sets which are optimal for a given functional, with respect to internal perturbations. The notion of shape subsolution was introduced by Bucur in [20] and had a basic role in the proof of the existence of optimal set for general spectral functionals. A particular attention was given a special

[1] w_μ indicates the energy state function associated to the measure μ.

class of domains known as *energy subsolution*, for which the cost functional depends on the torsion energy and the Lebesgue measure of the domain. In [20] it was shown that the energy subsolutions are necessarily bounded sets of finite perimeter and the proof was based on a technique introduced by Alt and Caffarelli in [1]. Similar results were obtained in the [59] and [26]. In [29], we investigated this notion obtaining a *density estimate*, which we used to prove a regularity result for the optimal set for the second eigenvalue λ_2 in a box, and a *three-phase monotonicity* formula of Cafarelli-Jerison-Kënig type, which allowed us to exclude the presence of triple points in some optimal partition problems.

In Chapter 5, we consider domains which are *shape supersolutions*, *i.e.* optimal sets with respect to external perturbations. This chapter contains the main regularity results concerning the state functions of the optimal sets. Our analysis is based on a regularity theorem for the quasi-minimizers of the Dirichlet integral, which is based on the technique developed by Briançon, Hayouni and Pierre (see [17] and also [77]) for the Lipschitz continuity of the state functions on the optimal sets for energy functionals. This result was then successfully applied, in an appropriate form, in the case of spectral functionals, to obtain the Lipschitz regularity of the corresponding eigenfunctions (see [28]).

The last section contains some of the main results from [59] and [58]. We investigate the supersolutions of functionals involving the perimeter, proving some general properties of these sets and also the Lipschitz continuity of their energy functions. This last result is the key step in the proof of the $C^{1,\alpha}$ regularity of the boundary of the optimal sets for spectral functionals with perimeter constraint, which is proved at the end of the chapter.

In Chapter 6 we consider various shape optimization problems involving spectral functionals. We present the recent results from [20]-[81] [25, 59] and [34]-[26], introducing the existence and regularity techniques involving the results from the previous chapters and simplifying some of the original proofs.

The last Chapter 7 is dedicated to the study of optimizations problems concerning one dimensional sets (graphs) in $\mathbb{R}^d$. The framework in this chapter significantly differs from the theory in the rest of the work. This is due to the fact that there is a lack of ambient functional space which hosts the functional spaces on the various *shapes*. With this Chapter we aim to keep the discussion open towards other problems which present similar difficulties as, for example, the optimization of the spectrum of the Neumann Laplacian.

The present monograph is a revised version, submitted for publication to "Edizioni della Normale", of the PhD thesis with the same title discussed on 8 November 2013 at Scuola Normale Superiore di Pisa and prepared under the joint supervision of Prof. Giuseppe Buttazzo and Prof. Dorin Bucur.

Pisa, 22 April 2014

Bozhidar Velichkov

Résumé of the main results

In this section we give a brief account on the main results in the present monograph.

The main result from Chapter 2 is the following existence Theorem, which is the non-linear variant of the classical Buttazzo-Dal Maso Theorem and was proved in [37]. Below, we state it in the framework of Cheeger's Sobolev spaces on metric measure spaces, but the main result is even more general and is discussed in Section 2.4.

Theorem 1 (Non-linear Buttazzo-Dal Maso theorem). *Consider a separable metric space (X, d) and a finite Borel measure m on X. Let $H^1(X, m)$ denote the Sobolev space on (X, d, m) and let $Du = g_u$ be the minimal generalized upper gradient of $u \in H^1(X, m)$. Under the assumption that the inclusion $H^1(X, m) \hookrightarrow L^2(X, m)$ is compact, we have that the problem*

$$\min\Big\{\mathcal{F}(\Omega)\,:\,\Omega\subset X,\ \Omega\ \mathit{Borel},\ |\Omega|\le c\Big\},$$

has solution, for every constant $c > 0$ and every functional $\mathcal{F}$ increasing and lower semi-continuous with respect to the strong-γ-convergence.[1]

This result was proved in [37] and naturally applies in many different frameworks as Finsler manifolds, Gaussian spaces of infinite dimension and Carnot-Caratheodory spaces.

In Chapter 3, we use some classical techniques to review the theory of the capacitary measures in $\mathbb{R}^d$ providing the reader with a self-contained exposition of the topic. One of our main contributions in this

[1] A typical example of such functionals is given by the eigenvalues of the Dirichlet Laplacian, variationally defined as

$$\lambda_k(\Omega) = \min_{K\subset H^1_0(\Omega)}\ \max_{u\in K}\frac{\int_X |Du|^2\,dm}{\int_X u^2\,dm},$$

where the minimum is over all k-dimensional subspaces K.

chapter is the generalization for capacitary measures of the concentration-compactness principle for quasi-open sets, a result from the paper of preparation [26].

Theorem 2. (Concentration-compactness principle for capacitary measures). *Suppose that μ_n is a sequence of capacitary measures in $\mathbb{R}^d$ such that the corresponding sequence of energy functions w_{μ_n} has uniformly bounded $L^1(\mathbb{R}^d)$ norms. Then, up to a subsequence, one of the following situations occur:*

(i1) (Compactness) *The sequence μ_n γ-converges to some $\mu \in \mathcal{M}^T_{\mathrm{cap}}(\mathbb{R}^d)$.*
(i2) (Compactness2) *There is a sequence $x_n \in \mathbb{R}^d$ such that $|x_n| \to \infty$ and $\mu_n(x_n + \cdot)$ γ-converges.*
(ii) (Vanishing) *The sequence μ_n does not γ-converge to the measure $\infty = I_\emptyset$, but the sequence of resolvents R_{μ_n} converges to zero in the strong operator topology of $\mathcal{L}(L^2(\mathbb{R}^d))$. Moreover, we have $\|w_{\mu_n}\|_{L^\infty} \to 0$ and $\lambda_1(\mu_n) \to +\infty$, as $n \to \infty$.*
(iii) (Dichotomy) *There are capacitary measures μ_n^1 and μ_n^2 such that:*

- $\mathrm{dist}\big(\{\mu_n^1 < \infty\}, \{\mu_n^2 < \infty\}\big) \to \infty$, *as* $n \to \infty$;
- $\mu_n \leq \mu_n^1 \wedge \mu_n^2$, *for every* $n \in \mathbb{N}$;
- $d_\gamma(\mu_n, \mu_n^1 \wedge \mu_n^2) \to 0$, *as* $n \to \infty$;
- $\|R_{\mu_n} - R_{\mu_n^1 \wedge \mu_n^2}\|_{\mathcal{L}(L^2)} \to 0$, *as* $n \to \infty$.

The results from Chapter 4, concerning the energy subsolutions, are from the recent paper [29]. Our main technical results, which are essential in the study of the qualitative properties of families of disjoint subsolutions (which naturally appear in the study of multiphase shape optimization problems) are a *density estimate* and a *three-phase monotonicity theorem* in the spirit of the two-phase formula by Caffarelli, Jerison and Kënig.

The following Theorem combines the results from Proposition 4.21 and Proposition 4.40, which were proved in [29].

Theorem 3 (Isolating an energy subsolution). *Suppose that $\Omega \subset \mathbb{R}^d$ is an energy subsolution. Then there exists a constant $c > 0$, depending only on the dimension, such that for every $x_0 \in \overline{\Omega}^M$, we have*

$$\limsup_{r \to 0} \frac{|\{w_\Omega > 0\} \cap B_r(x_0)|}{|B_r|} \geq c. \tag{1}$$

As a consequence, if the quasi-open sets Ω_1 and Ω_2 are two disjoint energy subsolutions, then there are open sets $D_1, D_2 \subset \mathbb{R}^d$ such that $\Omega_1 \subset D_1$, $\Omega_2 \subset D_2$ and $\Omega_1 \cap D_2 = \Omega_2 \cap D_1 = \emptyset$, up to sets of zero capacity.

As a consequence, we have the following (see Proposition 6.16):

Theorem 4 (Openness of the optimal set for λ_2). *Let $\mathcal{D} \subset \mathbb{R}^d$ be a bounded open set and Ω a solution of the problem*

$$\min\Big\{\lambda_2(\Omega) + m|\Omega| \,:\, \Omega \subset \mathcal{D},\ \Omega \text{ quasi-open}\Big\}.$$

Then there is an open set $\omega \subset \Omega$, which is a solution of the same problem.

A fundamental tool in the analysis of the optimal partitions is the following three-phase monotonicity lemma, which we proved in [29].

Theorem 5 (Three-phase monotonicity formula). *Let $u_i \in H^1(B_1), i = 1, 2, 3$, be three non-negative Sobolev functions such that $\Delta u_i \geq -1$, for each $i = 1, 2, 3$, and $\int_{\mathbb{R}^d} u_i u_j \, dx = 0$, for each $i \neq j$. Then there are dimensional constants $\varepsilon > 0$ and $C_d > 0$ such that, for every $r \in (0, 1)$, we have*

$$\prod_{i=1}^{3}\left(\frac{1}{r^{2+\varepsilon}}\int_{B_r}\frac{|\nabla u_i|^2}{|x|^{d-2}}\,dx\right) \leq C_d\left(1 + \sum_{i=1}^{3}\int_{B_1}\frac{|\nabla u_i|^2}{|x|^{d-2}}\,dx\right)^3.$$

We note that we do not assume that the functions u_i are continuous! This assumption was part of the two-phase monotonicity formula, proved in the original paper of Caffarelli, Jerison and Kenig, where can be dropped, as well.

In Chapter 5 we discuss a technique, developed in [28], for proving the regularity of the eigenfunctions associated to the optimal set for the k-th eigenvalue of the Dirichlet Laplacian. Our main result is the following theorem from [28].

Theorem 6 (Lipschitz continuity of the optimal eigenfunctions). *Let Ω be a solution of the problem*

$$\min\Big\{\lambda_k(\Omega) \,:\, \Omega \subset \mathbb{R}^d,\ \Omega \text{ quasi-open},\ |\Omega| = 1\Big\}.$$

Then there is an eigenfunction $u_k \in H^1_0(\Omega)$, corresponding to the eigenvalue $\lambda_k(\Omega)$, which is Lipschitz continuous on $\mathbb{R}^d$.

In the last section of Chapter 5 we study the properties of the measurable sets $\Omega \subset \mathbb{R}^d$ satisfying

$$P(\Omega) \leq P(\widetilde{\Omega}), \text{ for every measurable set } \widetilde{\Omega} \supset \Omega.$$

The results in this section are contained in [59], where we used them to prove the following Theorem, which can now be found in Chapter 6.

Theorem 7. (Existence and regularity for λ_k with perimeter constraint). *The shape optimization problem*

$$\min\Big\{\lambda_k(\Omega):\ \Omega\subset\mathbb{R}^d,\ \Omega\ open,\ P(\Omega)=1,\ |\Omega|<\infty\Big\},$$

has a solution. Moreover, any optimal set Ω is bounded, connected and its boundary $\partial\Omega$ is $C^{1,\alpha}$, for every $\alpha\in(0,1)$, outside a closed set of Hausdorff dimension at most $d-8$.

In Chapter 6 we prove **existence results for the following spectral optimization problems**, for every $k\in\mathbb{N}$.

1. Spectral optimization problems with internal constraint (see [25])

$$\min\Big\{\lambda_k(\Omega):\ \mathcal{D}^i\subset\Omega\subset\mathbb{R}^d,\ \Omega\text{ quasi-open},\ |\Omega|=1,\ |\Omega|<\infty\Big\};$$

2. Spectral optimization problems with perimeter constraint (see [59])

$$\min\Big\{\lambda_k(\Omega):\ \Omega\subset\mathbb{R}^d,\ \Omega\text{ open},\ P(\Omega)=1,\ |\Omega|<\infty\Big\};$$

3. Optimization problems for Schrödinger operators (for $k=1,2$ the result was proved in [34], while for generic $k\in\mathbb{N}$ the existence is proved in [26])

$$\min\Big\{\lambda_k(-\Delta+V):V:\mathbb{R}^d\to[0,+\infty]\text{ measurable},\int_{\mathbb{R}^d}V^{-1/2}\,dx=1\Big\};$$

4. Optimization problems for capacitary measures with torsion-energy constraint (see [26])

$$\min\Big\{\lambda_k(\mu):\ \mu\text{ capacitary measure in }\mathbb{R}^d,\ E(\mu)=-1\Big\},$$

where

$$E(\mu)=\min\Big\{\frac{1}{2}\int_{\mathbb{R}^d}|\nabla u|^2\,dx+\frac{1}{2}\int_{\mathbb{R}^d}u^2\,d\mu$$
$$-\int_{\mathbb{R}^d}u\,dx:\ u\in L^1(\mathbb{R}^d)\cap H^1(\mathbb{R}^d)\cap L^2(\mu)\Big\}.$$

In the last Chapter 7 we consider a spectral optimization problem, which was studied in [35]. More precisely we prove that the following problem

$$\min\Big\{\mathcal{E}(C):\ C\subset\mathbb{R}^d\text{ closed connected set},\ \mathcal{D}\subset C,\ \mathcal{H}^1(C)\le 1\Big\},$$

where $\mathcal{E}$ is the Dirichlet Energy of the one dimensional set C and $\mathcal{D}\subset\mathbb{R}^d$ is a finite set of points, has solution for some configurations of Dirichlet points $\mathcal{D}$ and might not admit a solution in some special cases (for example, when all the points in $\mathcal{D}$ are aligned).

Chapter 1
Introduction and Examples

1.1. Shape optimization problems

A *shape optimization problem* is a variational problem, in which the family of competitors consists of *shapes*, *i.e.* geometric objects that can be chosen to be metric spaces, manifolds or just domains in the Euclidean space. The shape optimization problems are usually written in the form

$$\min\Big\{\mathcal{F}(\Omega)\ :\ \Omega\in\mathcal{A}\Big\}, \tag{1.1}$$

where

- $\mathcal{F}$ is a *cost functional*,
- $\mathcal{A}$ is an *admissible family* (*set*, *class*) of shapes.

If there is a set $\Omega\in\mathcal{A}$ which realizes the minimum in (1.1), we call it an *optimal shape*, *optimal set* or simply a solution of (1.1). The theory of shape optimization concerns, in particular, the existence of optimal domains and their properties. These questions are of particular interest in the physics and engineering, where the cost functional $\mathcal{F}$ represents some energy we would like to minimize and the admissible class is the variety of shapes we are able to produce. We refer to the books [21, 71] and [72] for an extensive introduction to the shape optimization problems and their applications.

We are mainly interested in the class of shape optimization problems, where the admissible family of shapes consists of subsets of a given ambient space $\mathcal{D}$. In this case we will sometimes call the variables $\Omega\in\mathcal{A}$ *domains* instead of shapes. The set $\mathcal{D}$ is called *design region* and can be chosen to be a subset of $\mathbb{R}^d$, a differentiable manifold or a metric space. A typical example of an admissible class is the following:

$$\mathcal{A}=\Big\{\Omega\ :\ \Omega\subset\mathcal{D},\ \Omega\text{ open},\ |\Omega|\le c\Big\},$$

where $\mathcal{D}$ is a bounded open set in $\mathbb{R}^d$, $|\cdot|$ is the Lebesgue measure and c is a positive real number.

The cost functionals $\mathcal{F}$ we consider are defined on the admissible class of domains $\mathcal{A}$ through the solutions of some partial differential equation on each $\Omega \in \mathcal{A}$. The typical examples of cost functionals are:

- Energy functionals

$$\mathcal{F}(\Omega) = \int_\Omega g\big(x, u(x), \nabla u(x)\big)\, dx,$$

  where g is a given function and $u \in H_0^1(\Omega)$ is the weak solution of the equation

$$-\Delta u = f \quad \text{in} \quad \Omega\,, \qquad u \in H_0^1(\Omega),$$

  where f is a fixed function in $L^2(\mathcal{D})$ and $H_0^1(\Omega)$ is the Sobolev space of square integrable functions with square integrable distributional gradient on Ω.
- Spectral functionals

$$\mathcal{F}(\Omega) = F\big(\lambda_1(\Omega), \dots, \lambda_k(\Omega)\big),$$

  where $F : \mathbb{R}^k \to \mathbb{R}$ is a given function and $\lambda_k(\Omega)$ is the kth eigenvalue of the Dirichlet Laplacian on Ω, *i.e.* the kth smallest number such that the equation

$$-\Delta u_k = \lambda_k(\Omega) u_k \quad \text{in} \quad \Omega\,, \qquad u_k \in H_0^1(\Omega),$$

  has a non-trivial solution.

1.2. Why quasi-open sets?

In this section, we consider the shape optimization problem

$$\min\Big\{E(\Omega)\,:\;\Omega \subset \mathcal{D},\;\Omega \text{ open},\;|\Omega| = 1\Big\}, \tag{1.2}$$

where $\mathcal{D} \subset \mathbb{R}^d$ is a bounded open set (a box) of Lebesgue measure $|\mathcal{D}| \geq 1$ and $E(\Omega)$ is the *Dirichlet Energy* of Ω, *i.e.*

$$E(\Omega) = \min\left\{\frac{1}{2}\int_\Omega |\nabla u|^2\, dx - \int_\Omega u\, dx\,:\; u \in H_0^1(\Omega)\right\}. \tag{1.3}$$

In the terms of the previous section, we consider the shape optimization problem (1.1) with admissible set

$$\mathcal{A} = \Big\{\Omega : \ \Omega \subset \mathcal{D}, \ \Omega \text{ open}, \ |\Omega| = 1\Big\},$$

and cost functional

$$E(\Omega) = -\frac{1}{2}\int_\Omega w_\Omega \, dx, \tag{1.4}$$

where w_Ω is the weak solution of the equation

$$-\Delta w_\Omega = 1 \quad \text{in} \quad \Omega\,, \qquad w_\Omega \in H_0^1(\Omega). \tag{1.5}$$

Indeed, w_Ω is the unique minimizer in $H_0^1(\Omega)$ of the functional

$$J(u) = \frac{1}{2}\int_\Omega |\nabla u|^2 \, dx - \int_\Omega u \, dx,$$

and so

$$E(\Omega) = \frac{1}{2}\int_\Omega |\nabla w_\Omega|^2 \, dx - \int_\Omega w_\Omega \, dx. \tag{1.6}$$

On the other hand, using w_Ω as a test function in (1.5), we have that

$$\int_\Omega |\nabla w_\Omega|^2 \, dx = \int_\Omega w_\Omega \, dx, \tag{1.7}$$

which, together with (1.6), gives (1.4).

Remark 1.1. The functional $T(\Omega) = -E(\Omega)$ is called *torsion energy* or just *torsion*. We will call the function w_Ω *energy function* or sometimes *torsion function*.

Before we proceed, we recall some well-known properties of the energy functions.

- (Weak maximum principle) If $U \subset \Omega$ are open sets, then $0 \le w_U \le w_\Omega$. In particular, the Dirichlet Energy is decreasing with respect to inclusion
$$E(\Omega) \le E(U) \le 0.$$
- (Strong maximum principle) $w_\Omega > 0$ on Ω. Indeed, for any ball $B = B_r(x_0) \subset \Omega$, by the weak maximum principle, we have $w_\Omega \ge w_B$. On the other hand, w_B can be written explicitly as
$$w_B(x) = \frac{r^2 - |x - x_0|^2}{2d},$$
which is strictly positive on $B_r(x_0)$.

- (A priori estimate) The energy function w_Ω is bounded in $H_0^1(\Omega)$ by the constant depending only on the Lebesgue measure of Ω. Indeed, by (1.7) and the Hölder inequality, we have

$$\begin{aligned}\|\nabla w_\Omega\|_{L^2}^2 \le \|w_\Omega\|_{L^1} \le |\Omega|^{\frac{d+2}{2d}} \|w_\Omega\|_{L^{\frac{2d}{d-2}}} \\ \le C_d |\Omega|^{\frac{d+2}{2d}} \|\nabla w_\Omega\|_{L^2},\end{aligned} \tag{1.8}$$

where C_d is the constant in the Gagliardo-Nirenberg-Sobolev inequality in $\mathbb{R}^d$.

We now try to solve the shape optimization problem (1.12) by a direct method. Indeed, let Ω_n be a minimizing sequence for (1.12) and let, for simplicity, $w_n := w_{\Omega_n}$. By the estimate (1.8), we have

$$\|\nabla w_n\| \le C_d, \quad \forall n \in \mathbb{N}.$$

By the boundedness of $\mathcal{D}$, the inclusion $H_0^1(\mathcal{D}) \subset L^2(\mathcal{D})$ is compact and so, up to a subsequence, we may suppose that w_n converges to $w \in H_0^1(\mathcal{D})$ strongly in $L^2(\mathcal{D})$. Suppose that $\Omega = \{w > 0\}$ is an open set. Then, we have

- semicontinuity of the Dirichlet Energy

$$E(\Omega) \le \liminf_{n\to\infty} E(\Omega_n). \tag{1.9}$$

Indeed, since $w \in H_0^1(\Omega)$, we have that

$$\begin{aligned} E(\Omega) &\le \frac{1}{2}\int_\Omega |\nabla w|^2\,dx - \int_\Omega w\,dx \\ &\le \liminf_{n\to\infty}\left\{\frac{1}{2}\int_\Omega |\nabla w_n|^2\,dx - \int_\Omega w_n\,dx\right\} \\ &= \liminf_{n\to\infty} E(\Omega_n),\end{aligned}$$

- semicontinuity of the Lebesgue measure

$$|\Omega| \le \liminf_{n\to\infty} |\Omega_n|. \tag{1.10}$$

This follows by the Fatou Lemma and the fact that

$$\mathbb{1}_\Omega \le \liminf_{n\to\infty} \mathbb{1}_{\Omega_n}, \tag{1.11}$$

where $\mathbb{1}_\Omega$ is the characteristics function of Ω. Indeed, by the strong maximum principle, we have that

$$\Omega_n = \{w_n > 0\}.$$

On the other hand, we may suppose, again up to extracting a subsequence, that w_n converges to w almost everywhere. Thus, if $x \in \Omega$, then $w(x) > 0$ and so $w_n(x) > 0$ definitively, *i.e.* $x \in \Omega_n$ definitively, which proves (1.11).

Let $\widetilde{\Omega} \subset \mathcal{D}$ be an open set of unit measure, containing Ω. Then, we have that $\widetilde{\Omega} \in \mathcal{A}$ and, by the monotonicity of E and (1.9),

$$E(\widetilde{\Omega}) \le E(\Omega) \le \liminf_{n\to\infty} E(\Omega_n),$$

i.e. $\widetilde{\Omega}$ is an optimal domain for (1.12). In conclusion, we obtained that, under the assumption that $\{w > 0\}$ is an open set, the shape optimization problem (1.12) has a solution. Unfortunately, at the moment, since w is just a Sobolev function, there is no reason to believe that $\{w > 0\}$ is open. In fact the proof of this fact would require some regularity arguments which can be quite involved even in the simple case when the cost functional is the Dirichlet Energy E. Similar arguments applied to more general energy and spectral functionals can be complicated enough (if even possible) to discourage any attempt of providing a general theory of shape optimization.

An alternative approach is relaxing the problem to a wider class of admissible sets. The above considerations suggest that the class of *quasi-open* sets, *i.e.* the level sets of Sobolev functions, is a good candidate for a family, where optimal domains may exist. Indeed, it was first proved in [33] that the shape optimization problem

$$\min\Big\{E(\Omega)\,:\,\Omega \subset \mathcal{D},\ \Omega \text{ quasi-open},\ |\Omega| = 1\Big\}, \tag{1.12}$$

has a solution. After defining appropriately the Sobolev spaces and the PDEs on domains which are not open sets, we will see that the same proof works even in the general framework of a metric measure spaces and for a large class of cost functionals decreasing with respect to the set inclusion. For example, one may prove that there is a solution of the problem

$$\min\Big\{\lambda_k(\Omega)\,:\,\Omega \subset \mathcal{D},\ \Omega \text{ quasi-open},\ |\Omega| = 1\Big\}, \tag{1.13}$$

where $\lambda_k(\Omega)$ is variationally characterized as

$$\lambda_k(\Omega) = \min_{K \subset H_0^1(\Omega)} \max_{u \in K, u \neq 0} \frac{\int |\nabla u|^2 \, dx}{\int u^2 \, dx},$$

where the minimum is taken over all k-dimensional subspaces K of $H_0^1(\Omega)$. Indeed, if Ω_n is a minimizing sequence, then we consider the vectors $(u_1^n, \dots, u_k^n) \in \left(H_0^1(\Omega_n)\right)^k$ of eigenfunctions, orthonormal in L^2. We may suppose that for each $j = 1, \dots, k$ there is a function $u_j \in H_0^1(\mathcal{D})$ such that $u_j^n \to u^j$ in L^2. Arguing as in the case of the Dirichlet Energy, it is not hard to prove that the (quasi-open) set

$$\Omega = \bigcup_{j=1}^{k} \{u_j \neq 0\},$$

is a solution of (1.13).

1.3. Compactness and monotonicity assumptions in the shape optimization

In the previous section we sketched the proofs of the existence of an optimal domain for the problems (1.12) and (1.13). The essential ingredients for these existence results were the following assumptions:

- The compactness of the inclusion $H_0^1(\mathcal{D}) \subset L^2(\mathcal{D})$ in the design region $\mathcal{D}$;
- The monotonicity of the cost functional $\mathcal{F}$.

In Chapter 2 we prove a general existence result under the above assumptions, even in the case when $\mathcal{D}$ is just a metric space endowed with a finite measure. Nevertheless, non-trivial shape optimization problems can be stated without imposing these conditions. For example, by a standard symmetrization argument, the problems

$$\min \left\{E(\Omega) : \ \Omega \subset \mathbb{R}^d, \ \Omega \text{ quasi-open}, \ |\Omega| = 1\right\}, \tag{1.14}$$

$$\min \left\{\lambda_1(\Omega) : \ \Omega \subset \mathbb{R}^d, \ \Omega \text{ quasi-open}, \ |\Omega| = 1\right\}, \tag{1.15}$$

have solution which, in both cases, is a ball of unit measure. It is also easy to construct some artificial examples, in which the functional is not monotone and the domain is not compact, but there is still an optimal set. For instance, one may take

$$\min\left\{\lambda_1(\Omega) + |E(\Omega) - E(B)|^2 : \Omega \subset \mathbb{R}^d, \Omega \text{ quasi-open}, |\Omega| = 1\right\}, \tag{1.16}$$

where B is a ball of measure 1.

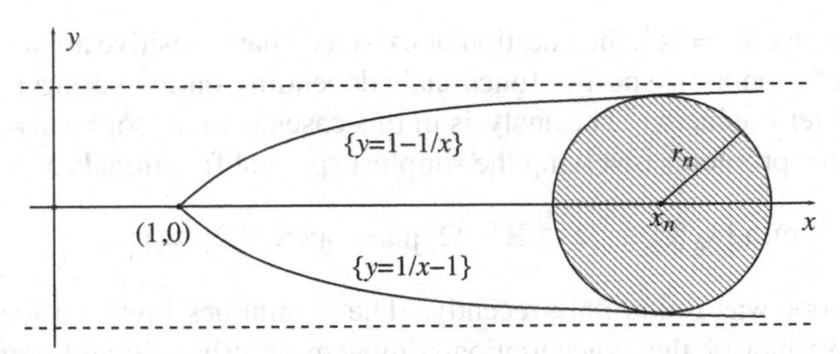

Figure 1.1. The convex and smooth design region from Example 1.2 with a minimizing sequence escaping at infinity.

In this section we investigate in which cases the compactness and monotonicity assumptions can be removed from the theory. In the framework of Euclidean space $\mathbb{R}^d$, the compactness assumption (more or less) corresponds to the assumption that $\mathcal{D} \subset \mathbb{R}^d$ has finite Lebesgue measure (see [22] for the conditions under which the inclusion of the Sobolev Space in $L^2(\mathcal{D})$ is compact). In general the existence does not hold in unbounded design regions $\mathcal{D}$ even for the simplest cost functionals and "nice" domains $\mathcal{D}$ (convex with smooth boundary).

Example 1.2. Let the design region $\mathcal{D} \subset \mathbb{R}^2$ be defined as follows (see Figure 1.1)

$$\mathcal{D} = \left\{(x, y) \in (1, +\infty) \times \mathbb{R} : \frac{1}{x} - 1 < y < 1 - \frac{1}{x}\right\}.$$

Then the shape optimization problem

$$\min\left\{\lambda_1(\Omega) : \Omega \subset \mathcal{D},\ \Omega \text{ quasi-open},\ |\Omega| = \pi\right\}, \tag{1.17}$$

does not have a solution. Since the ball of radius 1 is the minimizer for λ_1 in $\mathbb{R}^d$, we have that

$$\lambda_1(B_1) \le \inf\left\{\lambda_1(\Omega) : \Omega \subset \mathcal{D},\ \Omega \text{ quasi-open},\ |\Omega| \le \pi\right\}.$$

Moreover, the above inequality is, in fact, an equality since, by the rescaling property of λ_1 ($\lambda_1(t\Omega) = t^{-2}\lambda_1(\Omega)$), we have that

$$\lambda_1\big(B_{r_n}(x_n)\big) = r_n^2\lambda_1(B_1) \to \lambda_1(B_1), \quad \text{as} \quad n \to \infty,$$

where $B_{r_n}(x_n) \subset \mathcal{D}$ is a sequence of balls such that $r_n \to 1$ and $x_n \to \infty$, as $n \to \infty$. On the other hand, the ball of radius 1 is the unique minimizer for λ_1 in $\mathbb{R}^d$ and there is no ball of radius 1 contained in $\mathcal{D}$.

In the case $\mathcal{D} = \mathbb{R}^d$, the question of existence have positive answer in the case of monotone spectral functionals depending on the spectrum of the Dirichlet Laplacian. The analysis in this cases is more sophisticated and even for problems involving the simplest spectral functionals as

$$\min\left\{\lambda_k(\Omega):\ \Omega\subset\mathbb{R}^d,\ \Omega \text{ quasi-open},\ |\Omega|=1\right\}, \tag{1.18}$$

the proof was found only recently. The techniques involved are based on a variant of the concentration-compactness principle and arguments for the boundedness of the optimal set and can be applied essentially for functionals defined through the solutions of elliptic equations involving the Dirichlet Laplacian. In fact, for general monotone cost functionals, the existence in $\mathbb{R}^d$ does not hold.

Example 1.3. Let $a : \mathbb{R}^d \to (1, 2]$ be a smooth function such that $a(0) = 2$ and $a(x) \to 1$ as $x \to \infty$. Then, the shape optimization problem

$$\min\left\{\mathcal{F}(\Omega):\ \Omega\subset\mathbb{R}^d,\ \Omega \text{ quasi-open},\ |\Omega|=1\right\}, \tag{1.19}$$

does NOT have a solution, where the cost functional $\mathcal{F}$ is defined as

$$\mathcal{F}(\Omega) = -\frac{1}{2}\int_\Omega u\,dx,$$

where $u \in H_0^1(\Omega)$ is the weak solution of

$$-\operatorname{div}(a(x)\nabla u) = 1 \quad \text{in} \quad \Omega\,, \qquad u\in H_0^1(\Omega).$$

Indeed, since $a \geq 1$ and since the ball of unit measure B is the solution of (1.12) in the case $\mathcal{D} = \mathbb{R}^d$, we have

$$E(B) \leq \inf\left\{\mathcal{F}(\Omega):\ \Omega\subset\mathbb{R}^d,\ \Omega \text{ quasi-open},\ |\Omega|=1\right\}. \tag{1.20}$$

On the other hand, taking a sequence of balls of measure 1, which go to infinity, we obtain that there is an equality (1.20). Since, for every quasi-open set Ω of measure 1, we have

$$E(B) \leq E(\Omega) < \mathcal{F}(\Omega),$$

we conclude that the problem (1.19) does not have a solution.

The monotonicity of the cost functional seems to be an assumption even more difficult to drop. As the following example shows, even in the case of a bounded design region, the existence might not occur:

Example 1.4. Let $a_k, k \in \mathbb{N}$ be a sequence of real numbers converging to zero fast enough. For example $a_k = 2^{-2^{2^k}}$. Then the shape optimization problem

$$\min\Big\{\mathcal{F}(\Omega) : \ \Omega \subset \mathbb{R}^d, \ \Omega \text{ quasi-open}, \ |\Omega| = 1\Big\}, \qquad (1.21)$$

does NOT have a solution, where the cost functional F is given by

$$\mathcal{F}(\Omega) := \sum_{k=1}^{+\infty} a_k |\lambda_{k+1}(\Omega) - \lambda_k(\Omega)|.$$

Indeed, taking a minimizing sequence Ω_n such that each Ω_n consists of n different disjoint balls, it is not hard to check that $\mathcal{F}(\Omega_n) \to 0$. On the other hand, no set of positive measure can have spectrum of the Dirichlet Laplacian which consists of only one value.

Remark 1.5. We note that the choice of admissible set was crucial in the above example. In fact, with the convention $\lambda_k(\emptyset) = +\infty, \forall k \in \mathbb{N}$ and $\infty - \infty = 0$, we have that the empty set $\emptyset$ is a solution of

$$\min\Big\{\mathcal{F}(\Omega) : \ \Omega \subset \mathbb{R}^d, \ \Omega \text{ quasi-open}, \ |\Omega| \le 1\Big\}, \qquad (1.22)$$

where the cost functional $\mathcal{F}$ is as in (1.21).

1.4. Lipschitz regularity of the state functions

Once we obtain the existence of an optimal quasi-open set, a natural question concerns the regularity of this set. In particular, we expect that the optimal sets are open and that their boundaries are regular. In order to motivate these expectations we consider the following problem:

$$\min\Big\{\lambda_1(\Omega) + |\Omega| : \ \Omega \text{ open}, \ \Omega \subset \mathcal{D}\Big\}, \qquad (1.23)$$

where $\mathcal{D}$ is a bounded open set with smooth boundary or $\mathcal{D} = \mathbb{R}^d$.

Remark 1.6. One of the fundamental tools for understanding the behaviour of the optimal sets for spectra optimization problems is the *shape derivative* (for an introduction to this topic we refer to [72] and [71]). Let $k \in \mathbb{N}$ and let $\Omega \subset \mathcal{D}$ be an open set with smooth boundary $\partial\Omega \cap \mathcal{D}$. Consider a smooth vector field $V : \mathcal{D} \to \mathbb{R}^d$ with compact support in $\mathcal{D}$ and the following parametrized family of sets

$$\Omega_t := (Id + tV)(\Omega), \qquad t \in \mathbb{R}.$$

If the Dirichlet eigenvalue $\lambda_k(\Omega)$ is simple (*i.e.* of multiplicity one), we can express the shape derivative of λ_k along V as

$$\frac{d}{dt}\Big|_{t=0}\lambda_k(\Omega_t) = -\int_{\partial\Omega}|\nabla u_k|^2(V\cdot n)\,d\mathcal{H}^{d-1}, \tag{1.24}$$

where $u_k \in H_0^1(\Omega)$ is the kth eigenfunction on Ω, normalized in L^2 and $n(x)$ denotes the unit vector, normal to the surface $\partial\Omega$ in $x \in \partial\Omega$. The first variation of the Lebesgue measure $|\Omega_t|$ with respect to the field V is given by

$$\frac{d}{dt}\Big|_{t=0}|\Omega_t| = \int_{\partial\Omega} V\cdot n\,d\mathcal{H}^{d-1}.$$

Suppose now that $\Omega \subset \mathcal{D}$ is an open solution of (1.23) with smooth free boundary $\mathcal{D}\cap\partial\Omega$. Since the domain Ω is optimal, it is also connected and so the eigenvalue $\lambda_1(\Omega)$ is simple. Thus, we can apply the shape derivative from Remark 1.6 obtaining that

$$0 = \frac{d}{dt}\Big|_{t=0}\big(\lambda_1(\Omega_t) + |\Omega_t|\big) = \int_{\partial\Omega}\left(1 - |\nabla u_1|^2\right)V\cdot n\,d\mathcal{H}^{d-1},$$

for every smooth vector field V with compact support in $\mathcal{D}$. Since V is arbitrary we deduce the following optimality condition

$$|\nabla u_1|^2 = 1 \quad \text{on} \quad \partial\Omega\cap\mathcal{D}.$$

On the other hand, using the maximum principle and the regularity of $\mathcal{D}$, we have that

$$|\nabla u_1| \le \lambda_1(\Omega)\|u_1\|_{L^\infty}|\nabla w| \le C \quad \text{on} \quad \partial\Omega\cap\partial\mathcal{D},$$

where w solves the equation

$$-\Delta w = 1 \quad \text{in} \quad \mathcal{D}, \qquad w \in H_0^1(\mathcal{D}),$$

and C is a constant depending on $\mathcal{D}$ and $\lambda_1(\Omega)$. Thus

$$|\nabla u_1| \le \max\{C, 1\} \quad \text{on} \quad \partial\Omega,$$

and so, a standard P function argument shows that u_1 is Lipschitz continuous with constant depending on $\mathcal{D}$ and $\lambda_1(\Omega)$. Of course, this is not a rigorous argument, since we supposed already that $\partial\Omega\cap\mathcal{D}$ is smooth. Nevertheless, since the Lipschitz constant of u_1 does not depend on the regularity of $\partial\Omega$, it is natural to expect that there is a weaker form of the same argument that gives the Lipschitz continuity of u_1 (and also the openness of Ω).

The analogous argument in the case of higher eigenvalues is more complicated, since the expression (1.24) of the shape derivative does not hold in the case of multiple eigenvalues. On the other hand, it is expected (due to the numerical results in [85] and [7]) that the solutions of

$$\min\left\{\lambda_k(\Omega) + |\Omega| : \ \Omega \text{ quasi-open}, \ \Omega \subset \mathbb{R}^d\right\}, \qquad (1.25)$$

are such that $\lambda_k(\Omega) = \lambda_{k-1}(\Omega)$.[1] For the sake of simplicity, we suppose that the optimal set Ω^*, solution of (1.25), is such that

$$\lambda_{k-2}(\Omega^*) < \lambda_{k-1}(\Omega^*) = \lambda_k(\Omega^*) < \lambda_{k+1}(\Omega^*).$$

Suppose that Ω_δ is an open and regular set which solves the auxiliary problem[2]

$$\min\left\{(1-\delta)\lambda_k(\Omega)+\delta\lambda_{k-1}(\Omega)+2|\Omega| : \Omega \text{ quasi-open}, \Omega^* \subset \Omega \subset \mathbb{R}^d\right\}. \qquad (1.26)$$

Suppose that $\lambda_k(\Omega_\delta) = \lambda_{k-1}(\Omega_\delta)$. Then for any $\Omega \supset \Omega^*$ we have

$$\begin{aligned}\lambda_k(\Omega_\delta) + 2|\Omega_\delta| &= (1-\delta)\lambda_k(\Omega_\delta) + \delta\lambda_{k-1}(\Omega_\delta) + 2|\Omega_\delta| \\ &\leq (1-\delta)\lambda_k(\Omega) + \delta\lambda_{k-1}(\Omega) + 2|\Omega| \leq \lambda_k(\Omega) + 2|\Omega|,\end{aligned}$$

and so Ω_δ solves the problem

$$\min\left\{\lambda_k(\Omega) + 2|\Omega| : \ \Omega \text{ quasi-open}, \ \Omega^* \subset \Omega \subset \mathbb{R}^d\right\}. \qquad (1.27)$$

Using the optimality of Ω_δ and Ω^* we get

$$\lambda_k(\Omega_\delta) - \lambda_k(\Omega^*) \leq 2\big(|\Omega^*| - |\Omega_\delta|\big) \leq |\Omega^*| - |\Omega_\delta| \leq \lambda_k(\Omega_\delta) - \lambda_k(\Omega^*).$$

Thus, all the inequalities are equalities and so $|\Omega_\delta \Delta \Omega^*| = 0$, *i.e.* $\Omega_\delta = \Omega^*$.

Let now $\delta^* \in [0, 1]$ be the largest real number such that $\lambda_k(\Omega_{\delta^*}) = \lambda_{k-1}(\Omega_{\delta^*})$.[3] We now consider the main case $\delta^* \in (0, 1)$. Let $\delta_n > \delta^*$ be a sequence converging to δ^*. Then the sequence of Ω_{δ_n} converges to

[1] There is an argument due to Dorin Bucur that proves that there exists a solution Ω of (1.25) such that $\lambda_k(\Omega) = \lambda_{k-1}(\Omega)$.

[2] The idea to consider the functional $F_\delta(\Omega) = (1-\delta)\lambda_k(\Omega) + \delta\lambda_{k-1}(\Omega)$ was inspired by the recent work [84], where it was given a numerical evidence in the support of the conjecture that for small δ the optimal sets for λ_k are also optimal for F_δ.

[3] As we will see in Chapter 5, this condition is closed.

$\Omega_{\delta^*} = \Omega^*$ in L^1, *i.e.* $|\Omega_{\delta_n} \Delta \Omega^*| \to 0$. Up to a subsequence we may suppose that

$$\lambda_j(\Omega_{\delta_n}) \to \lambda_j(\Omega^*), \qquad \forall j = k-2, k-1, k, k+1.$$

Thus, we have

$$\lambda_{k-2}(\Omega_{\delta_n}) < \lambda_{k-1}(\Omega_{\delta_n}) < \lambda_k(\Omega_{\delta_n}) < \lambda_{k+1}(\Omega_{\delta_n}),$$

for each $n \in \mathbb{N}$. For n large enough we can suppose that the eigenvalues $\lambda_k(\Omega_{\delta_n})$ and $\lambda_{k-1}(\Omega_{\delta_n})$ are both simple and so, we can apply the shape derivative (1.24) to an *external* vector field V, *i.e.* such that $V \cdot n \geq 0$. Thus, we have

$$\begin{aligned} \frac{d}{dt}\Big|_{t=0^+} \Big[& (1-\delta_n)\lambda_k\big((Id+tV)(\Omega_{\delta_n})\big) + \delta_n \lambda_{k-1}\big((Id+tV)(\Omega_{\delta_n})\big) \\ & + 2\big|(Id+tV)(\Omega_{\delta_n})\big|\Big] \\ = (V \cdot n)\Big(& -(1-\delta_n)|\nabla u_k^n|^2 - \delta_n |\nabla u_{k-1}^n|^2 + 2\Big), \end{aligned}$$

where u_k^n and u_{k-1}^n are, respectively, the kth and $(k-1)$th eigenfunctions on Ω_{δ_n}, normalized in $L^2(\mathbb{R}^d)$. Since V is arbitrary, we have that

$$(1-\delta_n)|\nabla u_k^n|^2 + \delta_n |\nabla u_{k-1}^n|^2 \leq 2 \quad \text{on} \quad \partial\Omega_{\delta_n},$$

and so, both u_k^n and u_{k-1}^n are Lipschitz continuous. Moreover, the Lipschitz constant of u_k^n is uniform in n (even if $\delta^* = 0$). On the other hand, the infinity norm of the eigenfunctions can be estimated by a function depending only on λ_k and so, we have also the uniform estimate $\|u_k^n\|_{L^\infty} \leq C$, for every n. Thus u_k^n converge uniformly, as $n \to \infty$, to some bounded Lipschitz function $u : \mathbb{R}^d \to \mathbb{R}$.

It now remains to show that u is an eigenfunction on Ω^* relative to the eigenvalue $\lambda_k(\Omega^*)$. Since $\|\nabla u_n^k\|_{L^2} = \lambda_k(\Omega_{\delta_n})$, we have that $u \in H^1(\mathbb{R}^d)$ and that u_k^n converges to u weakly in $H^1(\mathbb{R}^d)$. We first note that $u = 0$ outside Ω^*, by the L^1 convergence of Ω_{δ_n} to Ω^*. Thus, since Ω^* is supposed to be regular, $u \in H_0^1(\Omega^*)$. Now it remains to check that u is a kth eigenfunction on Ω^*. Indeed, since $\Omega^* \subset \Omega_{\delta_n}$, we can use any $v \in H_0^1(\Omega^*)$ as a test function for u_k^n, *i.e.* we have

$$\int_{\mathbb{R}^d} \nabla u_k^n \cdot \nabla v \, dx = \lambda_k(\Omega_{\delta_n}) \int_{\mathbb{R}^d} u_k^n v \, dx,$$

and passing to the limit as $n \to \infty$, we obtain

$$\int_{\mathbb{R}^d} \nabla u \cdot \nabla v \, dx = \lambda_k(\Omega_{\delta_n}) \int_{\mathbb{R}^d} u v \, dx,$$

which concludes the proof that u is an eigenfunction on Ω^* with Lipschitz continuous extension on $\mathbb{R}^d$.

Chapter 2
Shape optimization problems in a box

In this chapter we define two different variational convergences on the family of domains contained in a given box. The term *box* is widely used in the shape optimization and classically refers to a bounded open set in $\mathbb{R}^d$. The theory of the weak-γ and the strong-γ-convergence[1] of sets in a box was developed in the Euclidean space (see, for example, [21] and the references therein). Nevertheless, as it was shown in [37], this is a theory that uses a purely variational techniques and it can be adapted to a much more general (non-linear) settings as those of measured metric spaces.

We start by introducing the Sobolev spaces and elliptic PDEs on a measured metric space together with some basic instruments as the weak and strong maximum principles. Since the analysis on metric spaces is a theme of intense research interest in the last years (see, for example, [68], or the more recent [4] and the references therein), we prefer to impose some minimal conditions on an abstractly defined Sobolev space instead of imposing more restrictive conditions on the metric space, which may later turn to be non-necessary.

2.1. Sobolev spaces on metric measure spaces

From now on (X, d, m) will denote a separable metric space (X, d) endowed with a σ-finite regular Borel measure m.

Let $L^2(X, m)$ be the Hilbert space of the real m-measurable functions $f : X \to \mathbb{R}$, with integrable square $\int_X |f|^2\, dm < +\infty$. Consider a linear subspace $H \subset L^2(X, m)$ such that:

(H1) H is a Riesz space ($u, v \in H \Rightarrow u \vee v, u \wedge v \in H$).

[1] The strong-γ-convergence is known in the literature as γ or also γ_{loc} convergence. Our motivation for introducing this new terminology is the fact that in the linear setting ($\mathbb{R}^d$) the strong-γ-convergence corresponds to the strong convergence of the corresponding resolvent operators. We reserve the term γ-convergence for an even stronger convergence, corresponding to the norm convergence of these operators (see Chapter 3).

Suppose that the application $D : H \to L^2(X, m)$ is such that:

(D1) $Du \geq 0$, for each $u \in H$,
(D2) $D(u + v) \leq Du + Dv$, for each $u, v \in H$,
(D3) $D(\alpha u) = |\alpha| Du$, for each $u \in H$ and $\alpha \in \mathbb{R}$,
(D4) $D(u \vee v) = Du \cdot \mathbb{1}_{\{u>v\}} + Dv \cdot \mathbb{1}_{\{u \leq v\}}$.

Remark 2.1. In the above hypotheses on H and D, we have that

$$D(u \wedge v) = Dv \cdot \mathbb{1}_{\{u>v\}} + Du \cdot \mathbb{1}_{\{u \leq v\}} \qquad \text{and} \qquad D(|u|) = Du.$$

Moreover, the quantity

$$\|u\|_H = \left(\|u\|^2_{L^2(m)} + \|Du\|^2_{L^2(m)} \right)^{1/2},$$

defined for $u \in H$, is a norm on the vector space H, which makes the inclusion $H \hookrightarrow L^2(X, m)$ continuous.

Remark 2.2. The main example we will keep in mind throughout this chapter is $X \subset \mathbb{R}^d$, an open set of finite Lebesgue measure, and $H = H^1_0(X)$, the classical Sobolev space on X. The operator D then is simply the modulus of the weak gradient, *i.e.* $Du = |\nabla u|$.

We furthermore assume that:

($\mathcal{H}1$) $(H, \|\cdot\|_H)$ is complete;
($\mathcal{H}2$) the norm of the gradient is lower semi-continuous with respect to the weak $L^2(X, m)$ convergence, *i.e.* if for the sequence $u_n \in H$ we have that

- u_n is bounded in H,
- u_n converges weakly in $L^2(X, m)$ to a function $u \in L^2(X, m)$,

then $u \in H$ and

$$\int_X |Du|^2 \, dm \leq \liminf_{n\to\infty} \int_X |Du_n|^2 \, dm. \tag{2.1}$$

Remark 2.3. If the embedding $H \hookrightarrow L^2(X, m)$ is compact, the condition ($\mathcal{H}2$) is equivalent to suppose that if u_n is a bounded sequence in H and strongly convergent in $L^2(X, m)$ to a function $u \in L^2(X, m)$, then we have that $u \in H$ and (2.1) holds.

From now on, with H we denote a linear subspace of $L^2(X, m)$ such that the conditions H1, D1, D2, D3, D4, $\mathcal{H}1$ **and** $\mathcal{H}2$ **are satisfied.**

Let now μ be a (not necessarily locally finite) **Borel measure on** X, **absolutely continuous with respect to** m, *i.e.* for every $E \subset X$ such that $m(E) = 0$, we have $\mu(E) = 0$. We will keep in mind two examples of such measures:

- $\mu = f\, dm$, for some measurable f;
- $\mu = \widetilde{I}_\Omega$, where $\Omega \subset X$ is a m-measurable set and

$$\widetilde{I}_\Omega(E) = \begin{cases} 0, & \text{if } m(E \setminus \Omega) = 0; \\ +\infty, & \text{if } m(E \setminus \Omega) > 0. \end{cases} \tag{2.2}$$

For a Borel measure μ as above, we define the space H_μ as

$$H_\mu = \left\{u \in H : \ u \in L^2(\mu)\right\}. \tag{2.3}$$

Remark 2.4. Equipped with the norm

$$\|u\|_{H_\mu} = \left(\|u\|_H^2 + \|u\|_{L^2(\mu)}^2\right)^{1/2}, \tag{2.4}$$

the space H_μ is Banach. Indeed, if $u_n \in H_\mu$ is Cauchy in H_μ, then u_n converges in H to $u \in H$, then u_n converges in $L^2(X, m)$ and so, we can suppose that u_n converges to u m-almost everywhere. Then u_n converges to u μ-almost everywhere and since u_n is Cauchy in $L^2(\mu)$, we have the claim.

Remark 2.5. We always have the inequality

$$\|u\|_H \le \|u\|_{H_\mu}.$$

If there is a constant $C > 0$ such that for every $u \in H_\mu$, we have

$$\|u\|_{H_\mu} \le C\|u\|_H,$$

then H_μ is a closed subspace of H.

Example 2.6. The space H_μ is not in general a closed subspace of H. In fact, suppose that the interval $X = (0, 1) \subset \mathbb{R}$ is equipped with the Euclidean distance and the Lebesgue measure. Take $H = H_0^1((0, 1))$, $Du = |u'|$ and let $\mu = \frac{dx}{x^3(1-x)^3}$. Then $C_c^\infty((0, 1)) \subset H_\mu$, and so H_μ is a dense subset of H. On the other hand the function $u(x) = x(1 - x)$ is such that $u \in H \setminus H_\mu$.

Example 2.7. If $\mu = \widetilde{I}_\Omega$, for some $\Omega \subset X$, then we have that $\|u\|_H = \|u\|_{H_\mu}$, for every $u \in H_\mu$. In particular, the space H_μ is a closed subspace of H, which we denote by $\widetilde{H}_0(\Omega)$ and can be characterized as

$$\widetilde{H}_0(\Omega) = \left\{u \in H : \ u = 0 \ m-\text{a.e. on } X \setminus \Omega\right\}.$$

Definition 2.8. Let μ be a Borel measure on (X, d, m), absolutely continuous with respect to m. We say that a function $u \in H$ is a solution of the elliptic boundary value problem

$$-D^2u + u + \mu u = f \quad \text{in} \quad H_\mu, \qquad u \in H_\mu, \tag{2.5}$$

where $f \in L^2(X, m)$, if u is a minimizer of the functional

$$J_{\mu,f}(u) = \begin{cases} \int_X \left(\frac{1}{2}|Du|^2 + \frac{1}{2}u^2 - fu\right) dm + \frac{1}{2}\int_X |u|^2 \, d\mu, & \text{if } u \in H, \\ +\infty, & \text{otherwise.} \end{cases}$$

Remark 2.9. If $\mu = \widetilde{I}_\Omega$, where $\Omega \subset X$, then we say that u is a solution of

$$-D^2u + u = f \quad \text{in} \quad \Omega, \qquad u \in \widetilde{H}_0(\Omega).$$

Lemma 2.10. *Suppose that μ is absolutely continuous with respect to m. Then for every sequence $u_n \in H_\mu$ such that:*

- *u_n is bounded in H_μ,*
- *u_n converges weakly in $L^2(X, m)$ to $u \in L^2(X, m)$,*

we have that $u \in H_\mu$ and

$$\|u\|_{H_\mu} \leq \liminf_{n\to\infty} \|u_n\|_{H_\mu}.$$

Proof. Under the assumptions of the Lemma, we have that the sequence u_n is bounded in $L^2(m + \mu)$. Thus it converges weakly in $L^2(m + \mu)$ to some $v \in L^2(m + \mu)$. Since $L^2(m + \mu) \subset L^2(m)$, we have that $v = u$. Now using (2.1) and the semi-continuity of the L^2 norm with respect to the weak L^2 convergence, we have the claim. □

Proposition 2.11. *Suppose that the Borel measure μ is absolutely continuous with respect to m. Then the problem* (2.5) *has a unique solution $w_{\mu,f} \in H_\mu$. Moreover, we have*

(i) $w_{\mu,tf} = tw_{\mu,f}$, *for every $t \in \mathbb{R}$;*
(ii) $\|w_{\mu,f}\|_{H_\mu}^2 = \int_X f w_{\mu,f} \, dm$;
(iii) *if $f \geq 0$, then $w_{\mu,f} \geq 0$.*

Proof. Suppose that u_n is a minimizing sequence for $J_{\mu,f}$ in H_μ. Since $J_{\mu,f}(0)=0$, we can assume that for each $n>0$

$$\frac{1}{2}\int_X (|Du_n|^2+u_n^2)\,dm+\frac{1}{2}\int_X u_n^2\,d\mu\leq\int_X fu_n\,dm\leq\|f\|_{L^2(m)}\|u_n\|_{L^2(m)},$$

and thus, we obtain

$$\|u_n\|_{L^2(m)}\leq\|u_n\|_{H_\mu}\leq 2\|f\|_{L^2(\mu)}.$$

Up to a subsequence we may suppose that u_n converges weakly to some $u\in L^2(m)$. By Lemma 2.10, we obtain that

$$J_{\mu,f}(u)\leq\liminf_{n\to\infty}J_{\mu,f}(u_n),$$

and so, $u\in H^1_\mu$ is a solution of (2.5).

Suppose now that $u,v\in H_\mu$ are two minimizers for $J_{\mu,f}$. Then

$$J_{\mu,f}\left(\frac{u+v}{2}\right)\leq\frac{J_{\mu,f}(u)+J_{\mu,f}(v)}{2}.$$

Moreover, by the strict convexity of the L^2 norm, we have $v=tu$. Since the function $j(t):=J_{\mu,f}(tu)$ is a polynomial of second degree in $t\in\mathbb{R}$ with positive leading coefficient, it has unique minimum in $\mathbb{R}$ and thus we have necessarily $t=1$.

To prove (i), we just note that for every $u\in H_\mu$ we have $J_{\mu,tf}(tu)=t^2J_{\mu,f}(u)$.

Point (ii) follows by minimizing the function $t\mapsto J_{\mu,f}(tw_{\mu,f})$, for $t\in\mathbb{R}$.

For (iii), we note that, in the case when $f\geq 0$, we have the inequality $J_{\mu,f}(|u|)\leq J_{\mu,f}(u)$, for each $u\in H_\mu$ and so we conclude by the uniqueness of the minimizer of $J_{\mu,f}$. □

Remark 2.12. From the proof of Proposition 2.11 we obtain, for any $f\in L^2(X,m)$ and $\mu<<m$, the estimates

$$\|w_{\mu,f}\|_{H_\mu}\leq\|f\|_{L^2(m)}$$
$$\text{and}\quad\left|J_{\mu,f}(w_{\mu,f})\right|=\frac{1}{2}\int_X fw_{\mu,f}\,dm\leq\frac{1}{2}\|f\|^2_{L^2(m)}. \tag{2.6}$$

For the solutions $w_{\mu,f}$ of (2.5), we have comparison principles, analogous to those in the Euclidean space $\mathbb{R}^d$.

Proposition 2.13. *Let μ be an absolutely continuous measure with respect to m. Then the solutions of (2.5) satisfy the following inequalities:*

(i) *If* $\mu \le \nu$ *and* $f \in L^2(m)$ *is a positive function, then* $w_{\nu,f} \le w_{\mu,f}$.
(ii) *If* $f, g \in L^2(X, m)$ *are such that* $f \le g$, *then* $w_{\mu,f} \le w_{\mu,g}$.

Proof.

(i) We write, for simplicity, $u = w_{\nu,f}$ and $U = w_{\mu,f}$. Note that we have $u \ge 0$ and $U \ge 0$. Consider the functions $u \vee U \in H_\mu$ and $u \wedge U \in H_\nu$. By the minimizing property of u and U, we have

$$J_{\nu,f}(u \wedge U) \ge J_{\nu,f}(u), \qquad J_{\mu,f}(u \vee U) \ge J_{\mu,f}(U).$$

We decompose the space as $X = \{u > U\} \cup \{u \le U\}$ to obtain

$$\begin{aligned}
&\int_{\{u>U\}} \left(\frac{1}{2}|DU|^2 + \frac{1}{2}U^2 - fU\right) dm + \frac{1}{2}\int_{\{u>U\}} U^2\, d\nu \\
&\ge \int_{\{u>U\}} \left(\frac{1}{2}|Du|^2 + \frac{1}{2}u^2 - fu\right) dm + \frac{1}{2}\int_{\{u>U\}} u^2\, d\nu, \\
&\int_{\{u>U\}\cap\omega} \left(\frac{1}{2}|Du|^2 + \frac{1}{2}u^2 - fu\right) dm + \frac{1}{2}\int_{\{u>U\}} u^2\, d\mu \\
&\ge \int_{\{u>U\}} \left(\frac{1}{2}|DU|^2 + \frac{1}{2}U^2 - fU\right) dm + \frac{1}{2}\int_{\{u>U\}} U^2\, d\mu.
\end{aligned} \tag{2.7}$$

Thus, we have

$$\int_{\{u>U\}} (u^2 - U^2)\, d\mu \ge \int_{\{u>U\}} (u^2 - U^2)\, d\nu,$$

and since $u^2 - U^2 > 0$ on $\{u > U\}$ and $\nu \ge \mu$, we have also the converse inequality and so

$$\int_{\{u>U\}} (u^2 - U^2)\, d\mu = \int_{\{u>U\}} (u^2 - U^2)\, d\nu.$$

Using again (2.7), we obtain that also

$$\begin{aligned}
&\int_{\{u>U\}} \left(\frac{1}{2}|DU|^2 + \frac{1}{2}U^2 - fU\right) dm \\
&= \int_{\{u>U\}} \left(\frac{1}{2}|Du|^2 + \frac{1}{2}u^2 - fu\right) dm,
\end{aligned}$$

and so

$$J_{\nu,f}(u \wedge U) = J_{\nu,f}(u) \qquad \text{and} \qquad J_{\mu,f}(u \vee U) = J_{\mu,f}(U).$$

By the uniqueness of the minimizers, we conclude that $u \le U$.

(ii) Let $u = w_{\mu,f}$ and $U = w_{\mu,g}$. As in the previous point, we consider the test functions $u \vee U$, $u \wedge U \in H_\mu$. Using the optimality of u and U, we have

$$J_{\mu,g}(u \vee U) \geq J_{\mu,g}(U), \qquad J_{\mu,f}(u \wedge U) \geq J_{\mu,f}(u).$$

We decompose the metric space X as $\{u > U\} \cup \{u \leq U\}$ to obtain

$$\int_{\{u>U\}} \left(\frac{1}{2}|Du|^2 + \frac{1}{2}u^2 - ug\right) dm + \frac{1}{2}\int_{\{u>U\}} u^2\, d\mu$$
$$\geq \int_{\{u>U\}} \left(\frac{1}{2}|DU|^2 + \frac{1}{2}U^2 - gU\right) dm + \frac{1}{2}\int_{\{u>U\}} U^2\, d\mu,$$

$$\int_{\{u>U\}} \left(\frac{1}{2}|DU|^2 + \frac{1}{2}U^2 - fU\right) dm + \frac{1}{2}\int_{\{u>U\}} U^2\, d\mu$$
$$\geq \int_{\{u>U\}} \left(\frac{1}{2}|Du|^2 + \frac{1}{2}u^2 - fu\right) dm + \frac{1}{2}\int_{\{u>U\}} u^2\, d\mu.$$

Then, we have

$$0 \geq \int_{\{u>U\}} \left(\frac{1}{2}|Du|^2 + \frac{1}{2}u^2 - fu\right) dm + \frac{1}{2}\int_{\{u>U\}} u^2\, d\mu$$
$$-\left(\int_{\{u>U\}} \left(\frac{1}{2}|DU|^2 + \frac{1}{2}U^2 - fU\right) dm + \frac{1}{2}\int_{\{u>U\}} U^2\, d\mu\right)$$
$$\geq \int_{\{u>U\}} g(u-U)\, dm - \int_{\{u>U\}} f(u-U)\, dm$$
$$= \int_{\{u>U\}} (g-f)(u-U)\, dm \geq 0.$$

Thus, we obtain the equality

$$\int_{\{u>U\}} \left(\frac{1}{2}|Du|^2 + \frac{1}{2}u^2 - fu\right) dm + \frac{1}{2}\int_{\{u>U\}} u^2\, d\mu$$
$$= \int_{\{u>U\}} \left(\frac{1}{2}|DU|^2 + \frac{1}{2}U^2 - fU\right) dm + \frac{1}{2}\int_{\{u>U\}} U^2\, d\mu,$$

and thus we have

$$J_{\mu,f}(u) = J_{\mu,f}(u \wedge U).$$

By the uniqueness of the minimizer of $J_{\mu,f}$, we conclude that $U \geq u$. □

Corollary 2.14. *Suppose that $\omega \subset \Omega$ and that $f \in L^2(X,m)$ is a positive function. Then we have $w_{\Omega,f} \geq w_{\omega,f}$, where $w_{\Omega,f}$ and $w_{\omega,f}$ are the solutions respectively of*

$$\begin{aligned} &-D^2 w_{\Omega,f} + w_{\Omega,f} = f \quad \text{in} \quad \Omega, \qquad w_{\Omega,f} \in \widetilde{H}_0(\Omega),\\ &-D^2 w_{\omega,f} + w_{\omega,f} = f \quad \text{in} \quad \omega, \qquad w_{\omega,f} \in \widetilde{H}_0(\omega). \end{aligned}$$

Proof. It is enough to note that $\widetilde{I}_\Omega \leq \widetilde{I}_\omega$ and then use Proposition 2.13 (a). □

The following lemma is similar to [51, Proposition 3.1].

Lemma 2.15. *Let μ be a measure on X, absolutely continuous with respect to m. For $u \in H_\mu$ and $\varepsilon > 0$ let u_ε be the unique solution of the equation*

$$-D^2 u_\varepsilon + u_\varepsilon + \mu u_\varepsilon + \varepsilon^{-1} u_\varepsilon = \varepsilon^{-1} u \quad \text{in} \quad H_\mu, \qquad u_\varepsilon \in H_\mu. \tag{2.8}$$

Then we have

(a) *u_ε converges to u in $L^2(X,m)$, as $\varepsilon \to 0$, and*

$$\|u - u_\varepsilon\|_{L^2(m)} \leq \varepsilon^{1/2} \|u\|_{H_\mu}; \tag{2.9}$$

(b) *$\|u_\varepsilon\|_{H_\mu} \leq \|u\|_{H_\mu}$, for every $\varepsilon > 0$, and*

$$\|u\|_{H_\mu} = \lim_{\varepsilon \to 0^+} \|u_\varepsilon\|_{H_\mu}; \tag{2.10}$$

(c) *if $u \geq 0$, then $u_\varepsilon \geq 0$;*
(d) *if $u \leq f$, then $u_\varepsilon \leq \varepsilon^{-1} C w_{\mu,f}$.*

Proof. We first note that u_ε is the minimizer of the functional

$$J_\varepsilon : L^2(X,m) \to \mathbb{R}$$

defined as

$$J_\varepsilon(v) = \int_X \left(|Dv|^2 + v^2\right) dm + \int_X v^2 \, d\mu + \frac{1}{\varepsilon} \int_X |v - u|^2 \, dm.$$

Since $J_\varepsilon(u_\varepsilon) \leq J_\varepsilon(u)$, we have

$$\|u_\varepsilon\|^2_{H_\mu} + \frac{1}{\varepsilon} \|u - u_\varepsilon\|^2_{L^2(m)} \leq \|u\|^2_{H_\mu},$$

and thus we obtain (a) and the inequality in (b). Since $u_\varepsilon \to u$ in $L^2(X,m)$ and u_ε is bounded in H_μ, we can apply Lemma 2.10 obtaining

$$\|u\|_{H_\mu} \leq \liminf_{\varepsilon\to 0^+} \|u_\varepsilon\|_{H_\mu} \leq \limsup_{\varepsilon\to 0^+} \|u_\varepsilon\|_{H_\mu} \leq \|u\|_{H_\mu},$$

which completes the proof of (b). Point (c) follows since $J_\varepsilon(|u_\varepsilon|) \leq J_\varepsilon(u_\varepsilon)$, whenever $u \geq 0$. To prove (d) we just apply the weak maximum principle (Proposition 2.13, (ii)) to the functions $\varepsilon^{-1}u \leq \varepsilon^{-1}C$. □

Remark 2.16. We note that if H_μ endowed with the norm $\|\cdot\|_{H_\mu}$ is a Hilbert space, then u_ε converges to u strongly in H_μ as $\varepsilon \to 0$. More generally, if H_μ is uniformly convex, then u_ε converges to u strongly in H_μ (see [16, Proposition III.30]).

We will refer to the following result as to the strong maximum principle for the solutions of (2.5).

Proposition 2.17. *Let μ be a measure on X, absolutely continuous with respect to m. Let $\psi \in L^2(X,m)$ be a strictly positive function on X such that for every $u \in H$ we have $\psi \wedge u \in H$. Then for every $u \in H_\mu$, we have that $\{u \neq 0\} \subset \{w_{\mu,\psi} > 0\}$, where $w_{\mu,\psi}$ is the solution of the equation*

$$-D^2 w_{\mu,\psi} + w_{\mu,\psi} + \mu w_{\mu,\psi} = \psi \quad in \quad H_\mu, \qquad w_{\mu,\psi} \in H_\mu.$$

Proof. Considering $|u|$ instead of u, we can restrict our attention only to non-negative functions. Moreover, by taking $u \wedge \psi$, we can suppose that $0 \leq u \leq \psi$. Consider the sequence u_ε of functions from Lemma 2.15. We have that $u_\varepsilon \leq \varepsilon^{-1} w_{\mu,\psi}$ and so

$$\{u_\varepsilon > 0\} \subset \{w_{\mu,\psi} > 0\}.$$

Passing to the limit as $\varepsilon \to 0$, we obtain

$$\{u > 0\} \subset \{w_{\mu,\psi} > 0\}.$$ □

Corollary 2.18. *Let ψ_1 and ψ_2 be two strictly positive functions satisfying the conditions of Proposition 2.17. Then we have*

$$\{w_{\mu,\psi_1} > 0\} = \{w_{\mu,\psi_2} > 0\}.$$

Definition 2.19. We say that H has the Stone property in $L^2(X,m)$, if there is a function $\psi \in L^2(X,m)$, strictly positive on X, such that for every $u \in H$ we have $\psi \wedge u \in H$.

Remark 2.20. If there is a function $\psi \in H$, strictly positive on X, then H has the Stone property in $L^2(X, m)$.

Remark 2.21. For a generic Riesz space $\mathcal{R}$, we say that $\mathcal{R}$ has the Stone property, if for every $u \in \mathcal{R}$, we have $u \wedge 1 \in \mathcal{R}$. If the constant 1 is in $L^2(X, m)$ and if H has the Stone property, then H has the Stone property in $L^2(X, m)$, in the sense of Definition 2.19.

Example 2.22. Let $X = \mathbb{R}^d$ and m be the Lebesgue measure. Then the Sobolev space $H^1_0(\Omega)$, for any (bounded or unbounded) set $\Omega \subset \mathbb{R}^d$, has the Stone property in $L^2(\mathbb{R}^d)$. In fact the Gaussian $e^{-|x|^2/2}$ is strictly positive Sobolev function on $\mathbb{R}^d$.

Definition 2.23. Suppose that the space H has the Stone property in $L^2(X, m)$. For every measure μ on X, absolutely continuous with respect to m, we define the set $\Omega_\mu \subset X$ as

$$\Omega_\mu = \{w_{\mu,\psi} > 0\}.$$

Remark 2.24. We note that, after Corollary 2.18, the definition of Ω_μ is independent on the choice of ψ.

Corollary 2.25. *Suppose that H has the Stone property in L^2 and let $\Omega \subset X$ be a Borel set. Then, setting $\mu = \widetilde{I}_\Omega$, we have*

$$\Omega_\mu \subset \Omega \qquad \text{and} \qquad \widetilde{H}_0(\Omega) = \widetilde{H}_0(\Omega_\mu).$$

Definition 2.26. Suppose that H satisfies the Stone property in $L^2(X,m)$. We say that the Borel set $\Omega \subset X$ is an energy set, if $\Omega = \Omega_\mu$,[2] where μ is the measure $\widetilde{I}_\Omega$.

Remark 2.27. For each $u \in H$ the set $\Omega = \{u > 0\}$ is an energy set. In fact, setting $\mu = \widetilde{I}_\Omega$, we have that $\{w_{\mu,\psi} > 0\} \subset \Omega = \{u > 0\}$, since $w_{\mu,\psi} \in H_\mu$. On the other hand, using Proposition 2.17, we have $\{u > 0\} \subset \{w_{\mu,\psi} > 0\}$.

2.2. The strong-γ and weak-γ convergence of energy domains

Throughout this section we will assume that H satisfies the properties H1, D1, D2, D3, D4, $\mathcal{H}1$ and $\mathcal{H}2$ and that H has the *Stone property* in $L^2(X, m)$. Moreover, we will need the further assumption that the inclusion $H \hookrightarrow L^2(X, m)$ is *locally compact*, *i.e.* every sequence $u_n \in H$ bounded in H admits subsequence for which there is a function $u \in H$ such that u_n converges to u in $L^2(B_R(x), m)$, for every ball $B_R(x) \subset X$.

[2] The equality is intended up to a set of zero m-measure, *i.e.* $m(\Omega \Delta \Omega_\mu) = 0$.

Under these assumptions, we introduce a suitable topology on the class of energy sets Ω, which involves the spaces $\widetilde{H}_0(\Omega)$ and the functionals defined on them as the first eigenvalue of the Dirichlet Laplacian, the Dirichlet Energy, etc.

2.2.1. The weak-γ-convergence of energy sets

From now on, for a given Borel set $\Omega \subset X$ and a function $f \in L^2(X, m)$ we will denote by $w_{\Omega,f}$ the solution of the problem

$$-D^2u + u = f \quad \text{in} \quad \Omega, \qquad u \in \widetilde{H}_0(\Omega),$$

i.e. the minimizer of the functional $J_{\Omega,f} := J_{\widetilde{I}_\Omega,f}$ in H.

Definition 2.28. Suppose that ψ is a Stone function in $L^2(X, m)$ for H. We say that a sequence of energy sets Ω_n weak-γ-converges to Ω if the sequence $\left(w_{\Omega_n,\psi}\right)_{n\geq 1}$ converges strongly in $L^2(X, m)$ to some $w \in L^2(X, m)$ and $\Omega = \{w > 0\}$.

Remark 2.29. We will prove later in Corollary 2.35 that the notion of the weak-γ-convergence is independent on the choice of ψ.

Remark 2.30. We first note that $w \in H$ and the set Ω from Definition 2.36 is an energy set. Indeed, since $w_n := w_{\Omega_n,\psi}$ satisfies

$$-D^2w_n + w_n = \psi \quad \text{in} \quad \Omega_n, \qquad w_n \in H_0(\Omega_n).$$

By the first estimate from (2.6) we have

$$\|w_n\|_H \leq \|\psi\|_{L^2(m)}, \; \forall\, n \in \mathbb{N}.$$

Thus, since $w_n \to w$, we have that $w \in H$ and

$$\|w\|_H \leq \liminf_{n\to\infty} \|w_n\|_H \leq \|\psi\|_{L^2(m)}.$$

Now, by Remark 2.27, $\Omega = \{w > 0\}$ is an energy set.

Remark 2.31. We note that the equation $w = w_{\mu,\psi}$, where $\mu = \widetilde{I}_\Omega$, does not necessarily hold. In the case $X = \mathbb{R}^d$ and $H = H^1(\mathbb{R}^d)$, we will see that w is of the form $w_{\mu,\psi}$, for some measure $\mu \geq \widetilde{I}_\Omega$.

Remark 2.32. If the inclusion $H \hookrightarrow L^2(X, m)$ is locally compact, then the family of energy sets is sequentially compact with respect to the weak-γ-convergence. Indeed, as we showed in Remark 2.30, the sequence $w_{\Omega_n,\psi}$ is bounded in H, for any choice of Ω_n. Moreover, $w_{\Omega_n,\psi} \leq w$, where w is the solution of

$$-D^2w + w = \psi \quad \text{in} \quad X, \qquad w \in H.$$

Thus, by the following Lemma 2.33, we have that $w_{\Omega_n,\psi}$ has a subsequence convergent in $L^2(X, m)$.

Lemma 2.33. *Suppose that the inclusion $H \hookrightarrow L^2(X,m)$ is locally compact. Let $w_n \in L^2(X,m)$ be a sequence strongly converging in $L^2(X,m)$ to $w \in L^2(X,m)$ and let $u_n \in H$ be a bounded sequence in H such that $|u_n| \le w_n$, for every $n \in \mathbb{N}$. Then up to a subsequence u_n converges strongly in $L^2(X,m)$ to some function $u \in H$.*

Proof. By assumption $(\mathcal{H}2)$, we have that u_n converges weakly in $L^2(X,m)$ to some $u \in H$. Thus, it is sufficient to check that the convergence is strong, *i.e.* that the sequence u_n is Cauchy in $L^2(X,m)$. Let $B_R(x) \subset X$ be a ball such that $\int_{X\setminus B_R(x)} w^2\,dm \le \varepsilon$. Then for n large enough we have

$$\int_{X\setminus B_R(x)} u_n^2\,dm \le \int_{X\setminus B_R(x)} w_n^2\,dm \le 2\varepsilon.$$

By hypothesis we have that up to a subsequence u_n converges to u in $L^2(B_R(x),m)$. Thus for $n,m \in \mathbb{N}$ large enough we get

$$\int_X |u_n - u_m|^2\,dm \le 8\varepsilon + \int_{X\setminus B_R(x)} |u_n - u_m|^2\,dm \le 9\varepsilon. \qquad \square$$

Proposition 2.34. *Suppose that the space H has the Stone property in $L^2(X,m)$ and that the inclusion $H \hookrightarrow L^2(X,m)$ is locally compact. Suppose that a sequence of energy sets Ω_n weak-γ-converges to Ω and suppose that $(u_n)_{n\ge 0} \subset H$ is a sequence bounded in H and strongly convergent in $L^2(X,m)$ to a function $u \in H$. If $u_n \in \widetilde{H}_0(\Omega_n)$ for every n, then $u \in \widetilde{H}_0(\Omega)$.*

Proof. For sake of simplicity, we set $w_n := w_{\Omega_n,\psi}$ and w to be the strong limit in $L^2(X,m)$ of w_n. Since $|u_n|$ also converges to $|u|$ in $L^2(X,m)$, we can suppose $u_n \ge 0$ for every $n \ge 1$. Moreover, since $u_n \wedge \psi$ converges to $u \wedge \psi$ in $L^2(X,m)$ and $\{u > 0\} = \{u \wedge \psi > 0\}$, we can suppose $u_n \le \psi$, for every $n \le 1$. For each $n \ge 1$ and every $\varepsilon > 0$ we define $u_{n,\varepsilon}$ to be the solution of

$$-D^2 u_{n,\varepsilon} + (1+\varepsilon^{-1})u_{n,\varepsilon} = \varepsilon^{-1}u_n \quad \text{in} \quad \Omega_n, \qquad u_{n,\varepsilon} \in \widetilde{H}_0(\Omega_n).$$

For every $\varepsilon > 0$, we have that $u_{n,\varepsilon}$ is bounded in H and $u_{n,\varepsilon} \le \varepsilon^{-1}w_n$. Since w_n converges to w in $L^2(X,m)$, we apply Lemma 2.33 to obtain that there is a function $u_\varepsilon \in H$ such that $u_{n,\varepsilon}$ converges strongly in $L^2(X,m)$ to u_ε. Moreover, we have $u_\varepsilon \le \varepsilon^{-1}w$ and so, $u_\varepsilon \in \widetilde{H}_0(\Omega)$. On the other hand, for every n and ε, we have

$$\|u_n - u_{n,\varepsilon}\|_{L^2(m)} \le \sqrt{\varepsilon}\|u_n\|_H \le \sqrt{\varepsilon}C,$$

and so passing to the limit in L^2, we have

$$\|u - u_\varepsilon\|_{L^2(m)} \leq \sqrt{\varepsilon}C,$$

which implies that $u_\varepsilon \to u$, strongly in $L^2(X, m)$ as $\varepsilon \to 0$, and so $u \in \widetilde{H}_0(\Omega)$. □

Corollary 2.35. *Suppose that H has the Stone property in $L^2(X, m)$ and that the inclusion $H \hookrightarrow L^2(X, m)$ is locally compact. Let φ and ψ be two Stone functions and let Ω_n be a sequence of energy sets such that $w_{\Omega_n,\varphi}$ converges in $L^2(X, m)$ to some $w_\varphi \in H$ and $w_{\Omega_n,\psi}$ converges in $L^2(X, m)$ to some $w_\psi \in H$. Then $\{w_\psi > 0\} = \{w_\varphi > 0\}$.*

Proof. Consider the function $\xi = \varphi \wedge \psi$. We note that ξ is a Stone function for H in $L^2(X, m)$. The sequence $w_{\Omega_n,\xi}$ is bounded in H and is such that $w_{\Omega_n,\xi} \leq w_{\Omega_n,\varphi}$. By Lemma 2.33, we can suppose that $w_{\Omega_n,\xi}$ converges in $L^2(X, m)$ to some w_ξ. Since $w_\xi \leq w_\varphi$, we have that $\{w_\xi > 0\} \subset \{w_\varphi > 0\}$. On the other hand, by Proposition 2.34, we have the converse inclusion, *i.e.* $\{w_\xi > 0\} = \{w_\varphi > 0\}$. Reasoning analogously, we have $\{w_\xi > 0\} = \{w_\psi > 0\}$ and so, we have the claim. □

2.2.2. The strong-γ-convergence of energy sets

Definition 2.36. Suppose that ψ is a Stone function in $L^2(X, m)$ for H. We say that a sequence of energy sets Ω_n strong-γ-converges Ω if the sequence $\left(w_{\Omega_n,\psi}\right)_{n\geq 1}$ converges strongly in $L^2(X, m)$ to the solution $w_{\Omega,\psi} \in L^2(X, m)$.

In what follows we will show that the definition of the strong-γ-convergence is independent on the choice of the function ψ (see Corollary 2.40). We start with two technical lemmas.

Lemma 2.37. *Suppose that H and D satisfy the assumptions* (H1), (D1), (D2), (D3), (D4), ($\mathcal{H}$1) *and* ($\mathcal{H}$2). *Suppose that $u_n \in H$ and $v_n \in H$ are two sequences converging strongly in $L^2(X, m)$ to $u \in H$ and $v \in H$, respectively. If we have*

$$\int_X |Du|^2 \, dm = \lim_{n\to\infty} \int_X |Du_n|^2 \, dm$$

$$\textit{and} \qquad \int_X |Dv|^2 \, dm = \lim_{n\to\infty} \int_X |Dv_n|^2 \, dm,$$

then also

$$\int_X |D(u \vee v)|^2\,dm = \lim_{n\to\infty}\int_X |D(u_n \vee v_n)|^2\,dm,$$

$$\int_X |D(u \wedge v)|^2\,dm = \lim_{n\to\infty}\int_X |D(u_n \wedge v_n)|^2\,dm.$$

Proof. Since we have that $u_n \wedge v_n \to u \wedge v$ and $u_n \vee v_n \to u \vee v$ in $L^2(X,m)$, we have

$$\begin{aligned}\int_X |D(u \vee v)|^2\,dm &\le \liminf_{n\to\infty}\int_X |D(u_n \vee v_n)|^2\,dm,\\ \int_X |D(u \wedge v)|^2\,dm &\le \liminf_{n\to\infty}\int_X |D(u_n \wedge v_n)|^2\,dm.\end{aligned} \tag{2.11}$$

On the other hand we have

$$\begin{aligned}&\|D(u\wedge v)\|^2_{L^2(m)} + \|D(u\vee v)\|^2_{L^2(m)}\\ &= \|Du\|^2_{L^2(m)} + \|Dv\|^2_{L^2(m)}\\ &= \lim_{n\to\infty}\left(\|Du_n\|^2_{L^2(m)} + \|Dv_n\|^2_{L^2(m)}\right)\\ &= \lim_{n\to\infty}\left(\|D(u_n\wedge v_n)\|^2_{L^2(m)} + \|D(u_n\vee v_n)\|^2_{L^2(m)}\right).\end{aligned} \tag{2.12}$$

Now the claim follows since by (2.12) both inequalities in (2.11) must be equalities. □

Lemma 2.38. *Suppose that the function $\psi \in L^2(X,m)$ is a Stone function for H and that the inclusion $H \hookrightarrow L^2(X,m)$ is locally compact. Suppose that the sequence $w_{\Omega_n,\psi}$ converges strongly in $L^2(X,m)$ to $w_{\Omega,\psi}$. Then, for every $v \in \widetilde{H}_0(\Omega)$, there is a sequence $v_n \in \widetilde{H}_0(\Omega_n)$ strongly converging to v in $L^2(X,m)$ and such that*

$$\int_X |Dv|^2\,dm = \lim_{n\to\infty}\int_X |Dv_n|^2\,dm. \tag{2.13}$$

Proof. We set for simplicity

$$w_n := w_{\Omega_n,\psi} \qquad \text{and} \qquad w =: w_{\Omega,\psi}.$$

We take for simplicity $v \ge 0$. The proof in the case when v changes sign is analogous. We first show that for $v \in \widetilde{H}_0(\Omega)$ the sequence $v_t =$

$v \wedge (tw) \in \widetilde{H}_0(\Omega)$ converges to v, strongly in $L^2(X, m)$ as $t \to +\infty$ and moreover the norm of the gradients converge

$$\int_X |Dv|^2\, dm = \lim_{t\to\infty} \int_X |Dv_t|^2\, dm.$$

Indeed, since $v_t \to v$ in $L^2(X, m)$, we have the semi-continuity

$$\int_X |Dv|^2\, dm \le \liminf_{t\to\infty} \int_X |Dv_t|^2\, dm.$$

For the other inequality, we note that $J_{\Omega,\psi}(w) \le J_{\Omega,\psi}(w \vee v)$, and thus

$$\begin{aligned}\int_{\{tw<v\}} &\left(\frac{t^2}{2}|Dw|^2 + \frac{t^2}{2}w^2 - tw\psi\right) dm \\ &\le \int_{\{tw<v\}} \left(\frac{1}{2}|Dv|^2 + \frac{1}{2}v^2 - v\psi\right) dm,\end{aligned} \tag{2.14}$$

which gives

$$\begin{aligned} t^2\int_{\{tw<v\}} |Dw|^2\, dm &\le \int_{\{tw<v\}} |Dv|^2\, dm + \int_{\{tw<v\}} (v^2 - t^2w^2)\, dm \\ &= \int_{\{tw<v\}} |Dv|^2\, dm + \left(\|v\|^2_{L^2(m)} - \|v_t\|^2_{L^2(m)}\right).\end{aligned} \tag{2.15}$$

Now since $|Dv_t| = |Dv|\mathbb{1}_{\{v<tw\}} + t|Dw|\mathbb{1}_{\{tw<v\}}$, we have

$$\int_X |Dv_t|^2\, dm \le \int_X |Dv|^2\, dm + \left(\|v\|^2_{L^2(m)} - \|v_t\|^2_{L^2(m)}\right), \tag{2.16}$$

which gives

$$\int_X |Dv|^2\, dm \ge \limsup_{t\to\infty} \int_X |Dv_t|^2\, dm.$$

Thus, by using a diagonal sequence argument, we can restrict our attention to functions $v \in \widetilde{H}_0(\Omega)$ such that $v \le tw$, for some $t > 0$. Up to substituting ψ by $t\psi$, we can assume $t = 1$. We now suppose $v \le w$ and define $v_n = v \wedge w_n \in \widetilde{H}_0(\Omega_n)$.

Since $w_n \to w$ in $L^2(X, m)$ and since w and w_n minimize $J_{\Omega,\psi}$ and $J_{\Omega_n,\psi}$, we get

$$\begin{aligned}\int_X |Dw_n|^2\, dm &= \int_X (w_n\psi - w_n^2)\, dm \xrightarrow[n\to\infty]{} \int_X (w\psi - w^2)\, dm \\ &= \int_X |Dw|^2\, dm.\end{aligned}$$

Now the claim follows by Lemma 2.37. □

Proposition 2.39. *Suppose that the function $\psi \in L^2(X, m)$ is a Stone function for H and that the inclusion $H \hookrightarrow L^2(X, m)$ is locally compact. Suppose that the sequence $w_{\Omega_n,\psi}$ converges strongly in $L^2(X, m)$ to $w_{\Omega,\psi}$. Then, for every function $f \in L^2(X, m)$, we have that $w_{\Omega_n,f}$ converges strongly in $L^2(X, m)$ to $w_{\Omega,f}$.*

Proof. We first note that, up to a subsequence, $w_{\Omega_n,f}$ converges to some $w \in H$. Moreover, since Ω_n weak-γ-converges to Ω, we have that $w \in \widetilde{H}_0(\Omega)$. We now prove that w minimizes the functional $J_{\Omega,f}$. Let $v \in \widetilde{H}_0(\Omega)$ and let $v_n \in \widetilde{H}_0(\Omega_n)$ be a sequence converging to v in $L^2(X, m)$ and such that

$$\int_X |Dv|^2\, dm = \lim_{n\to\infty} \int_X |Dv_n|^2\, dm.$$

We note that such a sequence exists by Lemma 2.38. Thus we have

$$J_{\Omega,f}(v) = \lim_{n\to\infty} J_{\Omega_n,f}(v_n) \geq \liminf_{n\to\infty} J_{\Omega_n,f}(w_{\Omega_n,f}) \geq J_{\Omega,f}(w),$$

which proves that w is the minimizer of $J_{\Omega,f}$. □

Corollary 2.40. *Suppose that the functions $\varphi, \psi \in L^2(X, m)$ are Stone function for H and that the inclusion $H \hookrightarrow L^2(X, m)$ is locally compact. Then the sequence $w_{\Omega_n,\varphi}$ converges strongly in $L^2(X, m)$ to $w_{\Omega,\varphi}$, if and only if, the sequence $w_{\Omega_n,\psi}$ converges strongly in $L^2(X, m)$ to $w_{\Omega,\psi}$.*

Before we continue with our next proposition we define, for every Borel set $\Omega \subset \mathbb{R}^d$, the operator $\|\cdot\|_{\widetilde{H}_0(\Omega)} : L^2(X, m) \to [0, +\infty]$ as

$$\|u\|_{\widetilde{H}_0(\Omega)} = \begin{cases} \|u\|_H, & \text{if } u \in \widetilde{H}_0(\Omega), \\ +\infty, & \text{otherwise.} \end{cases}$$

We also recall the definition of Γ-convergence of functionals:

Definition 2.41. Given a metric space (X, d) and sequence of functionals $F_n : X \to \mathbb{R} \cup \{+\infty\}$, we say that F_n Γ-converges to the functional $F : X \to \mathbb{R} \cup \{+\infty\}$, if the following two conditions are satisfied:

(a) **(the Γ-liminf inequality)** for every sequence x_n converging to $x \in X$, we have

$$F(x) \leq \liminf_{n\to\infty} F_n(x_n);$$

(b) **(the Γ-limsup inequality)** for every $x \in X$, there exists a sequence x_n converging to x, such that

$$F(x) = \lim_{n\to\infty} F_n(x_n).^3$$

[3] Due to the Γ-liminf inequality this property is equivalent to $F(x) \geq \limsup_{n\to\infty} F_n(x_n)$.

Proposition 2.42. *Suppose that H has the stone property in $L^2(X,m)$ and that the inclusion $H \hookrightarrow L^2(X,m)$ is locally compact. Then a sequence of energy sets $\Omega_n \subset X$ strong-γ-converges to the energy set Ω, if and only if, the sequence of operators $\|\cdot\|_{\widetilde{H}_0(\Omega_n)}$ Γ-converges in $L^2(X,m)$ to $\|\cdot\|_{\widetilde{H}_0(\Omega)}$.*

Proof. Suppose first that Ω_n strong-γ-converges to Ω. Let $u_n \in L^2(X,m)$ be a sequence strongly converging to $u \in L^2(X,m)$. Let u_n be such that $\lim_{n\to\infty} \|u_n\|_{\widetilde{H}_0(\Omega_n)} < +\infty$. Then $u_n \in \widetilde{H}_0(\Omega_n)$, for every $n \in \mathbb{N}$ and $\|u_n\|_H \leq C$. Then $u \in \widetilde{H}_0(\Omega)$ and by the semi-continuity of the norm H, we have

$$\|u\|_{\widetilde{H}_0(\Omega)} \leq \liminf_{n\to\infty} \|u_n\|_{\widetilde{H}_0(\Omega_n)}.$$

Let now $u \in \widetilde{H}_0(\Omega)$. Then, by Lemma 2.38, there is a sequence $u_n \in \widetilde{H}_0(\Omega_n)$ such that

$$\|u\|_{\widetilde{H}_0(\Omega)} = \lim_{n\to\infty} \|u_n\|_{\widetilde{H}_0(\Omega_n)},$$

which proves that $\|\cdot\|_{\widetilde{H}_0(\Omega_n)}$ Γ-converges in $L^2(X,m)$ to $\|\cdot\|_{\widetilde{H}_0(\Omega)}$.

Suppose now that the Γ-convergence holds and let $\psi \in L^2(X,m)$ be a Stone function for H. Since the functional $\Psi(u) := \int_X u\psi\, dm$ is continuous in $L^2(X,m)$, we have that the sequence of functionals

$$J_{\Omega_n,\psi}(u) = \frac{1}{2}\|u\|^2_{\widetilde{H}_0(\Omega_n)} - \Psi(u),$$

Γ-converges in $L^2(X,m)$ to $J_{\Omega,\psi}$. Thus the sequence of minima $w_{\Omega_n,\psi}$ converges in $L^2(X,m)$ to some $w \in H$, which is necessarily the minimizer of $J_{\Omega,\psi}$. □

2.2.3. From the weak-γ to the strong-γ-convergence

Let $\psi \in L^2(X,m)$ be a Stone function for H and let Ω_n be a sequence of energy sets such that $w_{\Omega_n,\psi}$ converges in $L^2(X,m)$ to w. In this subsection we investigate the relation between the functions w and $w_{\Omega,\psi}$, where $\Omega = \{w > 0\}$. We will mainly consider the case when m is a finite measure and ψ is a positive constant. Fixing $\psi = 1$, we will say that the sequence Ω_n strong-γ-converges to Ω, if $w = w_{\Omega,1}$. We will prove in Proposition 2.45 that in general the inequality $w \leq w_{\Omega,1}$ always holds. The equality does not always hold as some classical examples show (see [46] or [21]).

Lemma 2.43. *Suppose that the inclusion $H \hookrightarrow L^2(X,m)$ is locally compact and that ψ is a Stone function in $L^2(X,m)$. Consider a sequence Ω_n of energy sets, weak-γ-converging to the energy set Ω, and* *the sequence of functions $w_{\Omega_n,\psi}$ converging in $L^2(X,m)$ to w such that $\{w>0\}=\Omega$. Suppose that for each $n\ge 1$ we have that $\Omega\subset\Omega_n$. Then $w=w_{\Omega,\psi}$.*

Proof. For the sake of simplicity we set $w_n=w_{\Omega_n,\psi}$. For any set $E\subset X$, we consider the functional $J_E:L^2(X,m)\to\mathbb{R}$ defined as

$$J_E(u)=\int_X\left(\frac{1}{2}|Du|^2+\frac{1}{2}u^2-\psi u\right)dm+\int_X u^2\,d\widetilde{I}_E.$$

Since Ω_n is the unique minimizer of J_{Ω_n}, by the semi-continuity of the norm $\|D(\cdot)\|_{L^2(m)}$, we have

$$J_\Omega(w)\le\liminf_{n\to\infty}J_{\Omega_n}(w_n)\le\liminf_{n\to\infty}J_{\Omega_n}(w_{\Omega,\psi})=J_\Omega(w_{\Omega,\psi}),$$

where we used $w_{\Omega,\psi}$ as a test function in $\widetilde{H}_0(\Omega_n)$. Since $w_{\Omega,\psi}$ is the unique minimizer of J_Ω, we obtain $w=w_{\Omega,\psi}$. □

Lemma 2.44. *Let H and D satisfy the conditions* H1, $mD1$, $D2$, $D3$, D4, $\mathcal{H}1$ *and* $\mathcal{H}2$ *and suppose that*

(H2) *H has the Stone property,* i.e. *if $u\in H$, then $u\wedge 1\in H$;*
(D5) *for every $u\in H$ and $c\in\mathbb{R}$, $Du=0$ m-almost everywhere on the set $\{u=c\}$.*

Then we have:

(i) *If $u\in H$ and $\varepsilon>0$, then $(u-\varepsilon)^+\in H$;*
(ii) *If $u\in H$ and $\varepsilon>0$, then $D((u-\varepsilon)^+)=\mathbb{1}_{\{u>\varepsilon\}}Du$;*
(iii) *If $\Omega\subset X$ and $f\in L^2(X,m)$, then we have*

$$\left(w_{\Omega,f}-\varepsilon\right)^+=w_{\Omega_\varepsilon,(f-\varepsilon)}\le w_{\Omega_\varepsilon,f},$$

where $\Omega_\varepsilon=\{w_{\Omega,f}>\varepsilon\}$.

Proof. Claim (i) follows by the equality $(u-\varepsilon)^+=u-u\wedge\varepsilon$. For (ii) we note that, by (D5) $D((u-\varepsilon)^+)$ vanishes on $X\setminus\{u>\varepsilon\}$. On the other hand, we have

$$D(u-u\wedge\varepsilon)\le Du+D(u\wedge\varepsilon)\quad\text{and}\quad D(u)\le D(u-u\wedge\varepsilon)+D(u\wedge\varepsilon),$$

and since $D(u\wedge\varepsilon)=0$ on $\{u>\varepsilon\}$, we obtain (ii). To prove (iii), we set $w=w_{\Omega,f}$ and note that $w_\varepsilon:=(w_{\Omega,f}-\varepsilon)^+$ is the unique minimizer of

$$J(u)=\int_X\left(\frac{1}{2}|Du|^2+\frac{1}{2}(u+w\wedge\varepsilon)^2-f(u+w\wedge\varepsilon)\right)dm,$$
$$u\in\widetilde{H}_0(\Omega_\varepsilon).$$

Thus, w_ε satisfies the equation

$$-D^2 w_\varepsilon + w_\varepsilon = f - \varepsilon \quad \text{in} \quad \Omega_\varepsilon, \qquad w_\varepsilon \in \widetilde{H}_0(\Omega_\varepsilon). \qquad \square$$

In the next Proposition we will suppose that H satisfies also conditions (H2) and (D5) from Lemma 2.44. Under these assumptions we will prove a result resembling the weak maximum principle for weak-γ-limits. We note that in $\mathbb{R}^d$ this result is immediate due to the characterization of the limit $w = \lim_{n\to\infty} w_\Omega$.

Proposition 2.45. *Let $\psi \in L^2(X,m)$ be a Stone function for H. Suppose that the inclusion $H \hookrightarrow L^2(X,m)$ is locally compact and that H satisfies* (H1), (H2), (D1), (D2), (D3), (D4), (D5), $(\mathcal{H}1)$ *and* $(\mathcal{H}2)$. *Suppose that the sequence Ω_n of energy sets is such that $w_{\Omega_n,\psi}$ converges strongly in $L^2(X,m)$ to $w \in H$. Then, setting $\Omega = \{w > 0\}$, we have $w \le w_{\Omega,\psi}$.*

Proof. Consider, for $\varepsilon > 0$, the energy set $\Omega_n^\varepsilon = \{w_{\Omega_n,\psi} > \varepsilon\}$. By Lemma 2.44, we have

$$(w_{\Omega_n,\psi} - \varepsilon)^+ \le w_{\Omega_n^\varepsilon,\psi} \le w_{\Omega_n^\varepsilon \cup \Omega,\psi}. \tag{2.17}$$

Up to a subsequence, we may suppose that $w_{\Omega_n^\varepsilon \cup \Omega,\psi}$ converges strongly in $L^2(X,m)$ to some $w^\varepsilon \in H$. On the other hand, we note that $\left(w_{\Omega_n,\psi} > \varepsilon\right)^+$ converges in $L^2(X,m)$ to $(w-\varepsilon)^+$ and so, $v_n^\varepsilon \to v^\varepsilon$ strongly in $L^2(X,m)$, where

$$v_n^\varepsilon = 1 - \frac{1}{\varepsilon}(w_{\Omega_n,\psi} \wedge \varepsilon) \qquad \text{and} \qquad v^\varepsilon = 1 - \frac{1}{\varepsilon}(w \wedge \varepsilon).$$

Thus we obtain that $v_n^\varepsilon \wedge w_{\Omega_n^\varepsilon \cup \Omega,\psi}$ converges in $L^2(X,m)$ to $v^\varepsilon \wedge w^\varepsilon$. We now have

$$v_n^\varepsilon = 0 \ \text{on} \ \Omega_n^\varepsilon \qquad \text{and} \qquad w_{\Omega_n^\varepsilon \cup \Omega,\psi} = 0 \ \text{on} \ X \setminus (\Omega_n^\varepsilon \cup \Omega),$$

and thus we obtain that

$$v_n^\varepsilon \wedge w_{\Omega_n^\varepsilon \cup \Omega,\psi} = 0 \ \text{on} \ X \setminus \Omega.$$

Passing to the limit for $n \to \infty$, we have $v^\varepsilon \wedge w^\varepsilon \in \widetilde{H}_0(\Omega)$ and since $v^\varepsilon = 1$ on $X \setminus \Omega$, we deduce that $w^\varepsilon \in \widetilde{H}_0(\Omega)$. By Lemma 2.43, we have

$$w^\varepsilon \le w_{\{w^\varepsilon > 0\},\psi} \le w_{\Omega,\psi}. \tag{2.18}$$

On the other hand we have $w_{\Omega,\psi} \le w_{\Omega_n^\varepsilon \cup \Omega,\psi}$, for every $n \in \mathbb{N}$. Passing to the limit as $n \to \infty$ we get $w^\varepsilon \ge w_{\Omega,\psi}$, which together with (2.18) gives $w_\Omega = w^\varepsilon$. We now recall that after passing to the limit as $n \to \infty$ in (2.17), we have

$$(w - \varepsilon)^+ \le w_\varepsilon = w_{\Omega,\psi}.$$

Since $\varepsilon > 0$ is arbitrary, we obtain $w \le w_{\Omega,\psi}$. $\square$

Now we can prove the following result, which is analogous to [30, Lemma 4.10].

Proposition 2.46. *Suppose that H has the Stone property in $L^2(X, m)$, that the inclusion $H \hookrightarrow L^2(X, m)$ is locally compact and that H satisfies* (H1), (H2), (D1), (D2), (D3), (D4), (D5), $(\mathcal{H}1)$ *and* $(\mathcal{H}2)$. *Suppose that $(\Omega_n)_{n\geq 1}$ is a sequence of energy sets which weak-γ-converges to the energy set Ω. Then, there exists a sequence of energy sets $(\Omega'_n)_{n\geq 1}$ strong-γ-converging to Ω such that for each $n \geq 1$ we have the inclusion $\Omega_n \subset \Omega'_n$.*

Proof. Let $\psi \in L^2(X, m)$ be a Stone function for H. Consider, for each $\varepsilon > 0$, the sequence of minimizers $w_{\Omega_n \cup \Omega^\varepsilon, \psi}$, where $\Omega_\varepsilon = \{w_{\Omega,\psi} > \varepsilon\}$. We can suppose that for each (rational) $\varepsilon > 0$ the sequence is convergent in $L^2(X, m)$ to a positive function $w_\varepsilon \in H$.

Consider the function $v_\varepsilon = 1 - \frac{1}{\varepsilon}(w_{\Omega,\psi} \wedge \varepsilon)$, which is equal to 0 on Ω_ε and to 1 on $X \setminus \Omega$. Then we have that the sequence $w_{\Omega_n \cup \Omega^\varepsilon,\psi} \wedge v_\varepsilon \in \widetilde{H}_0(\Omega_n)$ converges to $w_\varepsilon \wedge v_\varepsilon$ strongly in $L^2(X, m)$ and is bounded in H. Then, since Ω_n weak-γ-converges to Ω, by Proposition 2.34, we have $w_\varepsilon \wedge v_\varepsilon \in \widetilde{H}_0(\Omega)$. Since $v_\varepsilon = 1$ on $X \setminus \Omega$, we have that also $w_\varepsilon \in \widetilde{H}_0(\Omega)$ and so, by Proposition 2.45, we have $w_\varepsilon \leq w_{\Omega,\psi}$. On the other hand, by the weak maximum principle and Lemma 2.44, we have

$$(w_{\Omega,\psi} - \varepsilon)^+ \leq w_{\Omega_\varepsilon,\psi} \leq w_{\Omega_n \cup \Omega_\varepsilon,\psi},$$

and thus, passing to the limit as $n \to \infty$, we obtain

$$(w_{\Omega,\psi} - \varepsilon)^+ \leq w_\varepsilon \leq w_\Omega,$$

from where we can conclude by a diagonal sequence argument. □

Remark 2.47. This last result is useful in the study of functionals defined on the family of energy sets $\mathcal{E}(X)$. More precisely, in the assumptions of Proposition 2.46, suppose that

$$\mathcal{F} : \mathcal{E}(X) \to [0, +\infty],$$

is a functional on the family of energy sets such that:

(J1) $\mathcal{F}$ is lower semi-continuous (shortly, lower semicontinuous) with respect to the strong-γ-convergence, that is

$$\mathcal{F}(\Omega) \leq \liminf_{n\to\infty} \mathcal{F}(\Omega_n) \quad \text{whenever} \quad \Omega_n \xrightarrow[n\to\infty]{\text{strong-}\gamma} \Omega.$$

(J2) $\mathcal{F}$ is monotone decreasing with respect to the inclusion, that is

$$\mathcal{F}(\Omega_1) \geq \mathcal{F}(\Omega_2) \quad \text{whenever} \quad \Omega_1 \subset \Omega_2.$$

Then $\mathcal{F}$ is lower semi-continuous with respect to the (weaker!) weak-γ-convergence. Indeed, suppose that Ω_n weak-γ-converges to Ω. By Proposition 2.46, there exists a sequence of energy sets $(\Omega'_n)_{n\geq 1}$ strong-γ-converging to Ω and such that $\Omega_n \subset \Omega'_n$. Thus we have

$$\mathcal{F}(\Omega) \leq \liminf_{n\to\infty} \mathcal{F}(\Omega'_n) \leq \liminf_{n\to\infty} \mathcal{F}(\Omega_n).$$

2.2.4. Functionals on the class of energy sets

In this subsection we analyse some of the functionals defined on the family $\mathcal{E}(X)$ of energy sets in X.

For a given positive m-measurable function $h : X \to [0, +\infty]$, we consider the mass of Ω with respect to h

$$M_h(\Omega) = \int_\Omega h\, dm.$$

If, for instance, h is constantly equal to 1, then $M_h(\Omega) = m(\Omega)$.

Lemma 2.48. *For every positive m-measurable function $h : X \to [0, +\infty]$, the functional $M_h : \mathcal{E}(X) \to [0, +\infty]$ is lower semi-continuous with respect to the weak-γ-convergence.*

Proof. Consider a weak-γ-converging sequence $\Omega_n \xrightarrow[n\to\infty]{\text{weak-}\gamma} \Omega$ and the function $w \in H$ such that $\{w > 0\} = \Omega$ and $w_{\Omega_n} \to w$ in $L^2(X, m)$. Up to a subsequence, we can assume that $w_{\Omega_n}(x) \to w(x)$ for m-almost every $x \in X$. Then $1\!\!1_\Omega \leq \liminf_{n\to\infty} 1\!\!1_{\Omega_n}$ and so, by Fatou lemma

$$M_h(\Omega) = \int_X 1\!\!1_\Omega h\, dm \leq \liminf_{n\to\infty} \int_X 1\!\!1_{\Omega_n} h\, dm = \liminf_{n\to\infty} M_h(\Omega_n). \qquad \square$$

Definition 2.49. For each Borel set $\Omega \in \mathcal{B}(X)$ the "first eigenvalue of the Dirichlet Laplacian" on Ω is defined as

$$\widetilde{\lambda}_1(\Omega) = \inf \left\{ \int_\Omega |Du|^2 dm \,:\, u \in \widetilde{H}_0(\Omega),\ \int_\Omega u^2 dm = 1 \right\}. \tag{2.19}$$

More generally, we can define $\widetilde{\lambda}_k(\Omega)$, for each $k > 0$, as

$$\widetilde{\lambda}_k(\Omega) = \inf_{K \subset \widetilde{H}_0(\Omega)} \sup \left\{ \int_\Omega |Du|^2 dm \,:\, u \in K,\ \int_\Omega u^2 dm = 1 \right\}, \tag{2.20}$$

where the infimum is over all k-dimensional linear subspaces K of $H_0(\Omega)$.

Definition 2.50. For each $f \in L^2(X,m)$ and $\Omega \subset X$ the Dirichlet Energy of Ω with respect to f is defined as

$$\widetilde{E}_f(\Omega) = \inf\left\{\frac{1}{2}\int_\Omega |Du|^2\,dm + \frac{1}{2}\int_\Omega u^2 d\,m - \int_\Omega uf\,dm : u \in \widetilde{H}_0(\Omega)\right\}. \quad (2.21)$$

Proposition 2.51. *Suppose that $\Omega \subset X$ is an energy set of positive measure such that the inclusion $\widetilde{H}_0(\Omega) \hookrightarrow L^2(X,m)$ is compact. Then there is a function $u_\Omega \in \widetilde{H}_0(\Omega)$ with $\|u_\Omega\|_{L^2} = 1$ and such that $\int_\Omega |Du|^2\,dm = \widetilde{\lambda}_1(\Omega)$. More generally, for each $k > 0$, there are functions $u_1, \ldots, u_k \in \widetilde{H}_0(\Omega)$ such that:*

(a) $\|u_j\|_{L^2(X,m)} = 1$, *for each* $j = 1, \ldots, k$,
(b) $\int_X u_i u_j\,dm = 0$, *for each* $1 \le i < j \le k$,
(c) $\int_X |Du|^2 dm \le \widetilde{\lambda}_k(\Omega)$, *for each* $u = \alpha_1 u_1 + \cdots + \alpha_k u_k$, *where* $\alpha_1^2 + \cdots + \alpha_k^2 = 1$.

Proof. Suppose that $(u_n)_{n\ge1} \subset \widetilde{H}_0(\Omega)$ is a minimizing sequence for $\widetilde{\lambda}_1(\Omega)$ such that $\|u_n\|_{L^2(X,m)} = 1$. Then $(u_n)_{n\ge1}$ is bounded with respect to the norm of H and so, there is a subsequence, still denoted in the same way, which strongly converges in $L^2(X,m)$ to some function $u \in H$:

$$u_n \xrightarrow[n\to\infty]{L^2(X,m)} u \in H.$$

We have that $\|u\|_{L^2} = 1$ and

$$\int_\Omega |Du|^2\,dm \le \liminf_{n\to\infty} \int_\Omega |Du_n|^2\,dm = \widetilde{\lambda}_1(\Omega).$$

Thus, u is the desired function. The proof in the case $k > 1$ is analogous. □

Proposition 2.52. *Suppose that H has the stone property in $L^2(X,m)$ and that the inclusion $H \hookrightarrow L^2(X,m)$ is compact. Then the functional $\widetilde{\lambda}_k : \mathcal{E}(X) \to \mathbb{R}$ defined by (6.40) is decreasing with respect to the set inclusion and lower semicontinuous with respect to the weak-γ-convergence.*

Proof. The monotonicity of $\widetilde{\lambda}_k$ with respect to the set inclusion holds since $\omega \subset \Omega$ implies $\widetilde{H}_0(\omega) \subset \widetilde{H}_0(\Omega)$.

We now prove the lower semi-continuity of $\widetilde{\lambda}_k$. Let $\Omega_n \xrightarrow[n\to\infty]{\text{weak-}\gamma} \Omega$, that is for some Stone function $\psi \in L^2(X,m)$ we have $w_{\Omega_n,\psi} \xrightarrow[n\to\infty]{L^2(X,m)} w$

with $w \in H$ and $\Omega = \{w > 0\}$. We can suppose that the sequence $\widetilde{\lambda}_k(\Omega_n)$ is bounded by some positive constant C_k. Let for each $n > 0$ the functions $u_1^n, \dots, u_k^n \in \widetilde{H}_0(\Omega_n)$ satisfy the conditions (a), (b) and (c) of Proposition 2.51. Then, we have that up to a subsequence we can suppose that u_j^n converges in $L^2(X, m)$ to some function $u_j \in H$. By Proposition 2.34, we have that $u_j \in \widetilde{H}_0(\Omega), \forall j = 1, \dots, k$. Consider the linear subspace $K \subset \widetilde{H}_0(\Omega)$ generated by $u_1, \dots, u_k$. Since $u_1, \dots, u_k$ are mutually orthogonal in $L^2(X, m)$, we have that $\dim K = k$ and so

$$\widetilde{\lambda}_k(\Omega) \leq \sup\left\{\int_\Omega |Du|^2\, dm : u \in K, \int_\Omega u^2\, dm = 1\right\}.$$

It remains to prove that for each $u \in K$ such that $\|u\|_{L^2(X,m)} = 1$, we have

$$\int_X |Du|^2\, dm \leq \liminf_{n\to\infty} \widetilde{\lambda}_k(\Omega_n).$$

In fact, we can suppose that $u = \alpha_1 u_1 + \dots + \alpha_k u_k$, where $\alpha_1^2 + \dots + \alpha_k^2 = 1$. Thus u is the strong limit in $L^2(X, m)$ of the sequence $u^n = \alpha_1 u_1^n + \dots + \alpha_k u_k^n \in \widetilde{H}_0(\Omega_n)$ and, by the semi-continuity of the norm of the gradient, we obtain

$$\int_X |Du|^2\, dm \leq \liminf_{n\to\infty} \int_X |Du^n|^2\, dm \leq \liminf_{n\to\infty} \widetilde{\lambda}_k(\Omega_n),$$

as required. □

Remark 2.53. If we drop the compactness assumption for inclusion $H \hookrightarrow L^2(X, m)$, then the semi-continuity of $\widetilde{\lambda}_k$ with respect to the weak-γ-convergence does not hold in general. For example consider $X = \mathbb{R}^d$ and $H = H^1(\mathbb{R}^d)$. Taking as a Stone function the Gaussian $\psi(x) = e^{-|x|^2/2}$, we have that the sequence of solutions[4] of

$$-\Delta w_n + w_n = \psi \quad \text{in} \quad B_1(x_n), \qquad w_n \in H_0^1(B_1(x_n)),$$

converges strongly to zero in $L^2(\mathbb{R}^d)$, as $x_n \to \infty$, since we have $\|w\|_{L^2} \leq \|\psi\|_{L^2(B_1(x_n))}$. Thus the sequence of unit balls $B_1(x_n)$ strong-γ-converges to the empty set, as $|x_n| \to \infty$ and so the semi-continuity does not hold:

$$\widetilde{\lambda}_1(B_1) = \liminf_{n\to\infty} \widetilde{\lambda}_1(B_1(x_n)) < \widetilde{\lambda}_1(\emptyset) = +\infty.$$

[4] In the Euclidean space $\mathbb{R}^d$ we have $H_0^1(B) = \widetilde{H}_0^1(B)$, for every ball B.

Proposition 2.54. *Suppose that H has the Stone property in $L^2(X, m)$ and that the inclusion $H \hookrightarrow L^2(X, m)$ is locally compact. Then, for every $f \in L^2(X, m)$, the functional $\widetilde{E}_f : \mathcal{E}(X) \to \mathbb{R}$ from Definition 2.50, is decreasing with respect to the set inclusion and lower semi-continuous with respect to the weak-γ-convergence.*

Proof. The fact that $\widetilde{E}_f$ is decreasing follows by the same argument as in Proposition 2.52. In order to prove the semi-continuity of $\mathcal{E}_f$, we consider a sequence Ω_n weak-γ-converging to Ω. Let now u_n be the solution of

$$-D^2 u_n + u_n = f \quad \text{in} \quad \Omega_n, \qquad u_n \in \widetilde{H}_0(\Omega_n).$$

Then we have that u_n is bounded in H. Moreover u_n is bounded from above and below by the solutions $u', u'' \in H$ of the equations

$$-D^2 u' + u' = |f| \quad \text{in} \quad X, \qquad u' \in H,$$
$$-D^2 u'' + u'' = -|f| \quad \text{in} \quad X, \qquad u' \in H.$$

Thus, u_n converges in $L^2(X, m)$ to some $u \in H$. By the weak-γ-convergence of Ω_n to Ω, we have that $u \in \widetilde{H}_0(\Omega)$ and by the semi-continuity of the $L^2(m)$-norm of Du, we have

$$\begin{aligned}\widetilde{E}_f(\Omega) &\leq \int_\Omega \left(\frac{1}{2}|Du|^2 + \frac{1}{2}u^2 - fu\right) dm \\ &\leq \liminf_{n\to\infty} \int_{\Omega_n} \left(\frac{1}{2}|Du_n|^2 + \frac{1}{2}u_n^2 - fu_n\right) dm = \liminf_{n\to\infty} \widetilde{E}_f(\Omega_n). \quad \square\end{aligned}$$

One can easily extend the above result to a much wider class functionals, depending on $w_{\Omega,f}$.

Proposition 2.55. *Suppose that H satisfies has the Stone property in $L^2(X, m)$, that the inclusion $H \hookrightarrow L^2(X, m)$ is locally compact and that satisfies the conditions* (H1), (H2), (D1), (D2), (D3), (D4), (D5), $(\mathcal{H}1)$ *and* $(\mathcal{H}2)$. *Let $j : X \times \mathbb{R} \to \mathbb{R}$ be a measurable function such that:*

(a) *$j(x, \cdot)$ is lower semi-continuous and decreasing for m-almost every $x \in X$;*
(b) *$j(x, s) \geq -\alpha(x)s - \beta s^2$, where $\beta \geq 0$ is a constant and $\alpha \in L^2(X, m)$ is a given function.*

Then for a given non-negative $f \in L^2(X, m)$, we have that the functional

$$\mathcal{F}_j(\Omega) = \int_X j(x, w_{\Omega,f})\, dx,$$

is decreasing with respect to the set inclusion and is lower semi-continuous with respect to the weak-γ-convergence.

Proof. Let $\omega \subset \Omega$. By the weak maximum principle, we get $w_{\omega,f} \leq w_{\Omega,f}$. Then

$$j(x, w_{\omega,f}(x)) \geq j(x, w_{\Omega,f}(x)), \quad \text{for every} \quad x \in X,$$

which proves the monotonicity part. For the lower semi-continuity we first notice that by Remark 2.47, it is sufficient to prove that F_j is lower semi-continuous with respect to the strong-γ-convergence. Consider a sequence Ω_n strong-γ-converging to Ω. By Proposition 2.39, we have that $w_{\Omega_n,f}$ converges in $L^2(X,m)$ to $w_{\Omega,f}$ and so, we have

$$j(x, w_{\Omega,f}(x)) \leq \liminf_{n\to\infty} j(x, w_{\Omega_n,f}(x)).$$

Since, for every $E \subset X$, we have

$$\begin{aligned} j(x, w_{E,f}(x)) &\geq j(x, w_{X,f}(x)) \\ &\geq -\alpha(x) w_{X,f}(x) - \beta w_{X,f}(x)^2 \in L^1(X,m), \end{aligned}$$

we can apply the Dominated Convergence Theorem, for the negative part of the function $j(x, w_{\Omega_n,f}(x))$, and the Fatou Lemma, for the positive part, obtaining the semi-continuity of $\mathcal{F}_j$. □

2.3. Capacity, quasi-open sets and quasi-continuous functions

Our main example of a couple $H \subset L^2(X,m), D : H \to L^2(X,m)$ is the Sobolev space $H = H^1(\mathbb{R}^d)$ and the modulus of the gradient $Du = |\nabla u|$. In this classical framework, we consider an open set $\Omega \subset \mathbb{R}^d$ and the Sobolev space $H_0^1(\Omega)$ on Ω. Denoting with $\widetilde{H}_0^1(\Omega) := \widetilde{H}_0(\Omega)$, we have that, in general, the spaces $\widetilde{H}_0^1(\Omega)$ and $H_0^1(\Omega)$ might be different. Thus also the functionals on the subsets Ω of $\mathbb{R}^d$, defined by minimizing a functional on $H_0^1(\Omega)$ or $\widetilde{H}_0^1(\Omega)$, might be different. In order to have a true extension of these functionals, classically defined for open sets Ω and the Sobolev spaces $H_0^1(\Omega)$, we need a new notion of a Sobolev space on a generic measurable set $\Omega \subset \mathbb{R}^d$. Classically, this definition is given through the notion of *capacity* and, as we will see below, can be extended to a very general setting.

In this section we give the notion of capacity in a very general setting, which is a natural continuation of the discussion in the previous sections; we then introduce the Sobolev spaces $H_0(\Omega)$ for a generic set Ω and show that the natural domains for these spaces are again the energy sets, introduced above. At the end of the section we discuss the questions concerning the shape optimization problems in the different frameworks of $H_0(\Omega)$ and $\widetilde{H}_0(\Omega)$.

Let $H \subset L^2(X, m)$ and $D : H \to L^2(X, m)$ satisfy the properties (H1), (H2), (D1), (D2), (D3), (D4), (D5), ($\mathcal{H}1$) and ($\mathcal{H}2$). We assume, furthermore, that

($\mathcal{H}3$) the linear subspace $H \cap C(X)$, where $C(X)$ denotes the set of real continuous functions on X, is dense in H with respect to the norm $\|\cdot\|_H$;
($\mathcal{H}4$) for every open set $\Omega \subset X$, there is a function $u \in H \cap C(X)$ such that $\{u > 0\} = \Omega$.

Remark 2.56. We note that ($\mathcal{H}4$) is equivalent to assume that for every ball $B_r(x) \subset X$ there is a function $u \in H \cap C(X)$ such that $\{u > 0\} = B_r(x)$.

Definition 2.57. We define the capacity (that depends on H and D) of an arbitrary set $\Omega \subset X$ as

$$\operatorname{cap}(\Omega) = \inf\left\{\|u\|_H^2 \ :\ u \in H,\ u \geq 1 \text{ in a neighbourhood of } \Omega\right\}. \quad (2.22)$$

We say that a property P holds quasi-everywhere (shortly q.e.), if the set on which it does not hold has zero capacity.

Remark 2.58. If $u \in H$ is such that $u \geq 0$ on X and $u \geq 1$ on $\Omega \subset X$, then $\|u\|_H^2 \geq m(\Omega)$. Thus, we have that $\operatorname{cap}(\Omega) \geq m(\Omega)$ and, in particular, if the property P holds q.e., then it also holds m-a.e.

It is straightforward to check that the capacity is an outer measure. More precisely, we have the following result.

Proposition 2.59.

(1) *If* $\omega \subset \Omega$, *then* $\operatorname{cap}(\omega) \leq \operatorname{cap}(\Omega)$.
(2) *If* $(\Omega_n)_{n\in\mathbb{N}}$ *is a family of disjoint sets, then*

$$\operatorname{cap}\left(\bigcup_{n=1}^{\infty} \Omega_n\right) \leq \sum_{n=1}^{\infty} \operatorname{cap}(\Omega_n).$$

(3) *For every* $\Omega_1, \Omega_2 \subset X$, *we have that*

$$\operatorname{cap}(\Omega_1 \cup \Omega_2) + \operatorname{cap}(\Omega_1 \cap \Omega_2) \leq \operatorname{cap}(\Omega_1) + \operatorname{cap}(\Omega_2).$$

(4) *If* $\Omega_1 \subset \Omega_2 \subset \cdots \subset \Omega_n \subset \ldots$, *then we have*

$$\operatorname{cap}\left(\bigcup_{n=1}^{\infty} \Omega_n\right) = \lim_{n\to\infty} \operatorname{cap}(\Omega_n).$$

Proof. Point (1) is a direct consequence of the definition; for a proof of point (2) see [62, Theorem 1, Section 4.7], while for the point (3) and (4) we refer to [62, Theorem 2, Section 4.7]. □

Remark 2.60. We note that the family of sets of zero capacity is closed with respect to the intersection and union of two sets, as well as, with respect to the denumerable unions.

Remark 2.61. Definition 2.58 coincides with the classical definition of capacity when $X = \mathbb{R}^d$ and $H = H^1(\mathbb{R}^d)$. For an introduction to the capacity in $\mathbb{R}^d$ we refer to [62] and [72].

Remark 2.62. We note that if $1 \in H$, then we simply have $\mathrm{cap}(\Omega) = m(\Omega)$. For example, this is the case when X is a compact differentiable manifold and H is the Sobolev space on X. Thus our definition is not satisfactory in all cases. For manifolds, for example it is natural to define the sets of capacity zero using the local charts and the definition in the Euclidean space, *i.e. we say that* $E \subset X$ *is of zero capacity* ($\mathrm{cap}(E) = 0$), *if for every* $r > 0$ *and every* $x \in X$ *we have* $\mathrm{cap}(\Omega \cap B_r(x);\, B_{2r}(x)) = 0$, *where*

$$\begin{aligned}&\mathrm{cap}\left(\Omega \cap B_r(x);\, B_{2r}(x)\right)\\&:= \inf\left\{\|u\|_H^2 : u \in \widetilde{H}_0(B_{2r}(x)),\, u \geq 1 \text{ in a neighbourhood of } \Omega \cap B_r(x)\right\}.\end{aligned}$$

Thus, one may define the capacity $\overline{\mathrm{cap}}$ as

$$\overline{\mathrm{cap}}(E) = \sup\left\{\mathrm{cap}\left(\Omega \cap B_r(x);\, B_{2r}(x)\right) : r \in (0, +\infty],\ x \in X\right\}.$$

In order to obtain the same results as below, one would need a further assumption on the space H. Namely that the existence of functions $\phi_{r,x} \in \widetilde{H}_0(B_{2r}(x))$ such that $\phi_{x,r} \equiv 1$, for every $x \in X$ and $r > 0$. Below we prefer to avoid this further technical complication and work with the capacity from Definition 2.57.

Definition 2.63. A function $u : X \to \mathbb{R}$ is said to be quasi-continuous if there exists a decreasing sequence of open sets $(\omega_n)_{n\geq 1}$ such that:

- $\mathrm{cap}(\omega_n) \xrightarrow[n\to\infty]{} 0$,
- On the complementary ω_n^c of ω_n the function u is continuous.

Definition 2.64. We say that a set $\Omega \subset X$ is quasi-open if there exists a sequence of open sets $(\omega_n)_{n\geq 1}$ such that

- $\Omega \cup \omega_n$ is open for each $n \geq 1$,
- $\mathrm{cap}(\omega_n) \xrightarrow[n\to\infty]{} 0$.

Remark 2.65. The sequence of open sets ω_n in both Definition 2.63 and Definition 2.64 can be taken to be decreasing.

The following two Propositions contain some of the fundamental properties of the quasi-continuous functions and the quasi-open sets.

Proposition 2.66. *Suppose that a function $u : X \to \mathbb{R}$ is quasi-continuous. Then*

(a) *the level set $\{u > 0\}$ is quasi-open;*
(b) *if $u \geq 0$ m-a.e., then $u \geq 0$ q.e. on X.*

Proof. See [72, Proposition 3.3.41] for a proof of (a) and [72, Proposition 3.3.30] for a proof of (b). □

Proposition 2.67.

(a) *For every function $u \in H$, there is a quasi-continuous function $\tilde{u}$ such that $u = \tilde{u}$ m-a.e.. We say that $\tilde{u}$ is a quasi-continuous representative of $u \in H$. If $\tilde{u}$ and $\tilde{u}'$ are two quasi-continuous representatives of $u \in H$, then $\tilde{u} = \tilde{u}'$ q.e.*
(b) *If $u_n \xrightarrow[n\to\infty]{H} u$, then there is a subsequence $(u_{n_k})_{k\geq 1} \subset H$ such that, for the quasi-continuous representatives of u_{n_k} and u, we have*
$$\tilde{u}_{n_k}(x) \xrightarrow[n\to\infty]{} \tilde{u}(x),$$
for q.e. $x \in X$.

Proof. See [72, Theorem 3.3.29] for a proof of (a), and [72, Proposition 3.3.33] for a proof of (b). □

Remark 2.68. We consider the following relations of equivalence on the Borel measurable functions
$$u \overset{cp}{\sim} v, \text{ if } u = v \text{ q.e.}, \quad u \overset{m}{\sim} v, \text{ if } u = v\ m\text{-a.e.}$$
We define the space
$$H^{cp} := \{u : X \to \mathbb{R} \ :\ u \text{ quasi-continuous}, u \in H\}/ \overset{cp}{\sim}, \tag{2.23}$$
and recall that
$$H := \{u : X \to \mathbb{R} \ :\ u \in H\}/ \overset{m}{\sim}. \tag{2.24}$$
Then the Banach spaces H^{cp} and H, both endowed with the norm $\|\cdot\|_H$, are isomorphic. In fact, in view of Proposition 2.66 and Proposition 2.67,

it is straightforward to check that the map $[u]_{cp} \mapsto [u]_m$ is a bijection, where $[u]_{cp}$ and $[u]_m$ denote the classes of equivalence of u related to $\overset{cp}{\sim}$ and $\overset{m}{\sim}$, respectively. In the sequel we will not make a distinction between H and H^{cp} and every function $u \in H$ will be identified with its quasi-continuous representative.

Proposition 2.69. *Let $\Omega \subset X$ be a quasi-open set. Then there is a (quasi-continuous) function $u \in H$ such that $\Omega = \{u > 0\}$ up to a set of zero capacity.*

Proof. Let ω_n be the sequence of open sets from Definition 2.63 and let $v_n \in H$ be such that $\omega_n \subset \{v_n = 1\}$ and $\|v_n\|_H^2 \leq 2\,\mathrm{cap}(\omega_n)$. Let $u_n \in H$ be such that $\{u_n > 0\} = \Omega \cup \omega_n$. Then $w_n = u_n \wedge (1 - v_n) \in H$ is such that $\{w_n > 0\} \subset \Omega$ and

$$\mathrm{cap}(\Omega \setminus \{w_n > 0\}) \leq \|v_n\|_H^2 \leq 2\,\mathrm{cap}(\omega_n).$$

After multiplying to an appropriate constant, we may suppose that $\|w_n\|_H \leq 2^{-n}$. Thus the limit $w = \sum_{n=1}^{\infty} w_n$ exists and $\{w > 0\} \subset \Omega$ q.e.. On the other hand

$$\mathrm{cap}(\Omega \setminus \{w > 0\}) \leq \mathrm{cap}(\Omega \setminus \{w_n > 0\}) \leq 2\,\mathrm{cap}(\omega_n),$$

and thus, passing to the limit as $n \to \infty$, we have the claim. □

Definition 2.70. For each $\Omega \subset X$ we define the space

$$H_0(\Omega) := \{u \in H \,:\, \mathrm{cap}(\{u \neq 0\} \setminus \Omega) = 0\}, \tag{2.25}$$

which, by Proposition 2.67 (b), is a closed linear subspace of H.

We define the function I_Ω on the m-measurable sets as

$$I_\Omega(E) = \begin{cases} 0, \text{ if } \mathrm{cap}(E \setminus \Omega) = 0, \\ +\infty, \text{ if } \mathrm{cap}(E \setminus \Omega) > 0. \end{cases} \tag{2.26}$$

Then I_Ω is a Borel measure on X. Moreover, if u and v are two nonnegative functions on X and $u = v$ quasi-everywhere on X, then we have that $\int_X u\, dI_\Omega = \int_X v\, dI_\Omega$. As a consequence the map

$$u \mapsto \int_X u^2\, dI_\Omega,$$

is well defined on H and so, we have the characterization

$$H_0(\Omega) = \left\{u \in H \,:\, u \in L^2(I_\Omega)\right\} = \left\{u \in H : \int_X u^2\, dI_\Omega < +\infty\right\}.$$

Thus, substituting I_Ω in place of the measure μ in Proposition 2.34, we have

Proposition 2.71. *Suppose that H has the Stone property in $L^2(X,m)$. Then for every $u \in H_0(\Omega)$, we have that* $\mathrm{cap}(\{w>0\}\setminus\{u\neq 0\})=0$, *where w is the minimizer in $H_0(\Omega)$ of the functional*

$$J_{\Omega,\psi}(u) := \frac{1}{2}\int_\Omega |Du|^2\,dm + \frac{1}{2}\int_\Omega u^2\,dm - \int_\Omega u\psi\,dm.$$

Remark 2.72. Proposition 2.71 suggests that the natural domains for the spaces $H_0(\Omega)$ are the quasi-open sets. Indeed, for every measurable set $\Omega \subset X$, there is a quasi-open set $\omega \subset \Omega$ such that $H_0(\omega) = H_0(\Omega)$.

Remark 2.73. We note that the inclusion $H_0(\Omega) \subset \widetilde{H}_0(\Omega)$ holds for each subset $\Omega \subset X$ and, in general, may be strict. For example, if $\Omega \subset \mathbb{R}^2$ is a square minus a horizontal line, *i.e.*

$$X = \mathbb{R}^2, \quad H = H^1(\mathbb{R}^2) \ \text{ and } \ \Omega = (-1,1)\times\{(-1,0)\cup(0,1)\} \subset \mathbb{R}^2,$$

then we have $H_0(\Omega) \neq \widetilde{H}_0(\Omega)$.

Proposition 2.74. *Suppose that H is uniformly convex and has the Stone property in $L^2(X,m)$. Let $\Omega \subset X$ be a given set. Then there is a quasi-open set ω such that $\omega \subset \Omega$ m-a.e. and*

$$H_0(\omega) = \widetilde{H}_0(\omega) = \widetilde{H}_0(\Omega). \tag{2.27}$$

Moreover, ω is unique up to a set of zero capacity.

Proof. Let w be (the quasi-continuous representative of) the solution of

$$-D^2 w + w = \psi \quad \text{in} \quad \Omega, \qquad w \in \widetilde{H}_0(\Omega),$$

where $\psi \in L^2(X,m)$ is the Stone function for H. Let $u \in \widetilde{H}_0(\Omega)$ be nonnegative and such that $u \leq \psi$ and let $u_\varepsilon \in \widetilde{H}_0(\Omega)$ be the sequence from Proposition 2.15 relative to the measure $\widetilde{I}_\Omega$. Since $u_\varepsilon \leq C\varepsilon^{-1}w$, we have that $\mathrm{cap}(\{u_\varepsilon>0\}\setminus\{w>0\})=0$. Moreover, by Remark 2.16, we have that u_ε converges strongly in H to u and so, $\mathrm{cap}(\{u>0\}\setminus\{w>0\})=0$, which proves that $\widetilde{H}_0(\Omega) \subset H_0(\{w>0\})$. Thus, we obtain the existence part by choosing $\omega=\{w>0\}$.

Suppose that $\omega=\{u>0\}$ and $\omega'=\{u'>0\}$ are two quasi-open sets satisfying (2.27). Then, $u' \in \widetilde{H}_0(\Omega) = H_0(\omega)$ and so, $\omega'=\{u'>0\}\subset\omega$ q.e. and analogously, $\omega \subset \omega'$ quasi-everywhere. □

Remark 2.75. One can substitute the uniform convexity assumption in Proposition 2.74 with the assumption that the space H is separable. If this is the case, consider a countable dense subset $(u_k)_{k=1}^{\infty} = \mathcal{A} \subset \widetilde{H}_0(\Omega)$. Then the desired quasi-open set is

$$\omega := \bigcup_{u \in \mathcal{A}} \{u \neq 0\} = \{w > 0\}, \quad \text{where} \quad w = \sum_{k=1}^{\infty} \frac{|u_k|}{2^k \|u_k\|_H}.$$

In fact, let $u \in \widetilde{H}_0(\Omega)$. Then, there is a sequence $(u_n)_{n \geq 1} \subset \mathcal{A}$ such that $u_n \xrightarrow[n\to\infty]{H} u$ and, by Proposition 2.67 (b), $u = 0$ q.e. on $X \setminus \omega$ and so, we have the existence of ω. The uniqueness follows as in Proposition 2.74.

Proposition 2.76. *Every quasi-open set is an energy set and every energy set is a quasi-open set, up to a set of measure zero.*

Proof. The first part of the claim follows since, by Proposition 2.69, every quasi-open set is of the form $u > 0$ for some $u \in H$. On the other hand, by Remark 2.27, the sets of the form $\{u > 0\}$ are energy sets. For the second part of the claim, we note that by the Definition of the energy set, we have that there is $w \in H$ such that $m(\Omega \Delta \{w > 0\}) = 0$. □

2.3.1. Quasi-open sets and energy sets from a shape optimization point of view

In this subsection we show that for a large class of shape optimization problems, working with energy sets or quasi-open sets makes no difference. This is the case when we consider spectral or energy optimization problems. The main reason for this fact is that the shape functionals are in fact not functionals on the sets Ω, but functionals on the Sobolev spaces $\widetilde{H}_0(\Omega)$ or $H_0(\Omega)$.

Suppose that F is a decreasing functional on the family of closed linear subspaces of H. Then we can define the functional $\widetilde{\mathcal{F}}$ on the family of Borel sets, by $\widetilde{\mathcal{F}}(\Omega) = F(\widetilde{H}_0(\Omega))$, and the functional $\mathcal{F}$ on the class of quasi-open sets, by $\mathcal{F}(\Omega) = F(H_0(\Omega))$. The following result shows that the shape optimization problems with measure constraint, related to $\mathcal{F}$ and $\widetilde{\mathcal{F}}$, are equivalent.

Theorem 2.77. *Suppose that H has the Stone property in $L^2(X, m)$ and that is separable or uniformly convex. Let F be a functional on the family of closed linear spaces of H, which is decreasing with respect to the inclusion. Then, we have that*

$$\begin{aligned} &\inf\Big\{F(\widetilde{H}_0(\Omega)) \;:\; \Omega \text{ Borel}, m(\Omega) \leq c\Big\} \\ &= \inf\Big\{F(H_0(\Omega)) \;:\; \Omega \text{ quasi-open}, m(\Omega) \leq c\Big\}. \end{aligned} \tag{2.28}$$

Moreover, if one of the infima is achieved, then the other one is also achieved.

Proof. We first note that by Corollary 2.25 and Proposition 2.76, the infimum in the right hand side of (2.28) can be considered on the family of quasi-open sets. Since F is a decreasing functional, we have that for each quasi-open $\Omega \subset X$

$$F(\widetilde{H}_0(\Omega)) \leq F(H_0(\Omega)).$$

On the other hand, by Proposition 2.76, there exists a quasi-open set ω such that $m(\omega) < m(\Omega)$ and $F(\widetilde{H}_0(\Omega)) = F(H_0(\omega))$ and so, we have that the two infima are equal.

Suppose now that Ω_{cp} is a solution of the problem

$$\min\Big\{F(H_0(\Omega)) \ : \ \Omega \text{ quasi-open}, m(\Omega) \leq c\Big\}.$$

Then we have that

$$F(\widetilde{H}_0(\Omega_{cp})) \leq F(H_0(\Omega_{cp})) = \inf\left\{F(\widetilde{H}_0(\Omega)) : \ \Omega \text{ Borel}, m(\Omega) \leq c\right\},$$

and so the infimum on the right hand side in (2.28) is achieved, too.

Let Ω_m be a solution of the problem

$$\min\Big\{F(\widetilde{H}_0(\Omega)) : \ \Omega \text{ Borel}, m(\Omega) \leq c\Big\},$$

and let $\tilde{\Omega}_m \subset \Omega_m$ a.e. such that $H_0(\tilde{\Omega}_m) = \widetilde{H}_0(\Omega_m)$. Then the infimum in the right hand side in (2.28) is achieved in $\tilde{\Omega}_m$. In fact, we have

$$F(H_0(\tilde{\Omega}_m)) = F(\widetilde{H}_0(\Omega_m)) = \inf\Big\{F(H_0(\Omega)) : \Omega \text{ quasi-open}, m(\Omega) \leq c\Big\},$$

which concludes the proof. □

Example 2.78. Typical examples of functionals satisfying the hypotheses of Theorem 2.77 are the eigenvalues λ_k defined variationally. Indeed, for any subspace $L \subset H$, we define

$$\Lambda_k(L) = \min_{S_k \subset L} \max_{u \in S_k \setminus \{0\}} \frac{\int_X |Du|^2 \, dm}{\int_X u^2 \, dm},$$

where the minimum is over the k-dimensional subspaces S_k of L. Thus, we have

$$\Lambda_k(\widetilde{H}_0(\Omega)) = \widetilde{\lambda}_k(\Omega) \qquad \text{and} \qquad \Lambda_k(H_0(\Omega)) = \lambda_k(\Omega),$$

where for each $\Omega \subset X$, we define

$$\lambda_k(\Omega) = \min_{S_k \subset H_0(\Omega)} \max_{u \in S_k \setminus \{0\}} \frac{\int_\Omega |Du|^2 dm}{\int_\Omega u^2 dm}, \tag{2.29}$$

where the minimum is over the k-dimensional subspaces S_k of $H_0(\Omega)$.

2.4. Existence of optimal sets in a box

In this section we apply the theory developed in Sections 2.1, 2.2 and 2.3. We state here a general Theorem in the abstract setting from these sections and then we will apply it to different situations.

Theorem 2.79. *Let (X, d) be a metric space and let m be a σ-finite Borel measure on X. Suppose that $H \subset L^2(X, m)$ has the Stone property in $L^2(X, m)$, that the inclusion $H \hookrightarrow L^2(X, m)$ is locally compact and that H satisfies the conditions* (H1), (H2), (D1), (D2), (D3), (D4), (D5), ($\mathcal{H}$1) *and* ($\mathcal{H}$2). *Let $\mathcal{F} : \mathcal{E}(X) \to \mathbb{R}$ be a functional on the family of energy sets $\mathcal{E}(X)$ and such that:*

- *$\mathcal{F}$ is decreasing with respect to the set inclusion;*
- *$\mathcal{F}$ is lower semicontinuous with respect to the strong-γ-convergence.*

Then, for every couple $A \subset B \subset X$ of energy sets, the shape optimization problem

$$\min \left\{ \mathcal{F}(\Omega) : \Omega \in \mathcal{E}(X), \ A \subset \Omega \subset B, \ \int_\Omega h \, dm \leq 1 \right\}, \tag{2.30}$$

has a solution for every m-measurable function $h : X \to [0, +\infty]$.

Proof. Let Ω_n be a minimizing sequence for (2.31). Then there is a set $\Omega \subset X$ such that Ω_n weak-γ-converges to Ω. We note that by the maximum principle we have $A \subset \Omega \subset B$. Moreover, in view of Lemma 2.48 and Remark 2.47, we have

$$\int_\Omega h \, dm \leq \liminf_{n \to \infty} \int_{\Omega_n} h \, dm \quad \text{and} \quad \mathcal{F}(\Omega) \leq \liminf_{n \to \infty} \mathcal{F}(\Omega_n),$$

which proves that Ω minimizes (2.31). □

Remark 2.80. We note that in the above Theorem one can take $A = \emptyset$ and also $B = X$.

Corollary 2.81. *Suppose that $H \subset L^2(X, m)$ satisfies the hypotheses of Theorem 2.79 and also conditions ($\mathcal{H}3$) and ($\mathcal{H}4$). Suppose, moreover, that H is separable or uniformly convex. Let $\mathcal{F}$ be a functional on the subspaces of H, decreasing with respect to the inclusion and such that the functional $\Omega \mapsto \mathcal{F}(\widetilde{H}_0(\Omega))$ is lower semicontinuous with respect to the strong-γ-convergence.*

Then, for every couple $A \subset B \subset X$ of quasi-open sets, the shape optimization problem

$$\min\Big\{\mathcal{F}\big(H_0(\Omega)\big) : \Omega \textit{ quasi-open}, \ A \subset \Omega \subset B, \ \int_\Omega h\,dm \leq 1\Big\}, \tag{2.31}$$

has a solution for every m-measurable function $h : X \to [0, +\infty]$.

2.4.1. The Buttazzo-Dal Maso Theorem

The first general result in the shape optimization was stated in the Eucldean setting. Indeed, taking $H = H^1(\mathbb{R}^d)$ and $Du = |\nabla u|$, we can define the weak-γ and the strong-γ-convergence as in Section 2.2. The following Theorem was proved in [33] and is now a consequence of Theorem 2.79.

Theorem 2.82. *Consider $\mathcal{D} \subset \mathbb{R}^d$ a bounded open set suppose that $\mathcal{F}$ is a functional on the quasi-open sets of $\mathbb{R}^d$, decreasing with respect to the set inclusion and lower semi-continuous with respect to the strong-γ-convergence. Then the shape optimization problem*

$$\min\Big\{\mathcal{F}(\Omega) : \Omega \textit{ quasi-open}, \ \Omega \subset \mathcal{D}, \ |\Omega| \leq c\Big\}, \tag{2.32}$$

has a solution.

Remark 2.83. In particular, the Buttazzo-Dal Maso theorem applies for functions depending on the spectrum of the Dirichlet Laplacian $\lambda_1(\Omega) \leq \lambda_2(\Omega) \leq \dots$ on Ω, which we recall are variationally characterized as

$$\lambda_k(\Omega) = \min_{S_k \subset H_0^1(\Omega)} \max_{u \in S_k \setminus \{0\}} \frac{\int_\Omega |\nabla u|^2\,dx}{\int_\Omega u^2\,dx}, \tag{2.33}$$

where the minimum is over the k-dimensional subspaces S_k of the Sobolev space $H_0^1(\Omega)$. Suppose that the function $F : \mathbb{R}^{\mathbb{N}} \to [0, +\infty]$ satisfies the following conditions:

(F1) If $z \in [0, +\infty]^{\mathbb{N}}$ and $(z_n)_{n \geq 1} \subset [0, +\infty]^{\mathbb{N}}$ is a sequence such that for each $j \in \mathbb{N}$

$$z_n^{(j)} \xrightarrow[n \to \infty]{} z^{(j)},$$

where $z_n^{(j)}$ indicates the jth component of z_n, then

$$F(z) \leq \liminf_{n\to\infty} F(z_n).$$

(F2) If $z_1^{(j)} \leq z_2^{(j)}$, for each $j \in \mathbb{N}$, then $F(z_1) \leq F(z_2)$.

Then the optimization problem

$$\min\Big\{F\big(\lambda_1(\Omega), \lambda_2(\Omega), \dots\big) \ : \ \Omega \subset \mathcal{D},\ \Omega \text{ quasi-open}, |\Omega| \leq c\Big\},$$

has a solution.

2.4.2. Optimal partition problems

In this subsection we recall a generalization of the Buttazzo-Dal Maso Theorem related to the partition problems. The existence of optimal partitions of quasi-open sets is a well-known result. We state it here for a class of functionals which may involve also the measures of the different regions. Following the terminology of [29], we call the optimization problems for this type of cost functionals *multiphase shape optimization problems*.

We consider a quasi-open set $\mathcal{D} \subset \mathbb{R}^d$ of finite Lebesgue measure and a functional $\mathcal{F}$ on the h-tuples of quasi-open subsets of $\mathcal{D}$ with the following properties:

($\mathcal{F}1$) $\mathcal{F}$ is decreasing with respect to the inclusion, *i.e.* if $\widetilde{\Omega}_i \subset \Omega_i$, for all $i = 1, \dots, h$, then

$$\mathcal{F}(\Omega_1, \dots, \Omega_h) \leq \mathcal{F}(\widetilde{\Omega}_1, \dots, \widetilde{\Omega}_h);$$

($\mathcal{F}2$) $\mathcal{F}$ is lower semi-continuous with respect to the strong-γ-convergence, *i.e.* if Ω_i^n strong-γ-converges to Ω_i, for every $i = 1, \dots, h$, then

$$\mathcal{F}(\Omega_1, \dots, \Omega_h) \leq \liminf_{n\to\infty} \mathcal{F}(\Omega_1^n, \dots, \Omega_h^n),$$

where the term strong-γ-convergence refers to the classical strong-γ-convergence in $\mathbb{R}^d$, *i.e.* the one defined through the space $H = H^1(\mathbb{R}^d)$.

Then we have the following result:

Theorem 2.84. *Let $\mathcal{D} \subset \mathbb{R}^d$ be a quasi-open set of finite Lebesgue measure let $\mathcal{F}$ be a decreasing and lower semicontinuous with respect to the* *strong-γ-convergence functional on the h-uples of quasi-open sets in $\mathcal{D}$. Then the multiphase shape optimization problem*

$$\min\Big\{\mathcal{F}(\Omega_1,\dots,\Omega_h)\ :\ \Omega_i\subset\mathcal{D}\ \textit{quasi-open},\ \forall i;\ \Omega_i\cap\Omega_j = \emptyset,\ \forall i\neq j\Big\},\tag{2.34}$$

has a solution.

Proof. Let $(\Omega_1^n,\dots,\Omega_h^n)$ be a minimizing sequence of disjoint quasi-open sets in $\mathcal{D}$. Then up to a subsequence, we may suppose that there are quasi-open sets $\Omega_1,\dots,\Omega_h\subset\mathcal{D}$ such that Ω_j^n weak-γ-converges to Ω_j, for each $j=1,\dots,h$. Let w_E denote the solution of

$$-\Delta w_E = 1\quad\text{in}\quad E,\qquad w_E\in H_0^1(E).$$

Then $w_{\Omega_j^n}$ converges in $L^2(\mathcal{D})$ to $w_j\in H_0^1(\Omega_j)$ such that $\{w_j>0\}=\Omega_j$. Thus, since $w_{\Omega_j^n}w_{\Omega_i^n}$ converges in L^1 to w_iw_j, we have that $|\{w_iw_j>0\}|=0$ and so $\mathrm{cap}(\Omega_i\cap\Omega_j)=\mathrm{cap}(\{w_iw_j>0\})=0$, which proves that Ω_i and Ω_j are disjoint when $i\neq j$. Thus the h-uple $(\Omega_1,\dots,\Omega_h)$ is an admissible competitor in (2.34) and so, by the semi-continuity of $\mathcal{F}$, we obtain the conclusion. □

Remark 2.85. We note that if $\mathcal{F}$ and $\mathcal{G}$ are two functionals on the h-uples of quasi-open sets in $\mathcal{D}$ satisfying $(\mathcal{F}1)$ and $(\mathcal{F}2)$, then the sum $\mathcal{F}+\mathcal{G}$ also satisfies $(\mathcal{F}1)$ and $(\mathcal{F}2)$.

We conclude this section noting that the following functionals satisfy $(\mathcal{F}1)$ and $(\mathcal{F}2)$:

(i) $\mathcal{F}(\Omega_1,\dots,\Omega_h)=\sum_{j=1}^h\lambda_{k_j}(\Omega_j)$, where $k_1,\dots,k_h\in\mathbb{N}$ are given natural numbers;

(ii) $\mathcal{F}(\Omega_1,\dots,\Omega_h)=\Big(\sum_{j=1}^h[\lambda_{k_j}(\Omega_j)]^p\Big)^{1/p}$, where $p\in\mathbb{N}$;

(iii) $\mathcal{F}(\Omega_1,\dots,\Omega_h)=\sum_{j=1}^h E_{f_j}(\Omega_j)$, where $f_1,\dots,f_h\in L^2(\mathcal{D})$ are given functions;

(iv) $\mathcal{F}(\Omega_1,\dots,\Omega_h)=\sum_{j=1}^h|\Omega_j|$.

2.4.3. Spectral drop in an isolated box

In the setting of the classical Buttazzo-Dal Maso Theorem the functionals we consider depend on the Dirichlet Laplacian. The kth Dirichlet eigenvalue and eigenfunction, for example, are a non trivial solution of the equation

$$-\Delta u_k = \lambda_k(\Omega)u_k \quad \text{in} \quad \Omega, \qquad u_k = 0 \quad \text{on} \quad \partial\Omega.$$

Thus in the shape optimization problem

$$\min\Big\{\lambda_k(\Omega) : \Omega \subset \mathcal{D},\ |\Omega| \le c\Big\},$$

we are in a situation where the box $\mathcal{D}$ has a boundary set to zero, *i.e.* $\partial\mathcal{D}$ is connected to the ground. In this case the box $\mathcal{D}$ has the role of a mechanical obstacle for the set Ω. A different situation occurs if we consider the set $\mathcal{D}$ to be isolated, *i.e.* the states of the system are described through the solutions of the problem

$$\begin{cases} -\Delta u_k = \lambda_k(\Omega;\mathcal{D})u_k & \text{in } \Omega, \\ u_k = 0 & \text{on } \partial\Omega \cap \mathcal{D}, \\ \dfrac{\partial u_k}{\partial n} = 0 & \text{on } \partial\mathcal{D} \cap \partial\Omega. \end{cases}$$

In this case the boundary $\partial\mathcal{D}$ is not only a mechanical obstacle, but also attracts the set Ω. This situation is similar to the classical liquid drop problem, where the functional on the set Ω is given through the relative perimeter $P(\Omega;\mathcal{D}) = \mathcal{H}^{d-1}(\partial\Omega \cap \mathcal{D})$.

Given a smooth bounded set $\mathcal{D} \subset \mathbb{R}^d$ and a (quasi-open) set $\Omega \subset \mathcal{D}$, we note that the relative eigenvalues $\lambda_k(\Omega;\mathcal{D})$ are variationally characterized as

$$\lambda_k(\Omega;\mathcal{D}) = \min_{S_k \subset H_0^1(\Omega;\mathcal{D})} \max_{u \in S_k \setminus \{0\}} \frac{\int_\Omega |\nabla u|^2\,dx}{\int_\Omega u^2\,dx},$$

where the minimum is over the k-dimensional subspaces S_k of $H_0^1(\Omega;\mathcal{D})$ and the Sobolev space $H_0^1(\Omega;\mathcal{D})$ is defined as

$$H_0^1(\Omega;\mathcal{D}) = \Big\{u \in H^1(\mathcal{D}) :\ u = 0 \text{ q.e. on } \mathcal{D} \setminus \Omega\Big\},$$

where we used the term *quasi-everywhere* in sense of the $H^1(\mathbb{R}^d)$-capacity. We have the following existence result.

Theorem 2.86. *Let $\mathcal{D} \subset \mathbb{R}^d$ be a smooth bounded open set in $\mathbb{R}^d$ and let F be an increasing and lower semi-continuous function on $\mathbb{R}^{\mathbb{N}}$. Then the shape optimization problem*

$$\min\Big\{F(\lambda_1(\Omega;\mathcal{D}), \lambda_2(\Omega;\mathcal{D}), \ldots) : \Omega \subset \mathcal{D},\ \Omega \textit{ quasi-open}, |\Omega| \le c\Big\}, \tag{2.35}$$

has a solution.

Proof. We start by noting that the inclusion $H^1(\mathcal{D}) \subset L^2(\mathcal{D})$ is compact. Thus, by Proposition 2.51, we have that the functional $\Omega \mapsto \lambda_k(\Omega; \mathcal{D})$ is lower semicontinuous with respect to the strong-γ-converges defined through the space $H = H^1(\mathcal{D})$. Thus, we have a solution of the problem 2.35 in the class of quasi-open sets with respect to the space $H^1(\mathcal{D})$. Now it is sufficient to note that these sets coincide with the quasi-open sets in $\mathbb{R}^d$, defined starting from the space $H^1(\mathbb{R}^d)$. Indeed, let $\Omega = \{u > 0\}$ for some $u \in H^1(\mathcal{D})$. Since $\mathcal{D}$ is regular, u admits an extension $\widetilde{u} \in H^1(\mathbb{R}^d)$ and thus $\Omega = \mathcal{D} \cap \{\widetilde{u} > 0\}$, which is a quasi-open set in the classical sense. □

2.4.4. Optimal periodic sets in the Euclidean space

In this subsection we consider an optimization problem for periodic sets in $\mathbb{R}^d$. We say that $\Omega \subset \mathbb{R}^d$ is t-periodic, if $\Omega = tv + \Omega$, for every vector with entire coordinates $v \in \mathbb{Z}^d$. Equivalently, we say that Ω is a set on the torus $\mathbb{T}_d = (S^1)^d$. For every $\Omega \subset \mathbb{T}_d$, we define

$$\lambda_k(\Omega; \mathbb{T}_d) = \min_{S_k \subset H^1_0(\Omega;\mathbb{T}_d)} \max_{u \in S_k \setminus \{0\}} \frac{\int_\Omega |\nabla u|^2\,dx}{\int_\Omega u^2\,dx},$$

where the minimum is over the k-dimensional subspaces S_k of $H^1_0(\Omega;\mathbb{T}_d)$, defined as

$$H^1_0(\Omega; \mathbb{T}_d) = \Big\{u \in H^1(\mathbb{T}_d) \,:\, u = 0 \text{ q.e. on } (0,1)^d \setminus \Omega\Big\},$$

where we used the term *quasi-everywhere* in sense of the space $H^1(\mathbb{R}^d)$ and $H^1(\mathbb{T}_d)$ is defined as

$$H^1(\mathbb{T}_d) = \Big\{u \in H^1\big((0,1)^d\big) \,:\, u(x_1, \dots, 0, \dots, x_d) = u(x_1, \dots, 1, \dots, x_d),\ \forall j = 1, \dots, d\Big\}.$$

Then, repeating the argument for Theorem 2.86, we have the following

Theorem 2.87. *Let F be an increasing and lower semi-continuous function on $\mathbb{R}^{\mathbb{N}}$. Then the shape optimization problem*

$$\min\Big\{F\big(\lambda_1(\Omega; \mathbb{T}_d), \lambda_2(\Omega; \mathbb{T}_d), \dots\big) \,:\, \Omega \subset \mathbb{T}_d,\ \Omega \textit{ quasi-open}, |\Omega \cap (0,1)^d| \le c\Big\},$$

has a solution, where the term quasi-open *is used in the classical sense given through the space $H^1(\mathbb{R}^d)$.*

2.4.5. Shape optimization problems on compact manifolds

Consider a differentiable manifold M of dimension d endowed with a Finsler structure, *i.e.* with a map $g : TM \to [0, +\infty)$ which has the following properties:

(1) g is smooth on $TM \setminus \{0\}$;
(2) g is 1-homogeneous, *i.e.* $g(x, \lambda X) = |\lambda| g(x, X), \forall \lambda \in \mathbb{R}$;
(3) g is strictly convex, *i.e.* the Hessian matrix with elements

$$g_{ij}(x) = \frac{1}{2} \frac{\partial^2}{\partial X^i \partial X^j} [g^2](x, X),$$

is positive definite for each $(x, X) \in TM$.

With these properties, the function $g(x, \cdot) : T_x M \to [0, +\infty)$ is a norm on the tangent space $T_x M$, for each $x \in M$. We define the gradient of a function $f \in C^\infty(M)$ as $Df(x) := g^*(x, df_x)$, where df_x stays for the differential of f at the point $x \in M$ and $g^*(x, \cdot) : T_x^* M \to \mathbb{R}$ is the co-Finsler metric, defined for every $\xi \in T_x^* M$ as

$$g^*(x, \xi) = \sup_{y \in T_x M} \frac{\xi(y)}{F(x, y)}.$$

The Finsler manifold (M, g) is a metric space with the distance:

$$d_g(x, y)$$
$$= \inf \left\{ \int_0^1 g(\gamma(t), \dot{\gamma}(t))\, dt : \gamma : [0, 1] \to M,\ \gamma(0) = x,\ \gamma(1) = y \right\}.$$

For any finite Borel measure m on M, we define $H := H_0^1(M, g, m)$ as the closure of the set of differentiable functions with compact support $C_c^\infty(M)$, with respect to the norm

$$\|u\| := \sqrt{\|u\|^2_{L^2(m)} + \|Du\|^2_{L^2(m)}}.$$

The functional λ_k is defined as in (2.33), on the class of quasi-open sets, related to the $H^1(M, g, m)$-capacity. Various choices for the measure m are available, according to the nature of the Finsler manifold M. For example, if M is an open subset of $\mathbb{R}^d$, it is natural to consider the Lebesgue measure $m = \mathcal{L}^d$. In this case, the non-linear operator associated to the functional $\int g^*(x, du_x)^2\, dx$ is called Finsler Laplacian. On the other hand, for a generic manifold M of dimension d, a canonical choice for

m is the Busemann-Hausdorff measure m_g, *i.e.* the d-dimensional Hausdorff measure with respect to the distance d_g. The non-linear operator associated to the functional $\int g^*(x, du_x)^2\, dm_g(x)$ is the generalization of the Laplace-Beltrami operator and its eigenvalues on the $\lambda_k(\Omega)$ on the set Ω are defined as in (2.33). In view of Theorem 2.79 and Corollary 2.81, we have the following existence results.

Theorem 2.88. *Given a compact Finsler manifold (M, g) with Busemann-Hausdorff measure m_g and an increasing and lower semi-continuous function F on $\mathbb{R}^{\mathbb{N}}$, we have that the problem*

$$\min\Big\{F(\lambda_1(\Omega), \lambda_2(\Omega), \dots) :\ m_g(\Omega) \le c,\ \Omega \text{ quasi-open}, \Omega \subset M\Big\},$$

has a solution for every $0 < c \le m_g(M)$.

Theorem 2.89. *Consider an open set $M \subset \mathbb{R}^d$ endowed with a Finsler structure g and the Lebesgue measure $\mathcal{L}^d$. Let F be an increasing and lower semi-continuous function on $\mathbb{R}^{\mathbb{N}}$. If the diameter of M with respect to the Finsler metric d_g is finite, then the following problem has a solution:*

$$\min\Big\{F\big(\lambda_1(\Omega), \lambda_2(\Omega), \dots\big) :\ |\Omega| \le c,\ \Omega \text{ quasi-open}, \Omega \subset M\Big\},$$

where $|\Omega|$ is the Lebesgue measure of Ω and $0 < c \le |M|$.

Remark 2.90. In [65] it was shown that if the Finsler metrics $g(x, \cdot)$ on $\mathbb{R}^d$ does not depend on $x \in \mathbb{R}^d$, then the solution of the optimization problem

$$\min\Big\{\lambda_1(\Omega)\ :\ |\Omega| \le c,\ \Omega \text{ quasi-open}, \Omega \subset \mathbb{R}^d\Big\},$$

is a ball (with respect to the Finsler distance d_g) of measure c. It is clear that it is also the case when in the hypotheses of Theorem 2.89 one considers $c > 0$ such that there is a ball of measure c contained in M. On the other hand , if c is big enough the solution is not, in general, the geodesic ball in M (see [71, Theorem 3.4.1]). If the Finsler metric is not constant in x, the solution will not be a ball even for small c. In this case it is natural to ask whether the optimal set gets close to the geodesic ball as $c \to 0$. In [86] this problem was discussed in the case when M is a Riemannian manifold. The same question for a generic Finsler manifold is still open.

2.4.6. Shape optimization problems in Gaussian spaces

Consider a separable Hilbert space $(\mathcal{H}, \langle\cdot,\cdot\rangle_{\mathcal{H}})$ with an orthonormal basis $(e_k)_{k\in\mathbb{N}}$. Suppose that $\mu = N_Q$ is a Gaussian measure on $\mathcal{H}$ with mean 0 and covariance operator Q (positive, of trace class) such that

$$Qe_k = \nu_k(Q)e_k,$$

where $0 < \cdots \leq \nu_n(Q) \leq \cdots \leq \nu_2(Q) \leq \nu_1(Q)$ is the spectrum of Q.

Denote with $\mathcal{E}(\mathcal{H})$ the space of all linear combinations of the functions on $\mathcal{H}$ which have the form $E_h(x) = e^{i\langle h,x\rangle}$ for some $h \in \mathcal{H}$, where for sake of simplicity we set $\langle\cdot,\cdot\rangle = \langle\cdot,\cdot\rangle_{\mathcal{H}}$. Then, the linear operator

$$\nabla : \mathcal{E}(\mathcal{H}) \subset L^2(\mathcal{H},\mu) \to L^2(\mathcal{H},\mu;\mathcal{H}), \qquad \nabla E_h = ihE_h,$$

is closable. We define the Sobolev space $W^{1,2}(\mathcal{H})$ as the domain of the closure of ∇. Thus, for any function $u \in W^{1,2}(\mathcal{H})$, we can define the gradient $\nabla u \in L^2(\mathcal{H},\mu;\mathcal{H})$ and we denote with $\nabla_k u \in L^2(\mathcal{H},\mu)$ the components of ∇u in $W^{1,2}(\mathcal{H})$, *i.e.*

$$\nabla_k u = \langle \nabla u, e_k\rangle.$$

We have the following integration by parts formula:

$$\int_{\mathcal{H}} \nabla_k u v\, d\mu + \int_{\mathcal{H}} u \nabla_k v\, d\mu = \frac{1}{\nu_k(Q)}\int_{\mathcal{H}} x_k u v\, d\mu.$$

If $\nabla_k u \in W^{1,2}(\mathcal{H})$, then we can test the above equation with $v = \nabla_k u$ to obtain

$$-\int_{\mathcal{H}} \nabla_k(\nabla_k u) v\, d\mu + \frac{1}{\nu_k(Q)}\int_{\mathcal{H}} x_k \nabla_k u v\, d\mu = \int_{\mathcal{H}} \nabla_k u \nabla_k v\, d\mu.$$

Summing over $k \in \mathbb{N}$, we get

$$\int_{\mathcal{H}} \left(-Tr[\nabla^2 u] + \langle Q^{-1}x, \nabla u\rangle\right) v\, d\mu = \int_{\mathcal{H}} \langle \nabla u, \nabla v\rangle\, d\mu,$$

where we used the notation

$$\langle Q^{-1}x, \nabla u\rangle := \sum_k \frac{1}{\nu_k(Q)} x_k \nabla_k u.$$

Definition 2.91. Given a Borel set $\Omega \subset \mathcal{H}$ and $\lambda \in \mathbb{R}$, we say that u is a weak solution of the equation

$$-Tr[\nabla^2 u] + \langle Q^{-1}x, \nabla u\rangle = \lambda u \quad \text{in } \Omega, \qquad u \in W_0^{1,2}(\Omega),$$

if $u \in W_0^{1,2}(\Omega)$ and

$$\int_{\mathcal{H}} \langle \nabla u, \nabla v \rangle \, d\mu = \lambda \int_{\mathcal{H}} uv \, d\mu, \quad \text{for every} \quad v \in W_0^{1,2}(\Omega).$$

By a well-known theorem from the functional analysis (see for example [57]), there is a self-adjoint operator A on $L^2(\Omega, \mu)$ such that for each $u, v \in Dom(A) \subset W_0^{1,2}(\Omega)$,

$$\int_{\mathcal{H}} Au \cdot v \, d\mu = \int_{\mathcal{H}} \langle \nabla u, \nabla v \rangle \, d\mu.$$

Then, by the compactness of the embedding $W_0^{1,2}(\Omega) \hookrightarrow L^2(\mu)$, A is a positive operator with compact resolvent. Keeping in mind the construction of A, we will write

$$A = -Tr[\nabla^2] + \langle Q^{-1}x, \nabla \rangle.$$

The spectrum of $-Tr[\nabla^2] + \langle Q^{-1}x, \nabla \rangle$ is discrete and consists of positive eigenvalues $0 \leq \lambda_1(\Omega) \leq \lambda_2(\Omega) \leq \ldots$ for which the usual min-max variational formulation holds.

Theorem 2.92. *Suppose that $\mathcal{H}$ is a separable Hilbert space with non-degenerate Gaussian measure μ. Then, for any $0 \leq c \leq 1$, the following optimization problem has a solution:*

$$\min\Big\{F\big(\lambda_1(\Omega), \lambda_2(\Omega), \ldots\big) \;:\; \Omega \subset \mathcal{H},\ \Omega \textit{ quasi-open},\ \mu(\Omega) = c\Big\},$$

where F is a decreasing and lower semicontinuous function on $\mathbb{R}^{\mathbb{N}}$.

Proof. Take $H := W^{1,2}(\mathcal{H})$ and $Du = \|\nabla u\|_{\mathcal{H}}$. The pair (H, D) satisfies the hypothesis $H1, \ldots, \mathcal{H}3$ and $\mathcal{H}4$. In fact, the norm $\|u\|^2 = \|u\|_{L^2}^2 + \|Du\|_{L^2}^2$ is the usual norm in $W^{1,2}(\mathcal{H})$ and with this norm $W^{1,2}(\mathcal{H})$ is a separable Hilbert space and the inclusion $H \hookrightarrow L^2(\mathcal{H}, \mu)$ is compact (see [55, Theorem 9.2.12]). Moreover, the continuous functions are dense in $W^{1,2}(\mathcal{H})$, by construction. Applying Proposition 2.52, Theorem 2.79 and Corollary 2.81 we obtain the conclusion. □

2.4.7. Shape optimization in Carnot-Caratheodory space

Consider a bounded open and connected set $\mathcal{D} \subset \mathbb{R}^d$ and C^∞ vector fields $Y_1, \ldots, Y_n$ defined on a neighbourhood U of $\overline{\mathcal{D}}$. We say that the vector fields satisfy the Hörmander's condition on U, if the Lie algebra generated by $Y_1, \ldots, Y_n$ has dimension d in each point $x \in U$.

We define the Sobolev space $W_0^{1,2}(\mathcal{D}; Y)$ on $\mathcal{D}$, with respect to the family of vector fields $Y = (Y_1, \dots, Y_n)$, as the closure of $C_c^\infty(\mathcal{D})$ with respect to the norm

$$\|u\|_Y = \left(\|u\|_{L^2}^2 + \sum_{j=1}^{n} \|Y_j u\|_{L^2}^2\right)^{1/2},$$

where the derivation $Y_j u$ is intended in sense of distributions. For $u \in W_0^{1,2}(\mathcal{D}; Y)$, we define

$$Yu=(Y_1u, \dots, Y_nu) \quad \text{and} \quad |Yu|=\left(|Y_1u|^2+\dots+|Y_nu|^2\right)^{1/2} \in L^2(\mathcal{D}).$$

Setting $Du := |Yu|$ and $H := W_0^{1,2}(\mathcal{D}; Y)$, we define, for any $\Omega \subset \mathcal{D}$, the kth eigenvalue $\lambda_k(\Omega)$ of the operator $Y_1^2 + \dots + Y_n^2$ as in (2.33).

Example 2.93. Consider the vector fields

$$X = \partial_x \qquad \text{and} \qquad Y = x\partial y.$$

We note that, since the commutator of X and Y is $[X, Y] = [\partial_x, x\partial_y] = \partial_y$, the vector fields X and Y satisfy the Hörmander condition in $\mathbb{R}^d$. Then operator $X^2 + Y^2$ is given by

$$X^2 + Y^2 = \partial_x^2 + x^2\partial_y^2,$$

and for every bounded $\Omega \subset \mathbb{R}^d$, $\lambda_k(\Omega)$ is defined as the kth biggest number such that the equation

$$-\left(\partial_x^2 + x^2\partial_y^2\right) u_k = \lambda_k(\Omega)u_k \quad \text{in} \quad \Omega, \qquad u_k \in W_0^{1,2}(\Omega; \{X, Y\}),$$

has a non-trivial weak solution.

Theorem 2.94. *Consider a bounded open set $\mathcal{D} \subset \mathbb{R}^d$ and a family $Y = (Y_1, \dots, Y_n)$ of C^∞ vector fields defined on an open neighbourhood U of the closure $\overline{\mathcal{D}}$ of $\mathcal{D}$ an suppose, moreover, that $Y_1, \dots, Y_n$ satisfy the Hörmander condition on U. Then for every increasing and lower semi-continuous function F on $\mathbb{R}^{\mathbb{N}}$, the following shape optimization problems has a solution:*

$$\min\left\{F\left(\lambda_1(\Omega), \lambda_2(\Omega), \dots\right) : \Omega \subset \mathcal{D},\ \Omega \text{ quasi-open}, |\Omega| \le c\right\}. \quad (2.36)$$

Proof. It is straightforward to check that the space $H := W_0^{1,2}(\mathcal{D}; Y)$ and the application $Du := |Yu|$ satisfy the assumptions of Theorem 2.79 and Corollary 2.81. Thus we only have to check the lower semi-continuity of λ_k with respect to the strong-γ-convergence. This follows by Proposition

2.52 since the inclusion $H \subset L^2(\mathcal{D})$ is compact. This last claim holds since $Y_1, \dots, Y_n$ satisfy the Hörmander condition on U. In fact, by the Hörmander Theorem (see [73]), there is some $\epsilon > 0$ and some constant $C > 0$ such that for any $\varphi \in C_c^\infty(\mathcal{D})$

$$\|\varphi\|_{H^\varepsilon} \leq C\left(\|\varphi\|_{L^2} + \sum_{j=1}^{k} \|Y_j\varphi\|_{L^2}\right),$$

where we set

$$\|\varphi\|_{H^\varepsilon} := \left(\int_{\mathbb{R}^d} |\widehat{\varphi}(\xi)|^2 (1 + |\xi|^2)^\varepsilon \, d\xi\right)^{1/2},$$

being $\widehat{\varphi}$ the Fourier transform of φ. Let $H_0^\varepsilon(\mathcal{D})$ be the closure of $C_c^\infty(\mathcal{D})$ with respect to the norm $\|\cdot\|_{H^\varepsilon}$. Since the inclusion $L^2(\mathcal{D}) \subset H_0^\varepsilon(\mathcal{D})$ is compact, we have the conclusion. □

2.4.8. Shape optimization in measure metric spaces

In this section we consider the framework, which inspired the general setting we introduced in the previous sections. We briefly recall the main definitions and results from [44] before we state our main existence result.

Definition 2.95. Let $u : X \to \overline{\mathbb{R}}$ be a measurable function. An upper gradient g for u is a Borel function $g : X \to [0, +\infty]$, such that for all points $x_1, x_2 \in X$ and all continuous rectifiable curves, $c : [0, l] \to X$ parametrized by arc-length, with $c(0) = x_1, c(l) = x_2$, we have

$$|u(x_2) - u(x_1)| \leq \int_0^l g(c(s))ds,$$

where the left hand side is intended as $+\infty$ if $|u(x_1)|$ or $|u(x_2)|$ is $+\infty$.

Following the original notation in [44], for $u \in L^2(X, m)$ we define the norms

$$|u|_{1,2} := \inf\left\{\liminf_{j\to\infty} \|g_j\|_{L^2}\right\} \qquad \text{and} \qquad \|u\|_{1,2} := \|u\|_{L^2} + |u|_{1,2},$$

where the infimum above is taken over all sequences (g_j), for which there exists a sequence $u_j \to u$ in L^2 such that, for each j, g_j is an upper gradient for u_j. We define the Sobolev space $H = H^1(X, m)$ as the class of functions $u \in L^2(X, m)$ such that the norm $\|u\|_{1,2}$ is finite. In [44, Theorem 2.7] it was proved that the space $H^1(X, m)$, endowed with the norm $\|\cdot\|_{1,2}$, is a Banach space. Moreover, in the same work, the following notion of a gradient was introduced.

Definition 2.96. The function $g \in L^2(X,m)$ is a generalized upper gradient of $u \in L^2(X,m)$, if there exist sequences $(g_j)_{j\geq 1} \subset L^2(X,m)$ and $(u_j)_{j\geq 1} \subset L^2(X,m)$ such that

$$u_j \to u \text{ in } L^2(X,m), \qquad g_j \to g \text{ in } L^2(X,m),$$

and g_j is an upper gradient for u_j, for every $j \geq 1$.

For each $u \in H^1(X,m)$ there exists a unique generalized upper gradient $g_u \in L^2(X,m)$, such that the following equality is satisfied:

$$\|u\|_{1,2} = \|u\|_{L^2} + \|g_u\|_{L^2}.$$

Moreover g_u is minimal in the sense that for every generalized upper gradient g of u, we have $g_u \leq g$. The function g_u is called *minimal generalized upper gradient of* u and is the metric space analogue of the modulus of the weak gradient $|\nabla u|$ in $\mathbb{R}^d$.

Under some mild conditions on the metric space X and the measure m, the minimal generalized upper gradient has a pointwise expression (see [44]). In fact, for any Borel function u, one can define

$$Lip\, u(x) = \liminf_{r\to 0} \sup_{d(x,y)=r} \frac{|u(x)-u(y)|}{r},$$

with the convention $Lip\, u(x) = 0$, whenever x is an isolated point. If the measure metric space (X,d,m) satisfies some standard assumptions (doubling and supporting a weak Poincaré inequality), then the function $Lip\, u$ is the minimal generalized upper gradient (see [44, Theorem 6.1] and also [4] for further analysis of g_u). Using the minimal generalized upper gradient one can consider elliptic boundary value problems on a metric space and thus define spectral and energy functionals on the subsets $\Omega \subset X$ as the Dirichlet Energy $E(\Omega)$ and the eigenvalue of the Dirichlet Laplacian $\lambda_k(\Omega)$ as in (2.33).

Theorem 2.97. *Consider a separable metric space (X,d) and a finite Borel measure m on X. Let $H^1(X,m)$ denote the Sobolev space on (X,d,m) and let $Du = g_u$ be the minimal generalized upper gradient of $u \in H^1(X,m)$. Under the assumption that the inclusion $H^1(X,m) \hookrightarrow L^2(X,m)$ is compact, the shape optimization problem*

$$\min\Big\{F\big(\lambda_1(\Omega),\lambda_2(\Omega),\dots\big) : \Omega \subset X,\ \Omega \text{ Borel},\ |\Omega| \leq c\Big\},$$

has solution, for every constant $c > 0$ and every increasing and lower semi-continuous function $F : \mathbb{R}^{\mathbb{N}} \to \mathbb{R}$.

Remark 2.98. There are various assumptions that can be made on the measure metric space (X, d, m) in order to have that the inclusion $H^1(X, m) \hookrightarrow L^2(X, m)$ is compact. A detailed discussion on this topic can be found in [68, Section 8]. For the sake of completeness, we state here a result from [68]:

Consider a separable metric space (X, d) of finite diameter equipped with a finite Borel measure m such that:

(a) *there exist constants $C_m > 0$ and $s > 0$ such that for each ball $B_{r_0}(x_0) \subset X$, each $x \in B_{r_0}(x_0)$ and $0 < r \leq r_0$, we have that*

$$\frac{m(B_r(x))}{m(B_{r_0}(x_0))} \geq C_m \frac{r^s}{r_0^s};$$

(b) *(X, d, m) supports a weak Poincaré inequality,* i.e. *there exist $C_P > 0$ and $\sigma \geq 1$ such that for each $u \in H^1(X, m)$ and each ball $B_r(x) \subset X$ we have*

$$\fint_{B_r(x)} \left| u(y) - \fint_{B_r(x)} u \, dm \right| dm(y) \leq C_P r \left(\fint_{B_{\sigma r}(x)} g_u^2 \, dm \right)^{1/2}.$$

Then, the inclusion $H^1(X, m) \hookrightarrow L^2(X, m)$ is compact.

Chapter 3
Capacitary measures

In this chapter we discuss one of the fundamental tools in the shape optimization. The capacitary measures generalize various situations involving PDEs in the Euclidean space $\mathbb{R}^d$, allowing us to threat at once problems concerning elliptic problems on domains, Schrödinger operators and operators involving traces of Sobolev functions on $(d-1)$-dimensional sets. In this setting we will use the following notations:

- d is the dimension of the space $\mathbb{R}^d$, which is endows with the norm

$$|x| = |(x_1, \dots, x_d)| = \sqrt{x_1^2 + \cdots + x_d^2};$$

- $B_r(x) := \{y \in \mathbb{R}^d : |x - y| < r\}$ will denote the ball of center x and radius r in $\mathbb{R}^d$; when $x = 0$, we will use the notation $B_r := B_r(x)$;
- for a real number $s > 0$ with $\mathcal{H}^s(E)$ we denote the s-Hausdorff measure on the set $E \subset \mathbb{R}^d$ (see [62]);
- $\mathcal{L}^d(E) = \mathcal{H}^d(E) = |E|$ is the Lebesgue measure of a measurable set $E \subset \mathbb{R}^d$;
- by ω_d we will denote the Lebesgue measure of the ball of radius 1 in $\mathbb{R}^d$. Thus we have

$$|B_r| = \omega_d r^d \quad \text{and} \quad \mathcal{H}^{d-1}(\partial B_r) = d\omega_d r^{d-1};$$

- we say that a property $\mathcal{P}$ holds *almost everywhere* (shortly *a.e.*), if the Lebesgue measure of the set where $\mathcal{P}$ does not hold is zero;
- the integral of a function f with respect to the Lebesgue measure on a measurable set $\Omega \subset \mathbb{R}^d$ is $\int_\Omega f\,dx$;
- for a set Ω endowed with a finite measure μ we will use the notation

$$\fint_\Omega f\,d\mu := \frac{1}{\mu(\Omega)} \int_\Omega f\,d\mu,$$

to indicate the mean value of the function $f : \Omega \to \mathbb{R}$.

- for $p \in [1, +\infty)$, with $L^p(\Omega)$ we denote the space of Lebesgue measurable functions $f : \Omega \to \mathbb{R}$ such that $\int_\Omega |f|^p\,dx < +\infty$, which is a Banach space endowed with the norm

$$\|f\|_{L^p(\Omega)} := \left(\int_\Omega |f|^p\,dx\right)^{1/p};$$

in the case when $\Omega = \mathbb{R}^d$ we will simply use the notations

$$L^p := L^p(\mathbb{R}^d) \qquad \text{and} \qquad \|\cdot\|_{L^p} := \|\cdot\|_{L^p(\mathbb{R}^d)};$$

- with $L^\infty(\Omega)$ we denote the space of Lebesgue measurable functions $f : \Omega \to \mathbb{R}$ such that

$$\|f\|_{L^\infty(\Omega)}$$
$$:= \inf\Big\{C > 0 : \ |f(x)| \le C \quad \text{almost everywhere on} \quad \Omega\Big\} < +\infty;$$

in the case when $\Omega = \mathbb{R}^d$ we will simply use the notations

$$L^\infty := L^\infty(\mathbb{R}^d) \qquad \text{and} \qquad \|\cdot\|_{L^\infty} = \|\cdot\|_{L^\infty} = \|\cdot\|_{L^\infty(\mathbb{R}^d)};$$

- for a measurable set $\Omega \subset \mathbb{R}^d$ we will denote with $P(\Omega)$ its *perimeter*, given by

$$P(\Omega) := |\nabla \mathbb{1}_\Omega|(\mathbb{R}^d),$$

where $\nabla \mathbb{1}_\Omega$ is the distributional gradient of the function $\mathbb{1}_\Omega : \mathbb{R}^d \to \mathbb{R}$ and $|\nabla \mathbb{1}_\Omega|(\mathbb{R}^d)$ is its total variation (see for example [67]).

3.1. Sobolev spaces in $\mathbb{R}^d$

We denote with $C_c^\infty(\mathbb{R}^d)$ the infinitely differentiable functions with compact support in $\mathbb{R}^d$. The spaces $H^1(\mathbb{R}^d)$ and $\dot{H}^1(\mathbb{R}^d)$ are the closures of $C_c^\infty(\mathbb{R}^d)$ with respect to the norms

$$\|u\|_{H^1} := \left(\int_{\mathbb{R}^d} |\nabla u|^2 + u^2\,dx\right)^{1/2}$$

$$\text{and} \qquad \|u\|_{\dot{H}^1} := \|\nabla u\|_{L^2} = \left(\int_{\mathbb{R}^d} |\nabla u|^2\,dx\right)^{1/2}.$$

We recall that if $d \ge 3$, the Gagliardo-Nirenberg-Sobolev inequality

$$\|u\|_{L^{2d/(d-2)}} \le C_d \|\nabla u\|_{L^2}, \qquad \forall u \in \dot{H}^1(\mathbb{R}^d), \tag{3.1}$$

holds, while in the cases $d \le 2$, we have respectively

$$\|u\|_{L^\infty} \le \left(\frac{r+2}{2}\right)^{2/(r+2)} \|u\|_{L^r}^{r/(r+2)} \|u'\|_{L^2}^{2/(r+2)}, \qquad \forall r \ge 1,\ \forall u \in \dot{H}^1(\mathbb{R}); \tag{3.2}$$

$$\|u\|_{L^{r+2}} \le \left(\frac{r+2}{2}\right)^{2/(r+2)} \|u\|_{L^r}^{r/(r+2)} \|\nabla u\|_{L^2}^{2/(r+2)}, \qquad \forall r \ge 1,\ \forall u \in \dot{H}^1(\mathbb{R}^2). \tag{3.3}$$

Thus, in any dimension we have

$$\|u\|_{H^1} \le C_d\Big(\|\nabla u\|_{L^2} + \|u\|_{L^1}\Big)$$

$$\text{and} \qquad H^1(\mathbb{R}^d) \cap L^1(\mathbb{R}^d) = \dot{H}^1(\mathbb{R}^d) \cap L^1(\mathbb{R}^d).$$

3.1.1. Concentration-compactness principle

In this section we recall a classical result due to P. L. Lions (see [79]). Our formulation is slightly different from the original one and is adapted to the use we will make of the concentration-compactness principle.

Definition 3.1. For every Borel measure μ on $\mathbb{R}^d$ we define the concentration function $Q_\mu : [0, +\infty) \to [0, +\infty]$ as

$$Q_\mu(r) = \sup_{x \in X} \mu(B_r(x)).$$

Remark 3.2. We note that Q_μ is nondecreasing, nonnegative and

$$\lim_{r \to +\infty} Q_\mu(r) = \|Q_\mu\|_{L^\infty} = \mu(\mathbb{R}^d).$$

The following lemma is elementary, but provides the compactness necessary for the concentration-compactness Theorem 3.4 below.

Lemma 3.3. *For every sequence of non-decreasing functions $Q_n : [0, +\infty) \to [0, 1]$, there is a subsequence converging pointwise to a non-decreasing function $Q : [0, +\infty) \to [0, 1]$.*

Theorem 3.4. *Consider a sequence $f_n \in L^1(\mathbb{R}^d)$ of positive functions uniformly bounded in $L^1(\mathbb{R}^d)$. Then, up to a subsequence, one of the following properties holds:*

(1) *There exists a sequence $(x_n)_{n\ge1} \subset \mathbb{R}^d$ with the property that for all $\epsilon > 0$ there is some $R > 0$ such that for all $n \in \mathbb{N}$ we have*

$$\int_{\mathbb{R}^d \setminus B_R(x_n)} f_n \, dx \le \epsilon.$$

(2) *For every $R > 0$ we have*

$$\lim_{n\to\infty}\left(\sup_{x\in\mathbb{R}^d}\int_{B_R(x)} f_n\,dx\right) = 0.$$

(3) *For every $\alpha > 1$, there is a sequence $x_n \in \mathbb{R}^d$ and an increasing sequence $R_n \to +\infty$ such that*

$$\lim_{n\to\infty}\int_{B_{\alpha R_n}(x_n)\setminus B_{R_n}(x_n)} f_n\,dx = 0,$$

$$\liminf_{n\to\infty}\int_{B_{R_n}(x_n)} f_n\,dx > 0 \quad \text{and} \quad \liminf_{n\to\infty}\int_{\mathbb{R}^d\setminus B_{\alpha R_n}(x_n)} f_n\,dx > 0.$$

Proof. We first note that, up to rescaling, we can suppose $\|f_n\|_{L^1} = 1$, for every $n \in \mathbb{N}$. Consider the concentration functions Q_n associated to the (probability) measure $f_n\,dx$. By Lemma 3.3, up to a subsequence, Q_n converges pointwise to some nondecreasing $Q : [0, +\infty) \to [0, 1]$. We first note that if $\lim_{t\to\infty} Q(t) = 0$, then $Q \equiv 0$ and so, (2) holds.

Suppose that $\lim_{t\to\infty} Q(t) = 1$. By the pointwise convergence of Q_n to Q, we have that for every $\varepsilon > 0$, there are $R_\varepsilon > 0$ and $n_\varepsilon \in \mathbb{N}$ such that $Q_n(R_\varepsilon) > (1-\varepsilon)$, for every $n \geq n_\varepsilon$. In particular, there is a sequence $y_n^\varepsilon \in \mathbb{R}^d$ such that

$$\int_{B_{R_\varepsilon}(y_n^\varepsilon)} f_n\,dx > 1-\varepsilon.$$

We note that the condition $\int f_n\,dx = 1$ implies $|y_n^{1/2} - y_n^\varepsilon| < R_{1/2} + R_\varepsilon$. Thus setting $x_n := y_n^{1/2}$ and $R = R_{1/2} + R_\varepsilon$, we have

$$\int_{B_R(x_n)} f_n\,dx \geq \int_{B_{R_\varepsilon}(y_n^\varepsilon)} f_n\,dx > 1-\varepsilon.$$

Suppose that $\lim_{t\to\infty} Q(t) =: l \in (0, 1)$ and fix $\varepsilon > 0$. Let $R_\varepsilon > 0$ be such that $l - \epsilon < Q(R_\varepsilon)$. In particular, we have $l - \epsilon < Q(R_\varepsilon) \leq Q(\alpha R_\varepsilon) \leq l$. Then, there exists $N = N(\varepsilon, \alpha) \in \mathbb{N}$ such that for each $n \geq N$, we have

$$l - \epsilon < Q_n(R_\varepsilon) \leq Q_n(\alpha R_\varepsilon) < l + \epsilon. \tag{3.4}$$

Thus, we can find a sequence y_k^ε such that for each $n \geq N$,

$$l - \epsilon < \int_{B_{R_\varepsilon}(y_k^\varepsilon)} f_n\,dx \leq \int_{B_{\alpha R_\varepsilon}(y_k^\varepsilon)} f_n\,dx \leq Q_n(\alpha R_\varepsilon) < l + \epsilon.$$

The conclusion follows by a diagonal sequence argument. $\square$

If the sequence $f_n \in L^1(\mathbb{R}^d)$ satisfies point (1) of the above Theorem, then it is *concentrated* in the dense of the following Definition.

Definition 3.5. We say that a sequence $f_n \in L^1(\mathbb{R}^d)$ has the concentration property if for every $\varepsilon > 0$ there is some $R_\varepsilon > 0$ such that

$$\int_{\mathbb{R}^d \setminus B_{R_\varepsilon}} |f_n|\, dx < \epsilon, \ \forall n \in \mathbb{N}.$$

Remark 3.6. If a sequence $f_n \in L^1(\mathbb{R}^d)$ has the concentration property and $g_n \in L^1(\mathbb{R}^d)$ is such that $|g_n| \leq C|f_n| + |f|$, for some $C > 0$ and some $f \in L^1(\mathbb{R}^d)$, then g_n also has the concentration property.

Remark 3.7. Since the inclusion $H^1(\mathbb{R}^d) \subset L^1_{\text{loc}}(\mathbb{R}^d)$ is compact, we have that if a sequence $u_n \in L^1(\mathbb{R}^d) \cap H^1(\mathbb{R}^d)$ is bounded in $L^1(\mathbb{R}^d) \cap H^1(\mathbb{R}^d)$ and has the concentration property, then there is a subsequence converging strongly in L^1.

3.1.2. Capacity, quasi-open sets and quasi-continuous functions

We define the *capacity* cap(E) of a measurable set $E \subset \mathbb{R}^d$, with respect to the Sobolev space $H^1(\mathbb{R}^d)$, as in Definition 2.57 (taking $H = H^1(\mathbb{R}^d)$), *i.e.*

$$\text{cap}(E) = \inf\left\{ \int_{\mathbb{R}^d} |\nabla u|^2 + u^2\, dx \ : \ u \in H^1(\mathbb{R}^d), \quad u \geq 1 \text{ in a neighbourhood of } E \right\}. \tag{3.5}$$

Remark 3.8. In dimension $d \geq 3$ one may define the capacity in an alternative way (see, for example, [62, Chapter 4.7]).

$$\widetilde{\text{cap}}(E) = \inf\left\{ \int_{\mathbb{R}^d} |\nabla u|^2\, dx \ : \ u \in H^1(\mathbb{R}^d), \quad u \geq 1 \text{ in a neighbourhood of } E \right\}. \tag{3.6}$$

For $d \geq 3$ the two quantities cap(E) and $\widetilde{\text{cap}}(E)$ are related by the inequality (3.7) below. Indeed, by definition we have $\widetilde{\text{cap}}(E) \leq \text{cap}(E)$, for every measurable $E \subset \mathbb{R}^d$. On the other hand, suppose that $u_n \in H^1(\mathbb{R}^d)$ is a sequence such that $\|\nabla u_n\|^2_{L^2}$ converges to $\widetilde{\text{cap}}(E)$. Since

$\|\nabla(0 \vee u_n \wedge 1)\|_{L^2} \leq \|\nabla u_n\|_{L^2}$, we may suppose that $0 \leq u_n \leq 1$. Thus, we have

$$\int_{\mathbb{R}^d} |\nabla u_n|^2 + u_n^2 \, dx \leq \int_{\mathbb{R}^d} |\nabla u_n|^2 + u_n^{\frac{2d}{d-2}} \, dx$$
$$\leq \int_{\mathbb{R}^d} |\nabla u_n|^2 \, dx + C_d \left(\int_{\mathbb{R}^d} |\nabla u_n|^2 \, dx \right)^{\frac{d}{d-2}},$$

which after passing to the limit as $n \to \infty$ gives

$$\widetilde{\text{cap}}(E) \leq \text{cap}(E) \leq \widetilde{\text{cap}}(E) + C_d \Big(\widetilde{\text{cap}}(E) \Big)^{\frac{d}{d-2}}. \tag{3.7}$$

In particular the sets of zero capacity defined through (3.5) and (3.6) are the same.

Remark 3.9. In dimension two, the above considerations are no more valid since the quantity defined in (3.6) is constantly zero. Indeed, for every function $u \in H^1(\mathbb{R}^2)$ and every scaling $u_t(x) := u(tx)$, defined for $t > 0$, we have

$$\int_{\mathbb{R}^2} |\nabla u_t|^2 \, dx = t^2 \int_{\mathbb{R}^2} |\nabla u|^2(tx) \, dx = \int_{\mathbb{R}^2} |\nabla u|^2 \, dx,$$

which in view of definition (3.6) gives that $\widetilde{\text{cap}}(E) = \widetilde{\text{cap}}(tE)$, for any $t > 0$. In particular $\widetilde{\text{cap}}(B_r) = \widetilde{\text{cap}}(B_1)$, for any ball $B_r \subset \mathbb{R}^2$. On the other hand, for $0 < r < 1$, we can use the radial test function $u(R) = \left[\frac{\log(R)}{\log(r)}\right]^+$ to obtain the bound

$$\widetilde{\text{cap}}(B_r) \leq \int_{\mathbb{R}^2} |\nabla u|^2 \, dx = 2\pi \int_r^1 \left[R \log^2(r)\right]^{-1} \, dR = \frac{2\pi}{|\log(r)|} \xrightarrow[r \to 0]{} 0,$$

which gives that $\widetilde{\text{cap}}(B_r) = 0$, for every $r > 0$. Then, using the monotonicity of $\widetilde{\text{cap}}$ and a standard approximation argument, we get that the value of $\widetilde{\text{cap}}$ is constantly zero on the subsets of $\mathbb{R}^2$.

Remark 3.10. Given an open set $\mathcal{D} \subset \mathbb{R}^d$ and a measurable set $E \subset \mathbb{R}^d$, one may define the capacity of E with respect to $\mathcal{D}$ in one of the following ways

$$\mathrm{cap}_{\mathcal{D}}(E) = \inf\left\{ \int_{\mathbb{R}^d} |\nabla u|^2 + u^2\,dx \,:\, u \in H^1(\mathcal{D}),\ u \geq 1 \right. \\ \left. \text{in a neighbourhood of } E \right\}, \tag{3.8}$$

$$\widetilde{\mathrm{cap}}_{\mathcal{D}}(E) = \inf\left\{ \int_{\mathbb{R}^d} |\nabla u|^2\,dx \,:\, u \in H_0^1(\mathcal{D}),\ u \geq 1 \right. \\ \left. \text{in a neighbourhood of } E \right\}. \tag{3.9}$$

Since the measure of $\mathcal{D}$ is finite, in any dimension $d \geq 1$, there is a constant $C_{\mathcal{D}} > 0$ such that

$$\widetilde{\mathrm{cap}}_{\mathcal{D}}(E) \leq \mathrm{cap}_{\mathcal{D}}(E) \leq C_{\mathcal{D}}\,\widetilde{\mathrm{cap}}_{\mathcal{D}}(E).$$

In is immediate to check[1] that in any dimension

$$\Big(\mathrm{cap}(E) = 0\Big) \Leftrightarrow \Big(\mathrm{cap}_{\mathcal{D}}(E) = 0\Big) \Leftrightarrow \Big(\widetilde{\mathrm{cap}}_{\mathcal{D}}(E) = 0\Big). \tag{3.10}$$

In particular, (3.10) shows that being of zero capacity is a local property. In fact an alternative way to define a set of zero capacity in $\mathbb{R}^d$ is the following:

$$\Big(\mathrm{cap}(E) = 0\Big) \\ \Leftrightarrow \Big(\mathrm{cap}_{B_{2r}(x)}\big(E \cap B_r(x)\big) = 0, \text{ for every ball } B_r(x) \subset \mathbb{R}^d\Big). \tag{3.11}$$

The advantage of this definition is that it can be easily extended to manifolds ot other settings, where the global definitions as (3.5) fail to provide a meaningful notion of zero capacity sets.[2]

In the following Proposition we list the main properties of the capacity in $\mathbb{R}^d$.

Proposition 3.11. *The following properties hold for the capacity in* $\mathbb{R}^d$*:*

(1) *If* $\omega \subset \Omega$, *then* $\mathrm{cap}(\omega) \leq \mathrm{cap}(\Omega)$.

[1] First for sets E, which are compactly included in $\mathcal{D}$, and then reasoning by approximation. The detailed proof can be found in [72, Proposition 3.3.17].

[2] On compact manifolds, for example, definition (3.5) gives precisely the measure of the sets E.

(2) *If* $(\Omega_n)_{n\in\mathbb{N}}$ *is a family of disjoint sets, then*

$$\mathrm{cap}\left(\bigcup_{n=1}^{\infty}\Omega_n\right)\le\sum_{n=1}^{\infty}\mathrm{cap}(\Omega_n).$$

(3) *For every* $\Omega_1, \Omega_2 \subset X$, *we have that*

$$\mathrm{cap}(\Omega_1\cup\Omega_2)+\mathrm{cap}(\Omega_1\cap\Omega_2)\le\mathrm{cap}(\Omega_1)+\mathrm{cap}(\Omega_2).$$

(4) *If* $\Omega_1 \subset \Omega_2 \subset \cdots \subset \Omega_n \subset \ldots$, *then we have*

$$\mathrm{cap}\left(\bigcup_{n=1}^{\infty}\Omega_n\right)=\lim_{n\to\infty}\mathrm{cap}(\Omega_n).$$

(5) *If* $K \subset \mathbb{R}^d$ *is a compact set, then we have*

$$\mathrm{cap}(K)=\inf\left\{\|\varphi\|_{H^1}^2\,:\,\varphi\in C_c^\infty(\mathbb{R}^d),\ \varphi\ge 1\ on\ K\right\}.$$

(6) *If* $A \subset \mathbb{R}^d$ *is an open set, then we have*

$$\mathrm{cap}(A)=\sup\left\{\mathrm{cap}(K)\,:\,K\ compact,\ K\subset A\right\}.$$

(7) *If* $\Omega \subset \mathbb{R}^d$ *is measurable, then*

$$\mathrm{cap}(\Omega)=\inf\left\{\mathrm{cap}(A)\,:\,A\ open,\ \Omega\subset A\right\}.$$

(8) *If* $K_1 \supset K_2 \supset \cdots \supset K_n \supset \ldots$ *are compact sets, then we have*

$$\mathrm{cap}\left(\bigcap_{n=1}^{\infty}K_n\right)=\lim_{n\to\infty}\mathrm{cap}(K_n).$$

Proof. The points (1), (2), (3) and (4) are the same as in Proposition 2.59. For the points (5), (6), (7) and (8), we refer to [72] and [62]. □

Analogously, we define the quasi-open sets and the quasi-continuous functions. We summarize the results from Section 2.3 in the following

Remark 3.12.

(1) For every Sobolev function $u \in H^1(\mathbb{R}^d)$, there is a unique, up to a set of zero capacity, quasi-continuous representative $\widetilde{u}$.
(2) If $\varphi : \mathbb{R}^d \to \mathbb{R}$ is a quasi-continuous function, then the level set $\{\varphi > 0\}$ is a quasi-open set.

(3) For every quasi-open set Ω there is a quasi-continuous function $u \in H^1(\mathbb{R}^d)$ such that $\Omega = \{u > 0\}$.
(4) If $u_n \in H^1(\mathbb{R}^d)$ converges strongly in $H^1(\mathbb{R}^d)$ to $u \in H^1(\mathbb{R}^d)$, then there is a subsequence of quasi-continuous representatives $\widetilde{u}_n$ which converges quasi-everywhere to the quasi-continuous representative $\widetilde{u}$.
(5) If $u : \mathbb{R}^d \to \mathbb{R}$ is quasi-continuous, then $|\{u \geq 0\}| = 0$, if and only if, $\mathrm{cap}(\{u \geq 0\}) = 0$.

Remark 3.13. From now on, we identify the Sobolev function $u \in H^1(\mathbb{R}^d)$ with its quasi-continuous representative $\widetilde{u}$.

All these results were already known in the general setting of Section 2.3. In $\mathbb{R}^d$ we can identify the precise representative $\widetilde{u}$ through the mean values of u (see [62, Section 4.8])

Theorem 3.14. *Let $u \in H^1(\mathbb{R}^d)$. Then, for quasi-every $x_0 \in \mathbb{R}^d$, we have*

$$\widetilde{u}(x_0) = \lim_{r\to 0} \fint_{B_r(x_0)} u\,dx. \tag{3.12}$$

3.2. Capacitary measures and the spaces H^1_μ

Definition 3.15. A Borel measure μ on $\mathbb{R}^d$ is called *capacitary*, if for every set $E \subset \mathbb{R}^d$ such that $\mathrm{cap}(E) = 0$ we have $\mu(E) = 0$.

Remark 3.16. If u_1 and u_2 are two positive Borel functions on $\mathbb{R}^d$ such that $\mathrm{cap}(\{u_1 \neq u_2\}) = 0$, then we have that $\int_{\mathbb{R}^d} u_1\,d\mu = \int_{\mathbb{R}^d} u_2\,d\mu$. In particular, a Sobolev function $u \in H^1(\mathbb{R}^d)$ is square integrable with respect to μ, *i.e.* $u \in L^2(\mu)$, if and only if its quasi-continuous representative $\widetilde{u}$, which is unique up to sets of zero capacity, is square integrable with respect to μ.

Let μ be a capacitary measure in $\mathbb{R}^d$. For a function $u \in H^1(\mathbb{R}^d)$, we define

$$\|u\|^2_{\dot{H}^1_\mu} := \int_{\mathbb{R}^d} |\nabla u|^2\,dx + \int_{\mathbb{R}^d} u^2\,d\mu, \tag{3.13}$$

$$\|u\|^2_{H^1_\mu} := \int_{\mathbb{R}^d} |\nabla u|^2\,dx + \int_{\mathbb{R}^d} u^2\,dx + \int_{\mathbb{R}^d} u^2\,d\mu = \|u\|^2_{\dot{H}^1_{\mu+1}}. \tag{3.14}$$

Definition 3.17. For every capacitary measure μ in $\mathbb{R}^d$, we define the space $H^1_\mu(\mathbb{R}^d)$ (or just H^1_μ) as

$$\begin{aligned} H^1_\mu(\mathbb{R}^d) &:= \left\{u \in H^1(\mathbb{R}^d) : \|u\|_{H^1_\mu} < +\infty\right\} \\ &= \left\{u \in H^1(\mathbb{R}^d) : \|u\|_{L^2(\mu)} < +\infty\right\}. \end{aligned} \tag{3.15}$$

Proposition 3.18. *For every capacitary measure μ the space H^1_μ endowed with the norm $\|\cdot\|_{H^1_\mu}$ is a Hilbert space. Moreover, H^1_μ is a Riesz space, has the Stone property and the functions in H^1_μ that have compact support are dense in H^1_μ.*

Proof. We first prove that H^1_μ is a Hilbert space (see also [33]). Indeed, let u_n be a Cauchy sequence with respect to $\|\cdot\|_{H^1_\mu}$. Then u_n converges to some $u \in H^1(\mathbb{R}^d)$ strongly in H^1 and thus quasi-everywhere. Since μ is absolutely continuous with respect to the capacity, we have that u_n converges to u μ-almost everywhere. On the other hand, u_n converges to some $v \in L^2(\mu)$ in $L^2(\mu)$ and so, up to a subsequence μ-almost everywhere. Thus $u = v$ in $L^2(\mu)$ and so $u \in H^1_\mu(\mathbb{R}^d) = H^1(\mathbb{R}^d) \cap L^2(\mu)$ is the limit of u_n in H^1_μ.

For the Riesz and the Stone properties of H^1_μ, we note that if $u, v \in H^1_\mu$, then also $u \wedge v \in H^1_\mu$ and $u \wedge 1 \in H^1_\mu$.

We now prove that the functions of compact support

$$H^1_{\mu,c} := \Big\{u \in H^1_\mu(\mathbb{R}^d) \,:\, \exists R > 0 \text{ such that } |\{u \neq 0\} \setminus B_R| = 0\Big\},$$

are dense in H^1_μ. We report the calculation here, since we will use this argument several times below. Consider the function $\eta_R(x) := \eta(x/R)$, where

$$\eta \in C^\infty_c(\mathbb{R}^d), \quad 0 \leq \eta \leq 1, \quad \eta = 1 \text{ on } B_1, \quad \eta = 0 \text{ on } \mathbb{R}^d \setminus B_2,$$

and let $u \in H^1_\mu$. Calculating the norm of $u - \eta_R u = (1 - \eta_R)u$, we have

$$\|(1-\eta_R)u\|^2_{H^1_\mu} = \int_{\mathbb{R}^d} |\nabla((1-\eta_R)u)|^2\,dx + \int_{\mathbb{R}^d} |(1-\eta_R)u|^2\,dx + \int_{\mathbb{R}^d} |(1-\eta_R)u|^2\,d\mu.$$

The last two terms converge to zero as $R \to \infty$ by the dominated convergence Theorem, while for the first one we note that $\|\nabla\eta_R\|_{L^\infty} = R^{-1}\|\nabla\eta\|_{L^\infty}$ and apply the Cauchy-Schwartz inequality obtaining

$$\begin{aligned}
&\int_{\mathbb{R}^d} |\nabla((1-\eta_R)u)|^2\,dx \\
&= \int_{\mathbb{R}^d} \left[(1-\eta_R)^2|\nabla u|^2 + |\nabla\eta_R|^2 u^2 + 2u\eta_R\nabla\eta_R \cdot \nabla u_R\right] dx \\
&\leq \int_{\mathbb{R}^d} (1-\eta_R)^2|\nabla u|^2\,dx + \left(2R^{-1} + R^{-2}\right)\|u\|_{H^1},
\end{aligned}$$

which proves the claim. □

Definition 3.19. We define the space $\dot{H}^1_\mu(\mathbb{R}^d)$ as the closure of the functions of compact support $H^1_{\mu,c} \subset H^1_\mu$ with respect to the norm $\|\cdot\|_{\dot{H}^1_\mu}$.

The following result is a consequence of the density of $H^1_{\mu,c}$ in both H^1_μ and $\dot{H}^1_\mu$.

Corollary 3.20. *Let μ be a capacitary measure in $\mathbb{R}^d$. Then the following are equivalent:*

(a) *$H^1_\mu \subset L^1(\mathbb{R}^d)$ and the injection $H^1_\mu \hookrightarrow L^1(\mathbb{R}^d)$ is continuous;*
(b) *$\dot{H}^1_\mu \subset L^1(\mathbb{R}^d)$ and the injection $\dot{H}^1_\mu \hookrightarrow L^1(\mathbb{R}^d)$ is continuous;*
(c) *$H^1_{\mu,c} \subset L^1(\mathbb{R}^d)$ and the injection $H^1_{\mu,c} \hookrightarrow L^1(\mathbb{R}^d)$ is continuous.*

Moreover, if one of (a), (b) and (c) holds, then we have that

$$H^1_\mu = H^1_\mu(\mathbb{R}^d) \cap L^1(\mathbb{R}^d) = \dot{H}^1_\mu(\mathbb{R}^d) \cap L^1(\mathbb{R}^d) = \dot{H}^1_\mu,$$

and the corresponding norms are equivalent.

Definition 3.21. We say that two capacitary measures μ and ν are equivalent, if

$$\mu(\Omega) = \nu(\Omega), \quad \forall \Omega \subset \mathbb{R}^d, \ \Omega \text{ quasi-open.}^3$$

Proposition 3.22. *Let μ and ν be capacitary measures. Then the following are equivalent:*

(a) *μ and ν are equivalent;*
(b) *for every non-negative quasi-continuous function $\varphi : \mathbb{R}^d \to \mathbb{R}^+$, we have*

$$\int_{\mathbb{R}^d} \varphi \, d\mu = \int_{\mathbb{R}^d} \varphi \, d\nu;$$

(c) *for every $u \in H^1(\mathbb{R}^d)$, we have*

$$\int_{\mathbb{R}^d} u^2 \, d\mu = \int_{\mathbb{R}^d} u^2 \, d\nu.$$

Proof. We first note that $(a) \Rightarrow (b)$ follows by the formula

$$\int_{\mathbb{R}^d} \varphi \, d\mu = \int_0^{+\infty} \mu(\{\varphi > t\}) \, dt.$$

Then (b) $\Rightarrow$ (c) holds since every $u \in H^1(\mathbb{R}^d)$ is quasi-continuous up to a set of zero capacity. Thus, we only have to prove that (c) $\Rightarrow$ (a).

[3] Recall that a quasi-open set $\Omega \subset \mathbb{R}^d$ is a set such that for every $\varepsilon > 0$ there is an open set $\omega_\varepsilon \subset \mathbb{R}^d$ such that $\Omega \cup \omega_\varepsilon$ is open and $\mathrm{cap}(\omega_\varepsilon) < \varepsilon$

Let $\Omega \subset \mathbb{R}^d$ be a quasi-open set. By Proposition 2.69, there is a function $u \in H^1(\mathbb{R}^d)$ such that $\{u > 0\} = \Omega$. Taking the positive part of u and then $u \wedge 1$, we can assume $0 \leq u \leq 1$ on $\mathbb{R}^d$. We now note that $u_\varepsilon = 1 \wedge (\varepsilon^{-1}u) \in H^1(\mathbb{R}^d)$ is decreasing in ε and converges pointwise to $\mathbb{1}_{\{u>0\}}$ as $\varepsilon \to 0$. Thus, we have

$$\mu(\Omega) = \lim_{\varepsilon\to\infty} \int_{\mathbb{R}^d} u_\varepsilon^2 \, d\mu = \lim_{\varepsilon\to\infty} \int_{\mathbb{R}^d} u_\varepsilon^2 \, d\nu = \nu(\Omega). \qquad \square$$

Remark 3.23. From now on we will identify the capacitary measure μ with its class of equivalence from Definition 3.21, which we will denote with $\mathcal{M}_{\text{cap}}(\mathbb{R}^d)$.

Remark 3.24. If μ, ν are two capacitary measures such that $\mu = \nu$, then $H^1_\mu = H^1_\nu$.

Definition 3.25. Let μ and ν be capacitary measures in $\mathbb{R}^d$. We will say that $\mu \geq \nu$, if

$$\mu(\Omega) \geq \nu(\Omega), \ \forall \Omega \subset \mathbb{R}^d, \ \Omega \text{ quasi-open}.$$

By the same argument as in the proof of Proposition 3.22, we have the following result.

Proposition 3.26. *Let μ and ν be capacitary measures. Then the following are equivalent:*

(a) $\mu \geq \nu$;

(b) *for every non-negative quasi-continuous function $\varphi : \mathbb{R}^d \to \mathbb{R}^+$, we have*

$$\int_{\mathbb{R}^d} \varphi \, d\mu \geq \int_{\mathbb{R}^d} \varphi \, d\nu;$$

(c) *for every $u \in H^1(\mathbb{R}^d)$, we have*

$$\int_{\mathbb{R}^d} u^2 \, d\mu \geq \int_{\mathbb{R}^d} u^2 \, d\nu.$$

Remark 3.27. If μ, ν are two capacitary measures such that $\mu \geq \nu$, then $H^1_\mu \subset H^1_\nu$.

Definition 3.28. Let μ and ν be capacitary measures in $\mathbb{R}^d$. We define the capacitary measure $\mu \vee \nu \in \mathcal{M}_{\text{cap}}(\mathbb{R}^d)$ as

$$(\mu \vee \nu)(E) := \max\Big\{\mu(A) + \nu(E \setminus A) : \ \forall \text{ Borel set } A \subset E\Big\},$$

for every Borel set $E \subset \mathbb{R}^d$.

Remark 3.29. It is straightforward to check that

$$\mu \le \mu \vee \nu \le \mu + \nu \qquad \text{and} \qquad H^1_{\mu\vee\nu} = H^1_\mu \cap H^1_\nu .$$

As we saw above, every capacitary measure $\mu \in \mathcal{M}_{\text{cap}}(\mathbb{R}^d)$ generates a closed subspace of H^1_μ. The classical Sobolev spaces $H^1_0(\Omega)$ can also be characterized through a specific capacitary measure. We give a precise definition of this concept below.

Definition 3.30. Given a Borel set $\Omega \subset \mathbb{R}^d$, we define the capacitary measures I_Ω and $\widetilde{I}_\Omega$ as

$$I_\Omega(E) = \begin{cases} 0, & \text{if } \operatorname{cap}(E \setminus \Omega) = 0, \\ +\infty, & \text{if } \operatorname{cap}(E \setminus \Omega) > 0, \end{cases}$$

$$\text{and} \qquad \widetilde{I}_\Omega(E) = \begin{cases} 0, & \text{if } |E \setminus \Omega| = 0, \\ +\infty, & \text{if } |E \setminus \Omega| > 0. \end{cases}$$

Remark 3.31. For every $\Omega \subset \mathbb{R}^d$, we have $I_\Omega \ge \widetilde{I}_\Omega$.

Remark 3.32. We note that for a function $u \in H^1(\mathbb{R}^d)$, we have

$$\left(u \in H^1_{I_\Omega}(\mathbb{R}^d)\right) \Leftrightarrow \left(\int_{\mathbb{R}^d} u^2\, dI_\Omega < +\infty\right) \Leftrightarrow \left(u \in H^1_0(\Omega)\right),$$

where for a generic set $\Omega \subset \mathbb{R}^d$, we define

$$H^1_0(\Omega) := \left\{u \in H^1(\mathbb{R}^d) :\ \operatorname{cap}\left(\{u \neq 0\} \setminus \Omega\right) = 0\right\}. \tag{3.16}$$

Analogously,

$$\left(u \in H^1_{\widetilde{I}_\Omega}(\mathbb{R}^d)\right) \Leftrightarrow \left(\int_{\mathbb{R}^d} u^2\, d\widetilde{I}_\Omega < +\infty\right) \Leftrightarrow \left(u \in \widetilde{H}^1_0(\Omega)\right),$$

where

$$\widetilde{H}^1_0(\Omega) := \left\{u \in H^1(\mathbb{R}^d) :\ \left|\{u \neq 0\} \setminus \Omega\right| = 0\right\}.$$

Remark 3.33. If $\Omega \subset \mathbb{R}^d$ is an open set, then the smooth functions with compact support in Ω, $C^\infty_c(\Omega)$ are dense in $H^1_0(\Omega)$, defined as in (3.16), with respect to the norm $\|\cdot\|_{H^1}$ (see [72, Theorem 3.3.42]). The analogous result for $\widetilde{H}^1_0(\Omega)$ is true under the additional assumption[4] that the boundary $\partial\Omega$ locally is a graph of a Lipschitz function.

[4] In Proposition 5.48 we will provide another more general condition.

3.3. Torsional rigidity and torsion function

Given a capacitary measure $\mu \in \mathcal{M}_{\text{cap}}(\mathbb{R}^d)$, we consider the functional

$$J_\mu : H^1(\mathbb{R}^d) \cap L^1(\mathbb{R}^d) \to \mathbb{R} \cup \{+\infty\},$$

$$J_\mu(u) := \frac{1}{2}\int_{\mathbb{R}^d} |\nabla u|^2\,dx + \frac{1}{2}\int_{\mathbb{R}^d} u^2\,d\mu - \int_{\mathbb{R}^d} u\,dx.$$

Definition 3.34. For a capacitary measure $\mu \in \mathcal{M}_{\text{cap}}(\mathbb{R}^d)$, we define

- the **torsional rigidity** (or the **torsion**) $T(\mu) \in [0, +\infty]$[5] of μ

$$T(\mu) := \max\Big\{ - J_\mu(u) : \ u \in H^1(\mathbb{R}^d) \cap L^1(\mathbb{R}^d)\Big\}$$
$$= \max\Big\{ - J_\mu(u) : \ u \in H^1_\mu(\mathbb{R}^d) \cap L^1(\mathbb{R}^d)\Big\};$$

- the **Dirichlet Energy** $E(\mu) \in [-\infty, 0]$ of μ

$$E(\mu) := -T(\mu) = \min\Big\{J_\mu(u) : \ u \in H^1(\mathbb{R}^d) \cap L^1(\mathbb{R}^d)\Big\}.$$

Definition 3.35. We say that the capacitary measure μ is of finite torsion if $T(\mu) < +\infty$.

Remark 3.36. Let μ and ν be capacitary measure such that $\mu \geq \nu$. Then we have $J_\mu \geq J_\nu$ and $T(\mu) \leq T(\nu)$. In particular, if $T(\nu) < +\infty$, then also $T(\mu) < +\infty$.

Remark 3.37. Every capacitary measure in a bounded open set is of finite torsion. Indeed, consider a bounded open set with smooth boundary $\Omega \subset \mathbb{R}^d$. Note that for every $u \in H^1_0(\Omega)$, we have (for the second inequality below, see [61, Theorem 1, Section 5.6])

$$\begin{aligned}\int_\Omega |u|\,dx &\leq |\Omega|^{\frac{1}{d}} \left(\int_{\mathbb{R}^d} |u|^{\frac{d}{d-1}}\,dx\right)^{\frac{d-1}{d}} \leq |\Omega|^{\frac{1}{d}} \int_{\mathbb{R}^d} |\nabla u|\,dx \\ &\leq |\Omega|^{\frac{2+d}{2d}} \left(\int_\Omega |\nabla u|^2\,dx\right)^{1/2}.\end{aligned} \tag{3.17}$$

In particular, for every capacitary measure $\mu \in \mathcal{M}_{\text{cap}}(\mathbb{R}^d)$ and every $u \in H^1_0(\Omega)$, we have

$$J_{I_\Omega \vee \mu}(u) \geq \frac{1}{2}\|\nabla u\|^2_{L^2} + \frac{1}{2}\int_{\mathbb{R}^d} u^2\,d\mu - |\Omega|^{\frac{2+d}{2d}}\|\nabla u\|_{L^2}. \tag{3.18}$$

---

[5] In the literature the torsion of μ is sometimes denoted by $P(\mu)$. In this monograph the we prefer the notation $T(\mu)$ since P is reserved for the perimeter.

Since $J_{I_\Omega\vee\mu}(0) = 0$, we can suppose that a minimizing sequence u_n for $J_{I_\Omega\vee\mu}$ is such that $J_{I_\Omega\vee\mu}(u_n) \leq 0$. By (3.18), we have $\|\nabla u_n\|_{L^2} \leq 2|\Omega|^{(2+d)/2d}$. Thus, the sequence u_n is bounded in $H_0^1(\Omega)$ and also in $H^1_{I_\Omega\vee\mu}$. By the compact inclusion $H_0^1(\Omega) \subset L^1(\Omega)$, we can suppose that u_n converges to some $u \in H^1_\mu \cap L^1(\Omega)$ both weakly in H^1_μ and strongly in $L^2(\Omega)$. Thus, u is a minimizer of $J_{I_\Omega\vee\mu}$ in $H^1(\mathbb{R}^d) \cap L^1(\mathbb{R}^d)$. Moreover, by the strict convexity of the functional, u is the unique minimizer of $J_{I_\Omega\vee\mu}$. Let $v \in H^1_\mu \cap H^1_0(\Omega) \cap L^1(\mathbb{R}^d)$. Using that for every $\varepsilon \in \mathbb{R}$, $J_{I_\Omega\vee\mu}(u) \leq J_{I_\Omega\vee\mu}(u + \varepsilon v)$ and taking the derivative for $\varepsilon = 0$, we obtain the Euler-Lagrange equation

$$\int_{\mathbb{R}^d} \nabla u \cdot \nabla v \, dx + \int_{\mathbb{R}^d} uv \, d\mu = \int_{\mathbb{R}^d} v \, dx. \tag{3.19}$$

In particular, taking $v = u$ in (3.19), we get

$$\int_{\mathbb{R}^d} |\nabla u|^2 \, dx + \int_{\mathbb{R}^d} u^2 \, d\mu = \int_{\mathbb{R}^d} u \, dx. \tag{3.20}$$

and thus, the Dirichlet Energy is given by

$$E(I_\Omega \vee \mu) = J_{I_\Omega\vee\mu}(u) = -\frac{1}{2}\int_{\mathbb{R}^d} u \, dx. \tag{3.21}$$

Let now $\mu \in \mathcal{M}_{\text{cap}}(\mathbb{R}^d)$ be a capacitary measure. For every $R > 0$, we consider the unique minimizer $w_R \in H^1(\mathbb{R}^d) \cap L^1(\mathbb{R}^d)$ of the functional $J_{I_{B_R}\vee\mu}$, which exists due to Remark 3.37. Reasoning as in Proposition 2.13, we have that the weak maximum principle holds, *i.e.* for every $R \geq r > 0$, we have $w_R \geq w_r$. Thus, the family of functions $\{w_R\}_{R>0}$ is increasing in $L^1(\mathbb{R}^d)$ and so it has a limit for almost every point in $\mathbb{R}^d$ as $R \to +\infty$.

Definition 3.38. Let $\mu \in \mathcal{M}_{\text{cap}}(\mathbb{R}^d)$ be a capacitary measure. The **torsion function**[6] w_μ of μ is the Lebesgue measurable function defined as

$$w_\mu := \lim_{R\to\infty} w_R = \sup_{R>0} w_R,$$

where w_R is the unique minimizer of the functional $J_{I_{B_R}\vee\mu} : H^1(\mathbb{R}^d) \cap L^1(\mathbb{R}^d) \to \mathbb{R} \cup \{+\infty\}$.

[6] In the literature it is also known as *energy function*.

Example 3.39. If $\Omega \subset \mathbb{R}^d$ is a bounded set and $\mu = I_\Omega$, then w_μ is the weak solution of the boundary value problem

$$-\Delta w = 1 \quad \text{in} \quad \Omega, \qquad w \in H_0^1(\Omega).$$

In particular, if Ω is the ball $B_R(x_0)$, then

$$w_\mu(x) = \frac{\left(R^2 - |x - x_0|^2\right)^+}{2d}.$$

Example 3.40. If $\mu = 0$, then $w_\mu \equiv +\infty$.

Example 3.41. If $\mu = I_S$, where $S \subset \mathbb{R}^2$ is the strip $S = \{(x, y) : x \in \mathbb{R},\ y \in (-1, 1)\}$, then

$$w_\mu(x, y) = \frac{(1 - y^2)^+}{2}.$$

This example shows that there are capacitary measures μ of infinite torsion whose torsion function w_μ is finite almost everywhere and even bounded (but not integrable).

The following result relates the integrability of w_μ to the finiteness of the torsion $T(\mu)$ and to the compact embedding of H^1_μ into $L^1(\mathbb{R}^d)$.

Theorem 3.42. *Let $\mu \in \mathcal{M}_{\text{cap}}(\mathbb{R}^d)$ and let w_μ be its torsion function. Then the following conditions are equivalent:*

(1) *The inclusion $H^1_\mu \subset L^1(\mathbb{R}^d)$ is continuous and there is a constant $C_\mu > 0$ such that*

$$\|u\|_{L^1} \le C_\mu\left(\|\nabla u\|^2_{L^2} + \|u\|^2_{L^2(\mu)}\right)^{1/2}, \qquad \text{for every } u \in H^1_\mu. \tag{3.22}$$

(2) *The inclusion $H^1_\mu \subset L^1$ is compact and* (3.22) *holds.*
(3) *The torsion function w_μ is in $L^1(\mathbb{R}^d)$.*
(4) *The torsion $T(\mu)$ is finite.*

Moreover, if the above conditions hold, then $w_\mu \in H^1_\mu \cap L^1(\mathbb{R}^d)$ is the unique minimizer of J_μ in H^1_μ and the constant from (3.22) *can be estimated by*

$$C_\mu^2 \le \int_{\mathbb{R}^d} w_\mu \, dx = 2T(\mu).$$

Proof. We first prove that (3) and (4) are equivalent.

(3) $\Rightarrow$ (4). Since the functions in $H^1_\mu \cap L^1$ with compact support are dense in $H^1_\mu \cap L^1$, we have

$$\begin{aligned}
&\inf\left\{J_\mu(u) : u \in H^1_\mu(\mathbb{R}^d) \cap L^1(\mathbb{R}^d)\right\} \\
&= \inf_{R>0}\left\{\inf\left\{J_\mu(u) : u \in H^1_{\mu\vee I_{B_R}}(\mathbb{R}^d) \cap L^1(\mathbb{R}^d)\right\}\right\} \\
&= \inf_{R>0} J_\mu(w_R) = \inf_{R>0}\left\{-\frac{1}{2}\int_{\mathbb{R}^d} w_R\,dx\right\} \\
&= -\frac{1}{2}\int_{\mathbb{R}^d} w_\mu\,dx > -\infty,
\end{aligned} \tag{3.23}$$

where the last equality is due to the fact that w_R is increasing in R and converges pointwise to w_μ. Moreover, we have that $w_\mu \in H^1_\mu \cap L^1(\mathbb{R}^d)$ and w_μ minimizes J_μ. Indeed, since w_R converges to w_μ in $L^1(\mathbb{R}^d)$ and w_R is uniformly bounded in H^1_μ by the inequality

$$\int_{\mathbb{R}^d} |\nabla w_R|^2\,dx + \int_{\mathbb{R}^d} w_R^2\,d\mu = \int_{\mathbb{R}^d} w_R\,dx \le \int_{\mathbb{R}^d} w_\mu\,dx,$$

we have that $w_\mu \in H^1_\mu$ and $J_\mu(w_\mu) \le \liminf_{R\to\infty} J_\mu(w_R)$.

(4) $\Rightarrow$ (3). By (3.23), we have that for every $R > 0$,

$$\int_{\mathbb{R}^d} w_R\,dx \le -2\inf\left\{J_\mu(u) : u \in H^1_\mu(\mathbb{R}^d) \cap L^1(\mathbb{R}^d)\right\} < +\infty.$$

Taking the limit as $R \to \infty$, and taking in consideration again (3.23), we obtain

$$\int_{\mathbb{R}^d} w_\mu\,dx = -2\inf\left\{J_\mu(u) : u \in H^1_\mu(\mathbb{R}^d) \cap L^1(\mathbb{R}^d)\right\} < +\infty. \tag{3.24}$$

Since the implication (2) $\Rightarrow$ (1) is clear, it is sufficient to prove that (1) $\Rightarrow$ (4) and (4) $\Rightarrow$ (2).

(1) $\Rightarrow$ (4). Let $u_n \in H^1_\mu$ be a minimizing sequence for J_μ such that $u_n \ge 0$ and $J_\mu(u_n) \le 0$, for every $n \in \mathbb{N}$. Then we have

$$\begin{aligned}
\frac{1}{2}\int_{\mathbb{R}^d} |\nabla u_n|^2\,dx + \frac{1}{2}\int_{\mathbb{R}^d} u_n^2\,d\mu &\le \int_{\mathbb{R}^d} u_n\,dx \\
&\le C\left(\int_{\mathbb{R}^d} |\nabla u_n|^2\,dx + \int_{\mathbb{R}^d} u_n^2\,d\mu\right)^{1/2},
\end{aligned}$$

and so u_n is bounded in $H^1_\mu(\mathbb{R}^d) \cap L^1(\mathbb{R}^d)$. Suppose that u is the weak limit of u_n in H^1_μ. Then

$$\|u\|_{H^1_\mu} \le \liminf_{n\to\infty} \|u_n\|_{H^1_\mu} \qquad \text{and} \qquad \int_{\mathbb{R}^d} u\,dx = \lim_{n\to\infty} \int_{\mathbb{R}^d} u_n\,dx,$$

where the last equality is due to the fact that the functional $\{u \mapsto \int u\,dx\}$ is continuous in H^1_μ. Thus, $u \in H^1_\mu \cap L^1(\mathbb{R}^d)$ is the (unique, due to the strict convexity of J_μ) minimizer of J_μ and so $E(\mu) = \inf J_\mu > -\infty$.

We now prove (3) $\Rightarrow$ (1). Since, $w_\mu \in H^1_\mu \cap L^1(\mathbb{R}^d)$ is the minimizer of J_μ in $H^1_\mu \cap L^1(\mathbb{R}^d)$, we have that the following Euler-Lagrange equation holds:

$$\int_{\mathbb{R}^d} \nabla w_\mu \cdot \nabla u\,dx + \int_{\mathbb{R}^d} w_\mu u\,d\mu = \int_{\mathbb{R}^d} u\,dx, \quad \forall u \in H^1_\mu(\mathbb{R}^d) \cap L^1(\mathbb{R}^d). \tag{3.25}$$

Thus, for every $u \in H^1_\mu(\mathbb{R}^d) \cap L^1(\mathbb{R}^d)$, we obtain

$$\begin{aligned} \|u\|_{L^1} &\le \Big(\|\nabla w_\mu\|^2_{L^2} + \|w_\mu\|^2_{L^2(\mu)}\Big)^{1/2} \Big(\|\nabla u\|^2_{L^2} + \|u\|^2_{L^2(\mu)}\Big)^{1/2} \\ &= \|w_\mu\|^{1/2}_{L^1} \Big(\|\nabla u\|^2_{L^2} + \|u\|^2_{L^2(\mu)}\Big)^{1/2}. \end{aligned} \tag{3.26}$$

Since $H^1_\mu(\mathbb{R}^d) \cap L^1(\mathbb{R}^d)$ is dense in $H^1_\mu(\mathbb{R}^d)$, we obtain (1).

(3) $\Rightarrow$ (2). Following [22, Theorem 3.2], consider a sequence $u_n \in H^1_\mu$ weakly converging to zero in H^1_μ and suppose that $u_n \ge 0$, for every $n \in \mathbb{N}$. Since the injection $H^1(\mathbb{R}^d) \hookrightarrow L^1_{\text{loc}}(\mathbb{R}^d)$ is locally compact, we only have to prove that for every $\varepsilon > 0$ there is some $R > 0$ such that $\int_{B^c_R} u_n\,dx \le \varepsilon$. Consider the function $\eta_R(x) := \eta(x/R)$ where

$$\eta \in C^\infty_c(\mathbb{R}^d), \quad 0 \le \eta \le 1, \quad \eta = 1 \text{ on } B_1, \quad \eta = 0 \text{ on } \mathbb{R}^d \setminus B_2.$$

Testing (3.25) with $(1-\eta_R)u_n$, we have

$$\begin{aligned} &\int_{\mathbb{R}^d} \Big[u_n \nabla w_\mu \cdot \nabla(1-\eta_R) + (1-\eta_R)\nabla w_\mu \cdot \nabla u_n)\Big]\,dx \\ &\qquad + \int_{\mathbb{R}^d} w_\mu (1-\eta_R) u_n\,d\mu = \int_{\mathbb{R}^d} (1-\eta_R) u_n\,dx, \end{aligned}$$

and using the identity $\|\nabla \eta_R\|_{L^\infty} = R^{-1}\|\nabla\eta\|_{L^\infty}$ and the Cauchy-Schwartz inequality, we have

$$\begin{aligned} \int_{B^c_{2R}} u_n\,dx &\le R^{-1}\|u_n\|_{L^2}\|\nabla w_\mu\|_{L^2} + \|\nabla u_n\|_{L^2}\|\nabla w_n\|_{L^2(B^c_R)} \\ &\qquad + \|u_n\|_{L^2(\mu)} \left(\int_{B^c_R} w^2_\mu\,d\mu\right)^{1/2}, \end{aligned}$$

which for R large enough gives the desired ε. □

Remark 3.43. In particular, by Theorem 3.42 the continuity of the inclusion $H^1_\mu \subset L^1(\mathbb{R}^d)$ is equivalent to the continuity of the inclusion $\dot{H}^1_\mu \subset L^1(\mathbb{R}^d)$. The norm of the injection operator $j_\mu : \dot{H}^1_\mu \hookrightarrow L^1(\mathbb{R}^d)$ can be calculated in terms of the torsion $T(\mu)$ and the torsion function w_μ. Indeed, by (3.26), we have that

$$\|u\|_{L^1} \le \|w_\mu\|_{L^1}^{1/2} \|u\|_{\dot{H}^1_\mu} = \big(2T(\mu)\big)^{1/2} \|u\|_{\dot{H}^1_\mu}, \qquad \forall u \in H^1_\mu. \quad (3.27)$$

On the other hand, for $u = w_\mu$, we have an equality in (3.27), which gives that the norm of j_μ is precisely $\big(2T(\mu)\big)^{1/2}$.

Example 3.44. Suppose that $\Omega \subset \mathbb{R}^d$ is a set of finite Lebesgue measure and $\mu = I_\Omega$ or $\mu = \widetilde{I}_\Omega$. Then the torsion function w_μ is in $L^1(\mathbb{R}^d)$ and so the inclusion $H^1_0(\Omega) \hookrightarrow L^1(\mathbb{R}^d)$ is compact.

Example 3.45. Suppose that μ is a capacitary measure and $V : \mathbb{R}^d \to [0, +\infty]$ is a measurable function such that

$$\int_{\mathbb{R}^d} V^{-1}\, dx < +\infty \qquad \text{and} \qquad \mu = V(x)\, dx.$$

Then the embedding $H^1_V \subset L^1(\mathbb{R}^d)$ is compact and the function w_μ is in $L^1(\mathbb{R}^d)$. Indeed, let w_n be a minimizing sequence for J_V in $H^1_V \cap L^1(\mathbb{R}^d)$. Since we can suppose $J_V(w_n) < 0$, we have

$$\frac{1}{2}\int_{\mathbb{R}^d} |\nabla w_n|^2 + w_n^2 V\, dx \le \int_{\mathbb{R}^d} w_n\, dx \le \left(\int_{\mathbb{R}^d} w_n^2 V\, dx\right)^{1/2} \left(\int_{\mathbb{R}^d} V^{-1}\, dx\right)^{1/2},$$

which proves that $\inf_n J_\mu(w_n) > -\infty$ and so, we can apply Theorem 3.42.

Remark 3.46. From now on we will denote the space of capacitary measures of finite torsion with $\mathcal{M}^T_{\text{cap}}(\mathbb{R}^d)$.

3.4. PDEs involving capacitary measures

Let $\Omega \subset \mathbb{R}^d$ be a bounded open set with smooth boundary and let $f \in L^2(\Omega)$. We recall that a function $u \in H^1_0(\Omega)$ is a weak solution of the equation

$$-\Delta u = f \quad \text{in} \quad \Omega, \qquad u \in H^1_0(\Omega), \qquad (3.28)$$

if it satisfies

$$\int_{\mathbb{R}^d} \nabla u \cdot \nabla v\, dx = \int_{\mathbb{R}^d} f v\, dx, \quad \text{for every} \quad v \in H^1_0(\Omega).$$

Equivalently, $u \in H_0^1(\Omega)$ solves (3.28) if it is the minimizer in $H_0^1(\Omega)$ of the functional

$$J_f(v) = \int_\Omega \frac{1}{2}|\nabla v|^2 - fv\,dx.$$

We generalize this concept for the class of capacitary measures (not necessarily of finite torsion).

Definition 3.47. Suppose that μ is a capacitary measure in $\mathbb{R}^d$, $\mu \in \mathcal{M}_{\text{cap}}(\mathbb{R}^d)$. Let $f \in L^p(\mathbb{R}^d)$ for some $p \in (1, +\infty]$. We will say that the function $u \in H_\mu^1$ is a (weak) solution of the equation

$$-\Delta u + \mu u = f \quad \text{in} \quad H_\mu^1, \qquad u \in H_\mu^1, \tag{3.29}$$

if u is the minimizer for the variational problem

$$\min\left\{J_{\mu,f}(u)\ :\ u \in H_\mu^1(\mathbb{R}^d) \cap L^{p'}(\mathbb{R}^d)\right\},$$

where the functional $J_{\mu,f} : H^1(\mathbb{R}^d) \cap L^{p'}(\mathbb{R}^d) \to \mathbb{R} \cup \{+\infty\}$ is defined as

$$J_{\mu,f}(u) := \frac{1}{2}\int_{\mathbb{R}^d} |\nabla u|^2\,dx + \frac{1}{2}\int_{\mathbb{R}^d} u^2\,d\mu - \int_{\mathbb{R}^d} uf\,dx. \tag{3.30}$$

Remark 3.48. If $u \in H_\mu^1 \cap L^{p'}(\mathbb{R}^d)$ is a solution of (3.29), then we have

$$\int_{\mathbb{R}^d} \nabla u \cdot \nabla v\,dx + \int_{\mathbb{R}^d} uv\,d\mu = \int_{\mathbb{R}^d} fv\,dx, \qquad \forall v \in H_\mu^1 \cap L^{p'}(\mathbb{R}^d).$$

Proposition 3.49 (Existence of weak solutions). *Let μ be a capacitary measure of finite torsion: $\mu \in \mathcal{M}_{\text{cap}}^T(\mathbb{R}^d)$. Let $f \in L^p(\mathbb{R}^d)$, where*

- $p \in [\frac{2d}{d+2}, +\infty]$, *if* $d \geq 3$;
- $p \in (1, +\infty]$, *if* $d = 2$;
- $p \in [1, +\infty]$, *if* $d = 1$.

Then there is a unique solution of the equation (3.29).

Proof. The existence follows by the compact injection $H_\mu^1 \hookrightarrow L^1(\mathbb{R}^d)$ and the Sobolev inequalities (3.1), (3.2) and (3.3). The uniqueness is a consequence of the strict convexity of $J_{\mu,f}$. □

If μ and f satisfy the hypotheses of Proposition 3.49, then we denote with $w_{\mu,f}$ the unique minimizer of $J_{\mu,f}$ in H_μ^1 and we will refer to it as to the solution of the equation (3.29). As in the metric case, we can compare the different solutions of (3.29) using the weak maximum principle.

Proposition 3.50 (Weak maximum principle). *Let $\mu \in \mathcal{M}_{\text{cap}}^T(\mathbb{R}^d)$ be a capacitary measure in $\mathbb{R}^d$ of finite torsion and let the exponent p be as in Proposition 3.49. Then the solutions of* (3.29) *satisfy the following inequalities:*

(i) *If $\mu \le \nu$ and $f \in L^p(\mathbb{R}^d)$ is a positive function, then $w_{\nu,f} \le w_{\mu,f}$.*
(ii) *If $f, g \in L^p(\mathbb{R}^d)$ are such that $f \le g$, then $w_{\mu,f} \le w_{\mu,g}$.*

Proof. We note that since $\mu \le \nu$, $T(\nu) \le T(\mu) < +\infty$ and so the solution $w_{\nu,f}$ exists. The rest of the proof follows by the same argument of Proposition 2.13. □

Some of the classical estimates for solution of PDEs on a bounded open set can be repeated in the framework of capacitary measures of finite torsion. In what follows, we obtain the classical estimate $\|u\|_{L^\infty} \le C\|f\|_{L^p}$, for $p > d/2$.

Lemma 3.51. *Let $\mu \in \mathcal{M}_{\text{cap}}^T(\mathbb{R}^d)$ be a capacitary measure of finite torsion. Let f be a non-negative function such that $f \in L^p(\mathbb{R}^d)$, for $p \in (d/2, +\infty]$, and let $u \in H_\mu^1$ be the solution of*

$$-\Delta u + \mu u = f \quad in \quad H_\mu^1, \qquad u \in H_\mu^1.$$

Then, there is a dimensional constant $C_d > 0$ such that, for every $t \ge 0$, we have

$$\|(u-t)^+\|_{L^\infty} \le \frac{C_d}{2/d - 1/p}\|f\|_{L^p}|\{u > t\}|^{2/d-1/p}.$$

More precisely, $C_d = \left(d\omega_d^{1/d}\right)^{-2}$, where ω_d is the volume of the unit ball in $\mathbb{R}^d$.

Proof. We start noticing that by the weak maximum principle, $u \ge 0$ on $\mathbb{R}^d$. For every $t \in [0, \|u\|_{L^\infty})$ and $\varepsilon > 0$, we consider the function

$$u_{t,\varepsilon} = u \wedge t + (u - t - \varepsilon)^+ \in H^1(\mathbb{R}^d).$$

Since $u_{t,\varepsilon} \le u$, we have that $u_{t,\varepsilon} \in H_\mu^1$ and so, we can use it as a test function for the functional $J_{\mu,f}$. Indeed the inequalities $J_{\mu,f}(u) \le J_{\mu,f}(u_{t,\varepsilon})$ and $u_{t,\varepsilon} \le u$ give

$$\frac{1}{2}\int_{\mathbb{R}^d} |\nabla u|^2\,dx - \int_{\mathbb{R}^d} fu\,dx \le \frac{1}{2}\int_{\mathbb{R}^d} |\nabla u_{t,\varepsilon}|^2\,dx - \int_{\mathbb{R}^d} fu_{t,\varepsilon}\,dx.$$

In particular, we get

$$\frac{1}{2}\int_{\{t<u\le t+\varepsilon\}}|\nabla u|^2\,dx \le \int_{\mathbb{R}^d} f\left(u-u_{t,\varepsilon}\right)dx \le \varepsilon\int_{\{u>t\}} f\,dx.$$

By the co-area formula (see [67, Chapter 1]) we have

$$\int_{\{u=t\}}|\nabla u|\,d\mathcal{H}^{d-1} \le 2\int_{\{u>t\}} f\,dx \le 2\|f\|_{L^p}|\{u>t\}|^{1/p'}.$$

Setting $\varphi:(0,+\infty)\to(0,+\infty)$ to be the monotone decreasing function $\varphi(t):=|\{u>t\}|$, we have that

$$\begin{aligned}\varphi'(t) &= -\int_{\{u=t\}}\frac{1}{|\nabla u|}\,d\mathcal{H}^{d-1} \le -\left(\int_{\{u=t\}}|\nabla u|\,d\mathcal{H}^{d-1}\right)^{-1} P(\{u>t\})^2\\ &\le -\frac{1}{2}\|f\|_{L^p}^{-1}\varphi(t)^{-1+1/p}\left(d\omega_d^{1/d}\right)^2\varphi(t)^{\frac{2(d-1)}{d}}\\ &= -\frac{1}{2}\|f\|_{L^p}^{-1}\left(d\omega_d^{1/d}\right)^2\varphi(t)^{\frac{d-2}{d}+\frac{1}{p}},\end{aligned}$$

where P is the De Giorgi perimeter (see [67] or [5]) and $d\omega_d^{1/d}$ is the sharp constant from the isoperimetric inequality $P(\Omega)\ge d\omega_d^{1/d}|\Omega|^{\frac{d-1}{d}}$ in $\mathbb{R}^d$. Setting $\alpha=\frac{d-2}{d}+\frac{1}{p}<1$ and $C=\frac{1}{2}\left(d\omega_d^{1/d}\right)^2\|f\|_{L^p}^{-1}$, we consider the ODE

$$y'=-Cy^\alpha,\qquad y(t_0)=y_0. \tag{3.31}$$

The solution of (3.31) is given by $y(t)=\left(y_0^{1-\alpha}-(1-\alpha)C(t-t_0)\right)^{\frac{1}{1-\alpha}}$. Since $\phi(t)\ge 0$, for every $t\ge 0$ and $y(t)\ge\phi(t)$, we have that there is some $t_{\max}$ such that $\phi(t)=0$, for every $t\ge t_{\max}$. Thus, taking $y_0=\phi(t_0)=|\{u>t_0\}|$, we have the estimate

$$\|(u-t_0)^+\|_{L^\infty}\le t_{\max}-t_0\le 2\frac{\left(d\omega_d^{1/d}\right)^{-2}}{2/d-1/p}\|f\|_{L^p}|\{u>t_0\}|^{2/d-1/p}. \qquad \square$$

Corollary 3.52. *Let $\mu\in\mathcal{M}_{\mathrm{cap}}^T(\mathbb{R}^d)$ be a capacitary measure of finite torsion and let w_μ be the corresponding torsion function. If $\mu\ge I_\Omega$, for some set $\Omega\subset\mathbb{R}^d$ of finite Lebesgue measure then we have the estimate*

$$\|w_\mu\|_{L^\infty}\le\frac{1}{d}\frac{|\Omega|^{2/d}}{|B_1|^{2/d}}, \tag{3.32}$$

where B_1 is the unit ball in $\mathbb{R}^d$.

Remark 3.53. We note that the estimate (3.32) is not sharp since, taking $\Omega = B_1$ and $\mu = I_{B_1}$, the torsion function is precisely $w_{B_1}(x) = \frac{1}{2d}(1 - |x|^2)^+$ and so, $\|w_{B_1}\|_{L^\infty} = \frac{1}{2d}$. A classical result due to Talenti (see [89]) shows that the (sharp) estimate

$$\|w_\mu\|_{L^\infty} \leq \frac{1}{2d}\frac{|\Omega|^{2/d}}{|B_1|^{2/d}}, \tag{3.33}$$

holds for every set Ω of finite measure and every $\mu \geq I_\Omega$.

Proposition 3.54 (Infinity estimate). *Let $\mu \in \mathcal{M}^T_{\text{cap}}(\mathbb{R}^d)$, $d \geq 2$, $p \in (d/2, +\infty]$ and $f \in L^p(\mathbb{R}^d)$. Then there is a unique minimizer $u \in H^1_\mu$ of the functional $J_{\mu,f} : H^1_\mu \to \mathbb{R}$. Moreover, u satisfies the inequality*

$$\|u\|_{L^\infty} \leq CT(\mu)^\alpha \|f\|_{L^p}, \tag{3.34}$$

for some constants C and α, depending only on the dimension d and the exponent p.

Proof. We first note that for any $v \in H^1_\mu$ such that $J_{\mu,f}(v) \leq 0$, we have

$$\int_{\mathbb{R}^d} |\nabla v|^2\,dx + \int_{\mathbb{R}^d} v^2\,dx \leq 2\int_{\mathbb{R}^d} fv\,dx \leq 2\|f\|_{L^p}\|v\|_{L^{p'}}.$$

On the other hand $p > d/2$ implies $p' < \frac{d}{d-2}$ and so $p' \in [1, \frac{2d}{d-2}]$. Thus, using (3.22) with $C = T(\mu)^{1/2}$ and an interpolation, we obtain

$$\int_{\mathbb{R}^d} |\nabla v|^2\,dx + \int_{\mathbb{R}^d} v^2\,dx \leq C_d T(\mu)^\alpha \|f\|^2_{L^p}, \tag{3.35}$$

which in turn implies the existence of a minimizer u of $J_{\mu,f}$, satisfying the same estimate.

In order to prove (3.34) it is sufficient to consider the case $f \geq 0$. In this case the solution is nonnegative $u \geq 0$ (since the minimizer is unique and $J_{\mu,f}(|u|) \leq J_{\mu,f}(u)$) and, by Lemma 3.51, we have that $u \in L^\infty$. We set $M := \|u\|_{L^\infty} < +\infty$ and apply again Lemma 3.51 to obtain

$$\frac{M^2}{2} = \int_0^M (M-t)\,dt \leq C\|f\|_{L^p}\int_0^M |\{u > t\}|^\beta\,dt \leq C\|f\|_{L^p} M^{1-\beta}\|u\|^\beta_{L^1},$$

where we set $\beta = 2/d - 1/p \leq 1$. Thus we obtain

$$M^{1+\beta} \leq C\|f\|_{L^p}\|u\|^\beta_{L^1}, \tag{3.36}$$

and using (3.35) with $v = u$, we get (3.34). □

Corollary 3.55. *Let $\mu \in \mathcal{M}_{\text{cap}}^T(\mathbb{R}^d)$ be a capacitary measure of finite torsion and let w_μ be the corresponding torsion function. Then $w_\mu \in L^\infty(\mathbb{R}^d)$ and*

$$\|w_\mu\|_{L^\infty} \le C_d \left(\int_{\mathbb{R}^d} w_\mu \, dx\right)^{\frac{2}{d+2}}, \tag{3.37}$$

for a dimensional constant $C_d > 0$.

3.4.1. Almost subharmonic functions

In this subsection we consider functions $u \in H^1(\mathbb{R}^d)$, which are subharmonic ub $\mathbb{R}^d$ up to some perturbation term $f \in L^p(\mathbb{R}^d)$:

$$\Delta u + f \ge 0 \quad \text{in} \quad \left[C_c^\infty(\mathbb{R}^d)\right]', \tag{3.38}$$

where the above inequality is intended in sense of distributions, *i.e.*

$$\int_{\mathbb{R}^d} -\nabla u \cdot \nabla \varphi + f\varphi \, dx \ge 0, \quad \text{for every} \quad \varphi \in C_c^\infty(\mathbb{R}^d) \quad \text{such that} \quad \varphi \ge 0.$$

We will show that under some reasonable hypotheses on f the function u is pointwise defined everywhere on $\mathbb{R}^d$, *i.e.* every point of $\mathbb{R}^d$ is a Lebesgue point for u. This result applies to the functions u that solve equations of the form

$$-\Delta u + \mu u = f \quad \text{in} \quad H_\mu^1, \qquad u \in H_\mu^1(\mathbb{R}^d). \tag{3.39}$$

In fact, we will show that if u is a positive solution of (3.39), then it satisfies the inequality (3.38).

We start our discussion recalling some general measure theoretic notions and results.

Definition 3.56. Consider a set E and a vector space $\mathcal{R}$ of real functions defined on E

(1) We say that $\mathcal{R}$ is a *Riesz space*, if for each $u, v \in \mathcal{R}, u \wedge v \in \mathcal{R}$.
(2) We denote with $\mathcal{R}_\sigma$ the class of functions $u : E \to \mathbb{R} \cup \{+\infty\}$ of the form $u = \sup_n u_n$ for a sequence of functions $u_n \in \mathcal{R}$.
(3) We say that a linear functional $L : \mathcal{R} \to \mathbb{R}$ is Daniell, if:
   - $L(u) \ge 0$, whenever $u \ge 0$;
   - for each increasing sequence of functions $u_n \in \mathcal{R}$ such that $u := \sup_n u_n \in \mathcal{R}$, we have $L(u) = \sup_n L(u_n)$.

Remark 3.57. We note that a positive linear functional $L : \mathcal{R} \to \mathbb{R}$ is Daniell if and only if, every decreasing sequence of functions $u_n \in \mathcal{R}$ such that $\inf_n u_n = 0$, we have $\inf_n L(u_n) = 0$.

Theorem 3.58 (Representation of Daniell functionals). *Let $\mathcal{R}$ be a Riesz space of real functions defined on the set E such that $1 \in R_\sigma$ and let L be a Daniell functional on $\mathcal{R}$. Then, there is a unique measure μ defined on the sigma-algebra of sets $\mathcal{E}$, generated by $\mathcal{R}$, such that*

$$\mathcal{R} \subset \mathcal{L}^1(\mu) \quad \text{and} \quad L(u) = \int_E u\,d\mu, \quad \text{for every} \quad u \in \mathcal{R}. \tag{3.40}$$

Proposition 3.59. *Let $p \in [1, +\infty]$, $f \in L^p(\mathbb{R}^d)$ and $u \in H^1(\mathbb{R}^d)$ be such that*

$$\Delta u + f \geq 0 \qquad \text{in} \qquad \left[H^1(\mathbb{R}^d) \cap L^{p'}(\mathbb{R}^d)\right]'.$$

Then, there is a Radon capacitary measure ν on $\mathbb{R}^d$ satisfying

$$-\int_{\mathbb{R}^d} \nabla u \cdot \nabla v\,dx + \int_{\mathbb{R}^d} f v\,dx = \int_{\mathbb{R}^d} v\,d\nu, \qquad \text{for every} \quad v \in H^1(\mathbb{R}^d) \cap C_c(\mathbb{R}^d). \tag{3.41}$$

Proof. Let L be the restriction of the operator $\Delta u + f : H^1(\mathbb{R}^d) \cap L^{p'}(\mathbb{R}^d) \to \mathbb{R}$ to the Riesz space $\mathcal{R} = C_c(\mathbb{R}^d) \cap H^1(\mathbb{R}^d)$. Then L is a positive functional. We will prove that L is also Daniell. Consider a decreasing sequence of functions $v_n \in \mathcal{R}$ such that $\inf_n v_n = 0$ and a function $g \in \mathcal{R}$ such that $g \geq \mathbb{1}_{\{v_1>0\}}$. Thus, we have that $0 \leq L(v_n) \leq L(\|v_n\|_{L^\infty} g) = \|v_n\|_{L^\infty} L(g)$. Thus it is sufficient to prove that $\|v_n\|_{L^\infty} \to 0$. Indeed, for every $\varepsilon \geq 0$, the sequence of sets $K_n := \{v_n \geq \epsilon\}$ is a decreasing sequence of compact sets with empty intersection and so, it is definitively constituted of empty sets.

Applying Daniell's Theorem 3.58, we have that there is a measure ν, on the σ-algebra generated by $\mathcal{R}$, such that (3.41) holds for any $v \in \mathcal{R}$. Since for every open set $A \subset \mathbb{R}^d$, there is a function $v \in \mathcal{R}$ such that $A = \{v > 0\}$, we have that ν is a Borel measure. Moreover, for every compact set $K \subset \mathbb{R}^d$, there is a function $\varphi \in H^1(\mathbb{R}^d) \cap C_c(\mathbb{R}^d)$ such that $\varphi \equiv 1$ on K. Thus, we have

$$\nu(K) \leq \int_{\mathbb{R}^d} \varphi\,d\nu = -\int_{\mathbb{R}^d} \nabla u \cdot \nabla\varphi\,dx + \int_{\mathbb{R}^d} \varphi f\,dx < +\infty,$$

which proves that ν is a Radon measure.

In order to prove that the measure ν is capacitary, it is sufficient to check that for every compact set $K \subset \mathbb{R}^d$ such that $\operatorname{cap}(K) = 0$, we have also $\nu(K) = 0$. Indeed, if $\operatorname{cap}(K) = 0$, then there is a sequence of functions $v_n \in C_c(\mathbb{R}^d) \cap H^1(\mathbb{R}^d)$ such that $v_n \geq 1$ on K and $\|v_n\|_{H^1} \to 0$ as $n \to \infty$. Thus, we have that

$$\mu(E) \leq \int_{\mathbb{R}^d} v_n\,d\mu = -\int_\Omega \nabla u \cdot \nabla v_n\,dx + \int_\Omega v_n f\,dx \xrightarrow[n\to+\infty]{} 0. \qquad \square$$

Theorem 3.60. *Assume that*

(a) $u \in H^1(\mathbb{R}^d) \cap L^\infty(\mathbb{R}^d)$;
(b) $f \in L^p(\mathbb{R}^d)$, *for some* $p \in (d/2, +\infty]$;
(c) $\Delta u + f \geq 0$ *on* $\mathbb{R}^d$ *in sense of distributions.*

Then

(i) *the function* $M_r : (0,1) \to \mathbb{R}$, *defined as*

$$M(r) := \fint_{\partial B_r(x_0)} u \, d\mathcal{H}^{d-1},$$

is of bounded variation.

(ii) Δu *is a signed Radon measure on* $\mathbb{R}^d$ *and the weak derivative of* M *is characterized by*

$$M'(r) = \frac{\Delta u\big(B_r\big)}{d\omega_d r^{d-1}}.$$

Proof. We will prove the above Theorem in three steps.

Step 1. We first prove (i) and (ii) under the additional hypothesis $u \in C^2(\mathbb{R}^d)$. Indeed, for each $0 < r < R < 1$, we have

$$\begin{aligned}\frac{\partial}{\partial r}\left[\fint_{\partial B_r} u \, d\mathcal{H}^{d-1}\right] &= \frac{\partial}{\partial r}\left[\fint_{\partial B_1} u(rx)\, d\mathcal{H}^{d-1}(x)\right] \\ &= \fint_{\partial B_1} \nabla u(rx)\cdot x \, d\mathcal{H}^{d-1}(x) = \fint_{\partial B_r} \nabla u(x)\cdot \frac{x}{r}\, d\mathcal{H}^{d-1}(x) \qquad (3.42) \\ &= \frac{1}{d\omega_d r^{d-1}} \int_{B_r} \Delta u(x)\, dx = \frac{\Delta u(B_r)}{d\omega_d r^{d-1}}.\end{aligned}$$

Moreover, $M' \in L^1\big((0,1)\big)$, since

$$M'(r) = \frac{\Delta u(B_r)}{d\omega_d r^{d-1}} \leq \frac{1}{d}\|\Delta u\|_{L^\infty(B_r)}.$$

Step 2. Proof of (i). We consider a function

$$\eta \in C_c^\infty(\mathbb{R}^d), \quad 0 \leq \eta \leq 1, \quad \eta = 1 \text{ on } B_1, \quad \eta = 0 \text{ on } \mathbb{R}^d \setminus B_2,$$

and, for every $r > 0$, we use the notation $\eta_r(x) := \eta(x/r)$ and $\phi_r(x) := r^{-d}\eta(x/r)$. Let $u_\varepsilon := u * \phi_\varepsilon$ and

$$M_\varepsilon(r) := \fint_{\partial B_r} u_\varepsilon \, d\mathcal{H}^{d-1}, \qquad \forall r \in (0,1).$$

Then we have $u_\varepsilon \in C^\infty(\mathbb{R}^d)$, $\|u_\varepsilon\|_{L^\infty} \le \|u\|_{L^\infty}$, $u_\varepsilon \xrightarrow[\varepsilon\to 0]{H^1(\mathbb{R}^d)} u$ and $M_\varepsilon \to M$ in $L^1\big((0,1)\big)$ and pointwise a.e. in $(0,1)$. Moreover, $M_\varepsilon \in BV\big((0,1)\big)$ and $\Delta u_\varepsilon + f \ge 0$. We now prove that the sequence M_ε is uniformly bounded in $BV\big((0,1)\big)$. Indeed, for any $\delta \in (0, 1/2)$ we have

$$\begin{aligned}
&\int_\delta^1 |M_\varepsilon'(r)|\,dr\\
&= \int_\delta^1 \frac{|\Delta u_\varepsilon(B_r)|}{d\omega_d r^{d-1}}\,dr\\
&\le \int_0^1 \frac{(\Delta u_\varepsilon + f)(B_r) + \int_{B_r}|f|(x)\,dx}{d\omega_d r^{d-1}}\,dr\\
&\le \int_\delta^1 \frac{\Delta u_\varepsilon(B_r)}{d\omega_d r^{d-1}}\,dr + 2\int_\delta^1 \frac{1}{d\omega_d r^{d-1}}\left(\int_{B_r}|f|\,dx\right)dr\\
&\le \fint_{\partial B_1} u_\varepsilon\,d\mathcal{H}^{d-1} - \fint_{\partial B_\delta} u_\varepsilon\,d\mathcal{H}^{d-1} + 2\int_\delta^1 \frac{\|f\|_{L^p}}{d\omega_d^{1/p}} r^{1-\frac{d}{p}}\,dr\\
&\le 2\|u\|_{L^\infty} + C_{d,p}\|f\|_{L^p},
\end{aligned} \tag{3.43}$$

where $C_{d,p}$ is a constant depending only on d and p. Passing to the limit as $\delta \to 0$ gives the uniform boundedness of M_ε in $BV\big((0,1)\big)$ and so, the claim.

Step 3. Proof of (ii). By Proposition 3.59 we have that $\nu := \Delta u + f$ is a Radon capacitary measure on $\mathbb{R}^d$. As a consequence, $\Delta u = \nu - f$ is a (signed) Radon capacitary measure on $\mathbb{R}^d$. Let u_ϵ be as in *Step* 2. Then we have that $\Delta u_\epsilon(B_r) \to \Delta u(B_r)$ for $\mathcal{L}^1$- almost every $r \in (0,1)$. In fact, since

$$|\Delta u|(B_R) \le \nu(B_R) + \int_{B_R} |f|\,dx < \infty, \qquad \forall R \in (0,1),$$

we have that for $\mathcal{L}^1$- almost every $r \in (0, R)$ the boundary ∂B_r is $|\Delta u|$-negligible. For those r, we have

$$\Delta u_\epsilon(B_r) = \int_{\mathbb{R}^d} \mathbb{1}_{B_r} * \phi_\epsilon\, d(\Delta u) \xrightarrow[\epsilon\to 0]{} \int_{\mathbb{R}^d} \mathbb{1}_{B_r}\, d(\Delta u),$$

where the passage to the limit is due to the dominated convergence theorem applied to the sequence $\int |\mathbb{1}_{B_r} * \phi_\epsilon - \mathbb{1}_{B_r}|\,d|\Delta u|$. In fact, for small enough ϵ, the integrand is bounded by $2\mathbb{1}_{B_{2r}}$ and $\mathbb{1}_{B_r} * \phi_\epsilon(x) \to \mathbb{1}_{B_r}(x)$, for every $x \notin \partial B_r$ and so, for $|\Delta u|$-almost every $x \in \mathbb{R}^d$. Moreover, it is immediate to check that

$$|\Delta u_\varepsilon|(B_r) \le (\Delta u_\varepsilon + f)(B_r) + \int_{B_r} |f|\,dx$$
$$\le (\Delta u)(B_{1+\varepsilon}) + 2\int_{B_{1+\varepsilon}} |f|\,dx < +\infty,$$

which shows that $M_\varepsilon'(r) \to \left(d\omega_d r^{d-1}\right)^{-1}\Delta u(B_r)$ in $L^1\big((\delta,1)\big)$, for every $\delta > 0$, which concludes the proof. □

Remark 3.61. If u satisfies the hypotheses (a), (b) and (c) of Theorem 3.60, then the function $M' \in L^1((0,1))$ and we have the estimate

$$\int_0^1 |M'(r)|\,dr \le 2\|u\|_{L^\infty} + C_{d,p}\|f\|_{L^p},$$

where $C_{d,p}$ is the constant, depending only on d and p, obtained in (3.43).

Remark 3.62. The conclusions of Theorem 3.60 hold also if we replace the condition (a) with the alternative assumption

(a') $u \in H^1(R^d)$ and $u \ge 0$.

Indeed, the only difference in the proof is in the last estimate of (3.43), where the term $2\|u\|_{L^\infty}$ should be replaced with $1 + \fint_{\partial B_1} u\,d\mathcal{H}^{d-1}$. In this case the L^1 norm of M' is estimated by

$$\int_0^1 |M'(r)|\,dr \le 1 + \fint_{\partial B_1} u\,d\mathcal{H}^{d-1} + C_{d,p}\|f\|_{L^p},$$

where $C_{d,p}$ is the constant from (3.43).

Remark 3.63. It is not hard to check that for a generic Sobolev function $u \in H^1(\mathbb{R}^d)$ the mean $M(r) := \fint_{\partial B_r} u\,d\mathcal{H}^{d-1}$ is continuous for $r \in (0,+\infty)$. Indeed, if $u \in C^1(\mathbb{R}^d)$, then for every $x \in \partial B_1$, we have

$$|u(Rx) - u(rx)| = \left|\int_r^R x\cdot\nabla u(sx)\,ds\right| \le (R-r)^{1/2}\left(\int_r^R |\nabla u|^2(sx)\,ds\right)^{1/2}.$$

Integrating for $x \in \partial B_1$, we have

$$|M(R) - M(r)| \le \fint_{\partial B_1} (R-r)^{1/2}\left(\int_r^R |\nabla u|^2(sx)\,ds\right)^{1/2} d\mathcal{H}^{d-1}$$
$$\le |R-r|^{1/2}\left(\fint_{\partial B_1}\int_r^R |\nabla u|^2(sx)\,ds\,d\mathcal{H}^{d-1}\right)^{1/2}$$
$$\le \frac{|R-r|^{1/2}}{(d\omega_d r^{d-1})^{1/2}}\|\nabla u\|_{L^2},$$

which, by approximation, continues to hold for every $u \in H^1(\mathbb{R}^d)$. In particular, we notice that the radially symmetric Sobolev functions are continuous.

Corollary 3.64. *In the hypotheses of Theorem* 3.60 *or Remark* 3.62, *we have that for every point $x_0 \in \mathbb{R}^d$, the limit*

$$\widetilde{u}(x_0) := \lim_{r\to 0} \fint_{\partial B_r(x_0)} u \, d\mathcal{H}^{d-1} \tag{3.44}$$

exists and $\widetilde{u} = u$ almost everywhere on $\mathbb{R}^d$. Moreover, for every $R > 0$, we have that

$$\fint_{\partial B_R(x_0)} u \, d\mathcal{H}^{d-1} - \widetilde{u}(x_0) = \int_0^R \frac{\Delta u(B_s(x_0))}{d\omega_d s^{d-1}} \, ds. \tag{3.45}$$

Proof. We note that

$$\fint_{\partial B_R(x_0)} u \, d\mathcal{H}^{d-1} - \fint_{\partial B_r(x_0)} u \, d\mathcal{H}^{d-1} = \int_r^R M'(s) \, ds \le \int_r^R |M'(s)| \, dx, \tag{3.46}$$

where $M'(s)$ is as in Theorem 3.60. Thus, by Remark 3.61 the limit (3.44) exists. Suppose now that $x_0 \in \mathbb{R}^d$ is a Lebesgue point for u. Then we have

$$\begin{aligned} u(x_0) &= \lim_{r\to 0} \fint_{B_r(x_0)} u \, dx = \lim_{r\to 0} \frac{1}{\omega_d r^d} \int_0^r d\omega_d s^{d-1} \left(\fint_{\partial B_s(x_0)} u \, d\mathcal{H}^{d-1} \right) ds \\ &= \lim_{r\to 0} \int_0^r \frac{d s^{d-1}}{r^d} \left(\fint_{\partial B_s(x_0)} u \, d\mathcal{H}^{d-1} \right) ds = \widetilde{u}(x_0), \end{aligned}$$

and so $u(x_0) = \widetilde{u}(x_0)$ for a.e. $x_0 \in \mathbb{R}^d$. The identity (3.45) follows after passing to the limit as $r \to 0$ in (3.46). □

The first part of Corollary 3.64 can be proved in an alternative way. For the sake of simplicity, we consider the case $f \in L^\infty(\mathbb{R}^d)$, which will be sufficient for our purposes.

Proposition 3.65. *Let $u \in H^1(\mathbb{R}^d)$ and $f \in L^\infty(\mathbb{R}^d)$. Suppose that $\Delta u + f \ge 0$ in sense of distributions on $\mathbb{R}^d$. Then every point $x_0 \in \mathbb{R}^d$ is a Lebesgue point for u and moreover, we have*

$$\lim_{r\to 0} \fint_{\partial B_r(x_0)} \big|u - u(x_0)\big| \, d\mathcal{H}^{d-1} = \lim_{r\to 0} \fint_{B_r(x_0)} \big|u - u(x_0)\big| \, dx = 0. \tag{3.47}$$

Proof. Since we have

$$\Delta u + \|f\|_{L^\infty} \geq \Delta u + f \geq 0,$$

we can restrict our attention to the case $f \equiv 1$. We now consider the function $v(x) := u(x) + \frac{|x|^2}{2d}$. We note that $\Delta v \geq 0$ and so, the function

$$r \mapsto \fint_{\partial B_r(x_0)} v \, d\mathcal{H}^{d-1},$$

is increasing in r. Thus, we may choose a representative of v such that for every point $x_0 \in \mathbb{R}^d$ the limit

$$v(x_0) = \lim_{r\to\infty} \fint_{\partial B_r(x_0)} v \, d\mathcal{H}^{d-1},$$

exists. Thus, we may suppose that for every point $x_0 \in \mathbb{R}^d$ we have

$$u(x_0) = \lim_{r\to\infty} \fint_{\partial B_r(x_0)} u \, d\mathcal{H}^{d-1}.$$

In order to prove (3.47) we write

$$\begin{aligned}
&\lim_{r\to 0} \fint_{\partial B_r(x_0)} \big|u - u(x_0)\big| \, d\mathcal{H}^{d-1} \\
&\leq \lim_{r\to 0} \fint_{\partial B_r(x_0)} \big|v - v(x_0)\big| \, d\mathcal{H}^{d-1} + \lim_{r\to 0} \fint_{\partial B_r(x_0)} \big||x|^2 - |x_0|^2\big| \, d\mathcal{H}^{d-1} \\
&\leq \lim_{r\to 0} \fint_{\partial B_r(x_0)} v \, d\mathcal{H}^{d-1} - v(x_0) + \lim_{r\to 0} \fint_{\partial B_r(x_0)} \big||x|^2 - |x_0|^2\big| \, d\mathcal{H}^{d-1},
\end{aligned}$$

and we note that by the definition of $v(x_0)$ the right-hand side converges to zero. The proof of the second equality in (3.47) is analogous. □

3.4.2. Pointwise definition, semi-continuity and vanishing at infinity for solutions of elliptic PDEs

In this section we investigate some of the fine properties of the solutions of the equation

$$-\Delta u + \mu u = f \quad \text{in} \quad H^1_\mu, \qquad u \in H^1_\mu,$$

where μ is a capacitary measure of finite torsion. Our results will depend strongly on the theory recalled in the previous section.

Lemma 3.66. *Let $\mu \in \mathcal{M}^T_{\text{cap}}(\mathbb{R}^d)$ be a capacitary measure of finite torsion. Suppose that $p \in [1, +\infty]$ is as in Proposition 3.49 and $f \in L^p(\mathbb{R}^d)$ is such that the solution u of the equation*

$$-\Delta u + \mu u = f \quad in \quad H^1_\mu, \qquad u \in H^1_\mu, \tag{3.48}$$

is non-negative on $\mathbb{R}^d$. Then the following inequality holds:

$$\Delta u + f \mathbb{1}_{\{u>0\}} \geq 0 \qquad in \qquad \left[C^\infty_c(\mathbb{R}^d)\right]'. \tag{3.49}$$

Proof. Let v be a non-negative function in $C^\infty_c(\Omega)$ or, more generally, in $H^1(\mathbb{R}^d) \cap L^1(\mathbb{R}^d) \cap L^\infty(\mathbb{R}^d)$. For each $n \geq 1$, consider the function $p_n : \mathbb{R} \to \mathbb{R}$ defined by

$$p_n(t) = \begin{cases} 0, & \text{if } t \leq 0, \\ nt, & \text{if } t \in [0, \frac{1}{n}], \\ 1, & \text{if } t \geq \frac{1}{n}. \end{cases} \tag{3.50}$$

Since p_n is Lipschitz, we have that $p_n(u) \in H^1(\mathbb{R}^d)$, $\nabla p_n(u) = p_n'(u)\nabla u$ and $v p_n(u) \in H^1(\mathbb{R}^d)$. Moreover, since $|p_n(u)| \leq n|u|$ and $v \in L^\infty(\mathbb{R}^d)$, we have that $v p_n(u) \in H^1_\mu$ and so we can use it to test the equation for u.

$$\begin{aligned} \int_{\mathbb{R}^d} f v p_n(u)\, dx &= \int_{\mathbb{R}^d} \nabla u \cdot \nabla\big(v p_n(u)\big)\, dx + \int_{\mathbb{R}^d} u v p_n(u)\, d\mu \\ &\geq \int_\Omega v p_n'(u)|\nabla u|^2\, dx + \int_\Omega p_n(u)\nabla u \cdot \nabla v\, dx \\ &\geq \int_\Omega p_n(u)\nabla u \cdot \nabla v\, dx. \end{aligned} \tag{3.51}$$

Since $p_n(u) \uparrow \mathbb{1}_{\{u>0\}}$, as $n \to \infty$, we obtain (3.49). □

Remark 3.67. It is sometimes convenient for the sign-changing solutions u of (3.48) to consider separately the positive and negative parts u_+ and u_-. Indeed, let $\mu \in \mathcal{M}^T_{\text{cap}}(\mathbb{R}^d)$ be a capacitary measure of finite torsion in $\mathbb{R}^d$ and let $f \in L^p(\mathbb{R}^d)$, where p is as in Proposition 3.49. Consider the solution u of the equation (3.48) and the capacitary measures

$$\mu_+ = \mu \vee I_{\{u>0\}} \qquad \text{and} \qquad \mu_+ = \mu \vee I_{\{u<0\}}.$$

We have that the positive and negative parts, $u_+ \in H^1_{\mu_+}$ and $u_- \in H^1_{\mu_-}$ of u are solutions respectively of

$$-\Delta u_+ + \mu_+ u_+ = f \ \text{ in } \ H^1_{\mu_+} \quad \text{and} \quad -\Delta u_- + \mu_- u_- = -f \ \text{ in } \ H^1_{\mu_-}.$$

Then, by Lemma 3.66 we have that

$$\Delta u_+ + f \mathbb{1}_{\{u>0\}} \geq 0 \qquad \text{and} \qquad \Delta u_- - f \mathbb{1}_{\{u<0\}} \geq 0,$$

in sense of distributions on $\mathbb{R}^d$. Thus, there are Radon capacitary measures ν_+ and ν_- on $\mathbb{R}^d$ such that

$$\nu_+ = \Delta u_+ + f \mathbb{1}_{\{u>0\}} \qquad \text{and} \qquad \nu_- = \Delta u_- - f \mathbb{1}_{\{u<0\}}.$$

Theorem 3.68. *Suppose that $\mu \in \mathcal{M}^T_{\text{cap}}(\mathbb{R}^d)$ is a capacitary measure of finite torsion and that $f \in L^p(\mathbb{R}^d)$, for some $p \in (d/2, +\infty]$. Let $u \in H^1_\mu$ be the solution of the equation*

$$-\Delta u + \mu u = f \quad in \quad H^1_\mu, \qquad u \in H^1_\mu.$$

Then Δu is a Radon measure on $\mathbb{R}^d$, every point $x_0 \in \mathbb{R}^d$ is a Lebesgue point for u and we have

$$u(x_0) = \lim_{r\to 0} \fint_{\partial B_r(x_0)} u \, d\mathcal{H}^{d-1} = \lim_{r\to 0} \fint_{B_r(x_0)} u \, dx.$$

Moreover, we have

$$\frac{d}{dr}\left[\fint_{\partial B_r(x_0)} u \, d\mathcal{H}^{d-1}\right] = \frac{\Delta u(B_r(x_0))}{d\omega_d r^{d-1}},$$

in sense of distributions on $(0, 1)$, and

$$\int_0^1 \frac{|\Delta u|(B_r(x_0))}{d\omega_d r^{d-1}} \, dr < +\infty,$$

where with $|\Delta u|$, we denote the total variation of the measure Δu.

Proof. It is sufficient to decompose u as in Remark 3.67 and then to apply Theorem 3.60 for u_+ and u_-. The integrability of the total variation of Δu follows by Remark 3.61 and the inequality

$$\begin{aligned} |\Delta u| &\leq |\Delta u_+| + |\Delta u_-| \leq (\nu_+ + |f|) + (\nu_- + |f|) \\ &\leq \Delta u_+ + \Delta u_- + 4|f|. \end{aligned}$$

□

Lemma 3.69. *Let $\mu \in \mathcal{M}^T_{\text{cap}}(\mathbb{R}^d)$ be a capacitary measure of finite torsion. Suppose that $p \in (d/2, +\infty]$ and $f \in L^p(\mathbb{R}^d)$. Then, there is a dimensional constant $C_d > 0$ such that the solution u of the equation*

$$-\Delta u + \mu u = f \quad in \quad H^1_\mu, \qquad u \in H^1_\mu,$$

satisfies the inequality

$$u(x_0) \leq \frac{C_d \|f\|_{L^p}}{2/d - 1/p} r^{2-\frac{d}{p}} + \fint_{B_r(x_0)} |u| \, dx, \tag{3.52}$$

for every $x_0 \in \mathbb{R}^d$.

Proof. We first note that by Remark 3.67, it is sufficient to prove the claim in the case when u is non-negative. Let $r > 0$ and let w be the solution of the problem

$$-\Delta w = |f| \quad \text{in} \quad B_r(x_0), \qquad w \in H_0^1(B_r(x_0)).$$

By Lemma 3.66, $u - w$ is subharmonic in $B_r(x_0)$, *i.e.*

$$\Delta(u - w) \geq 0 \qquad \text{in} \qquad \left[C_c^\infty(B_r(x_0))\right]'.$$

Thus, by the mean value property of the subharmonic functions and the infinity estimate from Lemma 3.51 we have

$$\begin{aligned} u(x_0) &\leq w(x_0) + \fint_{B_r(x_0)} (u - w) \, dx \leq w(x_0) + \fint_{B_r(x_0)} u \, dx \\ &\leq \frac{C_d \|f\|_{L^p}}{2/d - 1/p} \|B_r\|^{2/d - 1/p} + \fint_{B_r(x_0)} u \, dx, \end{aligned}$$

which proves the claim. □

Proposition 3.70. *Let* $\mu \in \mathcal{M}_{\text{cap}}^T(\mathbb{R}^d)$ *be a capacitary measure of finite torsion. Suppose that* $p \in (d/2, +\infty]$ *and* $f \in L^p(\mathbb{R}^d)$. *Then the solution* u *of the equation*

$$-\Delta u + \mu u = f \quad in \quad H_\mu^1, \qquad u \in H_\mu^1,$$

vanishes at infinity.

Proof. Suppose, that $x_n \in \mathbb{R}^d$ is a sequence such that $|x_n| \to \infty$ and $u(x_n) \geq \delta$ for some $\delta \geq 0$. For $r > 0$, by Lemma 3.69 we have

$$u(x_n) \leq \frac{C_d \|f\|_{L^p}}{2/d - 1/p} \|B_r\|^{2/d - 1/p} + \fint_{B_r(x_n)} u \, dx.$$

Passing to the limit as $n \to \infty$, we obtain

$$\delta \leq \frac{C_d}{2/d - 1/p} \|f\|_{L^p} \|B_r\|^{2/d - 1/p},$$

and since $r > 0$ is arbitrary, we conclude that $\delta = 0$. □

In a similar way we have the following semi-continuity result.

Proposition 3.71. *Let $\mu \in \mathcal{M}_{\text{cap}}^T(\mathbb{R}^d)$ be a capacitary measure of finite torsion. Suppose that $p \in (d/2, +\infty]$ and $f \in L^p(\mathbb{R}^d)$ is such that the solution u of the equation*

$$-\Delta u + \mu u = f \quad in \quad H_\mu^1, \qquad u \in H_\mu^1,$$

is non-negative on $\mathbb{R}^d$. Then u is upper semi-continuous, i.e.

$$u(x_0) = \lim_{r\to 0} \|u\|_{L^\infty(B_r(x_0))}, \quad for\ every \quad x_0 \in \mathbb{R}^d.$$

Proof. Suppose that $x_n \to x_0$ is such that $u(x_n) \geq (1-\varepsilon)\|u\|_{L^\infty(B_{1/n}(x_0))}$. For $r > 0$, by Lemma 3.69, we have

$$(1-\varepsilon)\|u\|_{L^\infty(B_{1/n}(x_0))} \leq u(x_n) \leq \frac{C_d\|f\|_{L^p}}{2/d - 1/p}\|B_r\|^{2/d-1/p} + \fint_{B_r(x_n)} u\,dx.$$

Passing to the limit as $n \to \infty$, we get

$$(1-\varepsilon)\|u\|_{L^\infty(B_{1/n}(x_0))} \leq \frac{C_d\|f\|_{L^p}}{2/d - 1/p}\|B_r\|^{2/d-1/p} + \fint_{B_r(x_0)} u\,dx.$$

Now, we pass to the limit for $r \to 0$ to obtain

$$(1-\varepsilon)\|u\|_{L^\infty(B_{1/n}(x_0))} \leq u(x_0),$$

which concludes the proof, since $\varepsilon > 0$ is arbitrary. □

3.4.3. The set of finiteness Ω_μ of a capacitary measure

In this subsection we introduce the notion of set of finiteness of a capacitary measure. Roughly speaking, we expect that whenever $u \in H_\mu^1$, $u = 0$ where $\mu = +\infty$ and so, it is supported on the set $\{\mu < +\infty\}$. The precise definition of this set will be given below through the torsion function w_μ.

Proposition 3.72. *Let μ be a capacitary measure in $\mathbb{R}^d$ and let w_μ be the torsion energy function for μ. For every $u \in H_\mu^1$, we have that* $\text{cap}\big(\{w_\mu > 0\} \setminus \{u \neq 0\}\big) = 0$.

Proof. As in Proposition 2.17, we can suppose that $0 \leq u \leq 1$. Since $\{w_\mu > 0\} = \bigcup_{R>0}\{w_R > 0\}$, where w_R are as in Definition 3.38, we have only to prove that $\text{cap}\big(\{u > 0\} \setminus \{w_R > 0\}\big) = 0$, for every $R > 0$. We first note that by the weak maximum principle $\{w_R > 0\} \subset B_R$ and so, we only have to prove that $\operatorname{cap}\big(\{u\eta_R > 0\} \setminus \{w_R > 0\}\big) = 0$, where $\eta_R(x) = \eta(x/R)$ and

$$\eta \in C_c^\infty(\mathbb{R}^d), \quad 0 \le \eta \le 1, \quad \{\eta > 0\} = B_1, \quad \eta = 1 \text{ on } B_{1/2}.$$

Setting $\mu_R = \mu \vee I_{B_R}$, we have that $w_R \in H^1_{\mu_R}$ and $\eta_R u \in H^1_{\mu_R}$. Reasoning as in Proposition 2.17 we consider the solution $u_\varepsilon \in H^1_{\mu_R}$ of

$$-\Delta u_\varepsilon + \mu_R u_\varepsilon + \varepsilon^{-1} u_\varepsilon = \varepsilon^{-1}\eta_R u \quad \text{in} \quad H^1_{\mu_R}.$$

By the weak maximum principle we have that $u_\varepsilon \le \varepsilon^{-1} w_R$. Moreover, by Lemma 2.15 and Remark 2.16, u_ε and converges to $\eta_R u$ strongly in H^1_μ as $\varepsilon \to 0$. Thus, $\operatorname{cap}\big(\{u\eta_R > 0\} \setminus \{w_R > 0\}\big) = 0$ and so, we have the claim. □

Definition 3.73. We define the set of finiteness Ω_μ of the capacitary measure μ as

$$\Omega_\mu := \{w_\mu > 0\}.$$

Proposition 3.74. *For every capacitary measure μ, we have $\mu \ge I_{\Omega_\mu}$.*

Proof. It is sufficient to check that for every $u \in H^1(\mathbb{R}^d)$, we have

$$\int_{\mathbb{R}^d} u^2 \, dI_{\Omega_\mu} \le \int_{\mathbb{R}^d} u^2 \, d\mu.$$

Indeed, let $u \in H^1_\mu$. Then $\operatorname{cap}(\{u \ne 0\} \setminus \Omega_\mu) = 0$ and thus $\int_{\mathbb{R}^d} u^2 \, dI_{\Omega_\mu} = 0$, which proves the claim. □

Example 3.75. If Ω is a quasi-open set and $\mu = I_\Omega$, then $\Omega_\mu = \Omega$.
Example 3.76. If $\mu = \widetilde{I}_\Omega$ for some $\Omega \subset \mathbb{R}^d$, then Ω_μ is such that $|\Omega_\mu \setminus \Omega| = 0$ and $\widetilde{H}^1_0(\Omega) = H^1_0(\Omega_\mu)$.

3.4.4. The resolvent associated to a capacitary measure μ

Let $\mu \in \mathcal{M}^T_{\text{cap}}(\mathbb{R}^d)$ be a capacitary measure of finite torsion and let $f \in L^2(\mathbb{R}^d)$. By Proposition 3.49 there is a unique solution of the equation

$$-\Delta u + \mu u = f \quad \text{in} \quad H^1_\mu, \qquad u \in H^1_\mu. \tag{3.53}$$

Moreover, by the Gagliardo-Nirenberg-Sobolev inequalities and the continuity of the inclusion $\dot{H}^1_\mu \hookrightarrow L^1(\mathbb{R}^d)$, there is a constant C_μ, depending on the dimension d and the torsion $T(\mu)$, such that

$$\int_{\mathbb{R}^d} u^2 \, dx \le C_\mu \left(\int_{\mathbb{R}^d} |\nabla u|^2 \, dx + \int_{\mathbb{R}^d} u^2 \, d\mu \right), \quad \text{for every} \quad u \in H^1_\mu(\mathbb{R}^d). \tag{3.54}$$

Thus, if u is a solution of (3.53), then using (3.54) and testing (3.53) with u itself, we obtain

$$\begin{aligned}\|u\|_{L^2}^2 &\le C_\mu\left(\int_{\mathbb{R}^d}|\nabla u^2|\,dx + \int_{\mathbb{R}^d}u^2\,d\mu\right)\\ &= C_\mu\int_{\mathbb{R}^d}uf\,dx \le C_\mu\|f\|_{L^2}\|u\|_{L^2},\end{aligned}$$

which finally gives the estimates

$$\|u\|_{L^2}\le C_\mu\|f\|_{L^2}\quad\text{and}\quad\int_{\mathbb{R}^d}|\nabla u^2|\,dx+\int_{\mathbb{R}^d}u^2\,d\mu\le C_\mu\|f\|_{L^2}^2. \tag{3.55}$$

Definition 3.77. We define the resolvent associated to the capacitary measure $\mu\in\mathcal{M}_{\text{cap}}^T(\mathbb{R}^d)$ as the (linear) operator $R_\mu : L^2(\mathbb{R}^d)\to L^2(\mathbb{R}^d)$ that associates to each function $f\in L^2(\mathbb{R}^d)$ the solution $u=R_\mu(f)$ of the equation (3.53).

In the rest of this subsection we will recall in a series of remarks the basic properties of the resolvent operator R_μ.

Remark 3.78. Given a capacitary measure $\mu\in\mathcal{M}_{\text{cap}}^T(\mathbb{R}^d)$, the resolvent operator $R_\mu : L^2(\mathbb{R}^d)\to L^2(\mathbb{R}^d)$ has the following properties:

- By the first estimate in (3.55), the operator R_μ is continuous and its norm is estimated by

$$\|R_\mu\|_{\mathcal{L}(L^2(\mathbb{R}^d),L^2(\mathbb{R}^d))}\le C_\mu.$$

- R_μ is a compact operator. Indeed, if f_n is a bounded sequence in $L^2(\mathbb{R}^d)$, then by the second estimate in (3.55) the sequence $R_\mu(f_n)$ is bounded in $H_\mu^1(\mathbb{R}^d)$ and by the compact inclusion $H_\mu^1(\mathbb{R}^d)\hookrightarrow L^2(\mathbb{R}^d)$ it has a subsequence that converges in $L^2(\mathbb{R}^d)$.
- R_μ is a self-adjoint operator on $L^2(\mathbb{R}^d)$. Indeed, if $f,g\in L^2(\mathbb{R}^d)$, then we have

$$\begin{aligned}\int_{\mathbb{R}^d}fR_\mu(g)\,dx &= \int_{\mathbb{R}^d}\nabla R_\mu(f)\cdot\nabla R_\mu(g)\,dx\\ &\quad+\int_{\mathbb{R}^d}R_\mu(f)R_\mu(g)\,d\mu=\int_{\mathbb{R}^d}gR_\mu(f)\,dx.\end{aligned}$$

- R_μ is a positive operator. Indeed, for every $f\in L^2(\mathbb{R}^d)$, we have

$$\int_{\mathbb{R}^d}fR_\mu(f)\,dx=\int_{\mathbb{R}^d}|\nabla R_\mu(f)|^2\,dx+\int_{\mathbb{R}^d}|R_\mu(f)|^2\,d\mu\ge 0.$$

Since for $\mu \in \mathcal{M}^T_{\text{cap}}(\mathbb{R}^d)$ the resolvent $R_\mu : L^2(\mathbb{R}^d) \to L^2(\mathbb{R}^d)$ is a compact positive self-adjoint operator its spectrum is real, positive and discrete, and its elements $\Lambda_k(\mu)$, $k \in \mathbb{N}$, can be ordered in a decreasing sequence as follows (see [57, Chapter 4]):

$$0 < \cdots \leq \Lambda_k(\mu) \leq \Lambda_{k-1}(\mu) \leq \cdots \leq \Lambda_1(\mu) = \|R_\mu\|_{\mathcal{L}(L^2(\mathbb{R}^d);L^2(\mathbb{R}^d))}.$$

For every $k \in \mathbb{N}$ we define

$$\lambda_k(\mu) := \frac{1}{\Lambda_k(\mu)}. \tag{3.56}$$

Thus $\lambda_k(\mu)$ can be ordered in an increasing sequence as follows:

$$0 < \lambda_1(\mu) \leq \lambda_2(\mu) \leq \cdots \leq \lambda_k(\mu) \leq \ldots.$$

Remark 3.79. For $\mu \in \mathcal{M}^T_{\text{cap}}(\mathbb{R}^d)$ the resolvent R_μ is a compact and self-adjoint operator and so there is a complete orthonormal system of eigenfunctions $\{u_k\}_{k\in\mathbb{N}} \subset L^2(\mathbb{R}^d)$, *i.e.*

- for every $k \in \mathbb{N}$, $u_k \in H^1_\mu$ satisfies the equation

$$-\Delta u_k + \mu u_k = \lambda_k(\mu) u_k \quad \text{in} \quad H^1_\mu, \qquad u_k \in H^1_\mu;$$

- $\displaystyle\int_{\mathbb{R}^d} u_i u_j \, dx = \delta_{ij}$, for every $i, j \in \mathbb{N}$;
- the linear combinations of u_k, $k \in \mathbb{N}$, are dense in $L^2(\mathbb{R}^d)$.

Remark 3.80. Let μ be a capacitary measure of finite torsion in $\mathbb{R}^d$. Then, by Proposition 3.54 the following equality holds for $p > d/2$:

$$\|R_\mu(f)\|_{L^\infty} \leq C\|f\|_{L^p}, \quad \text{for every} \quad f \in L^2(\mathbb{R}^d) \cap L^p(\mathbb{R}^d).$$

Thus R_μ can be extended to a continuous operator from $R_\mu : L^p(\mathbb{R}^d) \to L^\infty(\mathbb{R}^d)$ with norm depending only on the dimension d and the torsion $T(\mu)$.

Remark 3.81. Let μ be a capacitary measure of finite torsion in $\mathbb{R}^d$.

- If $d \leq 3$, then $d/2 < d$ and so, by Remark 3.80 R_μ extends to a continuous operator

$$R_\mu : L^2(\mathbb{R}^d) \to L^\infty(\mathbb{R}^d), \quad \text{for} \quad d = 1, 2, 3.$$

- If $d \geq 3$, then by the Gagliardo-Nirenberg-Sobolev inequality and the second inequality in (3.55) we have that R_μ is a continuous operator

$$R_\mu : L^2(\mathbb{R}^d) \to L^{2^*}(\mathbb{R}^d) \quad \text{and} \quad \|R_\mu\|_{\mathcal{L}(L^2(\mathbb{R}^d);L^{2^*}(\mathbb{R}^d))} \leq C_{d,\mu}, \tag{3.57}$$

where $2^* = \frac{2d}{d-2}$ and $C_{d,\mu}$ depends only on the dimension d and torsion $T(\mu)$.
- Suppose that the dimension d is 4 or 5. Then by (3.57) and Remark 3.80 we have that the composition $R_\mu^2 = R_\mu \circ R_\mu : L^2(\mathbb{R}^d) \to L^2(\mathbb{R}^d)$ can be extended to a continuous operator

$$R_\mu^2 : L^2(\mathbb{R}^d) \to L^\infty(\mathbb{R}^d),$$

with norm bounded by a constant depending on d and $T(\mu)$.
- In dimension $d > 3$ we can gain some integrability by interpolating between 2 and $d > d/2$. Indeed, let $p \in (2, d/2]$. Then since

$$R_\mu : L^2 \to L^2 \qquad \text{and} \qquad R_\mu : L^d \to L^\infty,$$

by the Riesz-Torin theorem we have

$$\|R_\mu(f)\|_{L^q} \leq C\|f\|_{L^p}, \qquad \text{where} \quad p \in [2, d] \quad \text{and} \quad q = p\left(1 + \frac{p-2}{d-p}\right), \tag{3.58}$$

where C depends only on the dimension d, the exponent p and the torsion $T(\mu)$.
- Suppose that $d \geq 6$ in which case we have $2^* = \frac{2d}{d-2} \in (2, d/2]$. Since the function $p \mapsto p\frac{p-2}{d-p}$ is increasing in p, by (3.58) and interpolation, we have that

$$\|R_\mu(f)\|_{L^{p+\alpha}} \leq C\|f\|_{L^p}, \qquad \text{for every} \quad p \in [2^*, d] \quad \text{and} \quad \alpha = \frac{8}{(d-2)(d-4)}. \tag{3.59}$$

Let k be the smallest natural number such that

$$k\alpha = \frac{8k}{(d-2)(d-4)} > \frac{d}{2} - \frac{2d}{d-2}.$$

Then we have

$$\begin{aligned}\|R_\mu^{k+2}(f)\|_{L^\infty} &\leq C_{k+1}\|R_\mu^{k+1}(f)\|_{L^{2^*+k\alpha}} \leq \dots \\ &\leq C_1\|R_\mu(f)\|_{L^{2^*}} \leq C\|f\|_{L^2},\end{aligned}$$

where the constants $C_{k+2}, \dots, C_1, C$ depend only on the dimension d and the torsion $T(\mu)$.

We summarize the results from Remark 3.81 in the following proposition.

Proposition 3.82. *Let $\mu \in \mathcal{M}^T_{\mathrm{cap}}(\mathbb{R}^d)$ be a capacitary measure of finite torsion. Then, there are constants $n \in \mathbb{N}$ and $C \in \mathbb{R}$, depending only on the dimension d and the torsion $T(\mu)$, such that the power of the resolvent $[R_\mu]^n$ is a continuous operator*

$$[R_\mu]^n : L^2(\mathbb{R}^d) \to L^\infty(\mathbb{R}^d) \quad \text{and} \quad \|[R_\mu]^n\|_{\mathcal{L}(L^2(\mathbb{R}^d);L^\infty(\mathbb{R}^d))} \le C.$$

Proposition 3.83. *Let $\mu \in \mathcal{M}^T_{\mathrm{cap}}(\mathbb{R}^d)$ be a capacitary measure of finite torsion. Then the normalized eigenfunctions u_k of the resolvent operator R_μ are bounded and*

$$\|u_k(\mu)\|_{L^\infty} \le C_d \lambda_k(\mu)^{d/4}, \tag{3.60}$$

where C_d is a dimensional constant.

Proof. We first prove that u_k is bounded. Let n and C be the constants from Proposition 3.82. Applying a power of the resolvent R_μ to the normalized eigenfunction $u_k \in H^1(\mu)$ we have

$$[R_\mu]^n(u_k) = \lambda_k(\mu)^{-n} u_k,$$

and by Proposition 3.82 we obtain

$$\|u_k\|_{L^\infty} \le C\lambda_k(\mu)^n \|u_k\|_{L^2} = C\lambda_k(\mu)^n,$$

where C is a constant depending on the capacitary measure μ.

We now note that for the positive and negative parts u_k^+ and u_k^- of u_k we have

$$\Delta u_k^+ + \lambda_k(\mu)\|u_k\|_{L^\infty} \ge \Delta u_k^+ + \lambda_k(\mu) u_k^+ \ge 0,$$

$$\Delta u_k^- + \lambda_k(\mu)\|u_k\|_{L^\infty} \ge \Delta u_k^- + \lambda_k(\mu) u_k^- \ge 0.$$

Setting for simplicity $u = u_k^+$ and $M = \lambda_k(\mu)\|u_k\|_{L^\infty}$, we get that for every $x_0 \in \mathbb{R}^d$ the function $x \mapsto u(x) + M\frac{|x-x_0|^2}{2d}$ is subharmonic and so, by the mean value inequality we have

$$\begin{aligned} u(x_0) &\le \fint_{B_R(x_0)} u\,dx + \frac{MR^2}{2d} \le \left(\fint_{B_R(x_0)} u^2\,dx\right)^{1/2} + \frac{MR^2}{2d} \\ &\le \frac{1}{(\omega_d R^d)^{1/2}} + \frac{MR^2}{2d}. \end{aligned}$$

Taking the minimum in $R \in (0, +\infty)$, we get

$$u(x_0) \leq C_d M^{\frac{d}{d+4}} = C_d \lambda_k(\mu)^{\frac{d}{d+4}} \|u_k\|_{L^\infty}^{\frac{d}{d+4}}.$$

Repeating the same argument for u_k^- and by the fact that x_0 is arbitrary, we get

$$\|u_k\|_{L^\infty} \leq C_d M^{\frac{d}{d+4}} = C_d \lambda_k(\mu)^{\frac{d}{d+4}} \|u_k\|_{L^\infty}^{\frac{d}{d+4}},$$

which gives (3.60). □

In the next subsection we will prove another estimate on the infinity norm of u_k which is due to Davies [56]. In particular, we will show that the constant C_d can be chosen to be independent even of the dimension d.

3.4.5. Eigenvalues and eigenfunctions of the operator $-\Delta + \mu$

Until now we studied the differential operator $-\Delta + \mu$ only implicitly, mainly through its resolvent R_μ. In this subsection we will give a precise definition to $-\Delta + \mu$ and its spectrum. In fact this will be an easy task since we already have the instruments necessary to identify it since we know its resolvent and also the quadratic form associated to it. Thus, we will simply define $-\Delta + \mu$ as the inverse of R_μ. Our main goal is the construction of the heat semigroup associated to this operator, which is a useful tool in the study of the properties of the eigenfunctions and eigenvalues associated to the capacitary measure μ.

We start our analysis with the following lemma.

Lemma 3.84. *Let $\mu \in \mathcal{M}_{\text{cap}}^T(\mathbb{R}^d)$ be a capacitary measure of finite torsion and let Ω_μ be its set of finiteness. Then the closure of the space H_μ^1 with respect to the norm $\|\cdot\|_{L^2}$ is precisely*

$$L^2(\Omega_\mu) := \Big\{ f \in L^2(\mathbb{R}^d) : \ f = 0 \ a.e. \ on \ \mathbb{R}^d \setminus \Omega_\mu \Big\}.$$

Proof. Denote with $\overline{H_\mu^1}$ the closure of H_μ^1 with respect to $\|\cdot\|_{L^2}$. Since $H_\mu^1 \subset L^2(\Omega_\mu)$, we obtain the inclusion $\overline{H_\mu^1} \subset L^2(\Omega_\mu)$. For the opposite one, consider an open set of finite measure $A \subset \mathbb{R}^d$ and a non-negative function $u \in H^1(\mathbb{R}^d)$ such that $A = \{u > 0\}$. Since $\Omega_\mu = \{w_\mu > 0\}$ by definition, we have that $\{w_\mu \wedge u > 0\} = \Omega_\mu \cap A$ and $w_\mu \wedge u \in H_\mu^1$. Now let $u_\varepsilon = 1 \wedge (\varepsilon^{-1}(w_\mu \wedge u))$. Then u_ε is an increasing sequence in ε converging pointwise to $\mathbb{1}_{A\cap\Omega_\mu}$. By the Fatou Lemma and the fact that A is arbitrary, we have that the characteristic functions of the Borel sets are in the closure of H_μ^1. By linearity and the density of the linear combinations of characteristic functions in $L^2(\Omega_\mu)$, we have the claim. □

Corollary 3.85. *Let $\mu \in \mathcal{M}^T_{\text{cap}}(\mathbb{R}^d)$ be a capacitary measure of finite torsion and let Ω_μ be its set of completeness. Then the resolvent operator $R_\mu : L^2(\Omega_\mu) \to L^2(\Omega_\mu)$ is injective.*

Proof. Suppose that $u, v \in L^2(\Omega_\mu)$ such that $R_\mu(u) = R_\mu(v)$. Then for every test function $\varphi \in H^1_\mu$ we have

$$\begin{aligned}\int_{\mathbb{R}^d} u\varphi\,dx &= \int_{\mathbb{R}^d} \nabla R_\mu(u)\cdot\nabla\varphi\,dx + \int_{\mathbb{R}^d} R_\mu(u)\varphi\,d\mu\\ &= \int_{\mathbb{R}^d} \nabla R_\mu(v)\cdot\nabla\varphi\,dx + \int_{\mathbb{R}^d} R_\mu(v)\varphi\,d\mu = \int_{\mathbb{R}^d} v\varphi\,dx.\end{aligned}$$

By the density of H^1_μ in $L^2(\Omega_\mu)$, we get that $u = v$. □

Definition 3.86. For a capacitary measure μ of finite torsion we define:

- the domain $Dom(-\Delta+\mu) \subset L^2(\Omega_\mu)$ as the image $Dom(-\Delta+\mu) = R_\mu(L^2(\Omega_\mu))$;
- the unbounded operator $-\Delta + \mu : Dom(-\Delta + \mu) \to L^2(\Omega_\mu)$ as the inverse of the map $R_\mu : L^2(\Omega_\mu) \to Dom(-\Delta + \mu)$.

Lemma 3.87. *Let $\mu \in \mathcal{M}^T_{\text{cap}}(\mathbb{R}^d)$ be a capacitary measure of finite torsion and let Ω_μ be its set of completeness. Then the operator $-\Delta + \mu$ with domain $Dom(-\Delta + \mu)$ is self-adjoint on the Hilbert space $L^2(\Omega_\mu)$.*

Proof. Let $(-\Delta + \mu)^*$ be the adjoint operator of $-\Delta + \mu$ and let $Dom((-\Delta+\mu)^*)$ be its domain. By the definition of an adjoint operator we have:

$$\begin{aligned}Dom((-\Delta + \mu)^*) = \Big\{&u \in L^2(\Omega_\mu) : \exists v \in L^2(\Omega_\mu) \;\text{ such that}\\ &\int_{\mathbb{R}^d} u(-\Delta+\mu)\varphi\,dx = \int_{\mathbb{R}^d} v\varphi\,dx,\quad \forall\varphi \in Dom(-\Delta+\mu)\Big\}.\end{aligned} \tag{3.61}$$

Taking $\psi \in L^2(\Omega_\mu)$ such that $R_\mu(\psi) = \varphi$, we get

$$\int_{\mathbb{R}^d} u\psi\,dx = \int_{\mathbb{R}^d} vR_\mu(\psi)\,dx = \int_{\mathbb{R}^d} \psi R_\mu(v)\,dx, \qquad \forall\psi \in L^2(\Omega_\mu).$$

Thus $u = R_\mu(v)$ and so, we obtain $Dom((-\Delta+\mu)^*) = Dom(-\Delta+\mu)$. Since by definition of the adjoint operator $(-\Delta + \mu)^*$ we have $(-\Delta + \mu)^*u = v$, where $u \in Dom((-\Delta + \mu)^*)$ and $v \in L^2(\Omega_\mu)$ is as in (3.61), we get that $(-\Delta + \mu)^* = -\Delta + \mu$. □

Remark 3.88. Let $\mu \in \mathcal{M}^T_{\text{cap}}(\mathbb{R}^d)$. We note that by construction we have that R_μ is the resolvent (in zero) of the unbounded self-adjoint operator $-\Delta + \mu$ on the Hilbert space $L^2(\Omega_\mu)$ and that the spectrum of $-\Delta + \mu$ is discrete and its elements are precisely

$$0 < \lambda_1(\mu) \le \lambda_2(\mu) \le \cdots \le \lambda_k(\mu) \le \dots,$$

where $\lambda_k(\mu)$ was defined in (3.56). Moreover, the following variational characterization holds for $\lambda_k(\mu)$:

$$\lambda_k(\mu) = \min_{S_k} \max_{u \in S_k \setminus \{0\}} \frac{\int_{\mathbb{R}^d} |\nabla u|^2\,dx + \int_{\mathbb{R}^d} u^2\,d\mu}{\int_{\mathbb{R}^d} u^2\,dx},$$

where the minimum is taken over all k-dimensional subspaces S_k of H^1_μ.

Since the operator $-\Delta + \mu$ is positive and self-adjoint, the Hille-Yoshida Theorem (see for example [60]) states that the operator $(\Delta - \mu)$ generates a strongly continuous semigroup T_μ on $L^2(\Omega_\mu)$, *i.e.* a family of operators $T_\mu(t) : L^2(\Omega_\mu) \to L^2(\Omega_\mu)$, for $t \in [0, +\infty)$, such that

- $T_\mu(t) : L^2(\Omega_\mu) \to L^2(\Omega_\mu)$ is continuous, for every $t \in [0, +\infty)$;
- $T_\mu(0) = Id$;
- $T_\mu(t) \circ T_\mu(s) = T_\mu(t + s)$, for every $t, s \in [0, +\infty)$;
- the map $t \mapsto T_\mu(t)u$ is continuous as a map from $[0, +\infty)$ to $L^2(\Omega_\mu)$ equipped with the strong topology, for every $u \in L^2(\Omega_\mu)$.

Example 3.89. If $\mu \equiv 0$ on $\mathbb{R}^d$, then the corresponding semigroup $T_0(t)$ can be defined through the heat kernel on $\mathbb{R}^d$ (see for example [61, Section 2.3]), *i.e.* for every $f \in L^2(\mathbb{R}^d)$ and every $t > 0$, we have

$$[T_0(t)f](x) = \frac{1}{(4\pi t)^{d/2}} \int_{\mathbb{R}^d} e^{-\frac{|x-y|^2}{4t}} f(y)\,dy.$$

Remark 3.90. Let $\mu \in \mathbb{R}^d$ be a generic capacitary measure. A classical result from the Theory of Semigroups (see for example [60]) states that a function $u \in Dom(-\Delta + \mu)$ if and only if the strong limit $\lim_{\varepsilon \to 0^+} \varepsilon^{-1}(T_\mu(\varepsilon)u - u)$ exists in $L^2(\Omega_\mu)$. If this is the case we have

$$(\Delta - \mu)u = \lim_{\varepsilon \to 0^+} \varepsilon^{-1}(T_\mu(\varepsilon)u - u).$$

Using this result and the semigroup property $T_\mu(t) \circ T_\mu(s) = T_\mu(t + s)$, it is straightforward to check that if $u \in Dom(-\Delta + \mu)$, then the application $t \mapsto T_\mu(t)u$ is Frechet differentiable as a map from $[0, +\infty)$ to $L^2(m)$ and its derivative is given by

$$\frac{d}{dt} T_\mu(t)u = T_\mu(t) \circ (\Delta - \mu)u = (\Delta - \mu) \circ T_\mu(t)u. \tag{3.62}$$

Remark 3.91. Suppose now that μ is a capacitary measure such that the inclusion $\dot{H}^1_\mu \subset L^2(\mathbb{R}^d)$ is compact. Let u_k be an eigenfunction for the operator R_μ, *i.e.* $R_\mu(u_k) = \Lambda_k(\mu)u_k$. Then $u_k \in Dom(-\Delta + \mu)$ and $(-\Delta + \mu)u_k = \lambda_k(\mu)u_k$. In particular, by (3.62), we have

$$\frac{d}{dt}T_\mu(t)u_k = T_\mu(t) \circ (\Delta - \mu)u_k = -\lambda_k(\mu)T_\mu(t)u_k,$$

and so, since $T_\mu(0)u_k = u_k$, we have

$$T_\mu(t)u_k = e^{-t\lambda_k(\mu)}u_k, \qquad \forall t \in [0, +\infty). \tag{3.63}$$

We now recall a classical result known as the Chernoff Product Formula (see [60, Theorem 5.2] and [60, Corollary 5.5]).

Theorem 3.92. *Let μ be a capacitary measure in $\mathbb{R}^d$ and let $f \in L^2(\Omega_\mu)$. Then we have*

$$T_\mu(t)f = \lim_{n\to\infty}\left[\frac{n}{t}R_{(\frac{n}{t}+\mu)}\right]^n(f), \quad \textit{for every} \quad t \in (0, +\infty), \tag{3.64}$$

where the limit on the right hand-side is strong in $L^2(\Omega_\mu)$.

A consequence of this formula is the following:

Corollary 3.93 (Weak maximum principle for semigroups). *Let μ be a capacitary measure in $\mathbb{R}^d$ and let $f \in L^2(\Omega_\mu)$. If $f \geq 0$, the for every $t \in [0, +\infty)$ we have $T_\mu(t)f \geq 0$. In particular, for every $f \in L^2(\Omega_\mu)$ and every $t \in [0, +\infty)$, we have $|T_\mu(t)f| \leq T_\mu(t)(|f|)$.*

Proof. It is sufficient to note that if $f \geq 0$, then the right hand-side of (3.64) is positive. □

In what follows we will need to compare the semigroups T_μ generated by different capacitary measures μ. In order to do that we extend the semigroup T_μ to the space $L^2(\mathbb{R}^d)$. Indeed, for the capacitary measure μ, we define the projection

$$P_\mu : L^2(\mathbb{R}^d) \to L^2(\Omega_\mu), \qquad P_\mu(u) := \mathbb{1}_{\Omega_\mu}u.$$

The one-parameter family of operators $\widetilde{T}_\mu(t) := T_\mu(t) \circ P_\mu : L^2(\mathbb{R}^d) \to L^2(\mathbb{R}^d)$ satisfies

- $\widetilde{T}_\mu(t) : L^2(\mathbb{R}^d) \to L^2(\Omega_\mu) \subset L^2(\mathbb{R}^d)$ is continuous, for every $t \in [0, +\infty)$;
- $\widetilde{T}_\mu(0) = P_\mu$;

- $\widetilde{T}_\mu(t) \circ \widetilde{T}_\mu(s) = \widetilde{T}_\mu(t+s)$, for every $t, s \in [0, +\infty)$;
- the map $t \to \widetilde{T}_\mu(t)u$ is continuous as a map from $[0, +\infty)$ to $L^2(\mathbb{R}^d)$ equipped with the strong topology, for every $u \in L^2(\mathbb{R}^d)$.

Proposition 3.94. *Let now μ and ν be capacitary measures in $\mathbb{R}^d$ such that $\mu \geq \nu$. Then for every nonnegative $f \in L^2(\mathbb{R}^d)$ and every $t \in [0, +\infty)$, we have $\widetilde{T}_\mu(t) f \leq \widetilde{T}_\nu(t) f$.*

Proof. We first note that $\mu \geq \nu$ implies $\Omega_\mu \subset \Omega_\nu$ and so, by Corollary f3.93, we have

$$\widetilde{T}_\nu(f \mathbb{1}_{\Omega_\nu}) \geq \widetilde{T}_\nu(f \mathbb{1}_{\Omega_\mu}).$$

Now using the approximation from Theorem 3.92, and the maximum principle for capacitary measures, we have that

$$\widetilde{T}_\nu(f \mathbb{1}_{\Omega_\mu}) \geq \widetilde{T}_\mu(f \mathbb{1}_{\Omega_\mu}),$$

which proves the claim. □

Corollary 3.95. *Suppose that μ is a capacitary measure such that the inclusion $\dot{H}^1_\mu \subset L^2(\mathbb{R}^d)$ is compact. Let $u_k \in L^2(\Omega_\mu)$ be an eigenfunction for the operator R_μ. Then we have the estimate*

$$\|u_k\|_{L^\infty} \leq e^{\frac{1}{8\pi}} \lambda_k(\mu)^{d/4} \|u_k\|_{L^2}. \tag{3.65}$$

Proof. By Remark 3.91, Corollary 3.93 and Proposition 3.94, we have

$$e^{-t\lambda_k(\mu)}|u_k| = |\widetilde{T}_\mu(t)u_k| \leq \widetilde{T}_\mu(t)|u_k| \leq T_0(|u_k|).$$

On the other hand, by Example 3.89, we have

$$|u_k| \leq \frac{e^{t\lambda_k(\mu)}}{(4\pi t)^{d/2}} \int_{\mathbb{R}^d} e^{-\frac{|x-y|^2}{4t}} |u_k(y)|\, dy \leq \frac{e^{t\lambda_k(\mu)}}{(4\pi t)^{d/2}} (2\pi t)^{d/4} \|u_k\|_{L^2}.$$

Now, choosing t appropriately, we have the claim. □

3.4.6. Uniform approximation with solutions of boundary value problems

Let $\mu \in \mathcal{M}^T_{\text{cap}}(\mathbb{R}^d)$ be a capacitary measure of finite torsion. For a positive real number $\varepsilon > 0$ and a function $f \in L^2(\mathbb{R}^d)$ we consider the variational problem

$$\min\left\{\int_{\mathbb{R}^d} |\nabla v|^2\, dx + \int_{\mathbb{R}^d} v^2\, d\mu + \frac{1}{\varepsilon}\int_{\mathbb{R}^d} |v-f|^2\, dx \,:\, v \in H^1_\mu(\mathbb{R}^d)\right\}. \tag{3.66}$$

By the compactness of the inclusion $H^1_\mu \hookrightarrow L^2(\mathbb{R}^d)$ and the strict convexity (in the variable $v \in H^1_\mu$) of the functional in (3.66), we have that there is a unique solution $u_\varepsilon \in H^1_\mu$ of the problem (3.66). Moreover, by the Euler-Lagrange for (3.66), the minimum u_ε is a solution of the equation

$$-\Delta u_\varepsilon + \mu u_\varepsilon + \frac{1}{\varepsilon} u_\varepsilon = \frac{1}{\varepsilon} f \quad \text{in} \quad H^1_\mu, \qquad u_\varepsilon \in H^1_\mu. \tag{3.67}$$

We denote with $Y_{\mu,\varepsilon} : L^2(\mathbb{R}^d) \to L^2(\mathbb{R}^d)$ the map that associates to every function $f \in L^2(\mathbb{R}^d)$ the solution u_ε of (3.67). Thus, $Y_{\mu,\varepsilon}$ is linear and continuous application, which can be expressed in terms of the resolvent operator as:

$$Y_{\mu,\varepsilon} := \frac{1}{\varepsilon} R_{\mu+\frac{1}{\varepsilon}} : L^2(\mathbb{R}^d) \to L^2(\mathbb{R}^d). \tag{3.68}$$

In fact, due to the fact that μ has finite torsion, the domains of the operators $-\Delta+\mu$ and $-\Delta+\mu+\frac{1}{\varepsilon}$ coincide, thus we have that $Y_{\mu,\varepsilon}(L^2(\mathbb{R}^d)) \subset Dom(-\Delta + \mu)$, *i.e.* the application of the map $Y_{\mu,\varepsilon}$ has a regularizing effect on f. Moreover, if we consider a function $u \in H^1_\mu$, then the regularized sequence $Y_{\mu,\varepsilon}(u)$ converges to u. More precisely, we have the following

Lemma 3.96. *Suppose that $\mu \in \mathcal{M}^T_{\text{cap}}(\mathbb{R}^d)$ is a capacitary measure of finite torsion. For every function $u \in H^1_\mu(\mathbb{R}^d)$ we have:*

(a) $\|Y_{\mu,\varepsilon}(u)\|_{\dot{H}^1_\mu} \le \|u\|_{\dot{H}^1_\mu}$, *for every* $\varepsilon > 0$;[7]
(b) $\|Y_{\mu,\varepsilon}(u) - u\|_{L^2} \le \varepsilon^{1/2}\|u\|_{\dot{H}^1_\mu}$, *for every* $\varepsilon > 0$;
(c) $Y_{\mu,\varepsilon}(u)$ *converges strongly in* H^1_μ *to* u *as* $\varepsilon \to 0$.

Proof. For sake of simplicity we ill use the notation $u_\varepsilon := Y_{\mu,\varepsilon}(u)$. We first test the optimality of u_ε in (3.66) against u, obtaining

$$\begin{aligned}\int_{\mathbb{R}^d} |\nabla u_\varepsilon|^2\,dx + \int_{\mathbb{R}^d} u_\varepsilon^2\,d\mu + \frac{1}{\varepsilon}\int_{\mathbb{R}^d} |u_\varepsilon - u|^2\,dx \le \int_{\mathbb{R}^d} |\nabla u|^2\,dx \\ + \int_{\mathbb{R}^d} u^2\,d\mu,\end{aligned} \tag{3.69}$$

which immediately gives both (a) and (b). For the proof (c) we first note that due to the uniform (in ε) bound $\|u_\varepsilon\|_{\dot{H}^1_\mu} \le \|u\|_{\dot{H}^1_\mu}$ and the $L^2(\mathbb{R}^d)$-convergence of u_ε to u, we have that u_ε converges to u weakly in H^1_μ. In

[7] We recall the notation $\|u\|_{\dot{H}^1_\mu} = \left(\int_{\mathbb{R}^d} |\nabla u|^2\,dx + \int_{\mathbb{R}^d} u^2\,d\mu\right)^{1/2}$.

order to show that the convergence is strong we estimate $\|u-u_\varepsilon\|_{\dot H^1_\mu}$ as follows:

$$\begin{aligned}
&\int_{\mathbb{R}^d}|\nabla(u_\varepsilon-u)|^2\,dx+\int_{\mathbb{R}^d}|u_\varepsilon-u|^2\,d\mu\\
&=2\left(\int_{\mathbb{R}^d}\nabla u\cdot\nabla(u_\varepsilon-u)\,dx+\int_{\mathbb{R}^d}u(u-u_\varepsilon)\,d\mu\right)\\
&\quad+\int_{\mathbb{R}^d}|\nabla u_\varepsilon|^2\,dx+\int_{\mathbb{R}^d}u_\varepsilon^2\,d\mu\\
&\quad-\left(\int_{\mathbb{R}^d}|\nabla u|^2\,dx+\int_{\mathbb{R}^d}u^2\,d\mu\right)\\
&\le 2\int_{\mathbb{R}^d}\nabla u\cdot\nabla(u_\varepsilon-u)\,dx+2\int_{\mathbb{R}^d}u(u-u_\varepsilon)\,d\mu.
\end{aligned}\tag{3.70}$$

Now by the weak convergence $u_\varepsilon\to u$ in H^1_μ, we obtain (c). ☐

In dimension $d\le 5$, using the Gagliardo-Nirenberg-Sobolev inequality we have that $u\in H^1(\mathbb{R}^d)$ implies $u\in L^p(\mathbb{R}^d)$, for some $p>d/2$. Thus we immediately obtain $Y_{\mu,\varepsilon}(u)\in L^\infty(\mathbb{R}^d)$, for $d\le 5$. In higher dimension ($d>5$) one can reason as in Remark 3.81, applying numerous times $Y_{\mu,\varepsilon}$ each time gaining some integrability, to obtain a function which is close to u in norm but bounded in L^∞.

Lemma 3.97. *There is a constant $M\in\mathbb{N}$, depending only on the dimension d, such that for every capacitary measure of finite torsion $\mu\in\mathcal{M}^T_{\text{cap}}(\mathbb{R}^d)$ and every function $u\in H^1_\mu(\mathbb{R}^d)$ we have:*

(i) $\|Y^M_{\mu,\varepsilon}(u)\|_{\dot H^1_\mu}\le\|u\|_{\dot H^1_\mu}$, *for every $\varepsilon>0$;*

(ii) $\|Y^M_{\mu,\varepsilon}(u)-u\|_{L^2}\le M\varepsilon^{1/2}\|u\|_{\dot H^1_\mu}$, *for every $\varepsilon>0$;*

(iii) $\|Y^M_{\mu,\varepsilon}(u)\|_{L^\infty}\le C\varepsilon^{-M}\|u\|_{L^2}$, *where the constant C depends on the dimension d and the torsion $T(\mu)$;*

(iv) $|Y^{M+1}_{\mu,\varepsilon}(u)|\le C\varepsilon^{-M-1}\|u\|_{L^2}w_\mu$, *where C is the constant from point (iii).*

Proof. Claim (i) is a direct consequence of Lemma 3.96 (a):

$$\|Y^M_{\mu,\varepsilon}(u)\|_{\dot H^1_\mu}\le\|Y^{M-1}_{\mu,\varepsilon}(u)\|_{\dot H^1_\mu}\le\cdots\le\|Y_{\mu,\varepsilon}(u)\|_{\dot H^1_\mu}\le\|u\|_{\dot H^1_\mu}.$$

For (ii) we apply numerous times the estimates from Lemma 3.96 (b):

$$\begin{aligned}
\|Y^M_{\mu,\varepsilon}(u)-u\|_{L^2}&\le\sum_{n=1}^M\|Y^n_{\mu,\varepsilon}(u)-Y^{n-1}_{\mu,\varepsilon}(u)\|_{L^2}\\
&\le\varepsilon^{1/2}\sum_{n=1}^M\|Y^{n-1}_{\mu,\varepsilon}(u)\|_{\dot H^1_\mu}\le M\varepsilon^{1/2}\|u\|_{\dot H^1_\mu}.
\end{aligned}$$

In order to prove (iii) we will show by induction that $Y^n_{\mu,\varepsilon}$ converges to u strongly in H^1_μ as $\varepsilon \to 0$. The base step $n = 1$ was proved in Lemma 3.96 (c). Let now $v_\varepsilon := Y^{n-1}_{\mu,\varepsilon}$ converges to u strongly in H^1_μ. It is sufficient to prove that $\|Y_{\mu,\varepsilon}(v_\varepsilon) - v_\varepsilon\|_{\dot{H}^1_\mu} \xrightarrow[\varepsilon\to 0]{} 0$. Since the difference $Y_{\mu,\varepsilon}(v_\varepsilon) - v_\varepsilon$ is bounded in H^1_μ and converges to zero in $L^2(\mathbb{R}^d)$, it also converges to zero weakly in H^1_μ Now using the estimate (3.70) for $u = v_\varepsilon$ we get

$$\begin{aligned}
&\lim_{\varepsilon\to 0} \|Y_{\mu,\varepsilon}(v_\varepsilon) - v_\varepsilon\|^2_{\dot{H}^1_\mu} \\
&\le 2 \limsup_{\varepsilon\to 0} \int_{\mathbb{R}^d} \nabla v_\varepsilon \cdot \nabla (Y_{\mu,\varepsilon}(v_\varepsilon) - v_\varepsilon)\, dx + \int_{\mathbb{R}^d} v_\varepsilon (Y_{\mu,\varepsilon}(v_\varepsilon) - v_\varepsilon)\, d\mu \\
&= 2 \limsup_{\varepsilon\to 0} \int_{\mathbb{R}^d} \nabla u \cdot \nabla (Y_{\mu,\varepsilon}(v_\varepsilon) - v_\varepsilon)\, dx + \int_{\mathbb{R}^d} u (Y_{\mu,\varepsilon}(v_\varepsilon) - v_\varepsilon)\, d\mu = 0.
\end{aligned}$$

For (iv) we first note that by the linearity of $Y_{\mu,\varepsilon}$ it is sufficient to prove the claim in the case $u \ge 0$. Now due to the representation (3.68), we have that $Y^n_{\mu,\varepsilon}(u) \ge 0$, for every $n \in \mathbb{N}$. On the other hand for a generic nonnegative function $f \in L^2(\mathbb{R}^d)$ we have

$$Y_{\mu,\varepsilon}(f) = \frac{1}{\varepsilon} R_{\mu+\frac{1}{\varepsilon}}(f) \le \frac{1}{\varepsilon} R_\mu(f),$$

and applying the above estimate to $f = u, \dots, R^{M-1}_\mu(u)$, we get

$$Y^M_{\mu,\varepsilon}(u) \le \varepsilon^{-M} R^M_\mu(u).$$

Now the claim follows by Proposition 3.82.

The last claim (v) follows by (iv) and the maximum principle. □

3.5. The γ-convergence of capacitary measures

The γ-convergence on the family of capacitary measures is a variational convergence which naturally appeared in the study of elliptic boundary value problems on variable domains. A great amount of literature was dedicated to the subject, starting from the pioneering works of De Giorgi, Dal Maso-Mosco, Chipot-Dal Maso and Cioranescu-Murat. Numerous applications were found to this theory, especially in the field of shape optimization, where a technique for proving existence of optimal domains was first introduced by Buttazzo and Dal Maso in [33]. In this section we give a self-contained introduction to the topic, following the ideas from [33, 51] and [19].

Definition 3.98. Let μ_n be a sequence of capacitary measures in $\mathbb{R}^d$. We say that μ_n γ-converges to the capacitary measure μ, if the sequence of energy functions w_{μ_n} converges to w_μ in $L^1(\mathbb{R}^d)$.

When the measures we consider correspond to domains in $\mathbb{R}^d$, we will sometimes use the following alternative terminology:

We say that the sequence of quasi-open sets $\Omega_n \subset \mathbb{R}^d$ γ-converges to the quasi-open set Ω, if the sequence of capacitary measures I_{Ω_n} γ-converges to I_Ω in sense of Definition 3.98.

Remark 3.99. The family $\mathcal{M}^T_{\text{cap}}(\mathbb{R}^d)$ of capacitary measures of finite torsion is a metric space with the metric $d_\gamma(\mu_1, \mu_2) = \|w_{\mu_1} - w_{\mu_2}\|_{L^1}$. On the ball $\{\mu \in \mathcal{M}_{\text{cap}}(\mathbb{R}^d) : \|w_\mu\|_{L^1} \leq 1\}$, this metric is equivalent to the distance $\|w_{\mu_1} - w_{\mu_2}\|_{L^p}$, for every $p \in (1, +\infty)$.

Remark 3.100. Classically, the term γ-convergence was used to indicate what we will call γ_{loc}-convergence, defined as follows: *The sequence of capacitary measures μ_n locally γ-converges (or γ_{loc}-converges) to the capacitary measure μ, if the sequence of energy functions $w_{\mu_n \vee I_\Omega}$ converges to $w_{\mu \vee I_\Omega}$ in $L^1(\mathbb{R}^d)$, for every bounded open set $\Omega \subset \mathbb{R}^d$.* The family of capacitary measures on $\mathbb{R}^d$, endowed with the γ_{loc} convergence, is metrizable (one can easily construct a metric using a sequence of balls B_n, for $n \to \infty$, and the distance d_γ from Remark 3.99). Moreover, it is a compact metric space.

3.5.1. Completeness of the γ-distance

In this subsection we prove that the metric space $(\mathcal{M}^T_{\text{cap}}(\mathbb{R}^d), d_\gamma)$ is complete. Essentially, there are two ways to approach this problem:

- The first one uses the classical result of the compactness with respect to the γ_{loc} convergence. In this case one has to prove that if $w_{\mu_n} \to w$ in L^1 and $\mu_n \to \mu$ in γ_{loc}, then $w = w_\mu$. This approach was used in [19], in the case $\mu_n = I_{A_n}$, and basically the same proof works in the general case. The further results on the γ-convergence rely on the analogous results for the γ_{loc} convergence.
- The second approach consists in constructing, given the limit function $w := \lim w_{\mu_n}$ in $L^1(\mathbb{R}^d)$, a capacitary measure μ such that $w = w_\mu$. This technique was introduced in [45] and was adopted in [51] (see also [72]). The results in [51] refer to the case of measures in a bounded open set $\Omega \subset \mathbb{R}^d$, but hold also in our case essentially with the same proofs.

We will prove the completeness of the γ-distance using the second approach.

Consider the set

$$\mathcal{K} = \left\{ w \in H^1(\mathbb{R}^d) \cap L^1(\mathbb{R}^d) : \Delta w + 1 \geq 0 \text{ in } \left[H^1(\mathbb{R}^d) \cap L^1(\mathbb{R}^d)\right]' \right\}.$$

Remark 3.101. We note that $\mathcal{K}$ is a closed convex set in $H^1(\mathbb{R}^d) \cap L^1(\mathbb{R}^d)$. Moreover, if $\mu \in \mathcal{M}^T_{\text{cap}}(\mathbb{R}^d)$, then by Lemma 3.66 we have

$$\Delta w_\mu + \mathbb{1}_{\{w_\mu > 0\}} \geq 0, \quad \text{as operator on} \quad H^1(\mathbb{R}^d) \cap L^1(\mathbb{R}^d),$$

and so $w_\mu \in \mathcal{K}$.

Theorem 3.102. *The space $\mathcal{M}^T_{\text{cap}}(\mathbb{R}^d)$ endowed with the metric d_γ is a complete metric space.*

Proof. Let μ_n be a sequence of capacitary measures, which is Cauchy with respect to the distance d_γ. Then the sequence $w_n := w_{\mu_n}$ converges in L^1 to some $w \in L^1(\mathbb{R}^d)$. Since, for every $n \in \mathbb{N}$, we have the identity

$$\int_{\mathbb{R}^d} |\nabla w_n|^2 \, dx + \int_{\mathbb{R}^d} w_n^2 \, d\mu_n = \int_{\mathbb{R}^d} w_n \, dx,$$

the sequence w_n is bounded in $H^1(\mathbb{R}^d)$. In particular, $w \in H^1(\mathbb{R}^d) \cap L^1(\mathbb{R}^d)$ and the converges $w_n \to w$ holds also weakly in $H^1(\mathbb{R}^d)$. By Remark 3.101, $w_n \in \mathcal{K}$ and passing to the limit $w \in \mathcal{K}$. Now, using the positivity of $\Delta w + 1$, by Proposition 3.59 we have there is a Radon capacitary measure ν on $\mathbb{R}^d$ such that $\Delta w + 1 = \nu$.

Following [51, Proposition 3.4], we define the measure μ as

$$\mu(E) = \begin{cases} \displaystyle\int_E \frac{1}{w} \, d\nu, & \text{if } \operatorname{cap}\left(E \setminus \{w > 0\}\right) = 0, \\ +\infty, & \text{if } \operatorname{cap}\left(E \setminus \{w > 0\}\right) > 0. \end{cases} \tag{3.71}$$

It is straightforward to check that the function μ, defined on the Borel sets in $\mathbb{R}^d$, is a measure. Moreover, since ν is capacitary, μ is also a capacitary measure. By construction we have

$$\int_{\mathbb{R}^d} w^2 \, d\mu = \int_{\mathbb{R}^d} w \, d\nu = \int_{\mathbb{R}^d} -|\nabla w|^2 + w \, dx < +\infty$$

and for every $u \in H^1(\mathbb{R}^d) \cap L^2(\mu) \cap L^1(\mathbb{R}^d)$ we have

$$\int_{\mathbb{R}^d} wu \, d\mu = \int_{\{w>0\}} u \, d\nu = \int_{\mathbb{R}^d} u \, d\nu = -\int_{\mathbb{R}^d} \nabla u \cdot \nabla w \, dx + \int_{\mathbb{R}^d} u \, dx.$$

Thus, w satisfies

$$-\Delta w + w\mu = 1 \quad \text{in} \quad H^1_\mu \cap L^1, \qquad w \in H^1_\mu \cap L^1$$

and so w minimizes the convex functional J_μ in $L^1 \cap H^1_\mu$. Finally, we obtain $w = w_\mu \in L^1(\mathbb{R}^d)$. □

3.5.2. The γ-convergence of measures and the convergence of the resolvents R_μ

In this section we relate the γ-convergence of a sequence of capacitary measures $\mu_n \in \mathcal{M}^T_{\text{cap}}(\mathbb{R}^d)$ to the convergence of the resolvent operators $R_{\mu_n} : L^2(\mathbb{R}^d) \to L^2(\mathbb{R}^d)$. We recall that a sequence $R_{\mu_n} \in \mathcal{L}(L^2(\mathbb{R}^d); L^2(\mathbb{R}^d))$ converges

- *in (operator) norm* to $R_\mu \in \mathcal{L}(L^2(\mathbb{R}^d); L^2(\mathbb{R}^d))$, if
$$\lim_{n\to\infty} \|R_{\mu_n} - R_\mu\|_{\mathcal{L}(L^2(\mathbb{R}^d);L^2(\mathbb{R}^d))} = 0;$$
- *strongly (in $L^2(\mathbb{R}^d)$)* to $R_\mu \in \mathcal{L}(L^2(\mathbb{R}^d); L^2(\mathbb{R}^d))$, if
$$\lim_{n\to\infty} \|R_{\mu_n}(f) - R_\mu(f)\|_{L^2(\mathbb{R}^d)} = 0, \quad \text{for every} \quad f \in L^2(\mathbb{R}^d).$$

Remark 3.103. By definition we have that if the sequence of resolvent operators R_{μ_n} converges in norm to R_μ, then it also converges strongly in $L^2(\mathbb{R}^d)$ to R_μ. The converse implication does not hold in general. Indeed consider the sequence of capacitary measures associated to a ball escaping at infinity, *i.e.*
$$\mu_n := I_{x_n+B_1} \in \mathcal{M}^T_{\text{cap}}(\mathbb{R}^d), \quad \text{where} \quad |x_n| \to +\infty.$$
Then the sequence of resolvents R_{μ_n} converges strongly to zero, while the norm remains constant
$$\|R_{\mu_n}\|_{\mathcal{L}(L^2(\mathbb{R}^d);L^2(\mathbb{R}^d))} = \|R_{\mu_1}\|_{\mathcal{L}(L^2(\mathbb{R}^d);L^2(\mathbb{R}^d))}, \quad \text{for every} \quad n \in \mathbb{N}.$$

In what follows we will prove that for a sequence of capacitary measures $\mu_n \in \mathcal{M}^T_{\text{cap}}(\mathbb{R}^d)$ the following implications hold:

the γ-convergence	$\mu_n \to \mu$
implies the norm convergence	$R_{\mu_n} \to R_\mu$,
the norm convergence	$R_{\mu_n} \to R_\mu$
implies the strong convergence	$R_{\mu_n} \to R_\mu$,
the strong convergence	$R_{\mu_n} \to R_\mu$
implies the Γ-convergence	$\|\cdot\|_{H^1_{\mu_n}} \to \|\cdot\|_{H^1_\mu}$.

Remark 3.104. Suppose that $\Omega_n \subset \mathbb{R}^d$ is a sequence of measurable sets of uniformly bounded measure. Then, for a given measurable set $\Omega \subset$

$\mathbb{R}^d$, the strong convergence of the resolvent operators $R_{\Omega_n} \to R_\Omega$ corresponds, in the terminology of Chapter 2, to the strong-γ-convergence of the domains Ω_n to Ω. Thus we will show that the γ-convergence of Ω_n implies the strong-γ-convergence. Precisely, we will show in Proposition 3.110 that for sequences of domains of uniformly bounded measure the γ-convergence is equivalent to the norm convergence of the resolvent operators.

We start by the following key lemma.

Lemma 3.105. *Suppose that the sequence $\mu_n \in \mathcal{M}^T_{\text{cap}}(\mathbb{R}^d)$ γ-converges to the capacitary measure μ. Let $f_n \in L^2(\mathbb{R}^d)$ be a sequence converging weakly in L^2 to $f \in L^2(\mathbb{R}^d)$. Then the sequence $R_{\mu_n}(f_n)$ converges strongly in $L^2(\mathbb{R}^d)$ to $R_\mu(f)$.*

Proof. We set for simplicity

$$w_n = w_{\mu_n}, \quad w = w_\mu \quad \text{and} \quad u_n = R_{\mu_n}(f_n).$$

We note that since

$$\limsup_{n\to\infty} \|f_n\|_{L^2} < +\infty \qquad \text{and} \qquad \|u_n\|^2_{\dot{H}^1_{\mu_n}} = \int_{\mathbb{R}^d} f_n u_n \, dx,$$

we have that $\|u_n\|_{H^1_{\mu_n}} \le C$, some constant C not depending on $n \in \mathbb{N}$. In particular, u_n is uniformly bounded in $H^1(\mathbb{R}^d) \cap L^1(\mathbb{R}^d)$.

Consider now the operators $Y_{\mu_n,\varepsilon}$, for some $\varepsilon > 0$, and the dimensional constant M from Lemma 3.97. We have that the sequence $u_{n,\varepsilon} := Y^{M+1}_{\mu_n,\varepsilon}(u_n)$ is uniformly bounded in $H^1(\mathbb{R}^d) \cap L^1(\mathbb{R}^d)$ and since $u_{n,\varepsilon} \le C_\varepsilon w_n$, for some constant C_ε, we have that $u_{n,\varepsilon}$ converges in $L^2(\mathbb{R}^d)$. Since $\|u_n - u_{n,\varepsilon}\|_{L^2} \le (M+1)\varepsilon^{1/2} C$, for every $n \in \mathbb{N}$, we have that u_n is Cauchy sequence in $L^2(\mathbb{R}^d)$ and so, it converges strongly in L^2 to some $u \in H^1(\mathbb{R}^d) \cap L^1(\mathbb{R}^d)$.

We now prove that $u = R_\mu(f)$. Indeed, for every $\varphi \in C^\infty_c(\mathbb{R}^d)$, we have

$$\begin{aligned}
\int_{\mathbb{R}^d} u_n \varphi \, dx &= \int_{\mathbb{R}^d} \nabla w_n \cdot \nabla(u_n \varphi) \, dx + \int_{\mathbb{R}^d} w_n u_n \varphi \, d\mu_n \\
&= \int_{\mathbb{R}^d} \big(u_n \nabla w_n \cdot \nabla \varphi - w_n \nabla u_n \cdot \nabla \varphi\big) \, dx \\
&\quad + \int_{\mathbb{R}^d} \nabla(w_n \varphi) \cdot \nabla u_n \, dx + \int_{\mathbb{R}^d} w_n u_n \varphi \, d\mu_n \\
&= \int_{\mathbb{R}^d} \big(u_n \nabla w_n \cdot \nabla \varphi - w_n \nabla u_n \cdot \nabla \varphi\big) \, dx + \int_{\mathbb{R}^d} w_n \varphi f_n \, dx.
\end{aligned}$$

Passing to the limit as $n \to \infty$, we obtain that u satisfies the identity

$$\int_{\mathbb{R}^d} u\varphi\,dx = \int_{\mathbb{R}^d} \big(u\nabla w \cdot \nabla\varphi - w\nabla u \cdot \nabla\varphi\big)\,dx + \int_{\mathbb{R}^d} w\varphi f\,dx. \quad (3.72)$$

On the other hand, $R_\mu(f)$ also satisfies (3.72) and so, taking $v = u - R_\mu(f)$, we have

$$\int_{\mathbb{R}^d} v\varphi\,dx = \int_{\mathbb{R}^d} \big(v\nabla w \cdot \nabla\varphi - w\nabla v \cdot \nabla\varphi\big)\,dx, \qquad \forall \varphi \in C_c^\infty(\mathbb{R}^d),$$

or, equivalently,

$$\int_{\mathbb{R}^d} v\varphi\,dx + \int_{\mathbb{R}^d} w\nabla v \cdot \nabla\varphi\,dx = \int_{\mathbb{R}^d} v\nabla w \cdot \nabla\varphi\,dx, \quad \forall \varphi \in C_c^\infty(\mathbb{R}^d). \quad (3.73)$$

Since $v \in L^1(\mathbb{R}^d) \cap L^2(\mathbb{R}^d)$ and $w|\nabla v| \in L^2(\mathbb{R}^d)$, we can estimate the left-hand side of (3.73) by $\|\nabla\varphi\|_{L^2}$ and thus we obtain

$$\int_{\mathbb{R}^d} v\nabla w \cdot \nabla\varphi\,dx \le C\|\nabla\varphi\|_{L^2}, \qquad \forall \varphi \in C_c^\infty(\mathbb{R}^d), \qquad (3.74)$$

and so the operator

$$\varphi \mapsto \int_{\mathbb{R}^d} v\nabla w \cdot \nabla\varphi\,dx,$$

can be extended to a continuous operator on $H^1(\mathbb{R}^d)$. We are not allowed to use $v_t := -t \vee v \wedge t$, as a test function in (3.72), obtaining

$$\begin{aligned} \int_{\mathbb{R}^d} v_t^2\,dx &\le \int_{\mathbb{R}^d} \frac{1}{2}\nabla w \cdot \nabla(v_t^2) - w|\nabla v_t|^2\,dx \\ &\le \frac{1}{2}\int_{\mathbb{R}^d} v_t^2\,dx - \int_{\mathbb{R}^d} w|\nabla v_t|^2\,dx, \end{aligned}$$

where we used the inequality $\Delta w + 1 \ge 0$ in $H^1(\mathbb{R}^d)$. In conclusion, we have

$$\frac{1}{2}\int_{\mathbb{R}^d} v_t^2\,dx + \int_{\mathbb{R}^d} w|\nabla v_t|^2\,dx \le 0,$$

which gives $v_t = 0$. Since $t > 0$ is arbitrary, we obtain $u = R_\mu(f)$, which concludes the proof. $\square$

Remark 3.106. A careful inspection of the proof of Lemma 3.105 shows that if $\mu_n \in \mathcal{M}_{\mathrm{cap}}(\mathbb{R}^d)$ γ-converges to $\mu \in \mathcal{M}_{\mathrm{cap}}(\mathbb{R}^d)$ and if $f_n \in L^2(\mathbb{R}^d)$ converges weakly in L^2 to $f \in L^2(\mathbb{R}^d)$, then $R_{\mu_n+t}(f_n)$ converges strongly in $L^2(\mathbb{R}^d)$ to $R_{\mu+t}(f)$, for every $t \ge 0$.

Proposition 3.107 (**γ implies convergence in norm**). *Let $\mu_n \in \mathcal{M}^T_{\text{cap}}(\mathbb{R}^d)$ be a sequence of capacitary measures γ-converging to $\mu \in \mathcal{M}^T_{\text{cap}}(\mathbb{R}^d)$. Then the sequence of operators $R_{\mu_n} \in \mathcal{L}(L^2(\mathbb{R}^d))$ converges to $R_\mu \in \mathcal{L}(L^2(\mathbb{R}^d))$ in norm.*

Proof. By definition of the convergence in norm, we have to show that

$$\lim_{n\to\infty} \Big\{ \sup \big\{ \|R_{\mu_n}(f) - R_\mu(f)\|_{L^2} : f \in L^2(\mathbb{R}^d),\ \|f\|_{L^2} = 1 \big\} \Big\} = 0,$$

or, equivalently, that for every sequence $f_n \in L^2(\mathbb{R}^d)$ with $\|f_n\|_{L^2} = 1$, we have

$$\lim_{n\to\infty} \|R_{\mu_n}(f_n) - R_\mu(f_n)\|_{L^2} = 0.$$

Let $f \in L^2(\mathbb{R}^d)$ be the weak limit of f_n in $L^2(\mathbb{R}^d)$. Then we have,

$$\begin{aligned}\lim_{n\to\infty} \|R_{\mu_n}(f_n) - R_\mu(f_n)\|_{L^2} &\le \limsup_{n\to\infty} \|R_{\mu_n}(f_n) - R_\mu(f)\|_{L^2} \\ &\quad + \limsup_{n\to\infty} \|R_\mu(f_n) - R_\mu(f)\|_{L^2}.\end{aligned}$$

The first term on the right-hand side is zero due to Lemma 3.105. The second term is zero due to the compactness of the inclusion $H^1_\mu \hookrightarrow L^2(\mathbb{R}^d)$. □

Since the convergence in norm implies the convergence of the spectrum, we obtain the following result.

Corollary 3.108. *The functional $\lambda_k : \mathcal{M}^T_{\text{cap}}(\mathbb{R}^d) \to [0, +\infty]$, which associates to each capacitary measure μ the kth eigenvalue $\lambda_k(\mu)$ of the operator $-\Delta + \mu$ in $L^2(\mathbb{R}^d)$, is continuous with respect to the γ-convergence.*

The convergence of the resolvents R_{μ_n} does not, in general, imply the γ-convergence of the measures μ_n. Indeed, we have the following example.

Example 3.109. Consider a sequence of sets $\Omega_n \subset \mathbb{R}^d$ with the following properties:

- each of the sets Ω_n is a *disjoint* union of n^{d+2} balls of equal radius;
- the radius of each ball in Ω_n is precisely $1/n$.

Then we have:

- the sequence of resolvent operators R_{Ω_n} converges to zero in norm:

$$\|R_{\Omega_n}\|_{\mathcal{L}(L^2(\mathbb{R}^d);L^2(\mathbb{R}^d))}=\frac{1}{\lambda_1(\Omega_n)}=\frac{1}{\lambda_1(B_{1/n})}=\frac{1}{n^2\lambda_1(B_1)}\xrightarrow[n\to\infty]{}0;$$

- the torsion $T(\Omega_n)$ remains constant:

$$T(\Omega_n)=n^{d+2}T(B_{1/n})=n^{d+2}\frac{\omega_d}{d(d+2)}(1/n)^{d+2}=\frac{\omega_d}{d(d+2)}.$$

We note that in the previous example the measure of Ω_n diverges. Precisely, we have

$$|\Omega_n|=n^{d+2}|B_{1/n}|=n^2\omega_d\xrightarrow[n\to+\infty]{}+\infty.$$

Thus the non equivalence seem to appear when the sequence of energy functions w_{μ_n} tends to distribute its mass uniformly on $\mathbb{R}^d$. In fact, the equivalence between the γ-convergence of μ_n and the norm convergence of R_{μ_n} holds, under the additional *non-dissipation* assumption $|\Omega_{\mu_n}|\le C$.

Proposition 3.110 (Convergence in norm and the non-dissipation of mass imply γ). *Let $\mu_n\in\mathcal{M}^T_{\text{cap}}(\mathbb{R}^d)$ be a sequence of capacitary measures and let Ω_{μ_n} be the corresponding sequence of sets of finiteness. If the measure of Ω_{μ_n} is uniformly bounded ($|\Omega_{\mu_n}|\le C$, for every $n\in\mathbb{N}$), then the sequence μ_n γ-converges to μ, if and only if, the sequence of resolvent operators $R_{\mu_n}\in\mathcal{L}(L^2(\mathbb{R}^d))$ converges to $R_\mu\in\mathcal{L}(L^2(\mathbb{R}^d))$ in norm.*

Proof. Suppose that $R_{\mu_n}\to R_\mu$ in the operator norm $\|\cdot\|_{\mathcal{L}(L^2(\mathbb{R}^d);L^2(\mathbb{R}^d))}$. We first show that $|\Omega_\mu|\le C$. Indeed, setting $\phi(x):=e^{-|x|^2}$, the strong convergence $R_{\mu_n}(\phi)\to R_\mu(\phi)$ gives that up to a subsequence $R_{\mu_n}(\phi)\to R_\mu(\phi)$ pointwise almost everywhere and so

$$\mathbb{1}_{\{R_\mu(\phi)>0\}}\le\liminf_{n\to+\infty}\mathbb{1}_{\{R_{\mu_n}(\phi)>0\}}\le\liminf_{n\to+\infty}\mathbb{1}_{\Omega_{\mu_n}},$$

which in turn implies

$$|\{R_\mu(\phi)>0\}|\le\liminf_{n\to+\infty}|\{R_{\mu_n}(\phi)>0\}|\le C.$$

Thus it is sufficient to show that $\{R_\mu(\phi)>0\}=\Omega_\mu\ \big(=\{w_\mu>0\}=\{w_\mu\wedge\phi>0\}\big)$, where the equalities in the parenthesis are due to the definition of Ω_μ and the strict positivity of f. Since $R_\mu(\phi) \in H^1_\mu \subset H^1_0(\Omega_\mu)$, we have the inclusion $\{R_\mu(\phi) > 0\} \subset \Omega_\mu$. For the opposite inclusion we consider the sequence $u_\varepsilon := Y_{\mu,\varepsilon}(w_\mu \wedge \phi)$, where $Y_{\mu,\varepsilon}$ is the operator from (3.68). By the maximum principle we have

$$u_\varepsilon \le \frac{1}{\varepsilon} R_{\mu+\frac{1}{\varepsilon}}(w_\mu \wedge \phi) \le \frac{1}{\varepsilon} R_\mu(w_\mu \wedge \phi) \le \frac{1}{\varepsilon} R_\mu(\phi),$$

and so $\{u_\varepsilon > 0\} \subset \{R_\mu(\phi) > 0\}$. On the other hand $u_\varepsilon \to w_\mu \wedge \phi$ strongly in H^1_μ (by Lemma 3.96) and so $\Omega_\mu = \{w_\mu \wedge \phi > 0\} \subset \{R_\mu(\phi) > 0\}$, which proves the equality $\Omega_\mu = \{R_\mu(\phi) > 0\}$ and the estimate $|\Omega_\mu| \le c$.

We now consider the sequence of characteristic functions $f_n = \mathbb{1}_{\Omega_{\mu_n} \cup \Omega_\mu}$. Since f_n is bounded in $L^2(\mathbb{R}^d)$, there is a function $f \in L^2(\mathbb{R}^d)$ such that $f_n \rightharpoonup f$ weakly in $L^2(\mathbb{R}^d)$. Since $f_n \equiv 1$ on Ω_μ, for every n, we have $f \equiv 1$ on Ω_μ. On the other hand, the norm convergence $R_{\mu_n} \to R_\mu$ implies the strong convergence in $L^2(\mathbb{R}^d)$ of $R_{\mu_n}(f_n)$ to $R_\mu(f)$. It is now sufficient to notice that $R_{\mu_n}(f_n) = w_{\mu_n}$ and $R_\mu(f) = w_\mu$. □

In view of Proposition 3.107 and by the definition of the strong convergence of operators, we have that every sequence $\mu_n \in \mathcal{M}^T_{\text{cap}}(\mathbb{R}^d)$, γ-converging to $\mu \in \mathcal{M}^T_{\text{cap}}(\mathbb{R}^d)$, is such that the sequence of reslovents R_{μ_n} converges strongly as operators in $L^2(\mathbb{R}^d)$ to R_μ. In what follows we study the relation between the strong convergence of the resolvent operators and the variational Γ-convergence (Definition 2.41) of the norms $\|\cdot\|_{H^1_\mu}$, which was originally used to define the a convergence on the class of capacitary measures. Before we continue we recall that given a capacitary measure $\mu \in \mathcal{M}^T_{\text{cap}}(\mathbb{R}^d)$ we can extend the associated norm $\|\cdot\|_{H^1_\mu}$ on H^1_μ to the entire space $L^2(\mathbb{R}^d)$ as follows:

$$\|u\|^2_{H^1_\mu} = \begin{cases} \displaystyle\int_{\mathbb{R}^d} |\nabla u|^2\,dx + \int_{\mathbb{R}^d} u^2\,d\mu + \int_{\mathbb{R}^d} u^2\,dx\,, & \text{if } \ u \in H^1_\mu, \\ +\infty\,, & \text{otherwise.} \end{cases}$$

The following result is classical and can be proved by a technique from the Γ-convergence Theory (see [53, Proposition 4.3] and [9, Corollary 3.13]). For sake of completeness, we give here a direct proof.

Proposition 3.111 (The strong convergence of R_{μ_n} implies the Γ-convergence of $\|\cdot\|_{H^1_\mu}$). *Let $\mu \in \mathcal{M}^T_{\text{cap}}(\mathbb{R}^d)$ and $\mu_n \in \mathcal{M}^T_{\text{cap}}(\mathbb{R}^d)$, for $n \in \mathbb{N}$ be capacitary measures of finite torsion such that the sequence of resolvent operators R_{μ_n} converges strongly in $L^2(\mathbb{R}^d)$ to R_μ. Then the sequence $\|\cdot\|_{H^1_{\mu_n}}$ Γ-converges in $L^2(\mathbb{R}^d)$ to $\|\cdot\|_{H^1_\mu}$.*

Proof. We first prove the "$\Gamma - \liminf$" inequality. Let $u_n \in H^1_{\mu_n}$ be a sequence converging to $u \in L^2(\mathbb{R}^d)$ strongly in $L^2(\mathbb{R}^d)$ and sequence of norms is bounded: $\|u_n\|_{H^1_{\mu_n}} \le C$, for a constant $C > 0$. For every $\varepsilon > 0$, consider the functions

$$u_n^\varepsilon := Y_{\mu_n,\varepsilon}(u_n) \qquad \text{and} \qquad u^\varepsilon := Y_{\mu,\varepsilon}(u),$$

where $Y_{\mu,\varepsilon} = \dfrac{1}{\varepsilon}R_{\mu+\frac{1}{\varepsilon}} = \dfrac{1}{\varepsilon}\big[1 + \dfrac{1}{\varepsilon}R_\mu\big]^{-1}R_\mu$ is the operator from (3.68). Since the norms of $Y_{\mu_n,\varepsilon}$ are bounded uniformly in n

$$\|Y_{\mu_n,\varepsilon}\|_{\mathcal{L}(L^2(\mathbb{R}^d);L^2(\mathbb{R}^d))} \le C_\varepsilon, \quad \text{for every} \quad n \in \mathbb{N},$$

and since $Y_{\mu_n,\varepsilon} \to Y_{\mu,\varepsilon}$ strongly as operators in $L^2(\mathbb{R}^d)$, we have that $u_n^\varepsilon \to u^\varepsilon$ strongly in $L^2(\mathbb{R}^d)$. Using $u_n^\varepsilon \in H^1_{\mu_n}$ as a test function in the equation

$$-\Delta u_n^\varepsilon + \mu_n u_n^\varepsilon = \frac{1}{\varepsilon}u_n - \frac{1}{\varepsilon}u_n^\varepsilon \quad \text{in} \quad H^1_{\mu_n}, \qquad u_n^\varepsilon \in H^1_{\mu_n},$$

we obtain the convergence of the $\|\cdot\|_{\dot{H}^1_{\mu_n}}$ norms:

$$\|u_n^\varepsilon\|^2_{\dot{H}^1_{\mu_n}} = \int_{\mathbb{R}^d} \frac{u_n^\varepsilon(u_n - u_n^\varepsilon)}{\varepsilon}\,dx \xrightarrow[n\to\infty]{} \int_{\mathbb{R}^d} \frac{u^\varepsilon(u - u^\varepsilon)}{\varepsilon}\,dx = \|u^\varepsilon\|^2_{\dot{H}^1_\mu}.$$

Using Lemma 3.96 (a), we get

$$\|u^\varepsilon\|_{\dot{H}^1_\mu} = \lim_{n\to\infty} \|u_n^\varepsilon\|_{\dot{H}^1_{\mu_n}} \le \liminf_{n\to\infty} \|u_n\|_{\dot{H}^1_{\mu_n}}.$$

On the other hand, by Lemma 3.96 (b), $\|u_n - u_n^\varepsilon\|_{L^2} \le C\sqrt{\varepsilon}$ and so passing to the limit, $\|u - u^\varepsilon\|_{L^2} \le C\sqrt{\varepsilon}$. Thus, u^ε converges in $L^2(\mathbb{R}^d)$ to u and is bounded in H^1_μ. As a consequence $u \in H^1_\mu$ and

$$\|u\|_{\dot{H}^1_\mu} \le \liminf_{\varepsilon\to 0^+} \|u^\varepsilon\|_{\dot{H}^1_\mu} \le \liminf_{n\to\infty} \|u_n\|_{\dot{H}^1_{\mu_n}},$$

which concludes the $\Gamma - \liminf$ inequality since $\|\cdot\|^2_{H^1} = \|\cdot\|^2_{\dot{H}^1_\mu} + \|\cdot\|^2_{L^2}$.

We now prove the "$\Gamma - \limsup$" inequality. by definition of the Γ-convergence, for every $u \in H^1_\mu$, we have to find a sequence $u_n \in H^1_{\mu_n}$ converging in $L^2(\mathbb{R}^d)$ to u and such that $\|u\|_{\dot{H}^1_\mu} = \lim_{n\to\infty} \|u_n\|_{\dot{H}^1_{\mu_n}}$. We first note that if $u = R_{\mu+t}(f)$, for some $f \in L^2(\mathbb{R}^d)$ and $t \ge 0$, then we may choose $u_n := R_{\mu_n+t}(f)$. Indeed, the strong convergence of $R_{\mu_n} \to R_\mu$ implies the strong convergence of $u_n = R_{\mu_n+t}(f) \to R_{\mu+t}(f) = u$ and testing with u_n the equation

$$-\Delta u_n + \mu_n u_n = f - t u_n \quad \text{in} \quad H^1_{\mu_n}, \qquad u_n \in H^1_{\mu_n},$$

we obtain

$$\|u_n\|^2_{\dot{H}^1_{\mu_n}} = \int_{\mathbb{R}^d} u_n(f - tu_n)\,dx \xrightarrow[n\to\infty]{} \int_{\mathbb{R}^d} u(f - tu)\,dx = \|u\|^2_{\dot{H}^1_\mu},$$

which completes the proof in the case $u = R_{\mu+t}(f)$. In the general case, it is sufficient to approximate in H^1_μ, the function $u \in H^1_\mu$ with functions of the form $R_{\mu+t}(f)$. Taking $u_\varepsilon = Y_{\mu,\varepsilon}(u)$, by Lemma 3.96, we have the strong convergence $u_\varepsilon \to u$ in $H^1_\mu(\mathbb{R}^d)$. Now the claim follows by a diagonal sequence argument. □

With the following example we show that the converse implication does not hold in general.

Example 3.112. We perform the following construction in $\mathbb{R}^d$, for $d \geq 3$.

- Let $f(x) = (1 + |x|)^{-\alpha}$, where $\frac{d}{2} < \alpha < \frac{d+2}{2}$. In particular we have that $f \in L^2(\mathbb{R}^d)$:

$$\int_{\mathbb{R}^d} f^2(x)\,dx = d\omega_d \int_0^{+\infty} \frac{r^{d-1}}{(1+r)^{2\alpha}}\,dr \leq d\omega_d \int_0^{+\infty} (1+r)^{-1-(2\alpha-d)}\,dr < +\infty.$$

- Let $x_n = (n, 0, \dots, 0) \in \mathbb{R}^d$ and let Ω_n be the half-ball centered in x_n:

$$\Omega_n := B_n(x_n) \cap \left\{(x^1, \dots, x^d) \in \mathbb{R}^d \,:\, x_1 > n\right\}.$$

In particular, we can obtain Ω_n by rescaling and translating Ω_1:

$$\Omega_n = x_n + n(-x_1 + \Omega_1). \tag{3.75}$$

The sequence of capacitary measures $\mu_n = I_{\Omega_n}$ has the following properties:

- The sequence of norms $\|\cdot\|_{H^1_{\mu_n}}$ Γ-converges in $L^2(\mathbb{R}^d)$ to $\|\cdot\|_{H^1_{L^\infty}}$, where

$$\|u\|_{H^1_{L^\infty}} = \begin{cases} 0, & \text{if } u \equiv 0, \\ +\infty, & \text{otherwise.} \end{cases}$$

- $\|\cdot\|_{H^1_{L^\infty}}$ is the norm associated to the capacitary measure $\mu = I_\emptyset$, defined by

$$\mu(E) = I_\emptyset(E) = \begin{cases} 0, & \text{if } \operatorname{cap}(E) = 0, \\ +\infty, & \text{otherwise.} \end{cases}$$

The Sobolev space H^1_μ contains only the constant zero ($H^1_\mu = \{0\}$) and the resolvent operator $R_\mu \equiv 0$, *i.e.* $R_\mu(\varphi) = 0$, for every $\varphi \in L^2(\mathbb{R}^d)$.

- The sequence $R_{\mu_n}(f)$ does not converge to $R_\mu(f) = 0$. Indeed, by the maximum principle and the fact that $f \geq \frac{1}{(1+2n)^\alpha}$ on Ω_n, we get

$$\|R_{\mu_n}(f)\|_{L^2} \geq \left\| R_{\mu_n}\left(\frac{1}{(1+2n)^\alpha}\right)\right\|_{L^2} = \frac{1}{(1+2n)^\alpha}\|w_{\Omega_n}\|_{L^2}$$
$$= \frac{n^{\frac{d+2}{2}}}{(1+2n)^\alpha}\|w_{\Omega_1}\|_{L^2} \to +\infty,$$

where we used the rescaling (3.75) to calculate the L^2 norm of the torsion function w_{Ω_n}.

The equivalence of the strong convergence of the resolvent operators R_{μ_n} and the Γ-convergence of the norms $\|\cdot\|_{H^1_{\mu_n}}$ does hold if an additional condition is imposed on the sequence μ_n.

Proposition 3.113. *Suppose that $\mu_n \in \mathcal{M}^T_{\text{cap}}(\mathbb{R}^d)$ is a sequence of capacitary measures of finite torsion such that the norms of the corresponding resolvent operators are uniformly bounded:*

$$\|R_{\mu_n}\|_{\mathcal{L}(L^2(\mathbb{R}^d);L^2(\mathbb{R}^d))} \leq C, \quad \textit{for every} \quad n \in \mathbb{N},$$

and let $\mu \in \mathcal{M}^T_{\text{cap}}(\mathbb{R}^d)$. Then the sequence of resolvents R_{μ_n} converges strongly in $L^2(\mathbb{R}^d)$ to R_μ, if and only if, the sequence $\|\cdot\|_{H^1_{\mu_n}}$ Γ-converges in $L^2(\mathbb{R}^d)$ to $\|\cdot\|_{H^1_\mu}$.

Proof. Let $f \in L^2(\mathbb{R}^d)$. We first prove that the sequence $u_n = R_{\mu_n}(f)$ converges in $L^2(\mathbb{R}^d)$. Indeed, by the bound on the resolvent we get

$$\|u_n\|_{L^2} \leq \|R_{\mu_n}\|_{\mathcal{L}(L^2(\mathbb{R}^d);L^2(\mathbb{R}^d))}\|f\|_{L^2} \leq C\|f\|_{L^2},$$

and by testing with u_n the equation

$$-\Delta u_n + \mu_n u_n = f \quad \text{in} \quad H^1_{\mu_n}, \qquad u_n \in H^1_{\mu_n}, \tag{3.76}$$

we obtain

$$\|u_n\|^2_{H^1} \leq \|u_n\|^2_{H^1_{\mu_n}} = \|u_n\|^2_{\dot{H}^1_{\mu_n}} + \|u_n\|^2_{L^2} \leq (C + C^2)\|f\|^2_{L^2}.$$

Thus, up to a subsequence u_n converges weakly in $L^2(\mathbb{R}^d)$ and strongly in $L^2_{\text{loc}}(\mathbb{R}^d)$ to a function $u \in L^2(\mathbb{R}^d)$. We now test the equation (3.76) with $(1-\eta_R)^2 u_n$, where $\eta_R(x) := \eta(x/R)$ and

$$\eta \in C_c^\infty(\mathbb{R}^d), \quad 0 \le \eta \le 1, \quad \eta = 1 \text{ on } B_1, \quad \eta = 0 \text{ on } \mathbb{R}^d \setminus B_2.$$

We obtain

$$\begin{aligned}
&\int_{\mathbb{R}^d} (1-\eta_R)^2 u_n f\,dx \\
&= \int_{\mathbb{R}^d} \nabla\big((1-\eta_R)^2 u_n\big)\cdot\nabla u_n\,dx + \int_{\mathbb{R}^d} (1-\eta_R)^2 u_n^2\,d\mu_n \\
&\ge \int_{\mathbb{R}^d} \left|\nabla\big((1-\eta_R)u_n\big)\right|^2 dx - \int_{\mathbb{R}^d} |\nabla(1-\eta_R)|^2 u_n^2\,dx \\
&\quad + \int_{\mathbb{R}^d} \big((1-\eta_R)u_n\big)^2\,d\mu_n \\
&\ge \frac{1}{C^2}\int_{\mathbb{R}^d} \big((1-\eta_R)u_n\big)^2\,dx - \int_{\mathbb{R}^d} |\nabla(1-\eta_R)|^2 u_n^2\,dx,
\end{aligned}$$

which gives the estimate

$$\begin{aligned}
\frac{1}{C^2}\int_{\mathbb{R}^d\setminus B_{2R}} u_n^2\,dx &\le \int_{\mathbb{R}^d} (1-\eta_R)^2 u_n f\,dx + \frac{\|\nabla\eta\|_{L^\infty}^2}{R^2}\int_{\mathbb{R}^d} u_n^2\,dx \\
&\le 2\int_{\mathbb{R}^d} (1-\eta_R)^2 u f\,dx + \frac{\|\nabla\eta\|_{L^\infty}^2}{R^2} C^2\|f\|_{L^2}^2,
\end{aligned}$$

for n large enough. Thus u_n converges to u strongly in $L^2(\mathbb{R}^d)$. Now the Γ-convergence of the norm gives

$$\begin{aligned}
J_{\mu,f}(u) &= \frac{1}{2}\|u\|_{H^1_\mu}^2 - \int_{\mathbb{R}^d} u\Big(f - \frac{1}{2}u\Big)\,dx \\
&\le \liminf_{n\to\infty}\left\{\frac{1}{2}\|u_n\|_{H^1_{\mu_n}}^2 - \int_{\mathbb{R}^d} u_n\left(f - \frac{1}{2}u_n\right)dx\right\} \\
&= \liminf_{n\to\infty} J_{\mu_n,f}(u_n)
\end{aligned}$$

On the other hand, for every $v \in H^1_\mu$, there is a sequence $v_n \in H^1_{\mu_n}$ such that

$$\begin{aligned}
J_{\mu,f}(v) &= \frac{1}{2}\|v\|_{H^1_\mu}^2 - \int_{\mathbb{R}^d} v\Big(f - \frac{1}{2}v\Big)\,dx \\
&= \lim_{n\to\infty}\left\{\frac{1}{2}\|v_n\|_{H^1_{\mu_n}}^2 - \int_{\mathbb{R}^d} v_n\left(f - \frac{1}{2}v_n\right)dx\right\} \ge \limsup_{n\to\infty} J_{\mu_n,f}(u_n),
\end{aligned}$$

where the last inequality is due to the fact that u_n minimizes $J_{\mu_n,f}$. Thus $J_{\mu,f}(u) \le J_{\mu,f}(v)$, for every $v \in H^1_\mu$, and so u minimizes $J_{\mu,f}$, *i.e.* $u = R_\mu(f)$. □

Remark 3.114. The hypothesis of Proposition 3.113 on the sequence R_{μ_n} is fulfilled in each of the following situations:

- The sequence of capacitary measures $\mu_n \in \mathcal{M}^T_{\text{cap}}(\mathbb{R}^d)$ is of uniformly bounded torsion: $T(\mu_n) \le C$, for every $n \in \mathbb{N}$.
- The sets of finiteness $\Omega_{\mu_n} \subset \mathbb{R}^d$ are of uniformly bounded measure: $|\Omega_{\mu_n}| \le C$, for every $n \in \mathbb{N}$.

In the case when the sequence μ_n has sets of finiteness of uniformly bounded measure, we can summarize the results from Propositions 3.107, 3.110, 3.111 and 3.113 in the following theorem.

Theorem 3.115. *Suppose that $\mu_n \in \mathcal{M}^T_{\text{cap}}(\mathbb{R}^d)$ is a sequence of capacitary measure of finite torsion and $\mu \in \mathcal{M}^T_{\text{cap}}(\mathbb{R}^d)$. Then* (i) $\Rightarrow$ (ii) $\Rightarrow$ (iii) $\Rightarrow$ (iv), *where*

(i) *The sequence μ_n γ-converges to μ.*
(ii) *The sequence of resolvent operators R_{μ_n} converges in the operator norm of $\mathcal{L}(L^2(\mathbb{R}^d); L^2(\mathbb{R}^d))$ to R_μ.*
(iii) *The sequence of resolvent operators R_{μ_n} converges strongly in $L^2(\mathbb{R}^d)$ to R_μ.*
(iv) *The sequence of norms $\|\cdot\|_{H^1_{\mu_n}}$ Γ-converges in $L^2(\mathbb{R}^d)$ to $\|\cdot\|_{H^1_\mu}$.*

If, moreover, the sequence of torsion functions w_{μ_n} is bounded from above by a function $w \in L^1(\mathbb{R}^d)$, then the claims above are equivalent: (i) $\Leftrightarrow$ (ii) $\Leftrightarrow$ (iii) $\Leftrightarrow$ (iv).

Proof. The implications (i) $\Rightarrow$ (ii) $\Rightarrow$ (iii) $\Rightarrow$ (iv) were proved respectively in the Propositions 3.107, 3.110, 3.111 and 3.113. In order to prove the equivalence under the additional hypothesis $w_{\mu_n} \le w \in L^1(\mathbb{R}^d)$, it is enough to prove that (iv) $\Rightarrow$ (i). Indeed, since

$$\int_{\mathbb{R}^d} |\nabla w_{\mu_n}|^2\,dx \le \int_{\mathbb{R}^d} w_{\mu_n}\,dx \le \int_{\mathbb{R}^d} w\,dx,$$

we have that w_{μ_n} converges (up to a subsequence) in $L^1_{\text{loc}}(\mathbb{R}^d)$. Due to the bound $w_{\mu_n} \le w$ it is concentrated and so, it converges strongly in $L^1(\mathbb{R}^d)$. By the completeness of the γ-distance we have that the $L^1(R^d)$ limit of w_{μ_n} is the torsion function $w_{\mu'}$ of a capacitary measure $\mu' \in \mathcal{M}^T_{\text{cap}}(\mathbb{R}^d)$.

Thus μ_n γ-converges to μ and so the sequence of norms $\|\cdot\|_{H^1_{\mu_n}}$ Γ-converges in $L^2(\mathbb{R}^d)$ to $\|\cdot\|_{H^1_{\mu'}}$, which together with (iv) gives

$$\|u\|_{H^1_{\mu'}} = \|u\|_{H^1_\mu}, \quad \text{for every} \quad u \in L^2(\mathbb{R}^d).$$

Thus $\mu \equiv \mu'$ and so μ_n γ-converges to μ. □

Remark 3.116. We note that if the capacitary measure $\nu \in \mathcal{M}^T_{\text{cap}}(\mathbb{R}^d)$ is such that $\nu \leq \mu_n$, for every capacitary measure $\mu_n \in \mathcal{M}^T_{\text{cap}}(\mathbb{R}^d)$, then by the maximum principle $w_{\mu_n} \leq w_\nu$ and so the four conditions from Theorem 3.115 are equivalent.

3.6. The γ-convergence in a box of finite measure

In this section we consider the case when the sequence of capacitary measures μ_n is uniformly bounded, *i.e.* when there is a capacitary measure ν in $\mathbb{R}^d$ such that $w_\nu \in L^1(\mathbb{R}^d)$ and $\mu_n \geq \nu$, for every $n \in \mathbb{N}$. For a generic capacitary measure $\nu \in \mathcal{M}_{\text{cap}}(\mathbb{R}^d)$ we will denote by $\mathcal{M}^{T,\nu}_{\text{cap}}$ the family of measures bounded from below by ν, *i.e.*

$$\mathcal{M}^{T,\nu}_{\text{cap}}(\mathbb{R}^d) := \Big\{\mu \in \mathcal{M}^T_{\text{cap}}(\mathbb{R}^d) : \mu \geq \nu\Big\}, \tag{3.77}$$

In the spacial case when ν is of the form $\nu = I_\Omega$ for a measurable set $\Omega \subset \mathbb{R}^d$ we will use the notation

$$\mathcal{M}^T_{\text{cap}}(\Omega) := \mathcal{M}^{T,\nu}_{\text{cap}}(\mathbb{R}^d) = \Big\{\mu \in \mathcal{M}^T_{\text{cap}}(\mathbb{R}^d) : \mu \geq I_\Omega\Big\}. \tag{3.78}$$

Theorem 3.117. *Let $\nu \in \mathcal{M}^T_{\text{cap}}(\mathbb{R}^d)$ be a capacitary measure of finite torsion. Then the family of capacitary measures $\mathcal{M}^{T,\nu}_{\text{cap}}(\mathbb{R}^d)$ equipped with the distance d_γ is a compact metric space.*

Proof. Let $\mu_n \in \mathcal{M}^{T,\nu}_{\text{cap}}(\mathbb{R}^d)$ be a given sequence of capacitary measures. Then by the maximum principle we get

$$w_{\mu_n} \leq w_\nu, \quad \text{for every} \quad n \in \mathbb{N}.$$

Now, reasoning as in Theorem 3.115, we get that up to a subsequence μ_n γ-converges to some $\mu \in \mathcal{M}^T_{\text{cap}}(\mathbb{R}^d)$ such that $w_\mu \leq w_\nu$.[8] Thus, it is sufficient to check that $\mu \geq \nu$, *i.e.* that for every non-negative $u \in H^1_\mu$, we have

$$\|u\|^2_{\dot{H}^1_\mu} = \int_{\mathbb{R}^d} |\nabla u|^2\,dx + \int_{\mathbb{R}^d} u^2\,d\mu \geq \int_{\mathbb{R}^d} |\nabla u|^2\,dx + \int_{\mathbb{R}^d} u^2\,d\nu = \|u\|^2_{\dot{H}^1_\nu}. \tag{3.79}$$

[8] We note that the inequality $w_\mu \leq w_\nu$ does not imply in general that $\mu \geq \nu$.

Indeed, by Theorem 3.115, the sequence of functionals $\|\cdot\|_{H^1_{\mu_n}}$ Γ-converges in $L^2(\mathbb{R}^d)$ to $\|\cdot\|_{H^1_\mu}$ and so, there is a sequence $u_n \in H^1_{\mu_n}$ such that u_n converges to u in $L^2(\mathbb{R}^d)$ and

$$\|u\|_{H^1_\mu} = \lim_{n\to\infty} \|u_n\|_{H^1_{\mu_n}} \geq \lim_{n\to\infty} \|u_n\|_{H^1_\nu} \geq \|u\|_{H^1_\nu},$$

where the last inequality is due to the semi-continuity od the norm $\|\cdot\|_{H^1_\nu}$ with respect to the strong $L^2(\mathbb{R}^d)$-convergence. □

In what follows we investigate the connection of the γ-convergence and the weak convergence of measures. In the particular case when the measures μ_n are absolutely continuous with respect to the Lebesgue measure, we have the following result.

Lemma 3.118. *Consider a measurable set $\Omega \subset \mathbb{R}^d$ and a fixed $p \in [1, +\infty)$ be fixed. Let $V_n \in L^1(\Omega)$ be a sequence weakly converging in $L^1(\Omega)$ to a function V. Setting $\mu_n = V_n dx + I_\Omega$ and $\mu = V dx + I_\Omega$, the sequence of functionals $\|\cdot\|_{H^1_{\mu_n}}$ Γ-converges in $L^2(\mathbb{R}^d)$ to the functional $\|\cdot\|_{H^1_\mu}$.*

Proof. We first prove the $\Gamma - \liminf$ inequality (Definition 2.41 (a)). Let $u_n \in H^1_{\mu_n}$ be a sequence converging in $L^2(\mathbb{R}^d)$ to some $u \in L^2(\mathbb{R}^d)$. By the lower semi-continuity of the $H^1(\Omega)$ norm we have $u \in h^1_0(\Omega)$ and

$$\int_\Omega |\nabla u|^2\, dx \leq \liminf_{n\to\infty} \int_\Omega |\nabla u_n|^2\, dx.$$

We now claim that the following inequality holds:

$$\int_\Omega V(x) u^2\, dx \leq \liminf_{n\to\infty} \int_\Omega V_n(x) u_n^2\, dx. \tag{3.80}$$

We will prove (3.80) in the case $p > 1$. For the limit case $p = 1$ we refer to [31]. Indeed, for any $t > 0$ we consider the functions $u_n^t := (-t) \vee u_n \wedge t$ and $u^t := (-t) \vee u \wedge t$. Since u_n converges strongly in $L^2(\Omega)$ to u, we have that $|u_n^t|^2$ converges strongly to $|u^t|^2$ in any $L^q(\Omega)$ and, in particular, for $q = p'$. Thus, we have

$$\int_\Omega V(x)|u^t|^2\, dx = \lim_{n\to\infty} \int_\Omega V_n(x)|u_n^t|^2\, dx \leq \liminf_{n\to\infty} \int_\Omega V_n(x) u_n^2\, dx.$$

Now passing to the limit as $t \to +\infty$, we obtain (3.80), which concludes the proof of the Γ − lim inf inequality

$$\|u\|^2_{H^1_\mu} \le \liminf_{n\to+\infty} \|u_n\|^2_{H^1_{\mu_n}}.$$

In order to prove the Γ − lim sup inequality (Definition 2.41 (b)) we construct, for every $u \in H^1_0(\Omega)$, a sequence $u_n \in H^1_{V_n}$ converging to u strongly in $L^2(\Omega)$ and such that

$$\begin{aligned}\limsup_{n\to\infty} &\left\{\int_\Omega |\nabla u_n|^2\,dx + \int_\Omega V_n(x)u_n^2\,dx\right\}\\ &\le \int_\Omega |\nabla u|^2\,dx + \int_\Omega V(x)u^2\,dx.\end{aligned}\tag{3.81}$$

For every $t > 0$ let $u^t = (u \wedge t) \vee (-t)$; then, by the weak convergence of V_n, for t fixed we have

$$\lim_{n\to\infty} \int_\Omega V_n(x)|u^t|^2\,dx = \int_\Omega V(x)|u^t|^2\,dx,$$

and

$$\lim_{t\to+\infty} \int_\Omega V(x)|u^t|^2\,dx = \int_\Omega V(x)|u|^2\,dx.$$

Then, by a diagonal argument, we can find a sequence $t_n \to +\infty$ such that

$$\lim_{n\to\infty} \int_\Omega V_n(x)|u^{t_n}|^2\,dx = \int_\Omega V(x)|u|^2\,dx.$$

Taking now $u_n = u^{t_n}$, and noticing that for every $t > 0$

$$\int_\Omega |\nabla u^t|^2\,dx \le \int_\Omega |\nabla u|^2\,dx,$$

we obtain (3.81) thus completing the proof. □

Theorem 3.119. *Let $\Omega \subset \mathbb{R}^d$ be a set of finite measure. Then, for every $p \in (1, +\infty)$ the set*

$$\mathcal{M}_{L^p}(\Omega) := \left\{\mu \in \mathcal{M}^T_{\text{cap}}(\mathbb{R}^d) : \mu = V dx + I_\Omega,\ V \ge 0,\ \int_\Omega V^p\,dx \le 1\right\},$$

is compact with respect to the γ-distance. Moreover, a sequence $\mu_n = V_n dx + I_\Omega \in \mathcal{M}_{L^p}(\Omega)$ γ-converges to $\mu = V dx + I_\Omega \in \mathcal{M}_{L^p}(\Omega)$, if and only if the corresponding sequence V_n converges to V weakly in L^p.

Proof. Consider a sequence $\mu_n = V_n dx + I_\Omega$. Then, up to a subsequence, V_n converges weakly in $L^p(\Omega)$ to a non-negative function $V \in L^p(\Omega)$ with $\int_\Omega V^p\,dx \le 1$. By Lemma 3.118, the sequence of functionals $\|\cdot\|_{H^1_{\mu_n}}$ Γ-converges in $L^2(\mathbb{R}^d)$ to the functional $\|\cdot\|_{H^1_\mu}$ associated to the capacitary measure $\mu = V dx + I_\Omega$. On the other hand, by the maximum principle we have $w_{\mu_n} \le w_\Omega$ and so, Theorem 3.115 gives that μ_n γ-converges to μ. □

In the case of weak* convergence of measures the statement of Theorem 3.119 is no longer true, as the following proposition shows.

Proposition 3.120. *Let $\Omega \subset \mathbb{R}^d$ $(d \ge 2)$ be a bounded open set and let $V, W \in L^1(\Omega)$ be two non-negative functions such that $V \ge W \ge 0$. Then, there is a sequence of non-negative functions $V_n \in L^1(\Omega)$, bounded in $L^1(\Omega)$, such that the sequence of measures $V_n\,dx$ converges weakly* in Ω to $V\,dx$ and the sequence $V_n dx + I_\Omega$ γ-converges to $W dx + I_\Omega$.*

Proof. For sake of simplicity, we will write w_μ instead of $w_{\mu+I_\Omega}$. Without loss of generality we can suppose $\int_\Omega (V - W)\,dx = 1$. Let μ_n be a sequence of probability measures on Ω weakly* converging to $(V - W)\,dx$ and such that each μ_n is a finite sum of Dirac masses. For each $n \in \mathbb{N}$ consider a sequence of positive functions $V_{n,m} \in L^1(\Omega)$ such that $\int_\Omega V_{n,m}\,dx = 1$ and such that the sequence of measures $V_{n,m}dx$ converges weakly* to μ_n as $m \to \infty$. Moreover, we note that we can choose $V_{n,m}$ to be a convex combination of functions of the form $|B_{1/m}|^{-1}\mathbb{1}_{B_{1/m}(x_j)}$.

We now prove that for fixed $n \in \mathbb{N}$, the sequence of capacitary measures $(V_{n,m} + W)\,dx$ γ-converges, as $m \to \infty$, to $W\,dx$ or, equivalently, that the sequence of torsion functions $w_{W+V_{n,m}}$ converges in L^2 to w_W, as $m \to \infty$. Indeed, by the weak maximum principle, we have

$$w_{W+I_{\Omega_{m,n}}} \le w_{W+V_{n,m}} \le w_W,$$

where $\Omega_{m,n} = \Omega \setminus \left(\bigcup_j B_{1/m}(x_j)\right)$ and so, we can estimate the distance between $W + V_{n,m}$ and W as follows:

$$d_\gamma\big(W, W + V_{n,m}\big) \le d_\gamma\big(W, W + I_{\Omega_{m,n}}\big) \tag{3.82}$$

$$= \int_\Omega \big(w_W - w_{W+I_{\Omega_{m,n}}}\big)\,dx = 2\big(E_1(W + I_{\Omega_{m,n}}) - E_1(W)\big) \tag{3.83}$$

$$\le \int_\Omega \Big(|\nabla w_m|^2 + W w_m^2 - 2w_m\Big)\,dx - \int_\Omega \Big(|\nabla w_W|^2 + W w_W^2 - 2w_W\Big)\,dx,$$

for a generic test function $w_m \in H_0^1(\Omega_{m,n})$. Since the single points have zero capacity in $\mathbb{R}^d$ $(d \geq 2)$ there exists a sequence $\phi_m \in H^1(\mathbb{R}^d)$ such that

$$\phi_m \equiv 1 \text{ on } B_{1/m}(0), \quad \phi_m \equiv 0 \text{ on } \mathbb{R}^d \setminus B_{1/sqrtm} \text{ and } \lim_{m\to\infty} \|\phi_m\|_{H^1} = 0.$$

Thus we may choose the test function w_m as the product

$$w_m(x) = w_W(x) \prod_j \big(1 - \phi_m(x - x_j)\big).$$

Now since $\phi_m \to 0$ strongly in $H^1(\mathbb{R}^d)$, it is easy to see that $w_m \to w_W$ strongly in $H^1(\Omega)$ and so, by (3.83), $d_\gamma\big(W, W + V_{n,m}\big) \to 0$, as $m \to \infty$. Since the weak convergence of probability measures and the γ-convergence are both induced by metrics, we can conclude by a diagonal sequence argument. □

Remark 3.121. When $d = 1$, a result analogous to Lemma 3.118 is that any sequence (μ_n) weakly* converging to μ is also γ-converging to μ. This is an easy consequence of the compact embedding of $H_0^1(\Omega)$ into the space of continuous functions on Ω.

We note that the hypothesis $V \geq W$ in Proposition 3.120 is necessary. Indeed, we have the following proposition, whose proof is contained in [36, Theorem 3.1] and we report it here for the sake of completeness.

Proposition 3.122 (Weak* limits are larger than the γ-limits). *Let $\mu_n \in \mathcal{M}_{\text{cap}}^T(\mathbb{R}^d)$ be a sequence of capacitary measures weakly* converging to a Borel measure ν and γ-converging to the capacitary measure $\mu \in \mathcal{M}_{\text{cap}}^T(\mathbb{R}^d)$. Then $\mu \leq \nu$ in $\mathbb{R}^d$.*

Proof. We note that it is enough to show that $\mu(K) \leq \nu(K)$, whenever $K \subset \mathbb{R}^d$ is a compact set. Let u be a nonnegative smooth function with compact support in $\mathbb{R}^d$ such that $u \leq 1$ in $\mathbb{R}^d$ and $u \equiv 1$ on K; we have

$$\mu(K) \leq \int_{\mathbb{R}^d} u^2\, d\mu \leq \liminf_{n\to\infty} \int_{\mathbb{R}^d} u^2\, d\mu_n = \int_{\mathbb{R}^d} u^2\, d\nu \leq \nu(\{u > 0\}).$$

Since u is arbitrary, we have the conclusion by the Borel regularity of ν. □

3.7. Concentration-compactness principle for capacitary measures

In this section we introduce one of the main tools for the study of shape optimization problems in $\mathbb{R}^d$. Since when we work in the whole Euclidean space, we don't have an a priori bound on the minimizing sequences of capacitary measures, as happens for example in a box. Thus, finding a convergent minimizing sequence becomes the main task in the of the existence of optimal solution. Since the γ-convergence of a sequence μ_n of capacitary measures is determined through the convergence of the corresponding energy functions w_{μ_n}, we can use the classical concentration-compactness principle of P.L.Lions to determine the behaviour of w_{μ_n}. At this point, we need to deduce the behaviour of the sequence μ_n from the behaviour of the sequence of energy functions. In order to do this we will need some preliminary technical results.

3.7.1. The γ-distance between comparable measures

The functional character of the distance d_γ makes quite technical the estimate on the distance between two capacitary measures. In this section, we collect various estimates on the distance between capacitary measures μ and ν which are comparable with respect to the order "$\leq$", *i.e.* when we have $\nu \leq \mu$ or $\mu \leq \nu$. In particular, we consider the most important cases, when the two measures differ outside a large ball (or a half-plane) or inside a small set. At the end we also give some estimates on the variation of eigenvalues and the resolvent operators with respect to the γ-distance.

Lemma 3.123. *Suppose that μ is a capacitary measure such that $w_\mu \in L^1(\mathbb{R}^d)$. Then, for every $R > 1$ and every $R_2 > R_1 > 1$ we have*

$$d_\gamma\big(\mu, \mu \vee I_{B_R}\big) \leq \int_{\mathbb{R}^d \setminus B_{R/2}} w_\mu \, dx + C R^{-2}, \tag{3.84}$$

$$d_\gamma\big(\mu, \mu \vee I_{B_R^c}\big) \leq \int_{B_{2R}} w_\mu \, dx + C R^{-2}, \tag{3.85}$$

$$d_\gamma\big(\mu, \mu \vee (I_{B_{R_1}} \wedge I_{B_{R_2}^c})\big) \leq \int_{B_{2R_2} \setminus B_{R_1/2}} w_\mu \, dx + C\big(R_1^{-2} + R_2^{-2}\big), \tag{3.86}$$

where the constant C depends only on $\|w_\mu\|_{L^1}$ and the dimension d.

Proof. We set for simplicity $w_R = w_{\mu \vee I_{B_R}}$ and $\eta_R(x) = \eta(x/R)$, where

$$\eta \in C_c^\infty(\mathbb{R}^d), \quad 0 \leq \eta \leq 1, \quad \eta = 1 \text{ on } B_1, \quad \eta = 0 \text{ on } \mathbb{R}^d \setminus B_2.$$

Then we have

$$
\begin{aligned}
&d_\gamma(\mu, \mu \vee I_{B_{2R}}) \\
&= \int_{\mathbb{R}^d} (w_\mu - w_{2R})\, dx \\
&= 2\big(J_\mu(w_{2R}) - J_\mu(w_\mu)\big) \le 2\big(J_\mu(\eta_R w_\mu) - J_\mu(w_\mu)\big) \\
&= \int_{\mathbb{R}^d} |\nabla(\eta_R w_\mu)|^2\, dx + \int_{\mathbb{R}^d} \eta_R^2 w_\mu^2\, d\mu - 2\int_{\mathbb{R}^d} \eta_R w_\mu\, dx + \int_{\mathbb{R}^d} w_\mu\, dx \\
&= \int_{\mathbb{R}^d} \big(w_\mu^2 |\nabla \eta_R|^2 + \nabla w_\mu \cdot \nabla(\eta_R^2 w_\mu)\big)\, dx \\
&\quad + \int_{\mathbb{R}^d} \eta_R^2 w_\mu^2\, d\mu - 2\int_{\mathbb{R}^d} \eta_R w_\mu\, dx + \int_{\mathbb{R}^d} w_\mu\, dx \\
&= \int_{\mathbb{R}^d} w_\mu^2 |\nabla \eta_R|^2\, dx + \int_{\mathbb{R}^d} \eta_R^2 w_\mu\, dx - 2\int_{\mathbb{R}^d} \eta_R w_\mu\, dx + \int_{\mathbb{R}^d} w_\mu\, dx \\
&= \int_{\mathbb{R}^d} w_\mu^2 |\nabla \eta_R|^2\, dx + \int_{\mathbb{R}^d} (1-\eta_R)^2 w_\mu\, dx \\
&\le \frac{\|\nabla \eta\|_{L^\infty}^2}{R^2} \|w_\mu\|_{L^2} + \int_{\mathbb{R}^d \setminus B_R} w_\mu\, dx,
\end{aligned}
$$

which proves (3.84). The estimates (3.85) and (3.86) are analogous. □

By a similar argument we have the following result, which is implicitly contained in [59, Lemma 3.7] in the case when $\mu = I_\Omega$.

Lemma 3.124 (Restriction to a half-space). *Suppose that μ is a capacitary measure in $\mathbb{R}^d$ such that $w_\mu \in L^1(\mathbb{R}^d)$. For the half-space $H = \{x \in \mathbb{R}^d : c + x \cdot \xi > 0\}$, where the constant $c \in \mathbb{R}$ and the unit vector $\xi \in \mathbb{R}^d$ are given, we have*

$$
\begin{aligned}
d_\gamma(\mu, \mu \vee I_H) \le \sqrt{8\|w_\mu\|_{L^\infty}} \int_{\partial H} w_\mu\, d\mathcal{H}^{d-1} - \int_{\mathbb{R}^d \setminus H} |\nabla w_\mu|^2\, dx \\
- \int_{\mathbb{R}^d \setminus H} w_\mu^2\, d\mu + 2\int_{\mathbb{R}^d \setminus H} w_\mu\, dx.
\end{aligned}
\tag{3.87}
$$

Proof. For sake of simplicity, set $w := w_\mu$, $M = \|w\|_{L^\infty}$, $c = 0$ and $\xi = (0, \dots, 0, -1)$. Consider the function

$$
v(x_1, \dots, x_d) = \begin{cases} M, & x_1 \le -\sqrt{M}, \\ \frac{1}{2}\Big(2M - (x_1 + \sqrt{2M})^2\Big), & -\sqrt{2M} \le x_1 \le 0, \\ 0, & 0 \le x_1. \end{cases}
\tag{3.88}
$$

Consider the function $w_H = w \wedge v \in H_0^1(H) \cap H_\mu^1$.

$$
\begin{aligned}
&d_\gamma(\mu, \mu \vee I_H)\\
&= \int_{\mathbb{R}^d} (w - w_{\mu\vee I_H})\,dx\\
&= 2\big(J_\mu(w_{\mu\vee I_H}) - J_\mu(w)\big) \le 2\big(J_\mu(w_H) - J_\mu(w)\big)\\
&\le \int_{\mathbb{R}^d} |\nabla(w_H)|^2 - |\nabla w|^2\,dx - \int_{\mathbb{R}^d\setminus H} w^2\,d\mu\\
&\quad + 2\int_{\mathbb{R}^d} (w - w_H)\,dx\\
&\le \int_{\{-\sqrt{2M}<x_1\le 0\}} |\nabla(w_H)|^2 - |\nabla w|^2\,dx - \int_{\mathbb{R}^d\setminus H} |\nabla w|^2\,dx\\
&\quad - \int_{\mathbb{R}^d\setminus H} w^2\,d\mu + 2\int_{\mathbb{R}^d} (w - w_H)\,dx\\
&\le 2\int_{\{-\sqrt{2M}<x_1\le 0\}} \nabla w_H \cdot \nabla(w_H - w)\,dx \qquad (3.89)\\
&\quad + 2\int_{\{-\sqrt{2M}<x_1\le 0\}} (w - w_H)\,dx\\
&\quad - \int_{\mathbb{R}^d\setminus H} |\nabla w|^2\,dx - \int_{\mathbb{R}^d\setminus H} w^2\,d\mu + 2\int_{\mathbb{R}^d\setminus H} w\,dx\\
&= 2\int_{\{-\sqrt{2M}<x_1\le 0\}} \nabla v \cdot \nabla(w_H - w)\,dx\\
&\quad + 2\int_{\{-\sqrt{2M}<x_1\le 0\}} (w - w_H)\,dx\\
&\quad - \int_{\mathbb{R}^d\setminus H} |\nabla w|^2\,dx - \int_{\mathbb{R}^d\setminus H} w^2\,d\mu + 2\int_{\mathbb{R}^d\setminus H} w\,dx\\
&= \sqrt{8M}\int_{\partial H} w\,d\mathcal{H}^{d-1} - \int_{\mathbb{R}^d\setminus H} |\nabla w|^2\,dx\\
&\quad - \int_{\mathbb{R}^d\setminus H} w^2\,d\mu + 2\int_{\mathbb{R}^d\setminus H} w\,dx.
\end{aligned}
$$

□

An analogous estimate allows us to prove the following

Lemma 3.125. *Suppose that μ is a capacitary measure such that $w_\mu \in L^1(\mathbb{R}^d)$. Then for every $\Omega \subset \mathbb{R}^d$, we have*

$$d_\gamma(\mu, \mu \vee I_{\Omega^c}) \le \|w_\mu\|_{L^\infty}^2 \operatorname{cap}(\Omega).$$

Proof. Suppose that $\mathrm{cap}(\Omega) > 0$ and let $\varphi \in H^1(\mathbb{R}^d)$ be a function such that

$$0 \le \varphi \le 1 \qquad \text{and} \qquad \mathrm{cap}(\Omega) \le \|\varphi\|_{H^1}^2 \le (1+\varepsilon)\,\mathrm{cap}(\Omega).$$

Then we have

$$\begin{aligned} d_\gamma(\mu, \mu \vee I_{\Omega^c}) &= \int_{\mathbb{R}^d} (w_\mu - w_{\mu\vee I_{\Omega^c}})\, dx = 2\big(J_\mu(w_{\mu\vee I_{\Omega^c}}) - J_\mu(w_\mu)\big) \\ &\le \int_{\mathbb{R}^d} \big|\nabla\big((1-\varphi)w_\mu\big)\big|^2\, dx + \int_{\mathbb{R}^d} (1-\varphi)^2 w_\mu^2\, d\mu \\ &\quad -2\int_{\mathbb{R}^d} (1-\varphi) w_\mu\, dx + \int_{\mathbb{R}^d} w_\mu\, dx \\ &= \int_{\mathbb{R}^d} |\nabla(1-\varphi)|^2 w_\mu^2\, dx + \int_{\mathbb{R}^d} \nabla w_\mu \cdot \nabla\big(w_\mu(1-\varphi)^2\big)\, dx \\ &\quad + \int_{\mathbb{R}^d} (1-\varphi)^2 w_\mu^2\, d\mu - 2\int_{\mathbb{R}^d} (1-\varphi) w_\mu\, dx + \int_{\mathbb{R}^d} w_\mu\, dx \\ &= \int_{\mathbb{R}^d} \big(|\nabla\varphi|^2 + \varphi^2\big) w_\mu^2\, dx \le (1+\varepsilon)\,\mathrm{cap}(\Omega)\|w_\mu\|_{L^\infty}^2, \end{aligned}$$

which, after letting $\varepsilon \to 0$, proves the claim. □

The following lemma is an estimate which appeared in [1] and [20] in the case $\mu = I_\Omega$.

Lemma 3.126 (Cutting off a ball). *Suppose that $\mu \in \mathcal{M}_{\mathrm{cap}}^T(\mathbb{R}^d)$ is a capacitary measure of finite torsion. Then there is a dimensional constant C_d such that, for every $B_r(x_0) \subset \mathbb{R}^d$, we have*

$$\begin{aligned} d_\gamma\big(\mu, \mu \vee I_{B_r(x_0)^c}\big) \le &-\int_{B_r} |\nabla w_\mu|^2\, dx - \int_{B_r} w_\mu^2\, d\mu + 2\int_{B_r} w_\mu\, dx \\ &+ C_d\left(r + \frac{\|w_\mu\|_{L^\infty(B_{2r}(x_0))}}{r}\right)\int_{\partial B_r} w_\mu\, d\mathcal{H}^{d-1}. \end{aligned}$$

Proof. Without loss of generality, we can suppose that $x_0 = 0$. We denote with A_r the annulus $B_{2r} \setminus \overline{B_r}$.

Let $\psi : A_1 \to \mathbb{R}^+$ be the solution of the equation

$$\Delta\psi = 0 \quad \text{on} \quad A_1, \qquad \psi = 0 \quad \text{on} \quad \partial B_1, \qquad \psi = 1 \quad \text{on} \quad \partial B_2.$$

With $\phi : A_1 \to \mathbb{R}^+$ we denote the solution of the equation

$$-\Delta\phi = 1 \quad \text{on} \quad A_1, \qquad \phi = 0 \quad \text{on} \quad \partial B_1, \qquad \phi = 0 \quad \text{on} \quad \partial B_2.$$

For an arbitrary $r > 0$, $\alpha > 0$ and $k > 0$, we have that the solution v of the equation

$$-\Delta v = 1 \quad \text{on} \quad A_r, \qquad v = 0 \quad \text{on} \quad \partial B_r, \qquad v = \alpha \quad \text{on} \quad \partial B_{2r},$$

is given by

$$v(x) = r^2\phi(x/r) + \alpha\psi(x/r), \tag{3.90}$$

and its gradient is of the form

$$\nabla v(x) = r(\nabla\phi)(x/r) + \frac{\alpha}{r}(\nabla\psi)(x/r). \tag{3.91}$$

Let v be as in (3.90) with $\alpha \geq \|w_\mu\|_{L^\infty(B_{2r})}$. Consider the function $w = w_\mu \mathbb{1}_{B_{2r}^c} + (w_\mu \wedge v)\mathbb{1}_{B_{2r}}$ and note that, by the choice of α, we have that $w \in H^1(\mathbb{R}^d)$.

$$\begin{aligned}
d_\gamma(\mu, \mu \vee I_{B_r^c}) &= \int_{\mathbb{R}^d} \left(w_\mu - w_{\mu\vee I_{B_r^c}}\right) dx \\
&= 2\left(J_\mu(w_{\mu\vee I_{B_r^c}}) - J_\mu(w_\mu)\right) \leq 2(J_\mu(w_r) - J_\mu(w_\mu)) \\
&= -\int_{B_r} |\nabla w_\mu|^2\,dx - \int_{B_r} w_\mu^2\,d\mu + 2\int_{B_r} w_\mu\,dx \\
&\quad + \int_{A_r\cap\{w_\mu > v\}} |\nabla v|^2 - |\nabla w_\mu|^2\,dx \\
&\quad + \int_{A_r\cap\{w_\mu > v\}} (v^2 - w_\mu^2)\,d\mu - 2\int_{A_r\cap\{w_\mu > v\}} (v - w_\mu)\,dx \\
&\leq -\int_{B_r} |\nabla w_\mu|^2\,dx - \int_{B_r} w_\mu^2\,d\mu + 2\int_{B_r} w_\mu\,dx \\
&\quad - \int_{A_r\cap\{w_\mu > v\}} |\nabla(v - w_\mu)|^2\,dx \\
&\quad + 2\int_{A_r\cap\{w_\mu > v\}} \nabla v\cdot\nabla(v - w_\mu)\,dx - 2\int_{A_r\cap\{w_\mu > v\}} (v - w_\mu)\,dx \\
&\leq -\int_{B_r} |\nabla w_\mu|^2\,dx - \int_{B_r} w_\mu^2\,d\mu + 2\int_{B_r} w_\mu\,dx \\
&\quad + 2\int_{\partial B_r} w_\mu |\nabla v|\,d\mathcal{H}^{d-1},
\end{aligned} \tag{3.92}$$

which, taking in consideration (3.91) and the choice of α , proves the claim. □

Our next result is the capacitary measure version of [19, Lemma 3.6].

Lemma 3.127. *Suppose that $\mu, \mu' \in \mathcal{M}^T_{\text{cap}}(\mathbb{R}^d)$ are capacitary measures of finite torsion such that $\mu' \geq \mu$. Then, we have*

$$\|R_\mu - R_{\mu'}\|_{\mathcal{L}(L^2)} \leq C\left[d_\gamma(\mu,\mu')\right]^{(d-1)/d^2},$$

where C is a constant depending only on the dimension d and the torsion $T(\mu)$ (but not on μ').

Proof. The proof follows the same argument as in [19, Lemma 3.6] and we report it here for the sake of completeness. Let $f \in L^p$, $f \geq 0$, for some $p \geq d \geq 2$. Then

$$\begin{aligned}
&\int_{\mathbb{R}^d} |R_\mu(f) - R_{\mu'}(f)|^p\,dx \\
&\leq \|R_\mu(f) - R_{\mu'}(f)\|_{L^\infty}^{p-1} \int_{\mathbb{R}^d} \big(R_\mu(f) - R_{\mu'}(f)\big)\,dx \\
&\leq C^{p-1}\|f\|_{L^p}^{p-1} \int_{\mathbb{R}^d} f(w_\mu - w_{\mu'})\,dx \\
&\leq C^{p-1}\|f\|_{L^p}^{p}\|w_\mu - w_{\mu'}\|_{L^{p'}},
\end{aligned} \tag{3.93}$$

and so, $R_\mu - R_{\mu'}$ is a linear operator from L^p to L^p such that

$$\|R_\mu - R_{\mu'}\|_{\mathcal{L}(L^p;L^p)} \leq C^{1-1/p}\|w_\mu - w_{\mu'}\|_{L^{p'}}^{1/p},$$

where, by Proposition 3.54, the constant C depends on the dimension d and the torsion $T(\mu) = \|w_\mu\|_{L^1}$. Since $R_\mu - R_{\mu'}$ is a self-adjoint operator in L^2, we can extend it to an operator on $L^{p'}$. Indeed, let $f \in L^2 \cap L^{p'}$, where $p' = p/(p-1)$. Since $L^{p'}$ is the dual of L^p and $L^2 \cap L^p$ and and $L^2 \cap L^{p'}$ are dense respectively in L^p and $L^{p'}$, we have

$$\begin{aligned}
&\|R_\mu(f) - R_{\mu'}(f)\|_{L^{p'}} \\
&= \sup\left\{\int_{\mathbb{R}^d} \big(R_\mu(f) - R_{\mu'}(f)\big)g\,dx \;:\; g \in L^2 \cap L^p,\ \|g\|_{L^p} = 1\right\}.
\end{aligned}$$

On the other hand, by the self-adjointness of $R_\mu - R_{\mu'}$ in L^2, for f and g as above, we have

$$\begin{aligned}
&\int_{\mathbb{R}^d} \big(R_\mu(f) - R_{\mu'}(f)\big)g\,dx = \int_{\mathbb{R}^d} \big(R_\mu(g) - R_{\mu'}(g)\big)f\,dx \\
&\leq \|R_\mu(g) - R_{\mu'}(g)\|_{L^p}\|f\|_{L^{p'}} \leq C^{1-1/p}\|w_\mu - w_{\mu'}\|_{L^{p'}}^{1/p}\|g\|_{L^p}\|f\|_{L^{p'}},
\end{aligned}$$

which gives that for every $f \in L^2 \cap L^{p'}$

$$\|R_\mu(f) - R_{\mu'}(f)\|_{L^{p'}} \leq C^{1-1/p}\|w_\mu - w_{\mu'}\|_{L^{p'}}^{1/p}\|f\|_{L^{p'}},$$

and so $R_\mu - R_{\mu'}$ can be extended to a linear operator on $L^{p'}$ such that

$$\|R_\mu - R_{\mu'}\|_{\mathcal{L}(L^{p'};L^{p'})} \leq C^{1-1/p}\|w_\mu - w_{\mu'}\|_{L^{p'}}^{1/p}.$$

By the classical Riesz-Thorin interpolation theorem we get

$$\begin{aligned}|R_\mu - R_{\mu'}\|_{\mathcal{L}(L^2)} &\leq C^{1-1/p}\|w_\mu - w_{\mu'}\|_{L^{p'}}^{1/p} \\ &\leq C^{1-1/p}\|w_\mu\|_\infty^{1/p^2}\|w_\mu - w_{\mu'}\|_{L^1}^{(p-1)/p^2}.\end{aligned}$$

Now using the L^∞ estimate on w_μ, and taking $p = d$, we have the claim. □

The following two results appeared respectively in [26] and [20]. We note that Lemma 3.128 is just a slight improvement of [20, Lemma 3], but is one of the crucial steps in the proof of existence of optimal measures for spectral-torsion functionals. We recall the notation $\Lambda_k(\mu) := 1/\lambda_k(\mu)$ for the kth eigenvalue of the resolvent operator R_μ associated to a capacitary measure μ.

Lemma 3.128. *Let $\mu \in \mathcal{M}_{\text{cap}}^T(\mathbb{R}^d)$ is a capacitary measure of finite torsion in $\mathbb{R}^d$. Then for every capacitary measure $\nu \geq \mu$ and every $k \in \mathbb{N}$, we have*

$$\Lambda_j(\mu) - \Lambda_j(\nu) \leq k^2 e^{\frac{1}{4\pi}}\lambda_k(\mu)^{\frac{d+4}{2}}\int_{\mathbb{R}^d}\big(R_\mu(w_\mu)w_\mu - R_\nu(w_\mu)w_\mu\big)\,dx. \quad (3.94)$$

Proof. Consider the orthonormal in $L^2(\mathbb{R}^d)$ family of eigenfunctions $u_1, \dots, u_k \in H_\mu^1$ corresponding to the compact self-adjoint operator $R_\mu : L^2(\mathbb{R}^d) \to L^2(\mathbb{R}^d)$. Let $P_k : L^2(\mathbb{R}^d) \to L^2(\mathbb{R}^d)$ be the projection

$$P_k(u) = \sum_{j=1}^k \left(\int_{\mathbb{R}^d} u u_j\,dx\right) u_j.$$

Consider the linear space $V = Im(P_k)$, generated by $u_1, \dots, u_k$ and the operators T_μ and T_ν on V, defined by

$$T_\mu = P_k \circ R_\mu \circ P_k \qquad \text{and} \qquad T_\nu = P_k \circ R_\nu \circ P_k.$$

It is immediate to check that $u_1, \dots, u_k$ and $\Lambda_1(\mu), \dots, \Lambda_1(\mu)$ are the eigenvectors and the corresponding eigenvalues of T_μ. On the other hand, for the eigenvalues $\Lambda_1(T_\nu), \dots, \Lambda_k(T_\nu)$ of T_ν, we have the inequality

$$\Lambda_j(T_\nu) \leq \Lambda_j(\nu), \quad \forall j = 1, \dots, k. \quad (3.95)$$

Indeed, by the min-max Theorem we have

$$\begin{aligned}
\Lambda_j(T_\nu) &= \min_{V_j\subset V}\ \max_{u\in V, u\perp V_j} \frac{\langle P_k\circ R_\nu\circ P_k(u), u\rangle_{L^2}}{\|u\|_{L^2}^2}\\
&= \min_{V_j\subset L^2}\ \max_{u\in V, u\perp V_j} \frac{\langle R_\nu(u), u\rangle_{L^2}}{\|u\|_{L^2}^2}\\
&\le \min_{V_j\subset L^2}\ \max_{u\in L^2, u\perp V_j} \frac{\langle R_\nu(u), u\rangle_{L^2}}{\|u\|_{L^2}^2} = \Lambda_j(\nu),
\end{aligned}$$

where with V_j we denotes a generic $(j-1)$-dimensional subspaces of $L^2(\mathbb{R}^d)$. Thus, we have the estimate

$$0 \le \Lambda_j(\mu) - \Lambda_j(\nu) \le \Lambda_j(T_\mu) - \Lambda_j(T_\nu) \le \|T_\mu - T_\nu\|_{\mathcal{L}(V)}, \quad (3.96)$$

and on the other hand

$$\begin{aligned}
\|T_\mu - T_\nu\|_{\mathcal{L}(V)} &= \sup_{u\in V} \frac{\langle (T_\mu - T_\nu)u, u\rangle_{L^2}}{\|u\|_{L^2}^2}\\
&= \sup_{u\in V} \frac{\langle (R_\mu - R_\nu)u, u\rangle_{L^2}}{\|u\|_{L^2}^2}\\
&= \sup_{u\in V} \frac{1}{\|u\|_{L^2}^2}\int_{\mathbb{R}^d} \big(R_\mu(u) - R_\nu(u)\big)u\,dx.
\end{aligned} \quad (3.97)$$

Let $u\in V$ be the function for which the supremum in the right hand side of (3.97) is achieved. We can suppose that $\|u\|_{L^2} = 1$, *i.e.* that there are real numbers $\alpha_1,\dots,\alpha_k$, such that

$$u = \alpha_1 u_1 + \dots + \alpha_k u_k, \qquad \text{where} \quad \alpha_1^2 + \dots + \alpha_k^2 = 1.$$

Thus, we have

$$\begin{aligned}
\|T_\mu - T_\nu\|_{\mathcal{L}(V)} &\le \int_{\mathbb{R}^d} \big|R_\mu(u) - R_\nu(u)\big|\cdot|u|\,dx\\
&\le \int_{\mathbb{R}^d} \left|\sum_{j=1}^k \alpha_j\big(R_\mu(u_j) - R_\nu(u_j)\big)\right|\cdot\left(\sum_{j=1}^k |u_j|\right)dx\\
&\le \int_{\mathbb{R}^d} \left(\sum_{j=1}^k \big|(R_\mu(u_j) - R_\nu(u_j)\big|\right)\cdot\left(\sum_{j=1}^k |u_j|\right)dx\\
&\le \int_{\mathbb{R}^d} \left(\sum_{j=1}^k \big(R_\mu(|u_j|) - R_\nu(|u_j|)\big)\right)\cdot\left(\sum_{j=1}^k |u_j|\right)dx,
\end{aligned} \quad (3.98)$$

where the last inequality is due to the linearity and the positivity of $R_\mu - R_\nu$. We now recall that by Corollary 3.95, we have $\|u_j\|_{L^\infty} \le e^{\frac{1}{8\pi}}\lambda_k(\mu)^{d/4}$, for every $j = 1, \dots, k$. By the weak maximum principle applied to u_j and w_μ, we have

$$|u_j| \le e^{\frac{1}{8\pi}}\lambda_k(\mu)^{\frac{d+4}{4}} w_\mu, \quad \text{for every} \quad 1 \le j \le k. \tag{3.99}$$

Using against the positivity of $R_\mu - R_\nu$ and substituting (3.99) in (3.98) we obtain the claim. □

Lemma 3.129. *Let μ be a capacitary measure such that $w_\mu \in L^1(\mathbb{R}^d)$. Then for every capacitary measure $\nu \ge \mu$ and every $k \in \mathbb{N}$, we have*

$$\Lambda_j(\mu) - \Lambda_j(\nu) \le C d_\gamma(\mu, \nu), \tag{3.100}$$

for every $0 < j \le k$, where C is a constant depending only on $\lambda_k(\mu)$ and the dimension d.

Proof. Reasoning as in Lemma 3.128, by (3.96) and (3.98), for each $j = 1, \dots, k$, we have

$$\begin{aligned}
\Lambda_j(\mu) - \Lambda_j(\nu) &\le \int_{\mathbb{R}^d} \left(\sum_{i=1}^k \big(R_\mu(|u_i|) - R_\nu(|u_i|)\big)\right) \cdot \left(\sum_{j=i}^k |u_i|\right) dx \\
&\le \left(\sum_{j=i}^k \|u_i\|_{L^\infty}\right)^2 \int_{\mathbb{R}^d} (w_\mu - w_\nu)\, dx
\end{aligned}$$

where $u_i \in H^1_\mu$ are the normalized eigenfunctions of $-\Delta + \mu$. Now the claim follows by the estimate from Corollary 3.95. □

3.7.2. The concentration-compactness principle

In this subsection, we finally state the version for capacitary measures of the concentration-compactness principle, which was proved in [26] and is based on the ideas for the analogous result for domains, originally proved in [19] for quasi-open sets. Our main tools for determining the behaviour of a sequence of capacitary measures are the estimates from the previous subsection.

In the theorem below we will use the notion of infimum of two capacitary measures μ and ν with disjoint sets of finiteness, *i.e.* $\mathrm{cap}(\Omega_\mu \cap \Omega_\nu) = 0$, namely

$$\mu \wedge \nu(E) = \begin{cases} \mu(\Omega_\mu \cap E) + \nu(\Omega_\nu \cap E), & \text{if } \mathrm{cap}\big(E \setminus (\Omega_\mu \cup \Omega_\nu)\big) = 0, \\ +\infty, & \text{if } \mathrm{cap}\big(E \setminus (\Omega_\mu \cup \Omega_\nu)\big) > 0. \end{cases}$$

Theorem 3.130. *Suppose that μ_n is a sequence of capacitary measures in $\mathbb{R}^d$ such that the corresponding sequence of energy functions w_{μ_n} has uniformly bounded $L^1(\mathbb{R}^d)$ norms. Then, up to a subsequence, one of the following situations occurs:*

(i1) *(Compactness) The sequence μ_n γ-converges to some $\mu \in \mathcal{M}_{\text{cap}}^T(\mathbb{R}^d)$.*
(i2) *(Compactness at infinity) There is a sequence $x_n \in \mathbb{R}^d$ such that $|x_n| \to \infty$ and $\mu_n(x_n + \cdot)$ γ-converges to some $\mu \in \mathcal{M}_{\text{cap}}^T(\mathbb{R}^d)$.*
(ii) *(Vanishing) The sequence μ_n does not γ-converge to the measure $\infty = I_\emptyset$, but the sequence of resolvents R_{μ_n} converges to zero in the strong operator topology of $\mathcal{L}(L^2(\mathbb{R}^d))$. Moreover, we have $\|w_{\mu_n}\|_{L^\infty} \to 0$ and $\lambda_1(\mu_n) \to +\infty$, as $n \to \infty$.*
(iii) *(Dichotomy) There are capacitary measures μ_n^1 and μ_n^2 such that:*

- $\text{dist}(\Omega_{\mu_n^1}, \Omega_{\mu_n^2}) \to \infty$, *as* $n \to \infty$;
- $\mu_n \leq \mu_n^1 \wedge \mu_n^2$, *for every* $n \in \mathbb{N}$;
- $d_\gamma(\mu_n, \mu_n^1 \wedge \mu_n^2) \to 0$, *as* $n \to \infty$;
- $\|R_{\mu_n} - R_{\mu_n^1 \wedge \mu_n^2}\|_{\mathcal{L}(L^2)} \to 0$, *as* $n \to \infty$;
- $\liminf_{n\to\infty} T(\mu_n^1) > 0$ *and* $\liminf_{n\to\infty} T(\mu_n^2) > 0$.

Proof. Consider the sequence $w_n := w_{\mu_n}$, which is bounded in $H^1(\mathbb{R}^d) \cap L^1(\mathbb{R}^d)$. We now apply the concentration compactness principle (Theorem 3.4) to the sequence w_n.

If the concentration (Theorem 3.4 (1)) occurs, then by the compactness of the embedding $H^1(\mathbb{R}^d) \subset L^1_{\text{loc}}(\mathbb{R}^d)$, up to a subsequence $w_n(\cdot + x_n)$ is concentrated in $L^1(\mathbb{R}^d)$ for some sequence $x_n \in \mathbb{R}^d$. If x_n has a bounded subsequence, then w_n converges (up to a subsequence) in $L^1(\mathbb{R}^d)$ and so, we have (i1). If $|x_n| \to \infty$, we directly obtain (i2).

Suppose now that the *vanishing* (Theorem 3.4 (2)) holds. We prove that (ii) holds. Let $\varphi \in C_c^\infty(\mathbb{R}^d)$ and let $\varepsilon > 0$. We choose $R > \varepsilon^{-d^2/2(d-1)}$ large enough and $N \in \mathbb{N}$ such that for every $n \geq N$, we have

$$\int_{B_R} w_n \, dx \leq \varepsilon^{d^2/(d-1)}.$$

By Lemma 3.123 and Lemma 3.127, we have

$$\|R_{\mu_n}(\varphi) - R_{\mu_n \vee I_{B_R}}(\varphi)\|_{L^2} \leq C\varepsilon\|\varphi\|_{L^2}$$

for some constant C, and by the vanishing property,

$$\|R_{\mu_n \vee I_{B_R}}(\varphi)\|_{L^2} \leq C\varepsilon\|\varphi\|_{L^2}.$$

Thus,

$$\|R_{\mu_n}(\varphi)\|_{L^2} \le \|R_{\mu_n}(\varphi) - R_{\mu_n \vee I_{B_R}}(\varphi)\|_{L^2} + \|R_{\mu_n \vee I_{B_R}}(\varphi)\|_{L^2} \le C\varepsilon\|\varphi\|_{L^2},$$

and we obtain the strong convergence in (ii).

We now prove that $\|w_n\|_{L^\infty} \to 0$. Suppose by contradiction that there is $\delta > 0$ and a sequence $x_n \in \mathbb{R}^d$ such that $w_n(x_n) > \delta$. Since $\Delta w_n + 1 \ge 0$ on $\mathbb{R}^d$ (by Lemma 3.66), we have that the function

$$x \mapsto w_n(x) - \frac{r^2 - |x - x_n|^2}{2d},$$

is subharmonic. Thus, choosing $r = \sqrt{d\delta}$, we have

$$\int_{B_r(x_n)} w_n \, dx \ge w_n(x_n) - \frac{r^2}{2d} \ge \delta/2,$$

which contradicts Theorem 3.4 (2).

Let $u_n \in H^1_{\mu_n}$ be the first, normalized in $L^2(\mathbb{R}^d)$, eigenfunction for the operator $-\Delta + \mu_n$. By Corollary 3.95, we have

$$-\Delta u_n + \mu_n u_n = \lambda_1(\mu_n) u_n \le \lambda_1(\mu_n)\|u_n\|_{L^\infty} \le e^{1/(8\pi)} \lambda_1(\mu_n)^{(d+4)/4}.$$

Suppose that the sequence $\lambda_1(\mu_n)$ is bounded. Then by the weak maximum principle (see [32, Proposition 3.4]) we have $u_n \le C w_n$, for some constant C. Thus, we have

$$1 = \int_{\mathbb{R}^d} u_n^2 \, dx \le C^2 \int_{\mathbb{R}^d} w_n^2 \, dx \le C^2 \|w_n\|_{L^\infty} \|w_n\|_{L^1} \to 0,$$

which is a contradiction.

Suppose that the *dichotomy* (Theorem 3.4 (3)) occurs. Choose $\alpha = 8$ and let $x_n \in \mathbb{R}^d$ and $R_n \to \infty$ be as in Theorem 3.4 (3). Setting

$$\mu_n^1 = \mu_n \vee I_{B_{2R_n}(x_n)} \qquad \text{and} \qquad \mu_n^2 = \mu_n \vee I_{B_{4R_n}(x_n)^c},$$

we have that μ_n^1 and μ_n^2 have disjoint sets of finiteness and it is immediate to check that

$$\mu_n^1 \wedge \mu_n^2 = \mu_n \vee I_{B_{2R_n}(x_n) \cup B^c_{4R_n}(x_n)}.$$

Since $R_n \to +\infty$, by the estimate (3.86) from Lemma 3.123, we obtain

$$\lim_{n\to\infty} d_\gamma(\mu_n, \mu_n^1 \wedge \mu_n^2) = 0.$$

By Lemma 3.127, we have

$$\liminf_{n\to\infty} \|R_{\mu_n} - R_{\mu_n^1 \wedge \mu_n^2}\|_{\mathcal{L}(L^2(\mathbb{R}^d))} \le C \lim_{n\to\infty} d_\gamma(\mu_n, \mu_n^1 \wedge \mu_n^2)^{(d-1)/d^2} = 0,$$

where C is a constant depending on the dimension and on $\sup_n T(\mu_n)$. For last claim of (iii) we note that by Theorem 3.4 (iii)

$$\liminf_{n\to\infty}\int_{B_{2R_n}(x_n)} w_n\,dx > 0 \qquad \text{and} \qquad \liminf_{n\to\infty}\int_{B^c_{4R_n}(x_n)} w_n\,dx,$$

and, on the other hand, by Lemma 3.123 we have

$$0 \le \int_{B_{2R_n}(x_n)} w_n\,dx - \int_{\mathbb{R}^d} w_{\mu_n^1}\,dx \le \int_{B_{8R_n}\setminus B_{R_n}} w_n\,dx + CR_n^{-2},$$
$$0 \le \int_{B^c_{4R_n}(x_n)} w_n\,dx - \int_{\mathbb{R}^d} w_{\mu_n^2}\,dx \le \int_{B_{8R_n}\setminus B_{R_n}} w_n\,dx + CR_n^{-2},$$

for some constant $C > 0$, which gives the claim since the right-hand side of both inequalities converges to zero as $n \to +\infty$. □

In the case when the measures μ_n have the specific forms $\mu_n = \widetilde{I}_{\Omega_n}$ or $\mu_n = I_{\Omega_n}$, we have the following result, which appeared for the first time in [19] and later in [24], where the perimeter was included as a variable. This result was also one of the fundamental tools in the proof of the existence of optimal sets for spectral functionals with perimeter constraint in [59].

Theorem 3.131. *Suppose that Ω_n is a sequence of measurable sets of uniformly bounded measure. Then, up to a subsequence, one of the following situations occur:*

(1a) *The sequence Ω_n γ-converges*[9] *to a capacitary measure $\mu \in \mathcal{M}^T_{\mathrm{cap}}(\mathbb{R}^d)$ and the sequence $1\!\!1_{\Omega_n} \in L^1(\mathbb{R}^d)$ is concentrated.*

(1b) *There is a sequence $x_n \in \mathbb{R}^d$ such that $|x_n| \to \infty$ and $x_n + \Omega_n$ γ-converges and the sequence $1\!\!1_{\Omega_n}(\cdot + x_n) \in L^1(\mathbb{R}^d)$ is concentrated.*

(2) *$\widetilde{\lambda}_1(\Omega_n) \to +\infty$, as $n \to \infty$.*

(3) *There are measurable sets Ω_n^1 and Ω_n^2 such that:*

- *$\mathrm{dist}(\Omega_n^1, \Omega_n^2) \to \infty$, as $n \to \infty$;*
- *$\Omega_n^1 \cup \Omega_n^2 \subset \Omega_n$, for every $n \in \mathbb{N}$;*
- *$d_\gamma\big(\widetilde{I}_{\Omega_n}, \widetilde{I}_{\Omega_n^1\cup\Omega_n^2}\big) \to 0$, as $n \to \infty$;*

[9] We recall that when we deal with sets Ω_n which are only measurable, the term γ-convergence refers to the sequence of capacitary measures $\widetilde{I}_{\Omega_n}$. On the other hand, we say that a sequence of quasi-open sets Ω_n γ-converges, if the sequence of measures I_{Ω_n} γ-converges.

- $\|R_{\Omega_n} - R_{\Omega_n^1 \cup \Omega_n^2}\|_{\mathcal{L}(L^2)} \to 0$, *as* $n \to \infty$;
- $\liminf_{n\to\infty} |\Omega_n^1| > 0$ *and* $\liminf_{n\to\infty} |\Omega_n^2| > 0$;
- *if* $P(\Omega_n) < +\infty$, *for every* $n \in \mathbb{N}$, *then*

$$\limsup_{n\to\infty} \big(P(\Omega_n^1) + P(\Omega_n^2) - P(\Omega_n)\big) = 0.$$

Proof. Let $w_n := w_{\Omega_n}$. By Corollary 3.52, we have $\|w_n\|_{L^1} \leq C$ for some universal constant C and so the sequences $\|w_n\|_{H^1}$ and $\|w_n\|_{L^\infty}$ are also bounded. We now apply the concentration compactness principle to the sequence of characteristic functions $\mathbb{1}_{\Omega_n}$.

If the *concentration* (Theorem 3.4 (1)) occurs, then the sequence $w_n \leq \|w_n\|_{L^\infty} \mathbb{1}_{\Omega_n}$ is also concentrated and so we have (1a) or (1b) as in Theorem 3.130.

If the *vanishing* (Theorem 3.4 (2)) occurs, then the vanishing holds also for the sequence $w_n \in L^1(\mathbb{R}^d)$. Thus, by Theorem 3.130 (ii) and the fact that $\|R_{\widetilde{I}_{\Omega_n}}\|_{\mathcal{L}(L^2(\mathbb{R}^d))} = \widetilde{\lambda}_1(\Omega_n)$, we obtain (2).

If the *dichotomy* (Theorem 3.4 (3)) occurs, then it holds also for the sequence $w_n \in L^1(\mathbb{R}^d)$. Thus, applying Theorem 3.130, we obtain all the claims in (3) but the last one. For the latter it is sufficient to note that one can take in Theorem 3.130 (iii), the sequence

$$\Omega_n^1 = \Omega_n \cap B_{R_n+\varepsilon}(x_n) \qquad \text{and} \qquad \Omega_n^2 = \Omega_n \setminus B_{8R_n-\varepsilon}(x_n),$$

for every $\varepsilon > 0$ small enough. Thus, choosing $\varepsilon > 0$ such that

$$\mathcal{H}^{d-1}\big(\partial^*\Omega_n \cap \partial B_{R_n+\varepsilon}(x_n)\big) = \mathcal{H}^{d-1}\big(\partial^*\Omega_n \cap \partial B_{8R_n-\varepsilon}(x_n)\big) = 0,$$

we have the claim. □

Remark 3.132. The same result holds if Ω_n is a sequence of quasi-open sets of uniformly bounded measure. In this case we apply Theorem 3.130 to the sequence of measures $\mu_n = I_{\Omega_n}$ and then proceed as in the proof of Theorem 3.131.

Chapter 4
Subsolutions of shape functionals

4.1. Introduction

In this chapter we consider domains (quasi-open or measurable sets) $\Omega \subset \mathbb{R}^d$, which are optimal for a given functional $\mathcal{F}$ only with respect to internal perturbations, *i.e.*

$$\mathcal{F}(\Omega) \leq \mathcal{F}(\omega), \text{ for every } \omega \subset \Omega. \tag{4.1}$$

We call the domains Ω satisfying (4.6) *subsolutions* for the functional $\mathcal{F}$. The subsolutions are a powerful tool in the study of many shape optimization problems. They naturally appear, for example, in the following situations:

- *Obstacle problems*. If $\mathcal{D} \subset \mathbb{R}^d$ is a given set (a box) and $\Omega \subset \mathcal{D}$ is a solution of the problem

$$\min\Big\{\mathcal{F}(\Omega)\,:\,\Omega \subset \mathcal{D}\Big\}, \tag{4.2}$$

  then Ω is a subsolution for $\mathcal{F}$.
- *Optimal partition problems*. If the domain $\mathcal{D} \subset \mathbb{R}^d$ is a given set (a box) and the couple (Ω_1, Ω_2) is a solution of the problem

$$\min\Big\{\mathcal{F}(\Omega_1) + \mathcal{F}(\Omega_2)\,:\,\Omega_1, \Omega_2 \subset \mathcal{D},\ \Omega_1 \cap \Omega_2 = \emptyset\Big\}, \tag{4.3}$$

  then each of the sets Ω_1 and Ω_2 is a subsolution for $\mathcal{F}$.
- *Change of the functional*. If the set $\Omega \subset \mathbb{R}^d$ is a solution of the problem

$$\min\left\{\mathcal{G}(\Omega)\,:\,\Omega \subset \mathbb{R}^d\right\}, \tag{4.4}$$

  and the functional $\mathcal{F}$ is such that

$$\mathcal{G}(\Omega) - \mathcal{G}(\omega) \geq \mathcal{F}(\Omega) - \mathcal{F}(\omega), \text{ for every } \omega \subset \Omega,$$

  then the sets Ω is a subsolution for $\mathcal{F}$.

This last case is particularly useful when the functional $\mathcal{G}$ depends in a non trivial way on the domain Ω. One may take for example $\mathcal{G}$ to be any function of the spectrum of Ω. In this case extracting information on the domain Ω, solution of (4.4), might be very difficult. Thus, it is convenient to search for a functional $\mathcal{F}$, which is easier to treat from the technical point of view.

If $\mathcal{F}$ is a decreasing functional with respect to the set inclusion, then every set $\Omega \subset \mathbb{R}^d$ is a subsolution for $\mathcal{F}$. Of course, we are interested in functionals which will allow us to extract some information on the subsolutions. Typical examples are the combinations of increasing and decreasing functionals as, for example, $\mathcal{F}(\Omega) = \lambda_1(\Omega) + |\Omega|$.

In many cases, the subsolution property (4.6) holds only for small perturbations of the domain Ω. In these cases, we will say that Ω is a local subsolution.

Definition 4.1 (Shape subsolutions in the class of Lebesgue measurable sets). Let $\mathcal{F}$ be a functional on the family $\mathcal{B}(\mathbb{R}^d)$ of Borel sets in $\mathbb{R}^d$ we will say that the set $\Omega \in \mathcal{B}(\mathbb{R}^d)$

- is a **local subsolution with respect to the Lebesgue measure**, if there is $\varepsilon > 0$ such that

$$\mathcal{F}(\Omega) \leq \mathcal{F}(\omega), \quad \forall \omega \subset \Omega \quad \text{such that} \quad |\Omega \setminus \omega| < \varepsilon.$$

- is a **local subsolution with respect to the distance** d_γ, if there is $\varepsilon > 0$ such that

$$\mathcal{F}(\Omega) \leq \mathcal{F}(\omega), \quad \forall \omega \subset \Omega \quad \text{such that} \quad d_\gamma(\widetilde{I}_\omega, \widetilde{I}_\Omega) < \varepsilon.$$

- is a **subsolution in** $D \subset \mathbb{R}^d$, if we have

$$\mathcal{F}(\Omega) \leq \mathcal{F}(\omega), \quad \forall \omega \subset \Omega \quad \text{such that} \quad \Omega \setminus \omega \subset D.$$

In this chapter we consider subsolutions for spectral and energy functionals. Before we start investigating the properties of these domains, we give an example of a well-studied functional, which suggests what can we expect from the shape subsolutions.

Example 4.2. Let $\mathcal{F}(\Omega) := P(\Omega)|\Omega|^{-1}$, for every measurable $\Omega \subset \mathbb{R}^d$, where with $P(\Omega)$ we denote the De Giorgi perimeter of Ω. If Ω is a (local with respect to the Lebesgue measure) shape subsolution for $\mathcal{F}$, then a standard argument gives that

1. Ω is a bounded set;
2. Ω has an internal density estimate.

Nevertheless, we cannot expect, in general, that Ω has any regularity property. Indeed, if Ω is the solution of

$$\min\Big\{\mathcal{F}(\Omega):\Omega\subset\mathcal{D}\Big\},\tag{4.5}$$

where $\mathcal{D}$ is a set with empty interior, then Ω is not even (equivalent to) an open set.

The notion of a shape subsolution with respect to a functional $\mathcal{F}$ depends on the domain of definition of $\mathcal{F}$. One can easily define shape subsolutions in the class of open sets, sets with smooth boundary, quasi-open sets, etc.

Definition 4.3 (Shape subsolutions in the class of quasi-open sets). Let $\mathcal{F}:\mathcal{A}_{\text{cap}}(\mathbb{R}^d)\to\mathbb{R}$ be a functional on the family of quasi-open sets $\mathcal{A}_{\text{cap}}(\mathbb{R}^d)$.

- We say that the quasi-open set Ω is a **shape subsolution for** $\mathcal{F}:\mathcal{A}_{\text{cap}}(\mathbb{R}^d)\to\mathbb{R}$, if

$$\mathcal{F}(\Omega)\le\mathcal{F}(\omega),\quad\forall\text{ quasi-open }\quad\omega\subset\Omega.\tag{4.6}$$

- We say that the quasi-open set Ω is a **local shape subsolution for** $\mathcal{F}:\mathcal{A}_{\text{cap}}(\mathbb{R}^d)\to\mathbb{R}$, **if there is $\varepsilon>0$ such that**

$$\mathcal{F}(\Omega)\le\mathcal{F}(\omega),\quad\forall\text{ quasi-open }\omega\subset\Omega\text{ such that }d_\gamma(\Omega,\omega)<\varepsilon.\tag{4.7}$$

Remark 4.4. Suppose that $\mathcal{F}$ is a functional on the class of Borel sets. If $\Omega\subset\mathbb{R}^d$ is a quasi-open set, which is a shape subsolution for $\mathcal{F}:\mathcal{B}(\mathbb{R}^d)\to\mathbb{R}$, then Ω is also a shape subsolution for the same functional restricted on the class of quasi-open set $\mathcal{F}:\mathcal{A}_{\text{cap}}(\mathbb{R}^d)\to\mathbb{R}$.

Remark 4.5. Suppose that the functional $\mathcal{F}:\mathcal{B}(\mathbb{R}^d)\to\mathbb{R}$ is of the form

$$\mathcal{F}(\Omega)=\Phi\big(H^1_0(\Omega)\big)+\mathcal{G}(\Omega),$$

where Φ is a functional on the closed subspaces of $H^1(\mathbb{R}^d)$ and $\mathcal{G}:\mathcal{B}(\mathbb{R}^d)\to\mathbb{R}$ is an increasing functional with respect to the set inclusion (defined up to sets of zero capacity). Let $\Omega\in\mathcal{B}(\mathbb{R}^d)$ be a shape subsolution for $\mathcal{F}$. Then, there is a quasi-open set $\omega\subset\Omega$ a.e. such that $\mathcal{F}(\omega)=\mathcal{F}(\Omega)$ and ω is a shape subsolution for $\mathcal{F}:\mathcal{A}_{\text{cap}}(\mathbb{R}^d)\to\mathbb{R}$. Indeed, there is a quasi-open set ω such that $\text{cap}(\omega\setminus\Omega)=0$ and $H^1_0(\Omega)=H^1_0(\omega)$. Now the claim follows by the definition of subsolution. An analogous result holds, if $\mathcal{F}$ is of the form

$$\mathcal{F}(\Omega)=\Phi\big(\widetilde{H}^1_0(\Omega)\big)+\mathcal{G}(\Omega),$$

for Φ is as above and $\mathcal{G}$ is an increasing functional with respect to the set inclusion (defined up to sets of zero measure). Indeed, it is sufficient to note that there is a quasi-open set ω such that $|\omega \setminus \Omega| = 0$ and $\widetilde{H}_0^1(\Omega) = \widetilde{H}_0^1(\omega) = H_0^1(\omega)$. Thus, ω is a subsolution for the functional $\mathcal{F}' : \mathcal{A}_{\text{cap}}(\mathbb{R}^d) \to \mathbb{R}$ defined as

$$\mathcal{F}'(\Omega) = \Phi\big(H_0^1(\Omega)\big) + \mathcal{G}(\Omega).$$

Remark 4.6 (Subsolutions in the space of capacitary measures). The notion of a subsolution can be extended in a natural way to the family of capacitary measures. Indeed, we say that the capacitary measure $\mu \in \mathcal{M}_{\text{cap}}^T(\mathbb{R}^d)$ is a subsolution for the functional $\mathcal{F} : \mathcal{M}_{\text{cap}}^T(\mathbb{R}^d) \to \mathbb{R}^d$, if we have

$$\mathcal{F}(\mu) \le \mathcal{F}(\nu), \quad \text{for every capacitary measure } \nu \ge \mu. \tag{4.8}$$

In this case the recovery of information on the set of finiteness Ω_μ can be easily reduced to the study of the shape subsolutions of the shape functional $\mathcal{G} : \mathcal{A}_{\text{cap}}(\mathbb{R}^d) \to \mathbb{R}$ defined as

$$\mathcal{G}(\Omega) := \mathcal{F}\big(\mu \vee I_\Omega\big).$$

Indeed, if the capacitary measure μ is a subsolution for $\mathcal{F}$, then the (quasi-open) set of finiteness Ω_μ is a shape subsolution for the functional $\mathcal{G}$, since for every quasi-open $\omega \subset \Omega_\mu$

$$\mathcal{G}(\Omega_\mu) = \mathcal{F}(\mu) \le \mathcal{F}\big(\mu \vee I_\omega\big) = \mathcal{G}(\omega).$$

4.2. Shape subsolutions for the Dirichlet Energy

We shall use throughout this section the notions of a measure theoretic closure $\overline{\Omega}^M$ and a measure theoretic boundary $\partial^M\Omega$ of a Lebesgue measurable set $\Omega \subset \mathbb{R}^d$, which are defined as:

$$\begin{aligned}
\overline{\Omega}^M &= \big\{x \in \mathbb{R}^d : \ |B_r(x) \cap \Omega| > 0, \quad \text{for every} \quad r > 0\big\},\\
\partial^M\Omega &= \big\{x \in \mathbb{R}^d : \ |B_r(x) \cap \Omega| > 0 \quad \text{and} \quad |B_r(x) \cap \Omega^c| > 0,\\
&\qquad\qquad\qquad\qquad\qquad\qquad\qquad \text{for every} \quad r > 0\big\}.
\end{aligned}$$

Moreover, for every $0 \le \alpha \le 1$, we define the set of points of density α as

$$\Omega_{(\alpha)} = \left\{x \in \mathbb{R}^d : \ \lim_{r\to 0} \frac{|B_r(x) \cap \Omega|}{|B_r|} = \alpha\right\}.$$

We recall that, if Ω has finite perimeter in sense of De Giorgi, *i.e.* the distributional gradient $\nabla\mathbb{1}_\Omega$ is a measure of finite total variation $|\nabla\mathbb{1}_\Omega|(\mathbb{R}^d)< +\infty$, then the generalized perimeter of Ω is given by

$$P(\Omega) = |\nabla\mathbb{1}_\Omega|(\mathbb{R}^d) = \mathcal{H}^{d-1}(\partial^*\Omega),$$

where $\partial^*\Omega$ is the reduced boundary of Ω (see for example [67]).

Let $\Omega \subset \mathbb{R}^d$ be a measurable set of finite Lebesgue measure $|\Omega| < +\infty$ and let $f \in L^2(\Omega)$ be a given function. We recall that the Sobolev space over Ω is defined as

$$H_0^1(\Omega) = \left\{u \in H^1(\mathbb{R}^d) : u = 0 \text{ q.e. on } \Omega^c\right\}.$$

The function $u \in H_0^1(\Omega)$ is a solution of the boundary value problem

$$-\Delta u = f \quad \text{in} \quad \Omega, \qquad u \in H_0^1(\Omega), \tag{4.9}$$

if u minimizes the functional $J_f : H_0^1(\Omega) \to \mathbb{R}$, where for every $v \in H_0^1(\Omega)$

$$J_f(v) := \frac{1}{2}\int_{\mathbb{R}^d} |\nabla u|^2\,dx - \int_{\mathbb{R}^d} uf\,dx.$$

We note that, for every $f \in L^2(\Omega)$, a solution u of (4.9) exists and is unique. Moreover, for every $v \in H_0^1(\Omega)$ we have

$$\int_{\mathbb{R}^d} \nabla u \cdot \nabla v\,dx = \int_{\mathbb{R}^d} vf\,dx,$$

and, taking $v = u$, we get

$$\min_{v\in H_0^1(\Omega)} J_f(v) = J_f(u) = -\frac{1}{2}\int_{\mathbb{R}^d} uf\,dx =: E_f(\Omega). \tag{4.10}$$

In the case when $f \equiv 1$, we denote with w_Ω the solution of (4.9) and with $E(\Omega)$ the quantity $E_1(\Omega)$. We call $E(\Omega)$ the *Dirichlet energy* and w_Ω the *energy* (or *torsion*) *function* of Ω. In the Remark below, we list a few properties of w_Ω which were proved in Section 3.4.

Remark 4.7. Suppose that $\Omega \subset \mathbb{R}^d$ is a set of finite measure and that $w_\Omega \in H_0^1(\Omega)$ is the energy function of Ω. Then we have

(a) w_Ω is bounded and

$$\|w_\Omega\|_{L^\infty} \le \frac{|\Omega|^{2/d}}{2d|B_1|^{2/d}},$$

where B_1 is the unit ball in $\mathbb{R}^d$.

(b) $\Delta w_\Omega + \mathbb{1}_{\{w_\Omega>0\}} \ge 0$ in sense of distributions on $\mathbb{R}^d$.
(c) Every point of $\mathbb{R}^d$ is a Lebesgue point for w_Ω.
(d) For every $x_0 \in \mathbb{R}^d$ and every $r > 0$, we have the inequalities

$$\begin{aligned} w_\Omega(x_0) &\le \frac{r^2}{2d} + \fint_{\partial B_r(x_0)} w_\Omega \, d\mathcal{H}^{d-1} \\ \text{and} \quad w_\Omega(x_0) &\le \frac{r^2}{2d} + \fint_{B_r(x_0)} w_\Omega \, dx. \end{aligned} \tag{4.11}$$

(e) w_Ω is upper semi-continuous on $\mathbb{R}^d$.
(f) $H_0^1(\Omega) = H_0^1\big(\{w_\Omega > 0\}\big)$.

Remark 4.8. Point (d) of Remark 4.7 in particular shows that the quasi-open sets are the natural domains for the Sobolev spaces. Indeed, we recall that for any measurable set Ω, the set $\{w_\Omega > 0\} \subset \Omega$ is quasi-open and such that $H_0^1(\Omega) = H_0^1\big(\{w_\Omega > 0\}\big)$. On the other hand, if Ω is quasi-open, then there is a function $u \in H_0^1(\Omega)$ such that $\Omega = \{u > 0\}$ up to a set of zero capacity. Since $u \in H_0^1(\{w_\Omega > 0\})$, we have that $\text{cap}(\{u > 0\} \setminus \{w_\Omega > 0\}) = 0$ and so the sets Ω and $\{w_\Omega > 0\}$ coincide quasi-everywhere.

Remark 4.9. From now on we identify w_Ω with its representative defined through the equality

$$w_\Omega(x_0) = \lim_{r\to 0} \fint_{B_r(x_0)} w_\Omega \, dx, \qquad \forall x_0 \in \mathbb{R}^d.$$

Thus, we identify every quasi-open set $\Omega \subset \mathbb{R}^d$ with its representative $\{w_\Omega > 0\}$. With this identification, we have the following simple observations:

- Let Ω be a quasi-open set, Then the measure theoretical and the topological closure of Ω coincide $\overline{\Omega} = \overline{\Omega}^M$. Indeed, we have $\overline{\Omega}^M \subset \overline{\Omega}$. On the other hand, if $x_0 \in \mathbb{R}^d \setminus \overline{\Omega}^M$, then there is a ball $B_r(x_0)$ such that $w_\Omega = 0$ on $B_r(x_0)$ and so, $x_0 \in \mathbb{R}^d \setminus \overline{\Omega}$. Thus we have also $\mathbb{R}^d \setminus \overline{\Omega}^M \subset \mathbb{R}^d \setminus \overline{\Omega}$, which proves the claim.
- Let Ω_1 and Ω_2 be two quasi-open sets. If $|\Omega_1 \cap \Omega_2| = 0$, then $\Omega_1 \cap \Omega_2 = \emptyset$. Indeed, we note that $\Omega_1 \cap \Omega_2 = \big\{x \in \mathbb{R}^d : w_{\Omega_1}(x) w_{\Omega_2}(x) > 0\big\}$. Since $|\Omega_1 \cap \Omega_2| = 0$, we have that $\int_{\mathbb{R}^d} w_1 w_2 \, dx = 0$. Note that every point of $x \in \mathbb{R}^d$ is a Lebesgue point for the product $w_1 w_2$, we have that $w_1 w_2 = 0$ everywhere on $\mathbb{R}^d$.
- Let Ω_1 and Ω_2 be two disjoint quasi-open sets. Then the measure theoretical and the topological common boundaries coincide

$$\partial\Omega_1 \cap \partial\Omega_2 = \overline{\Omega}_1 \cap \overline{\Omega}_2 = \overline{\Omega}_1^M \cap \overline{\Omega}_2^M = \partial^M \Omega_1 \cap \partial^M \Omega_2.$$

Following the original terminology from [20], we give the following:

Definition 4.10. We say that the quasi-open set $\Omega \in \mathcal{A}_{\rm cap}(\mathbb{R}^d)$ is an **energy subsolution** (with constant m) if Ω is a local subsolution for the functional $\mathcal{F}(\Omega) := E(\Omega) + m|\Omega|$, where $m > 0$ is a given constant, *i.e.* if there is $\varepsilon > 0$ such that

$$E(\Omega) + m|\Omega| \le E(\omega) + m|\omega|, \quad \forall \text{ quasi-open } \quad \omega \subset \Omega \quad \text{such that} \quad d_\gamma(\Omega, \omega) < \varepsilon. \tag{4.12}$$

Remark 4.11. For a pair of quasi-open sets $\Omega, \omega \subset \mathbb{R}^d$, we use the notation

$$d_\gamma(\Omega, \omega) := d_\gamma(I_\omega, I_\Omega) = \int_{\mathbb{R}^d} |w_\Omega - w_\omega|\, dx.$$

On the other hand, by the maximum principle we have $w_\Omega \ge w_\omega$, whenever $\omega \subset \Omega$ are quasi-open sets of finite measure. Thus, we have that

$$d_\gamma(\omega, \Omega) = \int_{\mathbb{R}^d} (w_\Omega - w_\omega)\, dx = 2\big(E(\omega) - E(\Omega)\big), \qquad \forall \omega \subset \Omega.$$

In particular, a set $\Omega \in \mathcal{A}_{\rm cap}(\mathbb{R}^d)$ is an energy subsolution, if and only if,

$$2m|\Omega \setminus \omega| \le d_\gamma(\omega, \Omega), \quad \forall \text{ quasi-open } \quad \omega \subset \Omega \quad \text{such that} \quad d_\gamma(\omega, \Omega) < \varepsilon. \tag{4.13}$$

Remark 4.12. If Ω is an energy subsolution with constant m and $m' \le m$, then Ω is also an energy subsolution with constant m'.

Remark 4.13. We recall that if $\Omega \subset \mathbb{R}^d$ is a quasi-open set of finite measure and $t > 0$ is a given real number, then we have

$$w_{t\Omega}(x) = t^2 w_\Omega(x/t) \qquad \text{and} \qquad E(t\Omega) = t^{d+2} E(\Omega).$$

Thus, if Ω is an energy subsolution with constants m and ε, then $\Omega' = t\Omega$ is an energy subsolution with constants $m' = 1$ and $\varepsilon' = \varepsilon t^{d+2}$, where $t = m^{-1/2}$.

Remark 4.14. If the energy subsolution $\Omega \subset \mathbb{R}^d$ is smooth, then writing the optimality condition for local perturbations of the domain Ω with smooth vector fields (see, for example, [72, Chapter 5]) we obtain

$$|\nabla w_\Omega|^2 \ge 2m \quad \text{on} \quad \partial\Omega.$$

Lemma 4.15. *Let $\Omega \subset \mathbb{R}^d$, for $d \geq 2$, be an energy subsolution with constant m and let $w = w_\Omega$. Then there exist constants C_d, depending only on the dimension d, and r_0, depending on the constant ε from Definition 4.10, such that for each $x_0 \in \mathbb{R}^d$ and each $0 < r < r_0$ we have the following inequality:*

$$\begin{aligned} &\frac{1}{2}\int_{B_r(x_0)} |\nabla w|^2\,dx + m\big|B_r(x_0)\cap\{w>0\}\big| \\ &\leq \int_{B_r(x_0)} w\,dx + C_d\left(r + \frac{\|w\|_{L^\infty(B_{2r}(x_0))}}{2r}\right)\int_{\partial B_r(x_0)} w\,d\mathcal{H}^{d-1}, \end{aligned} \tag{4.14}$$

Proof. Taking $\mu = I_\Omega$ in Lemma 3.125, we have that, for $r > 0$ small enough, the quasi-open set $\omega := \Omega \setminus \overline{B_r(x_0)}$ can be used to test (4.12). Now the conclusion follows by Lemma 3.126. □

Lemma 4.16. *Let $\Omega \subset \mathbb{R}^d$ be an energy subsolution with constant 1. Then there exist constants $C_d > 0$ (depending only on the dimension) and $r_0 > 0$ (depending on the dimension and on ε from Definition 4.10) such that for every $x_0 \in \mathbb{R}^d$ and $0 < r < r_0$ the following implication holds:*

$$\Big(\|w_\Omega\|_{L^\infty(B_r(x_0))} \leq C_d r\Big) \Rightarrow \Big(w_\Omega = 0 \text{ on } B_{r/2}(x_0)\Big). \tag{4.15}$$

Proof. Without loss of generality, we can assume that $x_0 = 0$ and we set $w := w_\Omega$. By the trace theorem for $W^{1,1}$ functions (see [5, Theorems 3.87 and 3.88]), we have that

$$\begin{aligned} \int_{\partial B_{r/2}} w\,d\mathcal{H}^{d-1} &\leq C_d\left(\frac{2}{r}\int_{B_{r/2}} w\,dx + \int_{B_{r/2}} |\nabla w|\,dx\right) \\ &\leq C_d\left(\frac{2}{r}\int_{B_{r/2}} w\,dx + \frac{1}{2}\int_{B_{r/2}} |\nabla w|^2\,dx + \frac{1}{2}\big|\{w>0\}\cap B_{r/2}\big|\right) \\ &\leq 2C_d\left(\frac{2}{r}\|w\|_{L^\infty(B_{r/2})} + \frac{1}{2}\right)\left(\frac{1}{2}\int_{B_{r/2}} |\nabla w|^2\,dx + \big|\{u>0\}\cap B_{r/2}\big|\right), \end{aligned} \tag{4.16}$$

where the constant $C_d > 0$ depends only on the dimension d.

We define the energy of w on the ball B_r as

$$E(w, B_r) = \frac{1}{2}\int_{B_r} |\nabla w|^2\,dx + \big|B_r \cap \{w>0\}\big|. \tag{4.17}$$

Combining (4.16) with the estimate from Lemma 4.14, we have

$$\begin{aligned} E(w, B_{r/2}) \leq & \int_{B_{r/2}} w\,dx + C_d\left(r + \frac{2}{r}\|w\|_{L^\infty(B_r)}\right)\int_{\partial B_{r/2}} w\,d\mathcal{H}^{d-1} \\ \leq & \left(\|w\|_{L^\infty(B_{r/2})} + C_d\left(\frac{2}{r}\|w\|_{L^\infty(B_{r/2})} + \frac{1}{2}\right)\right. \\ & \times \left.\left(r + \frac{1}{r}\|w\|_{L^\infty(B_r)}\right)\right) E(w, B_{r/2}), \end{aligned} \tag{4.18}$$

where the constants C_d depend only on the dimension d. The claim follows by observing that if

$$\|w\|_{L^\infty(B_r)} \leq cr,$$

for some small c and r, then by (4.18) we obtain $E(w, B_{r/2}) = 0$. □

Lemma 4.17. *Let μ be a capacitary measure in $\mathbb{R}^d$ such that $w_\mu \in L^1(\mathbb{R}^d)$. Suppose that there are constants $C > 0$ and $r_0 > 0$ such that for every $x_0 \in \mathbb{R}^d$ and $0 < r < r_0$ the following implication holds:*

$$\left(\|w_\mu\|_{L^\infty(B_r(x_0))} \leq Cr\right) \Rightarrow \left(w_\mu = 0 \text{ on } B_{r/2}(x_0)\right). \tag{4.19}$$

Then for every $0 < r < \min\{r_0, Cd/8\}$, the set $\Omega_\mu = \{w_\mu > 0\}$ can be covered with $N = C_d\|w_\mu\|_{L^1} r^{-d-1}$ balls of radius r, where C_d is a dimensional constant.

Proof. Suppose, by absurd that, for some $0 < r < R_0$, this is not the case and choose points $x_1, \ldots, x_N \in \mathbb{R}^d$ such that $x_1 \in \{w_\mu > 0\}$ and

$$x_{j+1} \in \{w_\mu > 0\} \setminus \left(\bigcup_{i=1}^{j} B_r(x_i)\right).$$

For each x_j, we have $\|w_\mu\|_{L^\infty(B_{r/4}(x_j))} > Cr/4$. For each $j = 1, \ldots, N$, consider $y_j \in B_{r/4}(x_j)$ such that

$$w(y_j) \geq Cr/8.$$

By construction we have that the balls $B_{r/4}(y_j)$ are disjoint for $j = 1, \ldots, N$. Since the function $w - \frac{r^2 - |\cdot - y_j|^2}{2d}$ is subharmonic in $B_r(y_j)$, we have the inequality

$$\int_{B_{r/4}(y_j)} \left(w(x) - \frac{r^2 - |x - y_j|^2}{2d}\right) dx \geq |B_{r/4}|\left(w(y_j) - \frac{r^2}{2d}\right),$$

and summing on j, we get

$$\|w\|_{L^1} \geq \sum_{j=1}^{N} \int_{B_{r/4}(y_j)} w\,dx \geq N|B_{r/4}| \left(\frac{Cr}{8} - \frac{r^2}{2d}\right) > N|B_{r/4}| \frac{Cr}{16}. \qquad \square$$

In other words, Lemma 4.16 says that in a point of $\overline{\Omega}^M$ (the measure theoretic closure of the energy subsolution Ω) the function w_Ω has at least linear growth. In particular, the maximum of w_Ω on $B_r(x)$ and the average on $\partial B_r(x)$ are comparable for $r > 0$ small enough.

Corollary 4.18. *Suppose that $\Omega \subset \mathbb{R}^d$ is an energy subsolution with $m = 1$ and let $w = w_\Omega$. Then there exists $r_0 > 0$, depending on the dimension and the constant ε from Definition 4.10, such that for every $x_0 \in \overline{\Omega}^M$ and every $0 < r < r_0$, we have*

$$2^{-d-2}\|w\|_{L^\infty(B_r(x_0))} \leq \fint_{\partial B_{2r}(x_0)} w\,d\mathcal{H}^{d-1} \leq \|w\|_{L^\infty(B_{2r}(x_0))}. \qquad (4.20)$$

Proof. Suppose that $x_0 = 0$ and consider the function $\varphi_{2r}(x) := \frac{(2r)^2 - |x|^2}{2d}$. By Remark 4.7 we have that $\Delta(w - \varphi_{2r}) \geq 0$ on $\mathbb{R}^d$ and $0 \leq \varphi_{2r} \leq 2r^2/d$ on B_{2r}. Comparing $w - \varphi_{2r}$ with the harmonic function on B_{2r} with boundary values w, we obtain that for every $x \in B_r$, we have

$$w(x) - \varphi_{2r}(x) \leq \frac{4r^2 - |x|^2}{d\omega_d 2r} \int_{\partial B_{2r}} \frac{w(y)}{|y - x|^d}\,d\mathcal{H}^{d-1}(y) \leq 2^d \fint_{\partial B_{2r}} w\,d\mathcal{H}^{d-1}.$$

For $0 < r < \min\left\{r_0, \frac{dC_d}{8}, 1\right\}$, where r_0 and C_d are the constants from Lemma 4.16, we choose $x_r \in B_r$ such that

$$w(x_r) > \frac{1}{2}\|w\|_{L^\infty(B_r)} > \frac{rC_d}{2}.$$

Then we have

$$\begin{aligned}\frac{\|w\|_{L^\infty(B_r)}}{2} &\leq w(x_r) \leq 2^d \fint_{\partial B_{2r}} w\,d\mathcal{H}^{d-1} + \frac{2r^2}{d} \\ &\leq 2^d \fint_{\partial B_{2r}} w\,d\mathcal{H}^{d-1} + \frac{\|w\|_{L^\infty(B_r)}}{4},\end{aligned}$$

which proves the claim. $\square$

Remark 4.19. In particular, there are constants c and r_0 such that if $x_0 \in \overline{\Omega}^M$, then for every $0 < r \leq r_0$, we have that

$$cr \leq \fint_{\partial B_r(x_0)} w_\Omega \, d\mathcal{H}^{d-1}.$$

Moreover, since $\int_{B_r} w_\Omega \, dx = \int_0^r \int_{\partial B_s} w_\Omega \, d\mathcal{H}^{d-1} \, ds$, we also have $cr \leq \fint_{B_r(x_0)} w_\Omega \, dx$.

As a consequence of Corollary 4.18, we can simplify (4.14). Precisely, we have the following result.

Corollary 4.20. *Suppose that $\Omega \subset \mathbb{R}^d$ is an energy subsolution with $m = 1$. Then there are constants $C_d > 0$, depending only on the dimension d, and r_0, depending on the dimension d and ε from Definition* 4.10, *such that for every $x_0 \in \overline{\Omega}^M$ and $0 < r < r_0$, we have*

$$\begin{aligned} &\frac{1}{2}\int_{B_r(x_0)} |\nabla w_\Omega|^2 \, dx + \left|\{w_\Omega > 0\} \cap B_r(x_0)\right| \\ &\leq C_d \frac{\|w_\Omega\|_{L^\infty(B_{2r}(x_0))}}{2r} \int_{\partial B_r(x_0)} w_\Omega \, d\mathcal{H}^{d-1}. \end{aligned} \tag{4.21}$$

Proof. We set for simplicity $w := w_\Omega$ and $x_0 = 0$. By Lemma 4.16 and Corollary 4.18, for $r > 0$ small enough, we have

$$\frac{1}{r}\|w\|_{L^\infty(B_r)} \geq C_d \qquad \text{and} \qquad \frac{1}{r}\fint_{\partial B_r} w \, d\mathcal{H}^{d-1} \geq 2^{-d-2} C_d. \tag{4.22}$$

Thus, for r as above, we have

$$\int_{B_r} w \, dx \leq |B_r| \frac{d2^{-d-2}C_d}{r} \|w\|_{L^\infty(B_r)} \leq \frac{1}{r}\|w\|_{L^\infty(B_r)} \int_{\partial B_r} w \, d\mathcal{H}^{d-1},$$

and so, it remains to apply the above estimate to (4.14). □

Relying on inequality (4.21) and Lemma 4.16 we get the following inner density estimate, which is much weaker than the density estimates from [1]. The main reason is that we work only with subsolutions and not with minimizers of a free boundary problem.

Proposition 4.21. *Suppose that $\Omega \subset \mathbb{R}^d$ is an energy subsolution. Then there exists a constant $c > 0$, depending only on the dimension, such that for every $x_0 \in \overline{\Omega}^M$, we have*

$$\limsup_{r \to 0} \frac{\left|\{w_\Omega > 0\} \cap B_r(x_0)\right|}{|B_r|} \geq c. \tag{4.23}$$

Proof. Without loss of generality, we can suppose that $x_0 = 0$ and by rescaling we can assume that $m = 1$. Let r_0 and C_d be as in Lemma 4.16 and let $0 < r < r_0$. By the Trace Theorem in $W^{1,1}(B_r)$, we have

$$\begin{aligned}
\int_{\partial B_r} w\,d\mathcal{H}^{d-1} &\le C_d\left(\int_{B_r}|\nabla w|\,dx + \frac{1}{r}\int_{B_r} w\,dx\right)\\
&\le C_d\left(\left(\int_{B_r}|\nabla w|^2\,dx\right)^{1/2}\big|\{w>0\}\cap B_r\big|^{1/2}\right.\\
&\qquad\left.+\frac{\|w\|_{L^\infty(B_r)}}{r}\big|\{w>0\}\cap B_r\big|\right) \qquad (4.24)\\
&\le C_d\left(\frac{\|w\|_{L^\infty(B_{2r})}}{2r}\int_{\partial B_r} w\,d\mathcal{H}^{d-1}\right)^{1/2}\big|\{w>0\}\cap B_r\big|^{1/2}\\
&\qquad + C_d\frac{\|w\|_{L^\infty(B_r)}}{r}\big|\{w>0\}\cap B_r\big|,
\end{aligned}$$

where the last inequality is due to Corollary 4.20 and C_d denotes a constant which depends only on the dimension d. Let

$$\begin{aligned}
X &= \left(\int_{\partial B_r} w\,d\mathcal{H}^{d-1}\right)^{1/2},\\
\alpha &= C_d\left(\frac{\|w\|_{L^\infty(B_{2r})}}{2r}\right)^{1/2}\big|\{w>0\}\cap B_r\big|^{1/2},\\
\beta &= C_d\frac{\|w\|_{L^\infty(B_r)}}{r}\big|\{w>0\}\cap B_r\big|.
\end{aligned}$$

Then, we can rewrite (4.24) as

$$X^2 \le \alpha X + \beta.$$

But then, since $\alpha, \beta > 0$, we have the estimate $X \le \alpha + \sqrt{\beta}$. Taking the square of both sides, we obtain

$$\begin{aligned}
\int_{\partial B_r} w\,d\mathcal{H}^{d-1} &\le C_d\big|\{w>0\}\cap B_r\big|\left(\frac{\|w\|_{L^\infty(B_{2r})}}{2r} + \frac{\|w\|_{L^\infty(B_r)}}{r}\right)\\
&\le 3C_d\big|\{w>0\}\cap B_r\big|\frac{\|w\|_{L^\infty(B_{2r})}}{2r}.
\end{aligned} \qquad (4.25)$$

By Corollary 4.18, we have that

$$\frac{\|w\|_{L^\infty(B_{r/2})}}{r/2} \le \frac{C_d\big|\{w>0\}\cap B_r\big|}{|B_r|}\frac{\|w\|_{L^\infty(B_{2r})}}{2r}, \qquad (4.26)$$

for some dimensional constant $C_d > 0$. We choose the constant c from (4.23) as $c = (2C_d)^{-1}$ and we argue by contradiction. Suppose, by absurd, that we have

$$\limsup_{r\to 0} C_d \frac{\big|\{w > 0\} \cap B_r\big|}{|B_r|} < \frac{1}{2}. \tag{4.27}$$

Setting, for $r > 0$ small enough,

$$f(r) := \frac{\|w\|_{L^\infty(B_r)}}{r},$$

and using (4.26), we have that for each $n \in \mathbb{N}$ the following inequality holds

$$f(r4^{-(n+1)}) \le \frac{C_d\big|\{w > 0\} \cap B_{2r4^{-(n+1)}}\big|}{|B_{2r4^{-(n+1)}}|} f(r4^{-n}), \tag{4.28}$$

and so

$$f(r4^{-(n+1)}) \le f(r) \prod_{k=0}^{n} \frac{C_d\big|\{w > 0\} \cap B_{2r4^{-(k+1)}}\big|}{|B_{2r4^{-(k+1)}}|}. \tag{4.29}$$

By equation (4.27), we have that $f(r4^{-n}) \to 0$, which is a contradiction with Lemma 4.16. □

Theorem 4.22. *Suppose that the quasi-open set $\Omega \subset \mathbb{R}^d$ is an energy subsolution with constant $m > 0$. Then, we have that:*

(i) *Ω is a bounded set and its diameter can be estimated by a constant depending on d, Ω, m and r_0;*
(ii) *Ω is of finite perimeter and*

$$\sqrt{2m}\,\mathcal{H}^{d-1}(\partial^*\Omega) \le |\Omega|; \tag{4.30}$$

(iii) *Ω is equivalent a.e. to a closed set. More precisely, $\Omega = \overline{\Omega}^M$ a.e., $\overline{\Omega}^M = \mathbb{R}^d \setminus \Omega_{(0)}$ and $\Omega_{(0)}$ is an open set. Moreover, if Ω is given through its canonical representative from Remark 4.9, then $\overline{\Omega} = \overline{\Omega}^M$.*

Proof. The first statements follows by Lemma 4.17. In order to prove (ii), we reason as in [20, Theorem 2.2]. Let $w = w_\Omega$ and consider the set $\Omega_\varepsilon = \{w > \varepsilon\}$. Since $w_{\Omega_\varepsilon} = (w - \varepsilon)^+$, we have that for small ε, the distance $d_\gamma(\Omega, \Omega_\varepsilon)$ is small, we can use Ω_ε as a competitor in (4.12) obtaining

$$\frac{1}{2}\int_{\mathbb{R}^d} |\nabla w|^2\,dx - \int_{\mathbb{R}^d} w\,dx + m|\Omega| \le E(\Omega) + m|\Omega| \le E(\Omega_\varepsilon) + m|\Omega_\varepsilon|$$
$$\le \frac{1}{2}\int_{\mathbb{R}^d} |\nabla (w-\varepsilon)^+|^2\,dx - \int_{\mathbb{R}^d} (w-\varepsilon)^+\,dx + m|\Omega_\varepsilon|.$$

In particular, we have

$$\begin{aligned}\varepsilon|\Omega| &\ge \int_{\mathbb{R}^d} w\,dx - \int_{\mathbb{R}^d} (w-\varepsilon)^+\,dx\\ &\ge \frac{1}{2}\int_{\{0<w\le\varepsilon\}} |\nabla w|^2\,dx + m|\Omega\setminus\Omega_\varepsilon|\\ &\ge \frac{1}{2}\big|\{0<w\le\varepsilon\}\big|^{-1}\left(\int_{\{0<w\le\varepsilon\}} |\nabla w|\,dx\right)^2 + m\big|\{0<w\le\varepsilon\}\big|\\ &\ge \sqrt{2m}\int_{\{0<w\le\varepsilon\}} |\nabla w|\,dx.\end{aligned}$$

By the co-area formula we have

$$\frac{1}{\varepsilon}\int_0^\varepsilon P\left(\{w>t\}\right)\,dt \le \sqrt{2m}|\Omega|,$$

for each $\varepsilon > 0$ small enough. Then, there is a sequence $(\varepsilon_n)_{n\ge1}$ converging to 0 and such that $P\left(\{w>\varepsilon_n\}\right) \le \sqrt{2m}|\Omega|$. Passing to the limit as $n\to\infty$, we obtain (ii).

For the third claim, it is sufficient to prove that $\Omega_{(0)}$ satisfies

$$\Omega_{(0)} = \mathbb{R}^d\setminus\overline{\Omega}^M = \left\{x\in\mathbb{R}^d : \text{exists } r>0 \text{ such that } |B_r(x)\cap\Omega| = 0\right\}, \quad (4.31)$$

where the second equality is just the definition of $\overline{\Omega}^M$. We note that $\Omega_{(0)}\subset\mathbb{R}^d\setminus\overline{\Omega}^M$ trivially holds for every measurable Ω. On the other hand, if $x\in\overline{\Omega}^M$, then, by Proposition 4.21, there is a sequence $r_n\to0$ such that

$$\lim_{n\to\infty}\frac{|B_{r_n}(x)\cap\Omega|}{|B_{r_n}|} \ge c > 0,$$

and so $x\notin\Omega_{(0)}$, which proves the opposite inclusion and the equality in (4.31).

$\square$

Remark 4.23. The second statement of Theorem 4.22 implies, in particular, that the energy subsolutions cannot be too small. Indeed, by the isoperimetric inequality, we have

$$c_d\sqrt{2m}|\Omega|^{\frac{d-1}{d}} \le \sqrt{2m}\mathcal{H}^{d-1}(\partial^*\Omega) \le |\Omega| \le C_d[\mathcal{H}^{d-1}(\partial^*\Omega)]^{\frac{d}{d-1}},$$

and so

$$c_d m^{\frac{d}{2}} \le |\Omega| \qquad \text{and} \qquad c_d m^{\frac{d-1}{2}} \le \mathcal{H}^{d-1}(\partial^*\Omega),$$

for some dimensional constant c_d.

4.3. Interaction between energy subsolutions

In this section we consider configurations of disjoint quasi-open sets $\Omega_1, \dots, \Omega_n$ in $\mathbb{R}^d$, each one being an energy subsolution. In particular, we will study the behaviour of the energy functions w_{Ω_i}, $i = 1, \dots, n$, around the points that belong to more than one of the measure theoretical boundaries $\partial^M\Omega_i$.

4.3.1. Monotonicity theorems

The Alt-Caffarelli-Friedman monotonicity formula is one of the most powerful tools in the study of the regularity of multiphase optimization problems as, for example, optimal partition problems for functionals involving some partial differential equation, a prototype being the multiphase Alt-Caffarelli problem

$$\min\left\{\sum_{i=1}^{m}\int_\Omega |\nabla u_i|^2 - f_i u_i + Q^2 \mathbb{1}_{\{u_i>0\}}\,dx : (u_1, \dots, u_m) \in \mathcal{A}(\Omega)\right\}, \tag{4.32}$$

where $\Omega \subset \mathbb{R}^d$ is a given (Lipschitz) bounded open set, $Q : \Omega \to \mathbb{R}$ is a measurable function, $f_1, \dots, f_m \in L^\infty(\Omega)$ and the admissible set $\mathcal{A}(\Omega)$ is given by

$$\begin{aligned}\mathcal{A}(\Omega) := \Big\{(u_1, \dots, u_m) \in \left[H^1(\Omega)\right]^m : \ u_i \ge 0,\ u_i = c \\ \text{on } \partial\Omega,\ u_i u_j = 0 \text{ a.e. on } \Omega, \forall i \ne j\Big\},\end{aligned} \tag{4.33}$$

where $c \ge 0$ is a given constant.

Remark 4.24.

- If $Q = 0$, then we have a classical optimal partition problem as the ones studied in [42, 47, 48, 49] and [69].
- If $c = 1, m = 1, f_1 = 0$ and $0 < a \le Q^2 \le b < +\infty$, then (4.32) reduces to the problem considered in [1].

- If $m = 1$, $Q \equiv 1$, $f_1 = f$ and $f_2 = -f$, then the solution of (4.32) is given by

$$u_1^* = u_+^* := \sup\{u^*, 0\}, \qquad u_2^* = u_-^* := \sup\{-u^*, 0\},$$

where $u^* \in H_0^1(\Omega)$ is a solution of the following problem, considered in [17],

$$\min\left\{\int_\Omega |\nabla u|^2 - fu\,dx + |\{u \neq 0\}| \,:\, u \in H_0^1(\Omega)\right\}.$$

- If, $Q \equiv 1$ and $f_1 = \cdots = f_m = f$, then (4.32) reduces to a problem considered in [29] and [12].

One of the main tools in the study of the Lipschitz continuity of the solutions $(u_1^*, \ldots, u_m^*)$ of the multiphase problem (4.32) is the monotonicity formula, which relates the behaviour of the different phases u_i^* in the points on the common boundary $\partial\{u_i^* > 0\} \cap \partial\{u_j^* > 0\}$, the main purpose being to provide a bound for the gradients $|\nabla u_i^*|$ and $|\nabla u_j^*|$ in these points. The following estimate was proved in [41], as a generalization of the monotonicity formula from [2], and was widely used (for example in [17] and also [28]) in the study of free-boundary problems.

Theorem 4.25 (Caffarelli-Jerison-Kenig). *Let $B_1 \subset \mathbb{R}^d$ be the unit ball in R^d and let $u_1, u_2 \in H^1(B_1)$ be non-negative and* continuous *functions such that*

$$\Delta u_i + 1 \geq 0, \quad for \quad i = 1, 2, \qquad and \qquad u_1 u_2 = 0 \quad on \quad B_1.$$

Then there is a dimensional constant C_d such that for each $r \in (0, 1)$ we have

$$\prod_{i=1}^{2}\left(\frac{1}{r^2}\int_{B_r} \frac{|\nabla u_i|^2}{|x|^{d-2}}\,dx\right) \leq C_d\left(1 + \sum_{i=1}^{2}\int_{B_1} \frac{|\nabla u_i|^2}{|x|^{d-2}}\,dx\right)^2. \tag{4.34}$$

The aim of this and the following subsections[1] 4.3.2, 4.3.3 and 4.3.4 is to show that the continuity assumption in Theorem 4.25 can be dropped (Theorem 4.30) and to provide the reader with a detailed proof of the multiphase version (Theorem 4.34 and Corollary 4.35) of Theorem 4.25, which was proved in [29]. We note that the proof of Theorem 4.30 follows precisely the one of Theorem 4.25 given in [41]. We report the

[1] The results in these sections are part of the note [93].

estimates, in which the continuity assumption was used, in Section 4.3.2 and we adapt them, essentially by approximation, to the non-continuous case.

A strong initial motivation was provided by the multiphase version of the Alt-Caffarelli-Friedman monotonicity formula, proved in [47] in the special case of sub-harmonic[2] functions u_i in $\mathbb{R}^2$, which avoids the continuity assumption and applies also in the presence of more phases. As a conclusion of the Introduction section, we give the proof of this result, which has the advantage of avoiding the technicalities, emphasising the presence of a stronger decay in the multiphase case and showing that the continuous assumption is unnecessary.

Theorem 4.26 (Alt-Caffarelli -Friedman; Conti-Terracini -Verzini). *Consider the unit ball $B_1 \subset \mathbb{R}^2$ and let $u_1, \dots, u_m \in H^1(B_1)$ be m non-negative subharmonic functions such that $\int_{\mathbb{R}^2} u_i u_j \, dx = 0$, for every choice of different indices $i, j \in \{1, \dots, m\}$. Then the function*

$$\Phi(r) = \prod_{i=1}^{m} \left(\frac{1}{r^m} \int_{B_r} |\nabla u_i|^2 \, dx \right) \tag{4.35}$$

is non-decreasing on $[0, 1]$. In particular,

$$\prod_{i=1}^{n} \left(\frac{1}{r^m} \int_{B_r} |\nabla u_i|^2 \, dx \right) \leq \left(\int_{B_1} |\nabla u_1|^2 \, dx + \dots + \int_{B_1} |\nabla u_m|^2 \, dx \right)^m. \tag{4.36}$$

Proof. The function Φ is of bounded variation and calculating its derivative we get

$$\frac{\Phi'(r)}{\Phi(r)} \geq -\frac{m^2}{r} + \sum_{i=1}^{m} \frac{\int_{\partial B_r} |\nabla u_i|^2 \, d\mathcal{H}^1}{\int_{B_r} |\nabla u_i|^2 \, dx}. \tag{4.37}$$

We now prove that the right-hand side is positive for every $r \in (0, 1)$ such that $u_i \in H^1(\partial B_r)$, for every $i = 1, \dots, m$, and $\int_{\partial B_r} u_i u_j \, d\mathcal{H}^1 = 0$, for every $i \neq j \in \{1, \dots, m\}$. We use the sub-harmonicity of u_i to calculate

$$\begin{aligned} \int_{B_r} |\nabla u_i|^2 \, dx &\leq \int_{\partial B_r} u_i \frac{\partial u_i}{\partial n} \, d\mathcal{H}^1 \\ &\leq \left(\int_{\partial B_r} u_i^2 \, d\mathcal{H}^1 \right)^{\frac{1}{2}} \left(\int_{\partial B_r} |\nabla_n u_i|^2 \, d\mathcal{H}^1 \right)^{\frac{1}{2}}, \end{aligned} \tag{4.38}$$

[2] The result in [47] is more general and applies to (non-linear) eigenfunctions.

and decomposing the gradient ∇u_i in the tangent and normal parts $\nabla_\tau u_i$ and $\nabla_n u_i$, we have

$$\begin{aligned}\int_{\partial B_r} |\nabla u_i|^2\, d\mathcal{H}^1 &= \int_{\partial B_r} |\nabla_n u_i|^2\, d\mathcal{H}^1 + \int_{\partial B_r} |\nabla_\tau u_i|^2\, d\mathcal{H}^1 \\ &\geq 2\left(\int_{\partial B_r} |\nabla_n u_i|^2\, d\mathcal{H}^1\right)^{\frac{1}{2}} \left(\int_{\partial B_r} |\nabla_\tau u_i|^2\, d\mathcal{H}^1\right)^{\frac{1}{2}}.\end{aligned} \tag{4.39}$$

Putting together (4.38) and (4.39), we obtain

$$\frac{\int_{\partial B_r} |\nabla u_i|^2\, d\mathcal{H}^1}{\int_{B_r} |\nabla u_i|^2\, dx} \geq 2\left(\frac{\int_{\partial B_r} |\nabla_\tau u_i|^2\, d\mathcal{H}^1}{\int_{\partial B_r} u_i^2\, d\mathcal{H}^1}\right)^{\frac{1}{2}} \geq 2\sqrt{\lambda_1(\partial B_r \cap \Omega_i)}, \tag{4.40}$$

where we use the notation $\Omega_i := \{u_i > 0\}$ and for an $\mathcal{H}^1$-measurable set $\omega \subset \partial B_r$ we define

$$\lambda_1(\omega) := \min\left\{\frac{\int_{\partial B_r} |\nabla_\tau v|^2\, d\mathcal{H}^1}{\int_{\partial B_r} v^2\, d\mathcal{H}^1} : v \in H^1(\partial B_r),\ \ \mathcal{H}^1\big(\{v \neq 0\} \setminus \omega\big) = 0\right\}.$$

By a standard symmetrization argument, we have $\lambda_1\big(\omega\big) \geq \left(\frac{\pi}{\mathcal{H}^1(\omega)}\right)^2$ and so, by (4.37) and the mean arithmetic-mean harmonic inequality, we obtain the estimate

$$\frac{\Phi'(r)}{\Phi(r)} \geq -\frac{m^2}{r} + \sum_{i=1}^m \frac{2\pi}{\mathcal{H}^1(\partial B_r \cap \Omega_i)} \geq 0,$$

which concludes the proof. □

4.3.2. The monotonicity factors

In this subsection we consider non-negative functions $u \in H^1(B_2)$ such that

$$\Delta u + 1 \geq 0 \qquad \text{weakly in} \qquad \left[H_0^1(B_2)\right]',$$

and we study the energy functional

$$A_u(r) := \int_{B_r} \frac{|\nabla u|^2}{|x|^{d-2}}\, dx,$$

for $r \in (0,1)$, which is precisely the quantity that appears in (4.56) and (4.70). We start with a lemma, which was first proved in [41, Remark 1.5].

Lemma 4.27. *Suppose that $u \in H^1(B_2)$ is a non-negative Sobolev function such that $\Delta u + 1 \geq 0$ on $B_2 \subset \mathbb{R}^d$. Then, there is a dimensional constant C_d such that*

$$\int_{B_1} \frac{|\nabla u|^2}{|x|^{d-2}}\,dx \leq C_d \left(1 + \int_{B_2 \setminus B_1} u^2\,dx\right). \tag{4.41}$$

Proof. Let $u_\varepsilon = \phi_\varepsilon * u$, where $\phi_\varepsilon \in C_c^\infty(B_\varepsilon)$ is a standard molifier. Then $u_\varepsilon \to u$ strongly in $H^1(B_2)$, $u_\varepsilon \in C^\infty(B_2)$ and $\Delta u_\varepsilon + 1 \geq 0$ on $B_{2-\varepsilon}$. We will prove (4.41) for u_ε. We note that a brief computation gives the inequality

$$\Delta(u_\varepsilon^2) = 2|\nabla u_\varepsilon|^2 + 2u_\varepsilon \Delta u_\varepsilon \geq 2|\nabla u_\varepsilon|^2 - 2u_\varepsilon \quad \text{in} \quad \left[H_0^1(B_{2-\varepsilon})\right]'. \tag{4.42}$$

We now choose a positive and radially decreasing function $\phi \in C_c^\infty(B_{3/2})$ such that $\phi = 1$ on B_1. By (4.42) we get

$$\begin{aligned}
2\int_{B_{3/2}} \frac{\phi(x)|\nabla u_\varepsilon|^2}{|x|^{d-2}}\,dx &\leq \int_{B_{3/2}} \phi(x)\frac{2u_\varepsilon + \Delta(u_\varepsilon^2)}{|x|^{d-2}}\,dx \\
&= \int_{B_{3/2}} 2\frac{\phi(x)u_\varepsilon}{|x|^{d-2}} + u_\varepsilon^2 \Delta\left(\frac{\phi(x)}{|x|^{d-2}}\right)dx \\
&= \int_{B_{3/2}} 2\frac{\phi(x)u_\varepsilon}{|x|^{d-2}} + u_\varepsilon^2 \frac{\Delta\phi(x)}{|x|^{d-2}} \\
&\quad + u_\varepsilon^2 \nabla\phi(x)\cdot\nabla(|x|^{2-d})\,dx - C_d u_\varepsilon^2(0) \\
&\leq 2\int_{B_{3/2}} \frac{\phi(x)u_\varepsilon}{|x|^{d-2}}\,dx + C_d \int_{B_2\setminus B_1} u_\varepsilon^2\,dx.
\end{aligned} \tag{4.43}$$

Thus, in order to obtain (4.41), it is sufficient to estimate the norm $\|u_\varepsilon\|_{L^\infty(B_1)}$ with the right hand side of (4.41). To do that, we first note that since $\Delta\big(u_\varepsilon(x) + |x|^2/2d\big) \geq 0$, we have

$$\max_{x\in B_1}\left\{u_\varepsilon(x) + |x|^2/2d\right\} \leq C_d + C_d \fint_{\partial B_r} u_\varepsilon\,d\mathcal{H}^{d-1}, \quad \forall r \in (3/2, 2-\varepsilon), \tag{4.44}$$

and, after integration in r and the Cauchy-Schwartz inequality, we get

$$\|u_\varepsilon\|_{L^\infty(B_1)} \leq C_d + C_d \left(\int_{B_2\setminus B_1} u_\varepsilon^2\,dx\right)^{1/2}, \tag{4.45}$$

which, together with (4.43), gives (4.41). □

Remark 4.28. For a non-negative function $u \in H^1(B_r)$, satisfying

$$\Delta u + 1 \geq 0 \qquad \text{in} \qquad \left[H_0^1(B_r)\right]', \tag{4.46}$$

we denote with $A_u(r)$ the quantity

$$A_u(r) := \int_{B_r} \frac{|\nabla u|^2}{|x|^{d-2}}\,dx < +\infty. \tag{4.47}$$

- The function $r \mapsto A_u(r)$ is bounded and increasing in r.
- A_u is differentiable almost everywhere and

$$\frac{d}{dr} A_u(r) = r^{2-d} \int_{\partial B_r} |\nabla u|^2\,d\mathcal{H}^{d-1}.$$

- The condition (4.46) holds also for the rescaled function $u_r(x) := r^{-2}u(rx)$ and we have

$$\begin{aligned} \int_{\partial B_1} |\nabla u_r|^2\,d\mathcal{H}^{d-1} &= \frac{1}{r^{d+1}} \int_{\partial B_r} |\nabla u|^2\,d\mathcal{H}^{d-1}, \\ \int_{B_1} \frac{|\nabla u_r|^2}{|x|^{d-2}}\,dx &= \frac{1}{r^4} \int_{B_r} \frac{|\nabla u|^2}{|x|^{d-2}}\,dx. \end{aligned} \tag{4.48}$$

The next result is implicitly contained in [41, Lemma 2.8] and it is the point in which the continuity of u_i was used. The inequality (4.49) is the analogue of the estimate (4.40), which is the main ingredient of the proof of Theorem 4.26.

Lemma 4.29. *Let $u \in H^1(B_2)$ be a non-negative function such that $\Delta u + 1 \geq 0$ on B_2. Then for Lebesgue almost every $r \in (0,1)$ we have the estimate*

$$\begin{aligned} \frac{1}{r^4} \int_{B_r} \frac{|\nabla u|^2}{|x|^{d-2}}\,dx \leq C_d &\left(1 + \frac{r^{-2}}{\sqrt{\lambda(u,r)}} \left(\fint_{\partial B_r} |\nabla u|^2\,d\mathcal{H}^{d-1}\right)^{\frac{1}{2}}\right) \\ &+ \frac{d\omega_d r^{-3}}{2\alpha(u,r)} \fint_{\partial B_r} |\nabla u|^2\,d\mathcal{H}^{d-1}, \end{aligned} \tag{4.49}$$

where

$$\begin{aligned} \lambda(u,r) := \min \Bigg\{ \frac{\int_{\partial B_r} |\nabla v|^2\,d\mathcal{H}^{d-1}}{\int_{\partial B_r} v^2\,d\mathcal{H}^{d-1}} \ :\ & v \in H^1(\partial B_r), \\ & \mathcal{H}^{d-1}\big(\{v \neq 0\} \cap \{u = 0\}\big) = 0 \Bigg\}, \end{aligned} \tag{4.50}$$

and $\alpha(u,r) \in \mathbb{R}^+$ *is the characteristic constant of* $\{u > 0\} \cap \partial B_r$, i.e. *the non-negative solution of the equation*

$$\alpha(u,r)\left(\alpha(u,r) + \frac{d-2}{r}\right) = \lambda(u,r). \tag{4.51}$$

Proof. We start by determining the subset of the interval (0, 1) for which we will prove that (4.49) holds. Let $u_\varepsilon := u * \phi_\varepsilon$, where ϕ_ε is a standard molifier. Then we have that:

(i) for almost every $r \in (0,1)$ the restriction of u to ∂B_r is Sobolev. *i.e.* $u_{|\partial B_r} \in H^1(\partial B_r)$;
(ii) for almost every $r \in (0,1)$ the sequence of restrictions $(\nabla u_\varepsilon)_{|\partial B_r}$ converges strongly in $L^2(\partial B_r;\mathbb{R}^d)$ to $(\nabla u)_{|\partial B_r}$.

We now consider $r \in (0,1)$ such that both (i) and (ii) hold. Using the scaling $u_r(x) := r^{-2}u(rx)$, we have that

$$\fint_{\partial B_r} |\nabla u|^2\, d\mathcal{H}^{d-1} = r^2 \fint_{\partial B_1} |\nabla u_r|^2\, d\mathcal{H}^{d-1},$$

$$\frac{1}{r^4}\int_{B_r} \frac{|\nabla u|^2}{|x|^{d-2}}\, dx = \int_{B_1} \frac{|\nabla u_r|^2}{|x|^{d-2}}\, dx,$$

$$\alpha(u_r,1) = r\alpha(u,r) \qquad \text{and} \qquad \lambda(u_r,1) = r^2\lambda(u,r).$$

Substituting in (4.49), we can suppose that $r = 1$ and set $\alpha := \alpha(u,1)$ and $\lambda := \lambda(u,1)$.

If $\mathcal{H}^{d-1}\big(\{u=0\}\cap\partial B_1\big)=0$, then $\lambda = 0$. Now if $\int_{\partial B_1} |\nabla u|^2\, d\mathcal{H}^{d-1} > 0$, then the inequality (4.49) is trivial. If on the other hand, $\int_{\partial B_1} |\nabla u|^2 d\mathcal{H}^{d-1} = 0$, then u is a constant on ∂B_1 and so, we may suppose that $u = 0$ on $\mathbb{R}^d \setminus B_1$, which again gives (4.49), by choosing C_d large enough. Thus, it remains to prove the Lemma in the case $\mathcal{H}^{d-1}\big(\{u=0\}\cap\partial B_1\big) > 0$.

We first note that since $\mathcal{H}^{d-1}\big(\{u=0\}\cap\partial B_1\big) > 0$, the constant λ defined in (4.50) is strictly positive. Using the restriction of u on ∂B_1 as a test function in (4.50) we get

$$\lambda \int_{\partial B_1} u^2\, d\mathcal{H}^{d-1} \le \int_{\partial B_1} |\nabla_\tau u|^2\, d\mathcal{H}^{d-1},$$

where ∇_τ is the tangential gradient on ∂B_1. In particular, we have

$$\lambda \int_{\partial B_1} u^2\, d\mathcal{H}^{d-1} \le \int_{\partial B_1} |\nabla_\tau u|^2\, d\mathcal{H}^{d-1} \le \int_{\partial B_1} |\nabla u|^2\, d\mathcal{H}^{d-1} =: B_u(1). \tag{4.52}$$

For every $\varepsilon > 0$, using the inequality

$$\Delta(u_\varepsilon^2) = 2u_\varepsilon \Delta u_\varepsilon + 2|\nabla u_\varepsilon|^2 \geq -2u_\varepsilon + 2|\nabla u_\varepsilon|^2,$$

and the fact that $\Delta\Big(u_\varepsilon + |x|^2/2d\Big) \geq 0$, we have

$$\begin{aligned} 2\int_{B_1} \frac{|\nabla u_\varepsilon|^2}{|x|^{d-2}}\,dx &\leq \int_{B_1} \frac{2u_\varepsilon + \Delta(u_\varepsilon^2)}{|x|^{d-2}}\,dx \\ &\leq C_d + C_d\Big(\int_{\partial B_1} u_\varepsilon^2\,d\mathcal{H}^{d-1}\Big)^{1/2} + \int_{B_1} \frac{\Delta(u_\varepsilon^2)}{|x|^{d-2}}\,dx. \end{aligned} \tag{4.53}$$

We now estimate the last term on the right-hand side.

$$\begin{aligned} \int_{B_1} \frac{\Delta(u_\varepsilon^2)}{|x|^{d-2}}\,dx &= \int_{B_1} \Delta(|x|^{2-d})u_\varepsilon^2\,dx \\ &\quad + \int_{\partial B_1} \left[\frac{\partial(u_\varepsilon^2)}{\partial n}|x|^{2-d} - \frac{\partial(|x|^{2-d})}{\partial n}u_\varepsilon^2\right] d\mathcal{H}^{d-1} \\ &\leq -d(d-2)\omega_d u_\varepsilon^2(0) + \int_{\partial B_1} 2u_\varepsilon \frac{\partial u_\varepsilon}{\partial n}\,d\mathcal{H}^{d-1} \\ &\quad + (d-2)\int_{\partial B_1} u_\varepsilon^2\,d\mathcal{H}^{d-1} \\ &\leq \int_{\partial B_1} 2u_\varepsilon \frac{\partial u_\varepsilon}{\partial n}\,d\mathcal{H}^{d-1} + (d-2)\int_{\partial B_1} u_\varepsilon^2\,d\mathcal{H}^{d-1}, \end{aligned} \tag{4.54}$$

where we used that $-\Delta(|x|^{2-d}) = d(d-2)\omega_d\delta_0$ (see for example [61, Section 2.2.1]). Since (ii) holds, we may pass to the limit in (4.53) and (4.54), as $\varepsilon \to 0$. Using (4.52) we obtain the inequality

$$\begin{aligned} 2\int_{B_1} \frac{|\nabla u|^2}{|x|^{d-2}}\,dx &\leq C_d + C_d\Big(\int_{\partial B_1} u^2\,d\mathcal{H}^{d-1}\Big)^{1/2} \\ &\quad +2\Big(\int_{\partial B_1} u^2\,d\mathcal{H}^{d-1}\Big)^{\frac12}\Big(\int_{\partial B_1}\Big|\frac{\partial u}{\partial n}\Big|^2 d\mathcal{H}^{d-1}\Big)^{\frac12} + (d-2)\int_{\partial B_1} u^2\,d\mathcal{H}^{d-1} \\ &\leq C_d + C_d\sqrt{\frac{B_u(1)}{\lambda}} + \frac{1}{\alpha}\int_{\partial B_1}\Big|\frac{\partial u}{\partial n}\Big|^2 d\mathcal{H}^{d-1} + \frac{\alpha + (d-2)}{\lambda}\int_{\partial B_1}\Big|\frac{\partial u}{\partial \tau}\Big|^2 d\mathcal{H}^{d-1} \\ &= C_d + C_d\sqrt{\frac{B_u(1)}{\lambda}} + \frac{B_u(1)}{\alpha}, \end{aligned}$$

where the last equality is due to the definition of α from (4.51). □

4.3.3. The two-phase monotonicity formula

In this subsection we prove the Caffarelli-Jerison-Kenig monotonicity formula for Sobolev functions. We follow precisely the proof given in [41], since the only estimates, where the continuity of u_i was used are now isolated in Lemma 4.27 and Lemma 4.29.

Theorem 4.30 (Two-phase monotonicity formula). *Let $B_1 \subset \mathbb{R}^d$ be the unit ball in $\mathbb{R}^d$ and $u_1, u_2 \in H^1(B_1)$ be two non-negative Sobolev functions such that*

$$\Delta u_i + 1 \geq 0, \quad \text{for } i = 1, 2, \qquad \text{and} \qquad u_1 u_2 = 0 \ \text{a.e. in } B_1. \tag{4.55}$$

Then there is a dimensional constant C_d such that for each $r \in (0,1)$ we have

$$\prod_{i=1}^{2} \left(\frac{1}{r^2} \int_{B_r} \frac{|\nabla u_i|^2}{|x|^{d-2}}\, dx \right) \leq C_d \left(1 + \sum_{i=1}^{2} \int_{B_1} \frac{|\nabla u_i|^2}{|x|^{d-2}}\, dx \right)^2. \tag{4.56}$$

For the sake of simplicity of the notation, for $i = 1, 2$ and u_1, u_2 as in Theorem 4.30, we set

$$A_i(r) := A_{u_i}(r) = \int_{B_r} \frac{|\nabla u_i|^2}{|x|^{d-2}}\, dx. \tag{4.57}$$

In the next Lemma we estimate the derivative (with respect to r) of the quantity that appears in the left-hand side of (4.56) from Theorem 4.30.

Lemma 4.31. *Let u_1 and u_2 be as in Theorem 4.30. Then there is a dimensional constant $C_d > 0$ such that the following implication holds: if $A_1(1/4) \geq C_d$ and $A_2(1/4) \geq C_d$, then*

$$\frac{d}{dr}\left[\frac{A_1(r)A_2(r)}{r^4} \right] \geq -C_d \left(\frac{1}{\sqrt{A_1(r)}} + \frac{1}{\sqrt{A_2(r)}} \right) \frac{A_1(r)A_2(r)}{r^4},$$

for Lebesgue almost every $r \in [1/4, 1]$.

Proof. We set, for $i = 1, 2$ and $r > 0$,

$$B_i(r) = \int_{\partial B_r} |\nabla u_i|^2\, d\mathcal{H}^{d-1}.$$

Since A_1 and A_2 are increasing functions, they are differentiable almost everywhere on $(0, +\infty)$. Moreover, $A_i'(r) = r^{2-d} B_i(r)$, for $i = 1, 2$, in sense of distributions and the function

$$r \mapsto r^{-4} A_1(r) A_2(r),$$

is differentiable a.e. with derivative

$$\frac{d}{dr}\left[\frac{A_1(r)A_2(r)}{r^4}\right]=\left(-\frac{4}{r}+\frac{r^{2-d}B_1(r)}{A_1(r)}+\frac{r^{2-d}B_2(r)}{A_2(r)}\right)\frac{A_1(r)A_2(r)}{r^4}.$$

Thus, it is sufficient to prove, that for almost every $r\in[1/4,1]$ we have

$$-\frac{4}{r}+\frac{r^{2-d}B_1(r)}{A_1(r)}+\frac{r^{2-d}B_2(r)}{A_2(r)}\geq -C_d\left(\frac{1}{\sqrt{A_1(r)}}+\frac{1}{\sqrt{A_2(r)}}\right). \quad (4.58)$$

Using the rescaling from (4.48), it is sufficient to prove (4.58) in the case $r=1$. We consider two cases:

(A) Suppose that $B_1(1)\geq 4A_1(1)$ or $B_2(1)\geq 4A_2(1)$. In both cases we have

$$-4+\frac{B_1(1)}{A_1(1)}+\frac{B_2(1)}{A_2(1)}\geq 0,$$

which gives (4.58).

(B) Suppose that $B_1(1)\leq 4A_1(1)$ and $B_2(1)\leq 4A_2(1)$. By Lemma 4.29 with the additional notation $\alpha_i:=\alpha(u_i,1)$ and $\lambda_i:=\lambda(u_i,1)$ we have

$$A_1(1)\leq C_d+C_d\sqrt{\frac{B_1(1)}{\lambda_1}+\frac{B_1(1)}{2\alpha_1}}\leq C_d+C_d\sqrt{\frac{A_1(1)}{\lambda_1}+\frac{B_1(1)}{2\alpha_1}}. \quad (4.59)$$

We now consider two sub-cases:

(B1) Suppose that $\alpha_1\geq 4$ or $\alpha_2\geq 4$. By (4.59), we get

$$A_1(1)\leq 2C_d\sqrt{\frac{A_1(1)}{\lambda_1}+\frac{B_1(1)}{\alpha_1}}.$$

Now since $\sqrt{\lambda_1}\geq\alpha_1\geq 4$ we obtain

$$4A_1(1)\leq 2C_d\sqrt{A_1(1)}+B_1(1)=A_1(1)\left(\frac{2C_d}{\sqrt{A_1(1)}}+\frac{B_1(1)}{A_1(1)}\right),$$

which gives (4.58).

(B2) Suppose that $\alpha_1\leq 4$ and $\alpha_2\leq 4$. Then for both $i=1,2$, we have $C_d\leq\sqrt{A_i/\lambda}$ and so, by (4.59)

$$2\alpha_i A_i(1)\leq C_d\sqrt{A_i(1)}+B_i(1).$$

Thus (4.58) reduces to $\alpha_1+\alpha_2\geq 2$, which was proved in [63] (see also [43]). □

The following is the discretized version of Lemma 4.31 and also the main ingredient in the proof of Theorem 4.30.

Lemma 4.32. *Let u_1 and u_2 be as in Theorem 4.30. Then there is a dimensional constant $C_d > 0$ such that the following implication holds: if for some $r \in (0,1)$*

$$\frac{1}{r^4}\int_{B_r}\frac{|\nabla u_1|^2}{|x|^{d-2}}\,dx \geq C_d \qquad \text{and} \qquad \frac{1}{r^4}\int_{B_r}\frac{|\nabla u_2|^2}{|x|^{d-2}}\,dx \geq C_d,$$

then we have the estimate

$$4^4 A_1(r/4)A_2(r/4) \leq \big(1+\delta_{12}(r)\big)A_1(r)A_2(r), \tag{4.60}$$

where

$$\delta_{12}(r) := C_d\left(\left(\frac{1}{r^4}\int_{B_r}\frac{|\nabla u_1|^2}{|x|^{d-2}}\,dx\right)^{-1/2}+\left(\frac{1}{r^4}\int_{B_r}\frac{|\nabla u_2|^2}{|x|^{d-2}}\,dx\right)^{-1/2}\right). \tag{4.61}$$

Proof. Using the rescaling $u_r(x) = r^{-2}u(rx)$, we can suppose that $r = 1$. We consider two cases:

(A) If $A_1(1) \geq 4^4 A_1(1/4)$ or $A_2(1) \geq 4^4 A_2(1/4)$, then

$$A_1(1)A_2(1) - 4^4 A_1(1/4)A_2(1/4) \geq A_1(1)\Big(A_2(1) - 4^4 A_2(1/4)\Big) \geq 0,$$

and so, we have the claim.

(B) Suppose that $A_1(1) \leq 4^4 A_1(1/4)$ or $A_2(1) \leq 4^4 A_2(1/4)$. Then $A_1(r) \geq C_d$ and $A_2(r) \geq C_d$, for every $r \in (1/4, 1)$ and so, we may apply Lemma 4.31

$$\begin{aligned}
&A_1(1)A_2(1) - 4^4 A_1(1/4)A_2(1/4)\\
&\geq -C_d\int_{1/4}^{1}\left(\frac{1}{\sqrt{A_1(r)}}+\frac{1}{\sqrt{A_2(r)}}\right)A_1(r)A_2(r)\,dr\\
&\geq -C_d\frac{3}{4}\left(\frac{1}{\sqrt{A_1(1/4)}}+\frac{1}{\sqrt{A_2(1/4)}}\right)A_1(1)A_2(1)\\
&\geq -C_d\frac{3}{4}\left(\frac{16}{\sqrt{A_1(1)}}+\frac{16}{\sqrt{A_2(1)}}\right)A_1(1)A_2(1),
\end{aligned}$$

where in the second inequality we used the monotonicity of A_1 and A_2. □

The following lemma corresponds to [41, Lemma 2.9] and its proof implicitely contains [41, Lemma 2.1] and [41, Lemma 2.3]. We state it here as a single separate result since it is only used in the proof of the two-phase monotonicity formula (Theorem 4.30).

Lemma 4.33. *Let u_1 and u_2 be as in Theorem* 4.30. *Then there are dimensional constants $C_d > 0$ and $\varepsilon > 0$ such that the following implication holds: if $A_1(1) \geq C_d$, $A_2(1) \geq C_d$ and $4^4 A_1(1/4) \geq A_1(1)$, then $A_2(1/4) \leq (1-\varepsilon) A_2(1)$.*

Proof. The idea of the proof is roughly speaking to show that if $A_1(1/4)$ is not too small with respect to $A_1(1)$, then there is a big portion of the set $\{u_1 > 0\}$ in the annulus $B_{1/2} \setminus B_{1/4}$. This of course implies that there is a small portion of $\{u_2 > 0\}$ in $B_{1/2} \setminus B_{1/4}$ and so $A_2(1/4)$ is much smaller than $A_2(1)$. We will prove the Lemma in two steps.

Step 1. *There are dimensional constants $C > 0$ and $\delta > 0$ such that if $A_1(1) \geq C$ and $4^4 A_1(1/4) \geq A_1(1)$, then $|\{u_1 > 0\} \cap B_{1/2} \setminus B_{1/4}| \geq \delta |B_{1/2} \setminus B_{1/4}|$.*

By Lemma 4.27 we have that

$$A_1(1/4) \leq C_d + C_d \int_{B_{1/2}\setminus B_{1/4}} u_1^2\,dx,$$

and by choosing $C > 0$ large enough we get

$$A_1(1/4) \leq C_d \int_{B_{1/2}\setminus B_{1/4}} u_1^2\,dx.$$

Now if $|\{u_1 > 0\} \cap B_{1/2} \setminus B_{1/4}| > 1/2 |B_{1/2} \setminus B_{1/4}|$, then there is nothing to prove. Otherwise, there is a dimensional constant C_d such that the Sobolev inequality holds

$$\left(\int_{B_{1/2}\setminus B_{1/4}} u_1^{\frac{2d}{d-2}}\,dx\right)^{\frac{d-2}{d}} \leq C_d \int_{B_{1/2}\setminus B_{1/4}} |\nabla u_1|^2\,dx \leq C_d A_1(1).$$

By the Hölder inequality, we get

$$\begin{aligned} A_1(1/4) &\leq C_d |\{u_1 > 0\} \cap B_{1/2} \setminus B_{1/4}|^{\frac{2}{d}} A_1(1) \\ &\leq C_d |\{u_1 > 0\} \cap B_{1/2} \setminus B_{1/4}|^{\frac{2}{d}} 4^4 A_1(1/4), \end{aligned}$$

which gives the claim[3] of *Step* 1 since $A_1(1/4) > 0$.

[3] In dimension 2 the argument is analogous.

Step 2. Let $\delta \in (0,1)$. Then there are constants $C > 0$ and $\varepsilon > 0$, depending on δ and the dimension, such that if $A_2(1) \geq C$ and $|\{u_2 > 0\} \cap B_{1/2} \setminus B_{1/4}| \leq (1-\delta)|B_{1/2} \setminus B_{1/4}|$, then $A_2(1/4) \leq (1-\varepsilon)A_2(1)$.

Since $|\{u_2 = 0\} \cap B_{1/2} \setminus B_{1/4}| \geq \delta |B_{1/2} \setminus B_{1/4}|$, there is a constant $C_\delta > 0$ such that

$$\int_{B_{1/2}\setminus B_{1/4}} u_2^2\,dx \leq C_\delta \int_{B_{1/2}\setminus B_{1/4}} |\nabla u_2|^2\,dx.$$

We can suppose that

$$\int_{B_{1/4}} |\nabla u_2|^2\,dx \geq \frac{1}{2}\int_{B_1} |\nabla u_2|^2\,dx \geq \frac{C}{2},$$

since otherwise the claim holds with $\varepsilon = 1/2$. Applying Lemma 4.27 we obtain

$$\begin{aligned}\int_{B_{1/4}} |\nabla u_2|^2\,dx &\leq C_d + C_d \int_{B_{1/2}\setminus B_{1/4}} u_2^2\,dx \\ &\leq C_d + C_d C_\delta \left(\int_{B_1} |\nabla u_2|^2\,dx - \int_{B_{1/4}} |\nabla u_2|^2\,dx\right) \\ &\leq \left(C_d C_\delta + \frac{1}{2}\right)\int_{B_1} |\nabla u_2|^2\,dx - C_d C_\delta \int_{B_{1/4}} |\nabla u_2|^2\,dx,\end{aligned} \tag{4.62}$$

where for the last inequality we chose $C > 0$ large enough. □

The proof of Theorem 4.30 continues exactly as in [41]. In what follows, for $i = 1, 2$, we adopt the notation

$$A_i^k := A_i(4^{-k}), \qquad b_i^k := 4^{4k} A_i(4^{-k}) \qquad \text{and} \qquad \delta_k := \delta_{12}(4^{-k}),$$

where A_i was defined in (4.57) and δ_{12} in (4.61).

Proof of Theorem 4.30. Let $M > 0$ be a fixed constant, larger than the dimensional constants in Lemma 4.31, Lemma 4.32 and Lemma 4.33.

Suppose that $k \in \mathbb{N}$ is such that

$$4^{4k} A_1^k A_2^k \geq M\big(1 + A_1^0 + A_2^0\big)^2. \tag{4.63}$$

Then we have

$$b_1^k = 4^{4k} A_1^k \geq M \qquad \text{and} \qquad b_2^k = 4^{4k} A_2^k \geq M. \tag{4.64}$$

Thus, applying Lemma 4.32 we get that if $k \in \mathbb{N}$ satisfies (4.63), then

$$4^4 A_1^{k+1} A_2^{k+1} \leq (1+\delta_k) A_1^k A_2^k. \tag{4.65}$$

We now denote with $S_1(M)$ the set

$$S_1(M) := \Big\{k \in \mathbb{N} : \ 4^{4k} A_1^k A_2^k \leq M\big(1 + A_1^0 + A_2^0\big)^2\Big\},$$

and with S_2 the set

$$S_2 := \Big\{k \in \mathbb{N} : \ 4^4 A_1^{k+1} A_2^{k+1} \leq A_1^k A_2^k\Big\}.$$

Let $L \in \mathbb{N}$ be such that $L \notin S_1(M)$ and let $l \in \{0, 1, \dots, L\}$ be the largest index such that $l \in S_1(M)$. Note that if $\{l+1, \dots, L-1\} \setminus S_2 = \emptyset$, then we have

$$4^{4L} A_1^L A_2^L \leq 4^{4(L-1)} A_1^{L-1} A_2^{L-1} \leq \dots \leq 4^{4(l+1)} A_1^{l+1} A_2^{l+1} \leq 4^4 4^{4l} A_1^l A_2^l,$$

which gives that $L \in S_1(4^4 M)$.

Repeating the proof of [41, Theorem 1.3], we consider the decreasing sequence of indices

$$l+1 \leq k_m < \dots < k_2 < k_1 \leq L,$$

constructed as follows:

- k_1 is the largest index in the set $\{l+1, \dots, L\}$ such that $k_1 \notin S_2$;
- k_{j+1} is the largest integer in $\{l+1, \dots, k_j - 1\} \setminus S_2$ such that

$$b_1^{k_{j+1}+1} \leq (1+\delta_{k_{j+1}}) b_1^{k_j} \qquad \text{and} \qquad b_2^{k_{j+1}+1} \leq (1+\delta_{k_{j+1}}) b_2^{k_j}. \tag{4.66}$$

We now conclude the proof in four steps.

Step 1. $4^{4L} A_1^L A_2^L \leq 4^{4(k_1+1)} A_1^{k_1} A_2^{k_1}$.

Indeed, since $\{k_1+1, \dots, L\} \subset S_2$, we have

$$\begin{aligned} 4^{4L} A_1^L A_2^L &\leq 4^{4(L-1)} A_1^{L-1} A_2^{L-1} \leq \dots \\ &\leq 4^{4(k_1+1)} A_1^{k_1+1} A_2^{k_1+1} \leq 4^4 4^{4k_1} A_1^{k_1} A_2^{k_1}. \end{aligned}$$

Step 2. $4^{4k_m} A_1^{k_m} A_2^{k_m} \leq 4^4 M\big(1 + A_1^0 + A_2^0\big)^2$.

Let $\tilde{k} \in \{l+1, \dots, k_m - 1\}$ be the smallest integer such that $\tilde{k} \notin S_2$. If no such $\tilde{k}$ exists, then we have

$$4^{4k_m} A_1^{k_m} A_2^{k_m} \leq \dots \leq 4^{4(l+1)} A_1^{l+1} A_2^{l+1} \leq 4^4 4^{4l} A_1^l A_2^l \leq 4^4 M\big(1 + A_1^0 + A_2^0\big)^2.$$

Otherwise, since k_m is the last index in the sequence constructed above, we have that

$$b_1^{\tilde{k}+1} > (1+\delta_{\tilde{k}})b_1^{k_m} \qquad \text{or} \qquad b_2^{\tilde{k}+1} > (1+\delta_{\tilde{k}})b_2^{k_m}.$$

Assuming, without loss of generality that the first inequality holds, we get

$$\begin{aligned} 4^{4k_m} A_1^{k_m} A_2^{k_m} &\le \frac{4^{4(\tilde{k}+1)} A_1^{\tilde{k}+1}}{1+\delta_{\tilde{k}}} A_2^{\tilde{k}+1} \le 4^{4\tilde{k}} A_1^{\tilde{k}} A_2^{\tilde{k}} \le \dots \le 4^4 4^{4l} A_1^l A_2^l \\ &\le 4^4 M\big(1 + A_1^0 + A_2^0\big)^2, \end{aligned}$$

where in the second inequality we used Lemma 4.32 and afterwards we used the fact that $\{l+1,\dots,\tilde{k}-1\} \subset S_2$.

Step 3. $4^{4k_j} A_1^{k_j} A_2^{k_j} \le (1+\delta_{k_{j+1}}) 4^{4k_{j+1}} A_1^{k_{j+1}} A_2^{k_{j+1}}$.

We reason as in *Step* 2 choosing $\tilde{k} \in \{k_{j+1}+1,\dots,k_j-1\}$ to be the smallest integer such that $\tilde{k} \notin S_2$. If no such $\tilde{k}$ exists, then $\{k_{j+1}+1,\dots,k_j-1\} \subset S_2$ and so we have

$$\begin{aligned} 4^{4k_j} A_1^{k_j} A_2^{k_j} &\le 4^{4(k_j-1)} A_1^{k_j-1} A_2^{k_j-1} \le \dots \le 4^{4(k_{j+1}+1)} A_1^{k_{j+1}+1} A_2^{k_{j+1}+1} \\ &\le (1+\delta_{k_{j+1}}) 4^{4k_{j+1}} A_1^{k_{j+1}} A_2^{k_{j+1}}, \end{aligned}$$

where the last inequality is due to Lemma 4.32. Suppose now that $\tilde{k}$ exists. Since k_j and k_{j+1} are consecutive indices, we have that

$$b_1^{\tilde{k}+1} > (1+\delta_{\tilde{k}})b_1^{k_j} \qquad \text{or} \qquad b_2^{\tilde{k}+1} > (1+\delta_{\tilde{k}})b_2^{k_j}.$$

As in *Step* 2, we assume that the first inequality holds. By Lemma 4.32 we have

$$\begin{aligned} 4^{4k_j} A_1^{k_j} A_2^{k_j} &\le \frac{4^{4(\tilde{k}+1)} A_1^{\tilde{k}+1}}{1+\delta_{\tilde{k}}} A_2^{\tilde{k}+1} \le 4^{4\tilde{k}} A_1^{\tilde{k}} A_2^{\tilde{k}} \le \dots \\ &\le 4^{4(k_{j+1}+1)} A_1^{k_{j+1}+1} A_2^{k_{j+1}+1} \le (1+\delta_{k_{j+1}}) 4^{4k_{j+1}} A_1^{k_{j+1}} A_2^{k_{j+1}}. \end{aligned}$$

which concludes the proof of *Step* 3.

Step 4. *Conclusion*. Combining the results of *Steps* 1, 2 and 3, we get

$$4^{4L} A_1^L A_2^L \le 4^8 M\big(1 + A_1^0 + A_2^0\big)^2 \prod_{j=1}^m (1+\delta_{k_j}). \tag{4.67}$$

We now prove that the sequences $b_1^{k_j}$ and $b_2^{k_j}$ can both be estimated from above by a geometric progression. Indeed, since $k_j \notin S_2$, we have

$$A_1^{k_j} A_2^{k_j} \leq 4^4 A_1^{k_j+1} A_2^{k_j+1} \leq 4^4 A_1^{k_j+1} A_2^{k_j}.$$

Thus $A_1^{k_j} \leq 4^4 A_1^{k_j+1}$ and analogously $A_2^{k_j} \leq 4^4 A_2^{k_j+1}$. Applying Lemma 4.33 we get

$$A_1^{k_j+1} \leq (1-\varepsilon) A_1^{k_j} \qquad \text{and} \qquad A_2^{k_j+1} \leq (1-\varepsilon) A_2^{k_j}.$$

Using again the fact that $k_j \notin S_2$, we obtain

$$A_1^{k_j} A_2^{k_j} \leq 4^4 A_1^{k_j+1} A_2^{k_j+1} \leq 4^4 A_1^{k_j+1} (1-\varepsilon) A_2^{k_j},$$

and so

$$\begin{aligned} b_1^{k_j} \leq (1-\varepsilon) b_1^{k_j+1} \quad &\text{and} \quad b_2^{k_j} \leq (1-\varepsilon) b_2^{k_j+1}, \\ &\text{for every} \ \ j = 1, \dots, m. \end{aligned} \tag{4.68}$$

By the construction of the sequence k_j, we have that for $i = 1, 2$

$$b_i^{k_j} \geq \frac{b_i^{k_{j+1}+1}}{1+\delta_{k_{j+1}}} \geq \frac{b_i^{k_{j+1}}}{(1+\delta_{k_{j+1}})(1-\varepsilon)} \geq \Big(1 - \frac{\varepsilon}{2}\Big)^{-1} b_i^{k_{j+1}},$$

where for the last inequality we choose M large enough such that $k \notin S_1(M)$ implies $\delta_k \leq \varepsilon/2$, where ε is the dimensional constant from Lemma 4.33. Setting $\sigma = (1-\varepsilon/2)^{1/2}$, we have that

$$b_i^{k_j} \geq \sigma^{-2} b_i^{k_{j+1}} \geq \dots \geq \sigma^{2(j-m)} b_i^{k_m} \geq M \sigma^{2(j-m)},$$

which by the definition of δ_{k_j} gives $\delta_{k_j} \leq \dfrac{C_d}{M} \sigma^{m-j} \leq C_d \sigma^{m-j}$, for $M > 0$ large enough, and

$$\begin{aligned} 4^{4L} A_1^L A_2^L &\leq \prod_{j=1}^{m} (1 + C_d \sigma^j) 4^8 M \big(1 + A_1^0 + A_2^0\big)^2 \\ &\leq \exp\Big(\sum_{j=1}^{m} \log(1 + C_d \sigma^j)\Big) 4^8 M \big(1 + A_1^0 + A_2^0\big)^2 \\ &\leq \exp\Big(C_d \sum_{j=1}^{m} \sigma^j\Big) 4^8 M \big(1 + A_1^0 + A_2^0\big)^2 \\ &\leq \exp\Big(\frac{C_d}{1-\sigma}\Big) 4^8 M \big(1 + A_1^0 + A_2^0\big)^2, \end{aligned} \tag{4.69}$$

which concludes the proof. ☐

4.3.4. Multiphase monotonicity formula

This subsection is dedicated to the multiphase version of Theorem 4.30, proved in [29]. The proof follows the same idea as in [41]. The major technical difference with respect to the two-phase case consists in the fact that we only need Lemma 4.32 and its three-phase analogue Lemma 4.38, while the estimate from Lemma 4.33 is not necessary.

Theorem 4.34 (Three-phase monotonicity formula). *Let $B_1 \subset \mathbb{R}^d$ be the unit ball in $\mathbb{R}^d$ and let $u_i \in H^1(B_1)$, $i = 1, 2, 3$, be three non-negative Sobolev functions such that*

$$\Delta u_i + 1 \geq 0, \quad \forall i = 1, 2, 3, \qquad \textit{and} \qquad u_i u_j = 0 \ \textit{a.e. in } B_1, \ \forall i \neq j.$$

Then there are dimensional constants $\varepsilon > 0$ and $C_d > 0$ such that for each $r \in (0, 1)$ we have

$$\prod_{i=1}^{3}\left(\frac{1}{r^{2+\varepsilon}}\int_{B_r}\frac{|\nabla u_i|^2}{|x|^{d-2}}\,dx\right) \leq C_d\left(1+\sum_{i=1}^{3}\int_{B_1}\frac{|\nabla u_i|^2}{|x|^{d-2}}\,dx\right)^3. \tag{4.70}$$

As a corollary, we obtain the following result.

Corollary 4.35 (Multiphase monotonicity formula). *Let $m \geq 2$ and $B_1 \subset \mathbb{R}^d$ be the unit ball in $\mathbb{R}^d$. Let $u_i \in H^1(B_1)$, $i = 1, \dots, m$, be m non-negative Sobolev functions such that*

$$\Delta u_i + 1 \geq 0, \quad \forall i = 1, \dots, m, \qquad \textit{and} \qquad u_i u_j = 0 \ \textit{a.e. in } B_1, \ \forall i \neq j.$$

Then there are dimensional constants $\varepsilon > 0$ and $C_d > 0$ such that for each $r \in (0, 1)$ we have

$$\prod_{i=1}^{m}\left(\frac{1}{r^{2+\varepsilon}}\int_{B_r}\frac{|\nabla u_i|^2}{|x|^{d-2}}\,dx\right) \leq C_d\left(1+\sum_{i=1}^{m}\int_{B_1}\frac{|\nabla u_i|^2}{|x|^{d-2}}\,dx\right)^m. \tag{4.71}$$

Remark 4.36. We note that the additional decay $r^{-\varepsilon}$ provided by the presence of a third phase is not optimal. Indeed, at least in dimension two, we expect that $\varepsilon = m-2$, where m is the number of phases involved. In our proof the constant ε cannot exceed $2/3$ in any dimension.

We now proceed with the proof of the three-phase formula. Before we start with the proof of Theorem 4.34 we will need some preliminary results, analogous to Lemma 4.31 and Lemma 4.32.

We recall that, for u_1, u_2 and u_3 as in Theorem 4.34, we use the notation

$$A_i(r) = \int_{B_r}\frac{|\nabla u_i|^2}{|x|^{d-2}}\,dx, \qquad \text{for } i = 1, 2, 3. \tag{4.72}$$

Lemma 4.37. *Let u_1, u_2 and u_3 be as in Theorem 4.34. Then there are dimensional constants $C_d > 0$ and $\varepsilon > 0$ such that if $A_i(1/4) \geq C_d$, for every $i = 1, 2, 3$, then*

$$\frac{d}{dr}\left[\frac{A_1(r)A_2(r)A_3(r)}{r^{6+3\varepsilon}}\right] \geq -C_d\left(\frac{1}{\sqrt{A_1(r)}}+\frac{1}{\sqrt{A_2(r)}}+\frac{1}{\sqrt{A_3(r)}}\right)\frac{A_1(r)A_2(r)A_3(r)}{r^{6+3\varepsilon}},$$

for Lebesgue almost every $r \in [1/4, 1]$.

Proof. We set, for $i = 1, 2, 3$ and $r > 0$,

$$B_i(r) = \int_{\partial B_r} |\nabla u_i|^2 \, d\mathcal{H}^{d-1}.$$

Since A_i, for $i = 1, 2, 3$, are increasing functions they are differentiable almost everywhere on $\mathbb{R}$ and $A_i'(r) = r^{2-d}B_i(r)$ in sense of distributions. Thus, the function

$$r \mapsto r^{-(6+3\varepsilon)}A_1(r)A_2(r)A_3(r),$$

is differentiable a.e. and we have

$$\frac{d}{dr}\left[\frac{A_1(r)A_2(r)A_3(r)}{r^{6+3\varepsilon}}\right] = \left(-\frac{6+3\varepsilon}{r}+\frac{r^{2-d}B_1(r)}{A_1(r)}+\frac{r^{2-d}B_2(r)}{A_2(r)}+\frac{r^{2-d}B_3(r)}{A_3(r)}\right)\frac{A_1(r)A_2(r)A_3(r)}{r^{6+3\varepsilon}}.$$

Thus, it is sufficient to prove that for almost every $r \in [1/4, 1]$ we have

$$\begin{aligned} &-\frac{6+3\varepsilon}{r}+r^{2-d}\left(\frac{B_1(r)}{A_1(r)}+\frac{B_2(r)}{A_2(r)}+\frac{B_3(r)}{A_3(r)}\right) \\ &\geq -C_d\left(\frac{1}{\sqrt{A_1(r)}}+\frac{1}{\sqrt{A_2(r)}}+\frac{1}{\sqrt{A_3(r)}}\right), \end{aligned} \tag{4.73}$$

and, by rescaling, we may assume that $r = 1$. We consider two cases.

(A) Suppose that there is some $i = 1, 2, 3$, say $i = 1$, such that $(6 + 3\varepsilon)A_1(1) \leq B_1(1)$. Then we have

$$-(6+3\varepsilon)+\frac{B_1(1)}{A_1(1)}+\frac{B_2(1)}{A_2(1)}+\frac{B_3(1)}{A_3(1)} \geq -(6+3\varepsilon)+\frac{B_1(1)}{A_1(1)} \geq 0,$$

which proves (4.73) and the lemma.

(B) Suppose that for each $i = 1, 2, 3$ we have $(6 + 3\varepsilon)A_i(1) \geq B_i(1)$. Since, for every $i = 1, 2, 3$ we have $A_i(1) \geq C_d$, by Lemma 4.29 with the additional notation $\alpha_i := \alpha(u_i, 1)$ and $\lambda_i := \lambda(u_i, 1)$ and by choosing $\varepsilon > 0$ small enough and then $C_d > 0$ large enough, we have

$$(2-\varepsilon)A_i(1) \leq C_d\sqrt{B_i(1)/\lambda_i} + B_i(1)/\alpha_i \leq C_d\sqrt{A_i(1)/\lambda_i} + B_i(1)/\alpha_i.$$

Moreover, $\alpha_i^2 \leq \lambda_i$, implies

$$(2 - \varepsilon)\alpha_i A_i(1) \leq C_d\sqrt{A_i(1)} + B_i(1). \tag{4.74}$$

Dividing both sides by $A_i(1)$ and summing for $i = 1, 2, 3$, we obtain

$$(2 - \varepsilon)(\alpha_1 + \alpha_2 + \alpha_3) \leq C_d \sum_{i=1}^{3} \frac{1}{\sqrt{A_i(1)}} + \sum_{i=1}^{3} \frac{B_i(1)}{A_i(1)},$$

and so, in order to prove (4.73), it is sufficient to prove that

$$\alpha_1 + \alpha_2 + \alpha_3 \geq \frac{6 + 3\varepsilon}{2 - \varepsilon}. \tag{4.75}$$

Let $\Omega_1^*, \Omega_2^*, \Omega_3^* \subset \partial B_1$ be the optimal partition of the sphere ∂B_1 for the characteristic constant α, *i.e.* the triple $\{\Omega_1^*, \Omega_2^*, \Omega_3^*\}$ is a solution of the problem

$$\min\Big\{\alpha(\Omega_1) + \alpha(\Omega_2) + \alpha(\Omega_3) : \ \Omega_i \subset \partial B_1\,, \\ \forall i;\ \mathcal{H}^{d-1}(\Omega_i \cap \Omega_j) = 0, \forall i \neq j\Big\}. \tag{4.76}$$

We recall that for a set $\Omega \subset \partial B_1$, the characteristic constant $\alpha(\Omega)$ is the unique positive real number such that $\lambda(\Omega) = \alpha(\Omega)(\alpha(\Omega) + d - 2)$, where

$$\lambda(\Omega) = \min\left\{\frac{\int_{\partial B_1} |\nabla v|^2 \mathcal{H}^{d-1}}{\int_{\partial B_1} v^2 \mathcal{H}^{d-1}} : v \in H^1(\partial B_1),\ \mathcal{H}^{d-1}\big(\{u \neq 0\} \setminus \Omega\big) = 0\right\}.$$

We note that, by [63], $\alpha(\Omega_i^*) + \alpha(\Omega_j^*) \geq 2$, for $i \neq j$ and so summing on i and j, we have

$$3 \leq \alpha(\Omega_1^*) + \alpha(\Omega_2^*) + \alpha(\Omega_3^*) \leq \alpha_1 + \alpha_2 + \alpha_3.$$

Moreover, the first inequality is strict. Indeed, if this is not the case, then $\alpha(\Omega_1^*) + \alpha(\Omega_2^*) = 2$, which in turn gives that Ω_1^* and Ω_2^* are two opposite hemispheres (see for example [43]). Thus $\Omega_3^* = \emptyset$, which is impossible[4] Choosing ε to be such that $\frac{6+3\varepsilon}{2-\varepsilon}$ is smaller than the minimum in (4.76), the proof is concluded. □

Lemma 4.38. *Let u_1, u_2 and u_3 be as in Theorem 4.34. Then, there are dimensional constants $C_d > 0$ and $\varepsilon > 0$ such that the following implication holds: if for some $r > 0$*

$$\frac{1}{r^4}\int_{B_r}\frac{|\nabla u_i|^2}{|x|^{d-2}}\,dx \ge C_d, \qquad \textit{for all} \quad i = 1, 2, 3,$$

then we have the estimate

$$4^{(6+3\varepsilon)}A_1(r/4)\,A_2(r/4)\,A_3(r/4) \le \big(1+\delta_{123}(r)\big)A_1(r)A_2(r)A_3(r), \tag{4.77}$$

where

$$\delta_{123}(r) := C_d\sum_{i=1}^{3}\left(\frac{1}{r^4}\int_{B_r}\frac{|\nabla u_i|^2}{|x|^{d-2}}\,dx\right)^{-1/2}. \tag{4.78}$$

Proof. We first note that the (4.77) is invariant under the rescaling $u_r(x) = r^{-2}u(xr)$. Thus, we may suppose that $r = 1$. We consider two cases:

(A) Suppose that for some $i = 1, 2, 3$, say $i = 1$, we have $4^{6+3\varepsilon}A_1(1/4) \le A_1(1)$. Then we have

$$4^{6+3\varepsilon}A_1(1/4)A_2(1/4)A_3(1/4) \le A_1(1)A_2(1)A_3(1).$$

(B) Suppose that for every $i = 1, 2, 3$, we have $4^{6+3\varepsilon}A_i(1/4) \ge A_i(1)$. Then $A_i(1/4) \ge C_d$ for some C_d large enough and so, we can apply Lemma 4.37, obtaining that

$$\begin{aligned}
&A_1(1)A_2(1)A_3(1) - 4^{6+3\varepsilon}A_1(1/4)A_2(1/4)A_3(1/4)\\
&\qquad\ge -C_d\int_{1/4}^{1}\left(\sum_{i=1}^{3}\frac{1}{\sqrt{A_i(r)}}\right)A_1(r)A_2(r)A_3(r)\,dr\\
&\qquad\ge -C_d\frac{3}{4}\left(\sum_{i=1}^{3}\frac{1}{\sqrt{A_i(1/4)}}\right)A_1(1)A_2(1)A_3(1)\\
&\qquad\ge -3C_d4^{2+\frac{3}{2}\varepsilon}\left(\sum_{i=1}^{3}\frac{1}{\sqrt{A_i(1)}}\right)A_1(1)A_2(1)A_3(1),
\end{aligned}$$

which gives the claim. □

[4] For example, it is in contradiction with the equality $\alpha(\Omega_1^*) + \alpha(\Omega_3^*) = 2$, which is also implied by the contradiction assumption.

We now proceed with the proof of the three-phase monotonicity formula. We present two different proofs: the first one repeats precisely the main steps of the proof of Caffarelli, Jerison and Kenig, while the second one follows a more direct argument.

Proof I of Theorem 4.34. For $i = 1, 2, 3$, we adopt the notation

$$A_i^k := A_i(4^{-k}), \quad b_i^k := 4^{4k} A_i(4^{-k}) \quad \text{and} \quad \delta_k := \delta_{123}(4^{-k}), \qquad (4.79)$$

where A_i was defined in (4.57) and δ_{123} in (4.78).

Let $M > 0$ and let

$$S_1(M) = \Big\{k \in \mathbb{N} : 4^{(6+3\varepsilon)k} A_1^k A_2^k A_3^k \le M\big(1 + A_1^0 + A_2^0 + A_3^0\big)^3\Big\},$$
$$S_2 = \big\{k \in \mathbb{N} : 4^{6+3\varepsilon} A_1^{k+1} A_2^{k+1} A_3^{k+1} \le A_1^k A_2^k A_3^k\big\}.$$

We first note that if $k \notin S_1$, then we have

$$\begin{aligned} M\big(1 + A_1^0 + A_2^0 + A_3^0\big)^3 &\le 4^{(6+3\varepsilon)k} A_1^k A_2^k A_3^k \\ &\le 4^{-(2-3\varepsilon)k} b_1^k 4^{4k} A_2^k A_3^k \\ &\le b_1^k C_d\big(1 + A_1^0 + A_2^0 + A_3^0\big)^2, \end{aligned}$$

where the last inequality is due to the two-phase monotonicity formula (Theorem 4.30). Choosing $M > 0$ big enough, we have that

$$\Big(k \notin S_1(M)\Big) \Rightarrow \Big(b_i^k \ge C_d, \ \forall i = 1, 2, 3\Big).$$

Fix $L \in \mathbb{N}$ and suppose that $L \notin S_1(M)$. Let $l \in \{0, \dots, L\}$ be the largest index such that $l \in S_1(M)$. We now consider two cases for the interval $[l + 1, L]$.

Case 1. If $\{l + 1, \dots, L\} \subset S_2$, then we have

$$\begin{aligned} 4^{(6+3\varepsilon)L} A_1^L A_2^L A_3^L &\le \cdots \le 4^{(6+3\varepsilon)(l+1)} A_1^{l+1} A_2^{l+1} A_3^{l+1} \\ &\le 4^{6+3\varepsilon} M \big(1 + A_1^0 + A_2^0 + A_3^0\big)^3, \end{aligned}$$

and so $L \in S_1(4^{6+3\varepsilon} M)$.

Case 2. If $\{l + 1, \dots, L\} \setminus S_2 \neq \emptyset$, then we choose k_1 to be the largest index in $\{l + 1, \dots, L\} \setminus S_2$. Then we define the sequence

$$l + 1 \le k_m < \cdots < k_1 \le L,$$

by induction as

$$k_{j+1} := \max\Big\{k \in \{l+1,\dots,k_j-1\}\setminus S_2 :\ b_i^{k_{j+1}+1} \le (1+\delta_{k_{j+1}})b_i^{k_j},\ \forall i=1,2,3\Big\}.$$

The proof now proceeds in four steps.

Step 1. $4^{(6+3\varepsilon)L}A_1^L A_2^L A_3^L \le 4^{(6+3\varepsilon)(k_1+1)}A_1^{k_1}A_2^{k_1}A_3^{k_1}$.

Indeed, since $\{k_1+1,\dots L\}\subset S_2$, we have

$$\begin{aligned}4^{(6+3\varepsilon)L}A_1^L A_2^L A_3^L &\le \dots \le 4^{(6+3\varepsilon)(k_1+1)}A_1^{k_1+1}A_2^{k_1+1}A_3^{k_1+1}\\ &\le 4^{6+3\varepsilon}4^{(6+3\varepsilon)k_1}A_1^{k_1}A_2^{k_1}A_3^{k_1}.\end{aligned}$$

Step 2. $4^{(6+3\varepsilon)k_m}A_1^{k_m}A_2^{k_m}A_3^{k_m} \le 4^{6+3\varepsilon}M\big(1+A_1^0+A_2^0+A_3^0\big)^3$.

Let $\tilde k \in \{l+1,\dots,k_m-1\}$ be the smallest index such that $\tilde k \notin S_2$. If no such $\tilde k$ exists, then we have

$$\begin{aligned}4^{(6+3\varepsilon)k_m}A_1^{k_m}A_2^{k_m}A_3^{k_m} &\le \dots \le 4^{(6+3\varepsilon)(l+1)}A_1^{l+1}A_2^{l+1}A_3^{l+1}\\ &\le 4^{6+3\varepsilon}4^{(6+3\varepsilon)l}A_1^{l}A_2^{l}A_3^{l}\\ &\le 4^{6+3\varepsilon}M\big(1+A_1^0+A_2^0+A_3^0\big)^3.\end{aligned}$$

Otherwise, since k_m is the last index in the sequence constructed above, there exists $i\in\{1,2,3\}$ such that

$$b_i^{\tilde k+1} > (1+\delta_{\tilde k})b_i^{k_m}. \tag{4.80}$$

Assuming, without loss of generality that $i=1$, we get

$$4^{(6+3\varepsilon)k_m}A_1^{k_m}A_2^{k_m}A_3^{k_m} = 4^{(-2+3\varepsilon)k_m}b_1^{k_m}4^{4k_m}A_2^{k_m}A_3^{k_m}$$

$$\le 4^{(-2+3\varepsilon)k_m}(1+\delta_{\tilde k})^{-1}b_1^{\tilde k+1}(1+\delta_{23}(4^{-k_m+1}))4^{4(k_m-1)}A_2^{k_m-1}A_3^{k_m-1} \tag{4.81}$$

$$\le 4^{(-2+3\varepsilon)(k_m-1)}(1+\delta_{\tilde k})^{-1}b_1^{\tilde k+1}4^{4(k_m-1)}A_2^{k_m-1}A_3^{k_m-1} \tag{4.82}$$

...

$$\le 4^{(-2+3\varepsilon)(\tilde k+1)}(1+\delta_{\tilde k})^{-1}b_1^{\tilde k+1}4^{4(\tilde k+1)}A_2^{\tilde k+1}A_3^{\tilde k+1} \tag{4.83}$$

$$= 4^{(6+3\varepsilon)(\tilde k+1)}(1+\delta_{\tilde k})^{-1}A_1^{\tilde k+1}A_2^{\tilde k+1}A_3^{\tilde k+1}$$

$$\le 4^{(6+3\varepsilon)\tilde k}A_1^{\tilde k}A_2^{\tilde k}A_3^{\tilde k} \le \dots \le 4^{(6+3\varepsilon)(l+1)}A_1^{l+1}A_2^{l+1}A_3^{l+1} \tag{4.84}$$

$$\le 4^{6+3\varepsilon}4^{(6+3\varepsilon)l}A_1^{l}A_2^{l}A_3^{l} \le 4^{6+3\varepsilon}M\big(1+A_1^0+A_2^0+A_3^0\big)^3, \tag{4.85}$$

where in order to obtain (4.81) we used (4.80) and the two-phase estimate from Lemma 4.32; for (4.82), we absorb the term that appears after applying Lemma 4.32, using that if M is large enough and $\varepsilon < 2/3$, then $\big(1+\delta_{23}(4^{-k_m+1})\big)4^{-2+3\varepsilon} \leq 1$; repeating the same estimate as above we obtain (4.83); for (4.84), we use the three-phase Lemma 4.38 and then the fact that $\{l+1,\dots,\tilde{k}\} \subset S_2$; for the last inequality (4.85) we just observed that $l \in S_1(M)$.

Step 3. $4^{(6+3\varepsilon)k_j} A_1^{k_j} A_2^{k_j} A_3^{k_j} \leq \big(1+\delta_{k_{j+1}}\big)4^{(6+3\varepsilon)k_{j+1}} A_1^{k_{j+1}} A_2^{k_{j+1}} A_3^{k_{j+1}}$.

We reason as in **Step 2** choosing $\tilde{k} \in \{k_{j+1}+1,\dots,k_j-1\}$ to be the smallest index such that $\tilde{k} \notin S_2$. If no such $\tilde{k}$ exists, then $\{k_{j+1}+1,\dots,k_j-1\} \subset S_2$ and so we have

$$\begin{aligned}4^{(6+3\varepsilon)k_j} A_1^{k_j} A_2^{k_j} A_3^{k_j} &\leq \dots \leq 4^{(6+3\varepsilon)(k_{j+1}+1)} A_1^{k_{j+1}+1} A_2^{k_{j+1}+1} A_3^{k_{j+1}+1}\\ &\leq \big(1+\delta_{k_{j+1}}\big)4^{(6+3\varepsilon)k_{j+1}} A_1^{k_{j+1}} A_2^{k_{j+1}} A_3^{k_{j+1}},\end{aligned}$$

where the last inequality is due to Lemma 4.32. Suppose now that $\tilde{k}$ exists. Since k_j and k_{j+1} are consecutive indices, there exists some $i \in \{1,2,3\}$ such that

$$b_i^{\tilde{k}+1} > (1+\delta_{\tilde{k}})b_i^{k_j}. \tag{4.86}$$

Without loss of generality we may assume that $i = 1$.

$$\begin{aligned}&4^{(6+3\varepsilon)k_j} A_1^{k_j} A_2^{k_j} A_3^{k_j} = 4^{(-2+3\varepsilon)k_j} b_1^{k_j} 4^{4k_j} A_2^{k_j} A_3^{k_j}\\ &\leq 4^{(-2+3\varepsilon)k_j}(1+\delta_{\tilde{k}})^{-1} b_1^{\tilde{k}+1}\,(1+\delta_{23}(4^{-k_j+1}))4^{4(k_j-1)} A_2^{k_j-1} A_3^{k_j-1} \quad (4.87)\\ &\leq 4^{(-2+3\varepsilon)(k_j-1)}(1+\delta_{\tilde{k}})^{-1} b_1^{\tilde{k}+1}\,4^{4(k_j-1)} A_2^{k_j-1} A_3^{k_j-1} \quad (4.88)\\ &\dots\\ &\leq 4^{(-2+3\varepsilon)(\tilde{k}+1)}(1+\delta_{\tilde{k}})^{-1} b_1^{\tilde{k}+1}\,4^{4(\tilde{k}+1)} A_2^{\tilde{k}+1} A_3^{\tilde{k}+1} \quad (4.89)\\ &= 4^{(6+3\varepsilon)(\tilde{k}+1)}(1+\delta_{\tilde{k}})^{-1} A_1^{\tilde{k}+1} A_2^{\tilde{k}+1} A_3^{\tilde{k}+1}\\ &\leq 4^{(6+3\varepsilon)\tilde{k}} A_1^{\tilde{k}} A_2^{\tilde{k}} A_3^{\tilde{k}} \leq \dots \leq 4^{(6+3\varepsilon)(k_{j+1}+1)} A_1^{k_{j+1}+1} A_2^{k_{j+1}+1} A_3^{k_{j+1}+1} \quad (4.90)\\ &\leq (1+\delta_{k_{j+1}})4^{(6+3\varepsilon)k_{j+1}} A_1^{k_{j+1}} A_2^{k_{j+1}} A_3^{k_{j+1}}, \quad (4.91)\end{aligned}$$

where for (4.87) we used (4.86) and Lemma 4.32; for (4.88) and (4.89), we use that for $M > 0$ large enough and $\varepsilon < 2/3$ we have $\big(1+\delta_{23}(4^{-k_m+1})\big)4^{-2+3\varepsilon} \leq 1$; for (4.90), we apply Lemma 4.38 and then the fact that $\{l+1,\dots,\tilde{k}\} \subset S_2$; for the last inequality (4.91) we use Lemma 4.38.

Step 4. *Conclusion.*

By the Steps 1, 2 and 3 we have that

$$4^{(6+3\varepsilon)L}A_1^L A_2^L A_3^L \le 4^{2(6+3\varepsilon)}M\left(1+A_1^0+A_2^0+A_3^0\right)^3\prod_{j=1}^m(1+\delta_{k_j}) \quad (4.92)$$

we now prove that for each $i=1,2,3$ the sequence $b_i^{k_j}$ is majorized by a geometric progression depending on M. Indeed, since $k_j\notin S_2$, we have

$$\begin{aligned}A_1^{k_j}A_2^{k_j}A_3^{k_j} &\le 4^{6+3\varepsilon}A_1^{k_j+1}A_2^{k_j+1}A_3^{k_j+1}\\ &\le 4^{-(2-3\varepsilon)}4^4A_1^{k_j+1}(1+\delta_{23}(4^{-k_j}))A_2^{k_j}A_3^{k_j}\\ &\le \sigma^2 4^4 A_1^{k_j+1}A_2^{k_j}A_3^{k_j},\end{aligned}$$

for some dimensional constant $\sigma<1$, where the second inequality is due to Lemma 4.32 and the last inequality is due to the choice of M large enough and $\varepsilon<2/3$. Thus we obtain

$$b_i^{k_j}\le\sigma^2 b_i^{k_j+1},\qquad \forall i=1,2,3 \quad\text{and}\quad \forall j=1,\dots,m. \quad (4.93)$$

for each $i=1,2,3$ and each $k_j\in S_3$. Now using the definition of the finite sequence k_j and (4.93), we deduce that for all $i=1,2,3$ and $j=2,\dots,m$ we have

$$b_i^{k_j}\le\sigma^2 b_i^{k_j+1}\le\sigma^2(1+\delta_{k_j})b_i^{k_{j-1}}\le\sigma b_i^{k_{j-1}},$$

and so, repeating the above estimate, we get

$$b_i^{k_j}\ge\sigma^{-1}b_i^{k_{j+1}}\ge\cdots\ge\sigma^{j-m}b_i^{k_m}\ge\sigma^{j-m}M,$$

and, by the definition (4.96) (and (4.78))of δ_{k_j},

$$\delta_{k_j}\le\frac{C_d}{M}\sigma^{\frac{m-j}{2}},\qquad \forall j=1,\dots,m. \quad (4.94)$$

By (4.92) and (4.94) and reasoning as in (4.69) we deduce

$$4^{(6+3\varepsilon)L}A_1^L A_2^L A_3^L\le\exp\left(\frac{C_d}{1-\sqrt{\sigma}}\right)4^{2(6+3\varepsilon)}M\left(1+A_1^0+A_2^0+A_3^0\right)^3, \quad (4.95)$$

which concludes the proof of Theorem 4.34. □

Proof II of Theorem 4.34. For $i = 1, 2, 3$, we adopt the notation

$$A_i^k := A_i(4^{-k}), \qquad b_i^k := 4^{4k} A_i(4^{-k}) \quad \text{and} \quad \delta_k := \delta_{123}(4^{-k}), \tag{4.96}$$

where A_i was defined in (4.57) and δ_{123} in (4.78).

Let $M > 0$ and let

$$S(M) = \left\{k \in \mathbb{N} : \ 4^{(6+3\varepsilon)k} A_1^k A_2^k A_3^k \le M\left(1 + A_1^0 + A_2^0 + A_3^0\right)^3\right\}.$$

We will prove that if $\varepsilon > 0$ is small enough, then there is M large enough such that for every $k \notin S(M)$, we have

$$4^{(6+3\varepsilon)k} A_1^k A_2^k A_3^k \le CM\left(1 + A_1^0 + A_2^0 + A_3^0\right)^3,$$

where C is a constant depending on d and ε.

We first note that if $k \notin S(M)$, then we have

$$\begin{aligned} M\left(1 + A_1^0 + A_2^0 + A_3^0\right)^3 &\le 4^{(6+3\varepsilon)k} A_1^k A_2^k A_3^k \\ &\le 4^{-(2-3\varepsilon)k} b_1^k 4^{4k} A_2^k A_3^k \\ &\le 4^{-(2-3\varepsilon)k} b_1^k C_d\left(1 + A_1^0 + A_2^0 + A_3^0\right)^2, \end{aligned}$$

and so $b_1^k \ge C_d^{-1} M 4^{(2-3\varepsilon)k}$, where C_d is the constant from Theorem 4.30. Thus, choosing $\varepsilon < 2/3$ and $M > 0$ large enough, we can suppose that, for every $i = 1, 2, 3$, $b_i^k > C_d$, where C_d is the constant from Lemma 4.38.

Suppose now that $L \in \mathbb{N}$ is such that $L \notin S(M)$ and let

$$l = \max\left\{k \in \mathbb{N} : \ k \in S(M) \cap [0, L]\right\} < L,$$

where we note that the set $S(M) \cap [0, L]$ is non-empty for large M, since for $k = 0, 1$, we can apply Theorem 4.30. Applying Lemma 4.38, for $k = l + 1, \dots, L - 1$ we obtain

$$\begin{aligned} 4^{(6+3\varepsilon)L} A_1^L A_2^L A_3^L &\le \left(\prod_{k=l+1}^{L-1} (1 + \delta_k)\right) 4^{(6+3\varepsilon)(l+1)} A_1^{l+1} A_2^{l+1} A_3^{l+1} \\ &\le \left(\prod_{k=l+1}^{L-1} (1 + \delta_k)\right) 4^{(6+3\varepsilon)(l+1)} A_1^l A_2^l A_3^l \\ &\le \left(\prod_{k=l+1}^{L-1} (1 + \delta_k)\right) 4^{6+3\varepsilon} M\left(1 + A_1^0 + A_2^0 + A_3^0\right)^2, \end{aligned} \tag{4.97}$$

where δ^k is the variable from Lemma 4.38.

Now it is sufficient to notice that for $k = l+1, \dots, L-1$, the sequence δ_k is bounded by a geometric progression. Indeed, setting $\sigma = 4^{-1+3\varepsilon/2} < 1$, we have that, for $k \notin S(M)$, $\delta_k \le C\sigma^k$, which gives

$$\begin{aligned}\prod_{k=l+1}^{L-1}(1+\delta_k) &\le \prod_{k=l+1}^{L-1}(1+C\sigma^k)\\ &= \exp\Big(\sum_{k=l+1}^{L-1}\log(1+C\sigma^k)\Big)\\ &\le \exp\Big(C\sum_{k=l-1}^{L+1}\sigma^k\Big) \le \exp\big(\tfrac{C}{1-\sigma}\big),\end{aligned} \tag{4.98}$$

which concludes the proof. □

4.3.5. The common boundary of two subsolutions. Application of the two-phase monotonicity formula.

We start our discussion with a result which is useful in multiphase shape optimization problems, since it allows to separate by an open set each quasi-open cell from the others.

Lemma 4.39. *Suppose that the disjoint quasi-open sets Ω_1 and Ω_2 are energy subsolutions. Then the corresponding energy function w_1 and w_2 vanish on the common boundary $\partial\Omega_1 \cap \partial\Omega_2 = \partial^M\Omega_1 \cap \partial^M\Omega_2$.*

Proof. Recall that, by Remark 4.9, we may suppose that $\Omega_i = \{w_i > 0\}$ and that, by Remark 4.7, every point $\mathbb{R}^d$ is a Lebesgue point for both w_1 and w_2.

Let $x_0 \in \partial^M\Omega_1 \cap \partial^M\Omega_2$. Then, for each $r > 0$ we have $\big|\{w_1 > 0\} \cap B_r(x_0)\big| > 0$ and so, by Proposition 4.21, there is a sequence $r_n \to 0$ such that

$$\lim_{n\to\infty}\frac{\big|\{w_1>0\}\cap B_{r_n}(x_0)\big|}{|B_{r_n}|} \ge c > 0. \tag{4.99}$$

Since $\big|\{w_1 > 0\} \cap \{w_2 > 0\}\big| = 0$, we have that

$$\limsup_{n\to\infty}\frac{\big|\{w_2>0\}\cap B_{r_n}(x_0)\big|}{|B_{r_n}|} \le 1-c < 1. \tag{4.100}$$

Since x_0 is a Lebesgue point for w_2, we have

$$\begin{aligned} w_2(x_0) &= \lim_{n\to\infty}\fint_{B_{r_n}(x_0)} w_2\,dx\\ &\le \limsup_{n\to\infty}\|w_2\|_{L^\infty(B_{r_n}(x_0))}\limsup_{n\to\infty}\frac{\big|\{w_2>0\}\cap B_{r_n}(x_0)\big|}{|B_{r_n}|}\\ &\le (1-c)\limsup_{n\to\infty}\|w_2\|_{L^\infty(B_{r_n}(x_0))} \le (1-c)w_2(x_0),\end{aligned}$$

where the last inequality is due to the upper semi-continuity of w_2 (see Remark 4.7). Thus, we conclude that $w_2(x_0) = 0$ and, analogously $w_1(x_0) = 0$. □

Proposition 4.40. *Suppose that the disjoint quasi-open sets Ω_1 and Ω_2 are energy subsolutions. Then there are open sets $D_1, D_2 \subset \mathbb{R}^d$ such that $\Omega_1 \subset D_1$, $\Omega_2 \subset D_2$ and $\Omega_1 \cap D_2 = \Omega_2 \cap D_1 = \emptyset$, up to sets of zero capacity.*

Proof. Define $D_1 = \mathbb{R}^d \setminus \overline{\Omega}_2^M$ and $D_2 = \mathbb{R}^d \setminus \overline{\Omega}_1^M$, which by the definition of a measure theoretic closure are open sets. As in Lemma 4.39, we recall that $\Omega_i = \{w_i > 0\}$ and that every point of Ω_i is a Lebesgue point for the energy function $w_i \in H_0^1(\Omega_i)$. Since $\Omega_i \subset \overline{\Omega}_i^M$, we have to show only that $\Omega_1 \subset D_1$ and $\Omega_2 \subset D_2$ or, equivalently, that $\Omega_1 \cap \overline{\Omega}_2^M = \Omega_2 \cap \overline{\Omega}_1^M = \emptyset$. Indeed, if this is not the case there is a point $x_0 \in \overline{\Omega}_2^M$ such that $w_1(x_0) > 0$, which is a contradiction with Lemma 4.39. □

4.3.6. Absence of triple points for energy subsolutions. Application of the multiphase monotonicity formula

This subsection is dedicated to the proof of the fact that no three energy subsolutions can meet in a single point. Our main tool will be the three-phase monotonicity formula from Theorem 4.34. We note that the monotonicity formula involves terms, which are basically of the form $\fint_{B_r} |\nabla w|^2\,dx$, while the condition that the subsolution property provides concerns the mean of the function, *i.e.* $\fint_{\partial B_r} w\,d\mathcal{H}^{d-1} \geq cr$. These two terms express in different ways the non-degeneracy of w on the boundary, but the connection between them raises some technical issues, which esentially concern the regularity of the free boundary.

Remark 4.41 (Application of the monotonicity formula). Let Ω_1, Ω_2 and Ω_3 be three disjoint quasi-open sets of finite measure in $\mathbb{R}^d$. Let $w_i \in H_0^1(\Omega_i)$, for $i = 1, 2, 3$, be the corresponding energy function and suppose that there is a constant $c > 0$ such that

$$\fint_{B_r(x_0)} |\nabla w_i|^2\,dx \geq c, \qquad \forall r \in (0,1),\ \forall x_0 \in \mathbb{R}^d,\ \forall i = 1, 2, 3. \tag{4.101}$$

Then, by Theorem 4.30, we have that for every $x_0 \in \partial^M \Omega_1 \cap \partial^M \Omega_2$, we have

$$\fint_{B_r(x_0)} |\nabla w_i|^2\,dx \leq \frac{C_d}{c}\left(1 + \int_{\mathbb{R}^d} w_1^2\,dx + \int_{\mathbb{R}^d} w_2^2\,dx\right)^2, \qquad \forall r \in (0,1) \text{ and } i = 1, 2. \tag{4.102}$$

Moreover, by the three-phase monotonicity formula, the set of triple points $\partial^M\Omega_1 \cap \partial^M\Omega_2 \cap \partial^M\Omega_3$ is empty. Indeed, if $x_0 \in \partial^M\Omega_1 \cap \partial^M\Omega_2 \cap \partial^M\Omega_3$, by Theorem 4.34 and the assumption (4.101), we would have

$$r^{-3\varepsilon}c^3 \leq \prod_{i=1}^{3}\left(\frac{1}{r^{d+\varepsilon}}\int_{B_r(x_0)}|\nabla w_i|^2\,dx\right) \leq C_d\left(1+\sum_{i=1}^{3}\int_{\mathbb{R}^d} w_i^2\,dx\right)^2,$$

which is false for $r > 0$ small enough.

Remark 4.42 (The two dimensional case). In dimension two, the energy subsolutions satisfy condition (4.101). Indeed, let $\Omega_1, \Omega_2 \subset \mathbb{R}^2$ be two disjoint energy subsolution with $m = 1$ and let $x_0 \in \partial^M\Omega_1 \cap \partial^M\Omega_2$. Setting $x_0 = 0$, by Corollary 5.47, we get that for each $0 < r \leq r_0$ the following estimates hold:

$$cr \leq \fint_{\partial B_r} w_1\,d\mathcal{H}^1 \qquad \text{and} \qquad cr \leq \fint_{\partial B_r} w_2\,d\mathcal{H}^1. \tag{4.103}$$

In particular, we get that $\partial B_r \cap \{w_1 = 0\} \neq \emptyset$ and $\partial B_r \cap \{w_2 = 0\} \neq \emptyset$. We now notice that for almost every $r \in (0, r_0)$ the restriction of w_1 and w_2 to ∂B_r are Sobolev functions. Thus, we have

$$2\pi c^2 r^3 \leq \frac{1}{|\partial B_r|}\left(\int_{\partial B_r} w_i\,d\mathcal{H}^1\right)^2 \leq \int_{\partial B_r} w_i^2\,d\mathcal{H}^1 \leq \frac{r^2}{\pi^2}\int_{\partial B_r}|\nabla w_i|^2\,d\mathcal{H}^1,$$

where $\lambda < +\infty$ a constant. Dividing by r^2 and integrating for $r \in [0, R]$, where $R < r_0$, we obtain that (4.101) for some constant $c > 0$.

In particular, we obtain that if $\Omega_1, \Omega_2, \Omega_3 \subset \mathbb{R}^2$ are three disjoint energy subsolutions then there are no triple points, *i.e.* the set $\partial^M\Omega_1 \cap \partial^M\Omega_2 \cap \partial^M\Omega_3$ is empty.

In higher dimension the inequality (4.101) on the common boundary points will be deduced by the following Lemma, which is implicitly contained in the proof of [1, Lemma 3.2].

Lemma 4.43. *For every $u \in H^1(B_r)$ we have the following estimate:*

$$\frac{1}{r^2}\big|\{u = 0\} \cap B_r\big|\left(\fint_{\partial B_r} u\,d\mathcal{H}^{d-1}\right)^2 \leq C_d\int_{B_r}|\nabla u|^2\,dx, \tag{4.104}$$

where C_d is a constant that depends only on the dimension d.

Proof. We report here the proof for the sake of completeness, and refer the reader to [1, Lemma 3.2]. We note that it is sufficient to prove the result in the case $u \geq 0$. Let $v \in H^1(B_r)$ be the solution of the problem

$$\min\left\{\int_{B_r} |\nabla v|^2\,dx \;:\; u - v \in H_0^1(B_r),\; v \geq u\right\}.$$

We note that v is superharmonic on B_r and harmonic on the quasi-open set $\{v > u\}$.

For each $|z| \leq \frac{1}{2}$, we consider the functions u_z and v_z defined on B_r as

$$u_z(x) := u\big((r - |x|)z + x\big) \qquad \text{and} \qquad v_z(x) := v\big((r - |x|)z + x\big).$$

Note that both u_z and v_z still belong to $H^1(B_r)$ and that their gradients are controlled from above and below by the gradients of u and v. We call S_z the set of all $|\xi| = 1$ such that the set $\left\{\rho : \frac{r}{8} \leq \rho \leq r,\ u_z(\rho\xi) = 0\right\}$ is not empty. For $\xi \in S_z$ we define

$$r_\xi = \inf\left\{\rho : \frac{r}{8} \leq \rho \leq r,\ u_z(\rho\xi) = 0\right\}.$$

For almost all $\xi \in S^{d-1}$ (and then for almost all $\xi \in S_z$), the functions $\rho \mapsto \nabla u_z(\rho\xi)$ and $\rho \mapsto \nabla v_x(\rho\xi)$ are square integrable. For those ξ, one can suppose that the equation

$$\begin{aligned}
&\big(u_z(\rho_2\xi) - v_x(\rho_2\xi)\big) - \big(u_z(\rho_1\xi) - v_x(\rho_1\xi)\big)\\
&= \int_{\rho_1}^{\rho_2} \xi \cdot \nabla\big(u_z(\rho\xi) - v_x(\rho\xi)\big)\,d\rho,
\end{aligned}$$

holds for all $\rho_1, \rho_2 \in [0, r]$. Moreover, we have the estimate

$$\begin{aligned}
v_z(r_\xi\xi) &= \int_{r_\xi}^{r} \xi \cdot \nabla(v_z - u_z)(\rho\xi)\,d\rho\\
&\leq \sqrt{r - r_\xi}\left(\int_{r_\xi}^{r} |\nabla(v_z - u_z)(\rho\xi)|^2\,d\rho\right)^{1/2}.
\end{aligned}$$

Since v is superharmonic we have that, by the Poisson's integral formula,

$$v(x) \geq c_d\frac{r - |x|}{r}\fint_{\partial B_r} u\,d\mathcal{H}^{d-1}.$$

Substituting $x = (r - r_\xi)z + r_\xi\xi$, we have

$$\begin{aligned}
v_z(r_\xi\xi) = v((r - r_\xi)z + r_\xi\xi) &\geq \frac{c_d}{2}\frac{r - r_\xi}{r}\fint_{\partial B_r} u\,d\mathcal{H}^{d-1}\\
&= \frac{c_d}{2}\frac{r - r_\xi}{r}\fint_{\partial B_r} u_z\,d\mathcal{H}^{d-1}.
\end{aligned}$$

Combining the two inequalities, we have

$$\frac{r-r_\xi}{r^2}\left(\fint_{\partial B_r} u\,d\mathcal{H}^{d-1}\right)^2 \le C_d \int_{r_\xi}^{r} |\nabla(v_z-u_z)(\rho\xi)|^2\,d\rho.$$

Integrating over $\xi \in S_z \subset S^{d-1}$, we obtain the inequality

$$\left(\int_{S_z}\frac{r-r_\xi}{r^2}\,d\xi\right)\left(\fint_{\partial B_r} u\,d\mathcal{H}^{d-1}\right)^2 \le C_d \int_{\partial B_1}\int_{r_\xi}^{r} |\nabla(v_z-u_z)(\rho\xi)|^2\,d\rho\,d\xi,$$

and, by the estimate that $\frac{r}{8} \le r_\xi \le r$, we have

$$\begin{aligned}\frac{1}{r^2}\big|\{u=0\}\cap B_r\setminus B_{r/4}(rz)\big|\left(\fint_{\partial B_r} u\,d\mathcal{H}^{d-1}\right)^2 &\le C_d \int_{B_r} |\nabla(v_z-u_z)|^2\,dx\\ &\le C_d \int_{B_r} |\nabla(v-u)|^2\,dx.\end{aligned}$$

Integrating over z, we obtain

$$\frac{1}{r^2}\big|\{u=0\}\cap B_r\big|\left(\fint_{\partial B_r} u\,d\mathcal{H}^{d-1}\right)^2 \le C_d \int_{B_r} |\nabla(u-v)|^2\,dx. \tag{4.105}$$

Now the claim follows by the fact that v is harmonic on $\{v-u>0\}$ and the calculation

$$\begin{aligned}\int_{B_r} |\nabla(u-v)|^2\,dx &= \int_{B_r} |\nabla u|^2 - |\nabla v|^2\,dx + 2\int_{B_r} \nabla v\cdot\nabla(v-u)\,dx\\ &\le \int_{B_r} |\nabla u|^2\,dx.\end{aligned}$$

□

Theorem 4.44. *Suppose that $\Omega_1, \Omega_2, \Omega_3 \subset \mathbb{R}^d$ are three mutually disjoint energy subsolutions. Then the set $\partial\Omega_1\cap\partial\Omega_2\cap\partial\Omega_3 = \partial^M\Omega_1\cap\partial^M\Omega_2\cap\partial^M\Omega_3$ is empty.*

Proof. Suppose for contradiction that there is a point $x_0 \in \partial^M\Omega_1\cap\partial^M\Omega_2\cap\partial^M\Omega_3$. Without loss of generality $x_0=0$. Using the inequality (4.26), we have

$$\prod_{i=1}^{3}\frac{\|w_i\|_{L^\infty(B_{r/2})}}{r/2} \le C_d\left(\prod_{i=1}^{3}\frac{|\{w_i>0\}\cap B_r|}{|B_r|}\right)\left(\prod_{i=1}^{3}\frac{\|w_i\|_{L^\infty(B_{2r})}}{2r}\right),$$

and reasoning as in Proposition 4.21, we obtain that there is a constant $c > 0$ and a decreasing sequence of positive real numbers $r_n \to 0$ such that

$$c \leq \prod_{i=1}^{3} \frac{\left|\{w_i > 0\} \cap B_{r_n}\right|}{|B_{r_n}|}, \qquad \forall n \in \mathbb{N},$$

Since $\left|\{w_i > 0\} \cap B_{r_n}\right| \leq |B_{r_n}|$, for each $i = 1, 2, 3$, we have

$$c \leq \frac{\left|\{w_i > 0\} \cap B_{r_n}\right|}{|B_{r_n}|}, \qquad \forall n \in \mathbb{N},$$

and since $\{w_1 > 0\}$, $\{w_2 > 0\}$ and $\{w_3 > 0\}$ are disjoint, we get

$$1 - 2c \leq \frac{\left|\{w_i = 0\} \cap B_{r_n}\right|}{|B_{r_n}|}, \qquad \forall n \in \mathbb{N}, \quad \forall i = 1, 2, 3.$$

Thus, we may apply Lemma 4.43 and then Lemma 4.16 and Corollary 4.18 , to obtain that there is a constant $\tilde{c} > 0$ such that for every $n \in \mathbb{N}$

$$\tilde{c} \leq \frac{\left|\{w_i = 0\} \cap B_{r_n}\right|}{|B_{r_n}|} \left(\frac{1}{r_n} \fint_{\partial B_{r_n}} u \, d\mathcal{H}^{d-1} \right)^2 \leq C_d \fint_{B_{r_n}} |\nabla w_i|^2 \, dx,$$

which proves that (4.101) holds for a sequence $r_n \to 0$. The conclusion follows as in Remark 4.41. □

Remark 4.45. Let $\Omega_1, \ldots, \Omega_h \subset \mathbb{R}^d$ be a family of disjoint energy subsolutions. Then we can classify the points in $\mathbb{R}^d$ in three groups, as follows:

- One-phase points

$$Z_1 = \left\{ x \in \mathbb{R}^d \, : \, \exists \Omega_i > 0 \text{ s.t. } x \notin \partial^M \Omega_j, \; \forall j \neq i \right\}.$$

- Internal double-phase points

$$Z_2^i = \Big\{ x \in \mathbb{R}^d \, : \, \exists i \neq j \text{ s.t. } x \in \partial^M \Omega_i \cap \partial^M \Omega_j; \\ \exists r > 0 \text{ s.t. } \left| B_r(x) \cap (\Omega_i \cup \Omega_j)^c \right| = 0 \Big\}.$$

- Boundary double-phase points

$$Z_2^b = \Big\{ x \in \mathbb{R}^d \, : \, \exists i \neq j \text{ s.t. } x \in \partial^M \Omega_i \cap \partial^M \Omega_j; \\ \left| B_r(x) \cap (\Omega_i \cup \Omega_j)^c \right| > 0, \; \forall r > 0 \Big\}.$$

4.4. Subsolutions for spectral functionals with measure penalization

In this section we investigate the properties of the local subsolutions for functionals of the form

$$\mathcal{F}(\Omega) = F\big(\lambda_1(\Omega), \dots, \lambda_k(\Omega)\big) + m|\Omega|,$$

i.e. we are interested in the quasi-opens sets $\Omega \subset \mathbb{R}^d$ such that

$$\begin{aligned} F\big(\lambda_1(\Omega), \dots, \lambda_k(\Omega)\big) + m|\Omega| \le F\big(\lambda_1(\omega), \dots, \lambda_k(\omega)\big) + m|\omega|, \\ \text{for every quasi-open } \omega \subset \Omega \text{ such that } d_\gamma(\omega, \Omega) < \varepsilon, \end{aligned} \tag{4.106}$$

where $m > 0$ and $\varepsilon > 0$ are constants and $f : \mathbb{R}^k \to \mathbb{R}$ is a given function. Many of the properties of the subsolutions Ω for the functionals descrived above are consequences of the results in the previous sections. Indeed, we have the following:

Theorem 4.46. *Suppose that Ω is a local subsolution, in sense of* (4.106), *for the functional*

$$\mathcal{F}(\Omega) := F\big(\lambda_1(\Omega), \dots, \lambda_k(\Omega)\big) + m|\Omega|,$$

where $m > 0$ and $F : \mathbb{R}^k \to \mathbb{R}$ is Lipschitz continuous in a neighbourhood of $\big(\lambda_1(\Omega), \dots, \lambda_k(\Omega)\big) \in \mathbb{R}^k$. Then Ω is an energy subsolution.

Proof. We first note that by Lemma 3.129, applied for $\mu = I_\Omega$ and $\nu = I_\omega$, we can find constants $\varepsilon > 0$ and $C > 0$ (depending on d, $|\Omega|$ and $\lambda_k(\Omega)$) such that

$$\begin{aligned} \lambda_j(\omega) - \lambda_j(\Omega) \le C d_\gamma(I_\Omega, I_\omega) = 2C\big(E(\omega) - E(\Omega)\big), \\ \forall j = 1, \dots, k. \end{aligned} \tag{4.107}$$

Thus, we can choose $\varepsilon > 0$ small enough such that

$$\begin{aligned} &F\big(\lambda_1(\omega), \dots, \lambda_k(\omega)\big) - F\big(\lambda_1(\Omega), \dots, \lambda_k(\Omega)\big) \\ &\le L \sum_{j=1}^{k} \big(\lambda_j(\omega) - \lambda_j(\Omega)\big) \\ &\le 2LCk\big(E(\omega) - E(\Omega)\big), \end{aligned} \tag{4.108}$$

where L is a local Lipschitz constant for f and C is a constant from (4.107). Now since Ω is a subsoluion for F, we have that it is also an energy subsolution with constant $m/(2LCk)$. □

Corollary 4.47. *Suppose that Ω is a local subsolution, in sense of* (4.106), *for the functional*

$$\mathcal{F}(\Omega) = F\big(\lambda_1(\Omega), \dots, \lambda_k(\Omega)\big) + m|\Omega|,$$

where $m > 0$ and $F : \mathbb{R}^k \to \mathbb{R}$ is Lipschitz continuous in a neighbourhood of $\big(\lambda_1(\Omega), \dots, \lambda_k(\Omega)\big) \in \mathbb{R}^k$. Then Ω is a bounded set of finite perimeter.

In the case $F(\lambda_1, \dots, \lambda_k) \equiv \lambda_1$, we can repeat some of the arguments obtaining some more precise results.

Theorem 4.48. *Suppose that the quasi-open set $\Omega \subset \mathbb{R}^d$ is a local (for the distance d_γ) subsolution for the functional $\lambda_1(\Omega) + m|\Omega|$. Then,*

(i) *$\lambda_1(\Omega) < \lambda_2(\Omega)$ and if u is the first eigenfunction on Ω, then $|\Omega \setminus \{u > 0\}| = 0$;*
(ii) *there are constants $r_0 > 0$ and $m > 0$ such that if $x \in \overline{\Omega}^M$, then for every $0 < r \leq r_0$ we have*

$$cr \leq \|u\|_{L^\infty(B_r(x))}, \tag{4.109}$$

where $u \in H^1_0(\Omega)$ is the first, normalized in L^2, eigenfunction on Ω;
(iii) *Ω has finite perimeter and we have the estimate*

$$\sqrt{m}\mathcal{H}^{d-1}(\partial^*\Omega) \leq \lambda_1(\Omega)|\Omega|^{1/2}; \tag{4.110}$$

(iv) *Ω is quasi-connected,* i.e. *if $A, B \subset \Omega$ are two quasi-open sets such that $A \cup B = \Omega$ and* $\mathrm{cap}(A \cap B) = 0$, *then* $\mathrm{cap}(A) = 0$ *or* $\mathrm{cap}(B) = 0$.

Proof. Let $u \in H^1_0(\Omega)$ be a first, normalized in $L^2(\Omega)$, eigenfunction on Ω. Then $\{u > 0\} \subset \Omega$

$$\lambda_1(\{u > 0\}) = \lambda_1(\Omega) = \int_\Omega |\nabla u^+|^2\, dx,$$

and so, we must have $|\Omega \setminus \{u > 0\}| = 0$. Now if $\widetilde{u}$ is another eigenfunction corresponding to $\lambda_1(\Omega)$ such that $\int_\Omega u\widetilde{u}\, dx = 0$, then $\widetilde{u}$ must change sign on Ω and so, taking $\widetilde{u}^+$ as first eigenfunction, we have

$$\lambda_1(\Omega) + m|\Omega| > \lambda_1(\{\widetilde{u} > 0\}) + m|\{\widetilde{u} > 0\}|,$$

which is a contradiction. Thus, we have (i).

In order to prove (ii), we reason as in Lemma 4.15 and Lemma 4.17. Indeed suppose $x_0 = 0, r > 0$ and let v be the solution of

$$-\Delta v = a \text{ in } B_{2r} \setminus \overline{B_r}, \quad v = 0 \text{ on } B_r \text{ and } v = \|u\|_{L^\infty(B_{2r})} \text{ on } B_{2r},$$

where a is a constant to be defined. Then, taking $u_r = u\mathbb{1}_{B_{2r}^c} + (u \wedge v)\mathbb{1}_{B_{2r}}$, for $r > 0$ small enough we have

$$\begin{aligned}
&\int_\Omega |\nabla u|^2\,dx + m\big|\{u > 0\} \setminus \{u_r > 0\}\big| \\
&\le \int_\Omega |\nabla u_r|^2\,dx + \left(\left(\int_\Omega u_r^2\,dx\right)^{-1} - 1\right)\int_\Omega |\nabla u_r|^2\,dx \\
&\le \int_\Omega |\nabla u_r|^2\,dx + 4\lambda_1(\Omega)\int_\Omega (u^2 - u_r^2)\,dx, \\
&\le \int_\Omega |\nabla u_r|^2\,dx + C\int_\Omega (u - u_r)\,dx,
\end{aligned}$$

where C is a constant depending only on the dimension d and $\lambda_1(\Omega)$ (we recall that $\|u\|_{L^\infty} \le C_d\lambda_1(\Omega)^{d/4}$, by Corollary 3.95). Now using the definition of u_r and taking $a = C$, we have

$$\begin{aligned}
&\int_{B_r} |\nabla u|^2\,dx + m\big|B_r \cap \{u > 0\}\big| \\
&\le \int_{\{v<u\}} \left(|\nabla v|^2 - |\nabla u|^2\right)dx + C\int_{\{v<u\}} (u - u_r)\,dx, \\
&\le \int_{\{v<u\}} \nabla v \cdot \nabla(v - u)\,dx + C\int_{\{v<u\}} (u - v)\,dx, \\
&= \int_{\partial B_r} u|\nabla v|\,d\mathcal{H}^{d-1} \le C_1\left(r + \frac{\|u\|_{L^\infty(B_{2r})}}{2r}\right)\int_{\partial B_r} u\,d\mathcal{H}^{d-1},
\end{aligned}$$

where C_1 is a constant depending only on the dimension d and $\lambda_1(\Omega)$. Now, reasoning a in Lemma 4.16 by the trace inequality and the boundedness of u, we obtain (ii).

In order to prove the bound (4.110), we follow the idea from [20]. Let u be the first, normalized in $L^2(\Omega)$, eigenfunction on Ω. Since $\lambda_1(\{u > 0\}) = \lambda_1(\Omega)$, we have that $|\{u > 0\}\Delta\Omega| = 0$. Consider the set $\Omega_\varepsilon = \{u > \varepsilon\}$. In order to use Ω_ε to test the (local) subminimality of Ω, we first note that Ω_ε γ-converges to Ω. Indeed, the family of torsion functions w_ε of Ω_ε is decreasing in ε and converges in L^2 to the torsion function w

of $\{u > 0\}$, as $\varepsilon \to 0$, since

$$\lambda_1(\Omega) \int_\Omega (w - w_\varepsilon) u \, dx = \int_\Omega \nabla w \cdot \nabla u \, dx - \int_{\Omega_\varepsilon} \nabla w_\varepsilon \cdot \nabla (u-\varepsilon)^+ \, dx$$
$$= \int_\Omega \left(u - (u-\varepsilon)^+\right) dx \to 0.$$

Now, using $(u-\varepsilon)^+ \in H_0^1(\Omega_\varepsilon)$ as a test function for $\lambda_1(\Omega_\varepsilon)$, we have

$$\lambda_1(\Omega) + m|\Omega| \le \lambda_1(\Omega_\varepsilon) + m|\Omega_\varepsilon| \le \frac{\int_\Omega |\nabla(u-\varepsilon)^+|^2 \, dx}{\int_\Omega |(u-\varepsilon)^+|^2 \, dx} + m|\Omega_\varepsilon|$$
$$\le \int_\Omega |\nabla(u-\varepsilon)^+|^2 \, dx$$
$$+\lambda_1(\Omega) \frac{\int_\Omega \left(u^2 - |(u-\varepsilon)^+|^2\right) dx}{\int_\Omega |(u-\varepsilon)^+|^2 \, dx} + m|\Omega_\varepsilon|$$
$$\le \int_\Omega |\nabla(u-\varepsilon)^+|^2 \, dx + \lambda_1(\Omega) \frac{2\varepsilon \int_\Omega u \, dx}{1 - 2\varepsilon \int_\Omega u \, dx} + m|\Omega_\varepsilon|$$
$$\le \int_\Omega |\nabla(u-\varepsilon)^+|^2 \, dx + \frac{2\varepsilon \lambda_1(\Omega)|\Omega|^{1/2}}{1 - 2\varepsilon \int_\Omega u \, dx} + m|\Omega_\varepsilon|.$$

Thus, we obtain

$$\begin{aligned} &\int_{\{0<u\le\varepsilon\}} |\nabla u|^2 \, dx + m\big|\{0 < u \le \varepsilon\}\big| \\ &\le 2\varepsilon \lambda_1(\Omega)|\Omega|^{1/2}\Big(1 - 2\varepsilon \int_\Omega u \, dx\Big)^{-1}. \end{aligned} \tag{4.111}$$

The mean quadratic-mean geometric and the Hölder inequalities give

$$\begin{aligned} 2m^{1/2} \int_{\{0<u\le\varepsilon\}} |\nabla u| \, dx &\le 2m^{1/2} \left(\int_{\{0<u\le\varepsilon\}} |\nabla u|^2 \, dx\right)^{1/2} \big|\{0 < u \le \varepsilon\}\big|^{1/2} \\ &\le 2\varepsilon \lambda_1(\Omega)|\Omega|^{1/2}\Big(1 - 2\varepsilon \int_\Omega u \, dx\Big)^{-1}. \end{aligned} \tag{4.112}$$

Using the co-area formula, we obtain

$$\frac{1}{\varepsilon} \int_0^\varepsilon \mathcal{H}^{d-1}\big(\partial^* \{u > t\}\big) \, dt \le m^{-1/2} \lambda_1(\Omega)|\Omega|^{1/2}\Big(1 - 2\varepsilon \int_\Omega u \, dx\Big)^{-1}, \tag{4.113}$$

and so, passing to the limit as $\varepsilon \to 0$, we obtain (4.110).

Let us now prove (iv). Suppose, by absurd that $\mathrm{cap}(A) > 0$ and $\mathrm{cap}(B) > 0$ and, in particular, $|A| > 0$ and $|B| > 0$. Since $\mathrm{cap}(A \cap B) = 0$, we have that $H_0^1(\Omega) = H_0^1(A) \oplus H_0^1(B)$ and so, $\lambda_1(\Omega) = \min\{\lambda_1(A), \lambda_1(B)\}$. Without loss of generality, we may suppose that $\lambda_1(\Omega) = \lambda_1(A)$. Then, we have

$$\lambda_1(A) + m|A| < \lambda_1(A) + m(|A| + |B|) = \lambda_1(\Omega) + m|\Omega|,$$

which is a contradiction with the subminimality of Ω. □

Remark 4.49. The claim (iv) from Theorem 4.48 gives a slightly stronger claim than that from the point (i) of the same Theorem. Indeed, we have that

$$\mathrm{cap}(\Omega \setminus \{u > 0\}) = 0,$$

where u is the first Dirichlet eigenfunction on Ω. We prove this claim in the following Lemma.

Lemma 4.50. *Suppose that $\Omega \subset \mathbb{R}^d$ is a quasi-open set of finite measure. If Ω is quasi-connected, then $\lambda_1(\Omega) < \lambda_2(\Omega)$ and $\Omega = \{u_1 > 0\}$, where u_1 is the first eigenvalue of the Dirichlet Laplacian on Ω.*

Proof. It is sufficient to prove that if $u \in H_0^1(\Omega)$ is a first eigenfunction of the Dirichlet Laplacian on Ω, then $\Omega = \{u > 0\}$. Indeed, let $\omega = \{u > 0\}$ and consider the torsion functions w_ω and w_Ω. We note that, by the weak maximum principle, we have $w_\omega \le w_\Omega$. Setting $\lambda = \lambda_1(\Omega)$, we have

$$\int_\Omega \lambda u w_\omega \, dx = \int_\Omega \nabla u \cdot \nabla w_\omega \, dx = \int_\Omega u \, dx,$$

$$\int_\Omega \lambda u w_\Omega \, dx = \int_\Omega \nabla u \cdot \nabla w_\Omega \, dx = \int_\Omega u \, dx.$$

Subtracting, we have

$$\int_\Omega u(w_\Omega - w_\omega) \, dx = 0, \tag{4.114}$$

and so, $w_\Omega = w_\omega$ on ω. Consider the sets $A = \Omega \cap \{w_\Omega = w_\omega\}$ and $B = \Omega \cap \{w_\Omega > w_\omega\}$. By construction, we have that $A \cup B = \Omega$ and $A \cap B = \emptyset$. Moreover, we observe that $A = \omega \neq \emptyset$. Indeed, one inclusion $\omega \subset A$, follows by (4.114), while the other inclusion follows, since by strong maximum principle for w_ω and w_Ω we have the equality

$$\Omega \cap \{w_\Omega = w_\omega\} = \{w_\Omega > 0\} \cap \{w_\Omega = w_\omega\} \subset \{w_\omega > 0\} = \omega.$$

By the quasi-connectedness of Ω, we have that $B = \emptyset$. Thus $w_\Omega = w_\omega$ and so, $\omega = \Omega$ up to a set of zero capacity. □

Remark 4.51. If Ω is a local subsolution for the functional $\lambda_1 + m|\cdot|$, then we have the estimate

$$\lambda_1(\Omega) \geq c_d m^{\frac{2}{d+2}}, \tag{4.115}$$

where c_d is a dimensional constant. In fact, by (4.110) and the isoperimetric inequality, we have

$$\lambda_1(\Omega)|\Omega|^{1/2} \geq \sqrt{m}P(\Omega) \geq c_d\sqrt{m}|\Omega|^{\frac{d-1}{d}},$$

and so

$$\lambda_1(\Omega) \geq c_d\sqrt{m}|\Omega|^{\frac{d-2}{2d}}.$$

By the Faber-Krahn inequality $\lambda_1(\Omega)|\Omega|^{2/d} \geq \lambda_1(B)|B|^{2/d}$, we obtain

$$\begin{aligned}\lambda_1(\Omega) &\geq c_d\sqrt{m}\left(|\Omega|^{\frac{2}{d}}\right)^{\frac{d-2}{4}} \geq c_d\sqrt{m}\left(\lambda_1(\Omega)^{-1}\lambda_1(B)|B|^{2/d}\right)^{\frac{d-2}{4}}\\ &\geq c_d\sqrt{m}\lambda_1(\Omega)^{-\frac{d-2}{4}}.\end{aligned}$$

Remark 4.52. Even if the subsolutions have some nice qualitative properties, their local behaviour might be very irregular. In fact, one may construct subsolutions for the first Dirichlet eigenvalue (and thus, energy subsolutions) with empty interior in sense of the Lebesgue measure, *i.e.* the set $\Omega_{(1)}$ of points of density 1 has empty interior. Consider a bounded quasi-open set $\mathcal{D}$ with empty interior as, for example,

$$\mathcal{D} = (0,1) \times (0,1) \setminus \left(\bigcup_{i=1}^{\infty} \overline{B}_{r_i}(x_i)\right) \subset \mathbb{R}^2,$$

where $\{x_i\}_{i\in\mathbb{N}} = \mathbb{Q}$ and r_i is such that

$$\sum_{i\in\mathbb{N}} \mathrm{cap}(\overline{B}_{r_i}(x_i)) < +\infty \qquad \text{and} \qquad \sum_{i\in\mathbb{N}} \pi r_i^2 < \frac{1}{2}.$$

Let $\Omega \subset \mathcal{D}$ be the solution of the problem

$$\min\left\{\lambda_1(\Omega) + |\Omega| : \Omega \subset \mathcal{D},\ \Omega \text{ quasi-open}\right\}.$$

Since, Ω is a global minimizer among all sets in $\mathcal{D}$, it is also a subsolution. On the other hand, $\mathcal{D}$ has empty interior and so does Ω.

4.5. Subsolutions for functionals depending on potentials and weights

In this subsection, we consider functionals depending on the spectrum of the Schrödinger operator $-\Delta + V$ for a fixed potential V. Indeed, let $\mathcal{F}$ be defined as

$$\mathcal{F}(\Omega) := F\big(\lambda_1^V(\Omega), \ldots, \lambda_k^V(\Omega)\big) + \int_\Omega h(x)\,dx, \tag{4.116}$$

where $V : \mathbb{R}^d \to [0, +\infty]$ and $h : \mathbb{R}^d \to [0, +\infty]$ are given Lebesgue measurable functions and where we used the notation

$$\lambda_k^V(\Omega) := \lambda_k(V dx + I_\Omega),$$

for the kth eigenvalue of the operator $-\Delta + (V + I_{L^\infty})$, associated to the capacitary measure $V dx + I_\Omega$. As in the previous sections, we say that Ω is a subsolution for F, if for every quasi-open set $\omega \subset \Omega$, we have $\mathcal{F}(\Omega) \leq \mathcal{F}(\omega)$. We note that Ω might have infinite Lebesgue measure and non-integrable torsion function w_Ω, even if the torsion function of $V dx + I_\Omega$ is integrable. Thus, the natural notion of local subsolution would concern the γ-distance between the measures $V dx + I_\Omega$ and $V dx + I_\omega$.

Definition 4.53. Suppose that Ω is a quasi-open set such that $\int_\Omega h(x)dx < +\infty$ and such that the capacitary measure $\mu = V dx + I_\Omega$ has integrable torsion function. We say that Ω is a local subsolution for the functional $\mathcal{F}$, if for every quasi-open $\omega \subset \Omega$ such that $(d_\gamma(V dx + I_\omega, V dx + I_\Omega) < \varepsilon$, we have $\mathcal{F}(\Omega) \leq \mathcal{F}(\omega)$.

For Ω such that $(V dx + I_\Omega) \in \mathcal{M}_{\text{cap}}^T(\mathbb{R}^d)$, we use the notation

$$\begin{aligned} E(\Omega; V) &= \min\left\{J_V(u) : u \in H_0^1(\Omega) \cap L^1(\Omega)\right\} \\ &= J_V(w_{\Omega,V}) = -\frac{1}{2}\int_{\mathbb{R}^d} w_{\Omega,V}\,dx, \end{aligned}$$

where

$$J_V(u) = \int_{\mathbb{R}^d} \left(\frac{1}{2}|\nabla u|^2 + \frac{1}{2}u^2 V - u\right) dx,$$

and $w_{\Omega,V}$ is the minimizer of J_V in $H_0^1(\Omega) \cap L^1(\Omega)$. As in the previous section, we can restrict our attention from the general functional $\mathcal{F}$ to the Dirichlet Energy $E(\Omega; V)$ with a volume term. Indeed, we have the following result.

Theorem 4.54. *Suppose that Ω is a local subsolution for the functional $\mathcal{F}$ given by* (4.116)*, where the function $F : \mathbb{R}^k \to \mathbb{R}$ is locally Lipschitz continuous. Then there is $\widetilde{m} > 0$ such that Ω is a local subsolution for the functional $E(\Omega; V) + \widetilde{m} \int_\Omega h(x)\,dx$.*

Proof. The claim follows by the same argument as in Theorem 4.46. □

We now prove that every local, in capacity, subsolution for the functional $E(\Omega; V) + m \int_\Omega h(x)\,dx$ is a bounded set. In order to do that we need to use appropriate perturbations of Ω as for example those from Lemma 4.16. On the other hand, using sets obtained by cutting off balls is rather complicated. In particular, we note that the estimate of the measure $|\{w_{\Omega;V} > 0\} \cap B_r|$ is a difficult or impossible task since we have no a priori argument that excludes the possibility that both V and h are strictly positive on the whole $\mathbb{R}^d$. Thus, instead of using perturbations with small balls, we will just test the subsolution Ω against sets of the form $\Omega \cap H_t$, where H_t is a half-space. This approach gives weaker results than these from Section 4.2, but the boundedness still holds.

Lemma 4.55. *Suppose that Ω is a local subsolution for the functional $E(\Omega; V) + m \int_\Omega h(x)\,dx$, where $m > 0$ and $V : \mathbb{R}^d \to [0, +\infty]$ and $h : \mathbb{R}^d \to [0, +\infty]$ are given measurable functions such that the torsion function $w_{\Omega,V}$ of $V\,dx + I_\Omega$ is integrable. If $h \geq V^{-\alpha}$, for some $\alpha \in [0, 1)$, then Ω is a bounded set.*

Proof. For each $t \in \mathbb{R}$, we set

$$
\begin{aligned}
H_t &= \{x \in \mathbb{R}^d : x_1 = t\},\\
H_t^+ &= \{x \in \mathbb{R}^d : x_1 > t\},\\
H_t^- &= \{x \in \mathbb{R}^d : x_1 < t\}.
\end{aligned}
\tag{4.117}
$$

We prove that there is some $t \in \mathbb{R}$ such that $|H_t^+ \cap \Omega| = 0$. For sake of simplicity, set $w := w_\Omega$ and $M = \|w\|_{L^\infty}$. By Lemma 3.124 and the subminimality of Ω, we have

$$
\begin{aligned}
&\frac{1}{2}\int_{H_t^+} |\nabla w|^2\,dx + \frac{1}{2}\int_{H_t^+} w^2 V\,dx + \int_{H_t^+} h\,dx\\
&\leq \sqrt{2M}\int_{H_t} w\,d\mathcal{H}^{d-1} + \int_{H_t^+} w\,dx,
\end{aligned}
\tag{4.118}
$$

for every $t \in \mathbb{R}$. By aim to prove that the right hand side is grater than a power of $\int_{H_t^+} w\,dx$. Indeed, we have

$$\begin{aligned}\int_{H_t^+} w^{2/p}\,dx &\le \left(\int_{H_t^+} w^2 V\,dx\right)^{1/p}\left(\int_{H_t^+} V^{-\alpha}\,dx\right)^{1/q}\\ &\le \frac{1}{p}\int_{H_t^+} w^2 V\,dx + \frac{1}{q}\int_{H_t^+} V^{-\alpha}\,dx,\end{aligned} \tag{4.119}$$

where $p \ge 1$ and $q \ge 1$ are such that

$$\begin{cases}\dfrac{1}{p}+\dfrac{1}{q}=1,\\ w^{2/p}=(w^2V)^{1/p}(V^{-\alpha})^{1/q},\end{cases}$$

i.e.

$$\frac{1}{p}+\frac{1}{q}=1 \qquad \text{and} \qquad \frac{1}{p}=\frac{\alpha}{q},$$

which gives

$$\frac{1}{q}=\frac{1}{1+\alpha} \qquad \text{and} \qquad \frac{1}{p}=\frac{\alpha}{1+\alpha},$$

and so,

$$\int_{H_t^+} w^{\frac{2\alpha}{\alpha+1}}\,dx \le \frac{\alpha}{1+\alpha}\int_{H_t^+} w^2 V\,dx + \frac{1}{1+\alpha}\int_{H_t^+} V^{-\alpha}\,dx. \tag{4.120}$$

On the other hand, by the Sobolev inequality, we have

$$\left(\int_{H_t^+} w^{\frac{2d}{d-2}}\,dx\right)^{\frac{d-2}{d}} \le C_d \int_{H_t^+} |\nabla w|^2\,dx.$$

Thus, we search for $\beta \in (0,1)$, $p \ge 1$ and $q \ge 1$ such that $1/p+1/q=1$ and

$$\left(\int_{H_t^+} w\,dx\right)^{\beta} \le \left(\int_{H_t^+} w^{\frac{2\alpha}{\alpha+1}}\,dx\right)^{\frac{1}{p}}\left(\int_{H_t^+} w^{\frac{2d}{d-2}}\,dx\right)^{\frac{1}{q}\frac{d-2}{d}}.$$

Thus, we have the system

$$\begin{cases}\dfrac{1}{p}+\dfrac{1}{q}=1,\\ \dfrac{1}{p\beta}+\dfrac{d-2}{d}\dfrac{1}{q\beta}=1,\\ \dfrac{2\alpha}{1+\alpha}\dfrac{1}{p\beta}+\dfrac{2}{q\beta}=1,\end{cases}$$

which gives

$$\frac{1}{p}=\frac{(1+\alpha)(d+2)}{2(d+1+\alpha)},\qquad \frac{1}{q}=\frac{d(1-\alpha)}{2(d+1+\alpha)},\qquad \beta=\frac{d+2\alpha}{d+1+\alpha}.$$

In conclusion, we get

$$\left(\int_{H_t^+} w\,dx\right)^\beta \leq C\sqrt{2M}\int_{H_t} w\,d\mathcal{H}^{d-1}+C\int_{H_t^+} w\,dx, \tag{4.121}$$

where C is a constant depending on α and the dimension d. Setting

$$\phi(t):=\int_{H_t^+} w\,dx,$$

we have that

$$\phi'(t)=-\int_{H_t} w\,d\mathcal{H}^{d-1},$$

and, by (4.121), we have

$$\phi(t)^\beta \leq -C\sqrt{2M}\phi'(t)+C\phi(t),$$

which gives that ϕ vanishes in a finite time. Repeating this argument in any direction and using that $\{w>0\}=\Omega$, we obtain that Ω is bounded. □

4.6. Subsolutions for spectral functionals with perimeter penalization

In this section we consider subsolutions for functionals of the form

$$\mathcal{F}(\Omega)=F\big(\widetilde{\lambda}_1(\Omega),\dots,\widetilde{\lambda}_k(\Omega)\big)+mP(\Omega), \tag{4.122}$$

where $m>0$, $\widetilde{\lambda}_k(\Omega):=\lambda_k(\widetilde{I}_\Omega)$ is the kth eigenvalue of the Laplacian on Ω with zero boundary conditions a.e. outside Ω, $F:\mathbb{R}^k\to\mathbb{R}$ is a given function and $P(\Omega)$ is the perimeter of the measurable set Ω in sense of De Giorgi. Since the perimeter is not an increasing functional with respect to the set inclusion, defining the subsolution using quasi-open or measurable sets is not equivalent. In this section, we choose to work with measurable sets, since in the shape optimization problems concerning the perimeter the existence results are easier to state in the class of measurable sets than in the class of quasi-open sets. Thus, we have

Definition 4.56. We say that the measurable set Ω is a local subsolution for the functional $\mathcal{F}$, if Ω has finite measure and for each measurable $\omega\subset\Omega$ such that $d_\gamma(\widetilde{I}_\Omega,\widetilde{I}_\omega)<\varepsilon$, we have $\mathcal{F}(\Omega)\leq\mathcal{F}(\omega)$.

As in the previous sections, we have

Theorem 4.57. *Suppose that the measurable set Ω is a local subsolution for the functional $\mathcal{F}$ from (4.122), where $F : \mathbb{R}^k \to \mathbb{R}$ is locally Lipschitz continuous. Then Ω is a local subsolution for the functional $\widetilde{E}(\Omega) + \widetilde{m}P(\Omega)$.*

Proof. See the proof of Theorem 4.46. □

As one may expect, all the subsolutions for functionals of the form $\mathcal{F}$, with locally Lipschitz F, are bounded sets. Indeed, we have the following:

Lemma 4.58. *Suppose that the measurable set $\Omega \subset \mathbb{R}^d$ is a subsolution for the functional $\widetilde{E}(\Omega) + mP(\Omega)$. Then Ω is a bounded set.*

Proof. We reason as in Lemma 4.55. For each $t \in \mathbb{R}$, we set

$$\begin{aligned} H_t &= \{x \in \mathbb{R}^d : x_1 = t\}, \\ H_t^+ &= \{x \in \mathbb{R}^d : x_1 > t\}, \\ H_t^- &= \{x \in \mathbb{R}^d : x_1 < t\}. \end{aligned} \tag{4.123}$$

We prove that there is some $t \in \mathbb{R}$ such that $|H_t^+ \cap \Omega| = 0$. For sake of simplicity, set $w := w_\Omega$ and $M = \|w\|_{L^\infty}$. By Lemma 3.124 and the subminimality of Ω, we have

$$\begin{aligned} &\frac{1}{2}\int_{H_t^+} |\nabla w|^2\,dx + m\big(P(\Omega; H_t^+) - \mathcal{H}^{d-1}(H_t \cap \Omega)\big) \\ &\le \sqrt{2M}\int_{H_t} w\,d\mathcal{H}^{d-1} + \int_{H_t^+} w\,dx, \end{aligned} \tag{4.124}$$

for every $t \in \mathbb{R}$. Using again the boundedness of w, we get

$$m\big(P(\Omega, H_t^+) - P(H_t^+, \Omega)\big) \le \sqrt{2}M^{3/2}\mathcal{H}^{d-1}(H_t \cap \Omega) + M|\Omega \cap H_t^+|. \tag{4.125}$$

On the other hand, by the isoperimetric inequality, for almost every t we have

$$|\Omega \cap H_t^+|^{\frac{d-1}{d}} \le C_d P(\Omega \cap H_t^+) = C_d\left(\mathcal{H}^{d-1}(H_t \cap \Omega) + P(\Omega, H_t^+)\right) \tag{4.126}$$

Putting together (4.125) and (4.126) we obtain

$$|\Omega \cap H_t^+|^{\frac{d-1}{d}} \le C_1\left(\mathcal{H}^{d-1}(H_t^+ \cap \Omega) + |\Omega \cap H_t^+|\right), \tag{4.127}$$

where C_1 is some constant depending on the dimension d, the constant m and the norm M. Setting $\phi(t) = |\Omega \cap H_t^+|$, we have that $\phi(t) \to 0$ as $t \to +\infty$ and $\phi'(t) = -\mathcal{H}^{d-1}(H_t \cap \Omega)$. Chosing $T = T(\Omega)$ such that

$$C_1\phi(t) \le \frac{1}{2}\phi(t)^{\frac{d-1}{d}} \qquad \forall t \ge T,$$

equation (4.127) gives

$$\phi'(t) \le -2C_1\phi(t)^{1-1/d} \qquad \forall t \ge T,$$

which implies that $\phi(\bar{t})$ vanishes for some $\bar{t} \in \mathbb{R}$. Repeating this argument in any direction, we obtain that Ω is bounded. □

4.7. Subsolutions for spectral-energy functionals

In this section we consider subsolutions for the functional, defined on the family of quasi-open sets in $\mathbb{R}^d$,

$$\mathcal{F}(\Omega) = F\big(\lambda_{1,\mu}(\Omega), \dots, \lambda_{k,\mu}(\Omega)\big) - E_\mu(\Omega), \qquad (4.128)$$

where $F : \mathbb{R}^k \to \mathbb{R}$ is a given function, μ is a capacitary measure such that $w_\mu \in L^1(\mathbb{R}^d)$ and we use the notation

$$\lambda_{k,\mu}(\Omega) := \lambda_k(\mu \vee I_\Omega).$$

For $f \in L^p(\mathbb{R}^d)$, where $p \in [2, \infty]$, we set

$$E_{\mu,f}(\Omega) = \min\bigg\{\frac{1}{2}\int_{\mathbb{R}^d} |\nabla u|^2\,dx + \frac{1}{2}\int_{\mathbb{R}^d} u^2\,d\mu - \int_{\mathbb{R}^d} uf\,dx \,:\, u \in H^1_\mu \cap H^1_0(\Omega)\bigg\},$$

i.e. $E_{\mu,f}(\Omega) = -\frac{1}{2}\int_{\mathbb{R}^d} f w_{\mu,f,\Omega}\,dx$, where the function $w_{\mu,f,\Omega}$ is the solution of the equation

$$-\Delta w + \mu w = f \quad \text{in} \quad H^1_\mu \cap H^1_0(\Omega), \qquad w \in H^1_\mu \cap H^1_0(\Omega).$$

In order to simplify the notation, we set $E_\mu(\Omega) := E_{\mu,1}(\Omega)$.

Since the above functionals are defined with respect to the measure μ, without any restriction on the quasi-open sets Ω, the definition of *local* subsolution depends on the measure μ.

Definition 4.59. We say that the quasi-open set $\Omega \subset \mathbb{R}^d$ is a subsolution for the functional $\mathcal{F}$, locally with respect to the measure $\mu \in \mathcal{M}_{\text{cap}}^T(\mathbb{R}^d)$, if there is an $\varepsilon > 0$ such that

$$\mathcal{F}(\Omega) \leq \mathcal{F}(\omega), \text{ for every quasi-open set } \omega \subset \Omega$$
$$\text{such that } d_\gamma(\mu \vee I_\omega, \mu \vee I_\Omega) < \varepsilon.$$

Theorem 4.60. *Suppose that μ is a capacitary measure such that $w_\mu \in L^1(\mathbb{R}^d)$ and let $\Omega \subset \mathbb{R}^d$ be a quasi-open set, local subsolution for $\mathcal{F}$ as in* (4.128) *with respect to μ. If $F : \mathbb{R}^k \to \mathbb{R}$ is locally Lipschitz, then Ω is a local subsolution for the functional $E_{\mu,f}(\Omega) - E_\mu(\Omega)$, where $f = cw_\mu$, for some constant $c > 0$ depending on μ and F.*

Proof. The claim follows from Lemma 3.128, by the argument as in Theorem 4.46. □

In the rest of this subsection we prove that the local subsolutions for the functionals of the form (4.128) are bounded sets. We need the following comparison principle "at infinity" for solutions of PDEs involving capacitary measures.

Lemma 4.61. *Consider a capacitary measure of finite torsion $\mu \in \mathcal{M}_{\text{cap}}^T(\mathbb{R}^d)$. Suppose that $u \in H^1_\mu$ is a solution of*

$$-\Delta u + \mu u = f \quad \text{in} \quad H^1_\mu, \qquad u \in H^1_\mu,$$

where $f \in L^1(\mathbb{R}^d) \cap L^\infty(\mathbb{R}^d)$ and $\lim_{x\to\infty} f(x) = 0$. Then, there is some $R > 0$, large enough, such that $u \leq w_\mu$ on $\mathbb{R}^d \setminus B_R$.

Proof. Set $v = u - w_\mu$. We will prove that the set $\{v > 0\}$ is bounded. Taking v^+ instead of v and $\mu \vee I_{\{v>0\}}$ instead of μ, we note that it is sufficient to restrict our attention to the case $v \geq 0$ on $\mathbb{R}^d$. We will prove the Lemma in four steps.

Step 1. There are constants $R_0 > 0$, $C_d > 0$ and $\delta > 0$ such that

$$\left(\int_{\mathbb{R}^d} v^2\varphi^{2(1+\delta)}\right)^{\frac{1}{1+\delta}} \leq C_d \int_{\mathbb{R}^d} |\nabla\varphi|^2 v^2\,dx, \quad \forall \varphi \in W_0^{1,\infty}(B^c_{R_0}). \tag{4.129}$$

For any $\varphi \in W^{1,\infty}(\mathbb{R}^d)$, we have that $v\varphi^2 \in H^1_\mu$ and so we may use it as a test function in

$$-\Delta v + \mu v = f - 1 \quad \text{in} \quad H^1_\mu, \qquad v \in H^1_\mu,$$

obtaining the identity

$$\begin{aligned}&\int_{\mathbb{R}^d} |\nabla(\varphi v)|^2\,dx + \int_{\mathbb{R}^d} \varphi^2 v^2\,d\mu\\ &= \int_{\mathbb{R}^d} |\nabla\varphi|^2 v^2\,dx + \int_{\mathbb{R}^d} v\varphi^2(f-1)\,dx, \quad \forall \varphi \in W^{1,\infty}(\mathbb{R}^d).\end{aligned} \tag{4.130}$$

Let $R_0 > 0$ be large enough such that $1 - f > \frac{4}{d+4}$. Then for any $\varphi \in W_0^{1,\infty}(\mathbb{R}^d \setminus B_{R_0})$, we use the Hölder, Young and the Sobolev's inequalities together with (4.130) to obtain

$$\begin{aligned}\left(\int_{\mathbb{R}^d} v^2\varphi^{\frac{2d+8}{d+2}}\,dx\right)^{\frac{d+2}{d+4}} &\le \left(\int_{\mathbb{R}^d} (\varphi v)^{\frac{2d}{d-2}}\,dx\right)^{\frac{d-2}{d+4}} \left(\int_{\mathbb{R}^d} v\varphi^2\,dx\right)^{\frac{4}{d+4}}\\ &\le \frac{d}{d+4}\left(\int_{\mathbb{R}^d} (\varphi v)^{\frac{2d}{d-2}}\,dx\right)^{\frac{d-2}{d}} + \frac{4}{d+4}\int_{\mathbb{R}^d} v\varphi^2\,dx\\ &\le C_d\left(\int_{\mathbb{R}^d} |\nabla(\varphi v)|^2\,dx + \int_{\mathbb{R}^d} v\varphi^2(1-f)\,dx\right)\\ &\le C_d\int_{\mathbb{R}^d} |\nabla\varphi|^2 v^2\,dx,\end{aligned} \tag{4.131}$$

where C_d is a dimensional constant.

Step 2. There is some $R_1 > 0$ such that the function $M(r) := \fint_{\partial B_r} v^2\,d\mathcal{H}^{d-1}$ is decreasing and convex on the interval $(R_1, +\infty)$. We first note that, for $R > 0$ large enough, $\Delta v \ge (1-f)\chi_{\{v>0\}} \ge 0$ as an element of $H^{-1}(B_R^c)$. Since $\Delta(v^2) = 2v\Delta v + 2|\nabla v|^2$, we get that the function $U := v^2$ is subharmonic on $\mathbb{R}^d \setminus B_R$. Now, the formal derivation of the mean M gives

$$M'(r) = \fint_{\partial B_r} \nu \cdot \nabla U\,d\mathcal{H}^{d-1},$$

where ν_r is the external normal to ∂B_r. Let $R_1 > 0$ be such that $1 \ge f$ on $\mathbb{R}^d \setminus B_{R_1}$. Then for any $R_1 < r < R < +\infty$ we have

$$\begin{aligned}d\omega_d\left(R^{d-1}M'(R) - r^{d-1}M'(r)\right) &= \int_{\partial B_R} \nu_R \cdot \nabla U\,d\mathcal{H}^{d-1}\\ &\quad - \int_{\partial B_r} \nu_r \cdot \nabla U\,d\mathcal{H}^{d-1}\\ &= \int_{B_{R_2}\setminus B_{R_1}} \Delta U\,dx \ge 0.\end{aligned}$$

If we have that $M'(r) > 0$ for some $r > R_1$, then $M'(R) > 0$ for each $R > r$ and so M is increasing on $[r, +\infty)$, which is a contradiction with the fact that v (and so, M) vanishes at infinity. Thus, $M'(r) \le 0$, for all $r \in (R_1, +\infty)$ and so for every $R_1 < r < R < +\infty$, we have

$$R^{d-1}\big(M'(R) - M'(r)\big) \ge R^{d-1}M'(R) - r^{d-1}M'(r) \ge 0,$$

which proves that $M'(r)$ is also increasing.

Step 3. *There are constants $R_2 > 0$, $C > 0$ and $0 < \delta < 1/(d-1)$ such that the mean value function $M(r)$ satisfies the differential inequality*

$$M(r) \le C\big(r|M'(r)| + M(r)\big)^{\frac{d-1}{2}\delta}|M'(r)|^{1-\frac{d-2}{2}\delta}, \quad \forall r \in (R_2, +\infty). \quad (4.132)$$

We first test the inequality (4.129) with radial functions of the form $\varphi(x) = \phi(|x|)$, where

$$\phi(r) = 0, \text{ for } r \le R,$$
$$\phi(r) = \frac{r-R}{\varepsilon(R)}, \text{ for } R \le r \le R + \varepsilon(R),$$
$$\phi(r) = 1, \text{ for } r \ge R + \varepsilon(R),$$

where $R > 0$ is large enough and $\varepsilon(R) > 0$ is a given constant. As a consequence, we obtain

$$\left(\int_{R+\varepsilon(R)}^{+\infty} r^{d-1}M(r)\,dr\right)^{\frac{1}{1+\delta}} \le C_d\varepsilon(R)^{-2}\int_R^{R+\varepsilon(R)} r^{d-1}M(r)\,dr. \quad (4.133)$$

By *Step* 2, we have that for R large enough:

- M is monotone, *i.e.* $M(r) \le M(R)$ for $r \ge R$;
- M is convex $M(r) \ge M'(R)(r-R) + M(R)$ for $r \ge R$.

We now consider take $\varepsilon(R) = \frac{1}{2}\frac{M(R)}{|M'(R)|}$, *i.e.* $2\varepsilon(R)$ is exactly the distance between $(R, 0)$ and the intersection point of the x-axis with the line tangent to the graph of M in $(R, M(R))$ (see Figure 4.1). With this choice of $\varepsilon(R)$ we estimate both sides of (4.133), obtaining

$$\big(R+\varepsilon(R)\big)^{\frac{d-1}{1+\delta}}\left(\frac{1}{4}M(R)\varepsilon(R)\right)^{\frac{1}{1+\delta}} \le C_d\big(R+\varepsilon(R)\big)^{d-1}\varepsilon(R)^{-2}M(R), \quad (4.134)$$

which, after substituting $\varepsilon(R)$ with $\frac{1}{2}\frac{M(R)}{|M'(R)|}$ gives (4.132).

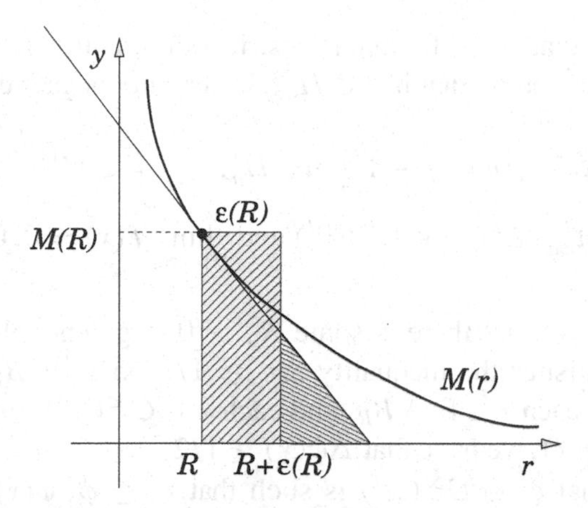

Figure 4.1. We estimate the integral $\int_R^{R+\varepsilon(R)} M(r)\,dr$ by the area of the rectangle on the right, while for the integral $\int_{R+\varepsilon(R)}^{+\infty} M(r)\,dr$ is bounded from below by the area of the triangle on the right.

Step 4. *Each non-negative (differentiable a.e.) function* $M(r)$, *which vanishes at infinity and satisfies the inequality* (4.132) *for some* $\delta > 0$ *small enough, has compact support.*

Let $r \in (R_2, +\infty)$, where R_2 is as in *Step 3*. We have two cases:

$$(a)\text{ If }\quad r|M'(r)| \geq M(r), \quad\text{ then }\quad M(r) \leq C_1 r^{\frac{(d-1)\delta}{2}} |M'(r)|^{1+\frac{\delta}{2}};$$

$$(b)\text{ If }\quad r|M'(r)| \leq M(r), \quad\text{ then }\quad M(r) \leq C_2 |M'(r)|^{1+\frac{\delta}{2}\left(1-\frac{(d-1)\delta}{2}\right)}.$$

Choosing δ small enough, we get that in both cases M satisfies the differential inequality

$$M(r)^{1-\delta_1} \leq -C r^{\delta_2} M'(r), \tag{4.135}$$

for appropriate constants $C > 0$ and $0 < \delta_1, \delta_2 < 1$. After integration, we have

$$C' - C'' r^{1-\delta_2} \geq M(r)^{\delta_1}, \tag{4.136}$$

for some constants $C', C'' > 0$, which concludes the proof. □

Below, we give an alternative and shorter proof of Lemma 4.61 which uses the notion of a viscosity solution.

Alternative proof of Lemma 4.61. Set $v = u - w_\mu$. We will prove that the set $\{v > 0\}$ is bounded. Taking v^+ instead of v and $\mu \vee I_{\{v>0\}}$ instead

of μ, we note that it is sufficient to restrict our attention to the case $v \geq 0$ on $\mathbb{R}^d$. We now prove that if $v \in H^1(\mathbb{R}^d)$ is a nonnegative function such that

$$-\Delta v + \mu v = f - 1 \quad \text{in} \quad H^1_\mu, \qquad v \in H^1_\mu, \tag{4.137}$$

where $\mu \in \mathcal{M}^T_{\text{cap}}(\mathbb{R}^d)$, $f \in L^\infty(\mathbb{R}^d)$ and $\lim_{|x|\to\infty} f(x) = 0$, then $\{v > 0\}$ is bounded.

We first prove that there is some $R_0 > 0$ large enough such that the function v satisfies the inequality $\Delta v \geq 1/2$ on $\mathbb{R}^d \setminus B_{R_0}$ in viscosity sense, *i.e.* for each $x \in \mathbb{R}^d \setminus B_{R_0}$ and each $\varphi \in C^\infty(\mathbb{R}^d)$, satisfying $v \leq \phi$ and $\varphi(x) = v(x)$, we have that $\Delta\varphi(x) \geq 1/2$.

Suppose that $\varphi \in C^\infty(\mathbb{R}^d)$ is such that $v \leq \phi$, $\varphi(x) = v(x)$ and $\Delta\varphi(x) < 1/2 - \varepsilon$. By modifying φ and considering $\varepsilon/2$ instead of ε, we may suppose that, for $\delta > 0$ small enough, $\{v + \delta > \varphi\} \subset B^c_{R_0}$ and $\Delta\varphi < 1/2 - \varepsilon$ on the set $\{v + \delta > \varphi\}$. Now taking $(v - \varphi + \delta)^+ \in H^1_\mu$ as a test function in (4.137), we get that

$$\begin{aligned}
\int_{\mathbb{R}^d} (f - 1)(v - \varphi + \delta)^+ \, dx &= \int_{\mathbb{R}^d} \nabla v \cdot \nabla (v - \varphi + \delta)^+ \, dx \\
&\quad + \int_{\mathbb{R}^d} v(v - \varphi + \delta)^+ \, d\mu \\
&\geq \int_{\mathbb{R}^d} \nabla\varphi \cdot \nabla (v - \varphi + \delta)^+ \, dx \\
&= -\int_{\mathbb{R}^d} (v - \varphi + \delta)^+ \Delta\varphi \, dx \\
&> \left(-\frac{1}{2} + \varepsilon\right) \int_{\mathbb{R}^d} (v - \varphi + \delta)^+ \, dx,
\end{aligned}$$

which gives a contradiction, once we choose $R_0 > 0$ large enough such that $f < 1/4$ on $\mathbb{R}^d \setminus B_{R_0}$.

For $r \in (R_0, +\infty)$, we consider the function $M(r) = \sup_{\partial B_r} v$. Then $M : (R_0, +\infty) \to \mathbb{R}$ satisfies the inequality

$$M''(r) + \frac{d-1}{r} M'(r) \geq \frac{1}{2}, \qquad \text{in viscosity sense.} \tag{4.138}$$

Indeed, let $r \in (R_0, +\infty)$ and $\phi \in C^\infty(\mathbb{R})$ be such that $\phi(r) = M(r)$ and $\phi \geq M$. Then, taking a point $x_0 \in \partial B_r$ such that $v(x) = M(r)$ (which exists due to the upper semi-continuity of v) and the function $\varphi(x) := \phi(|x|)$, we have that $\varphi \in C^\infty(\mathbb{R}^d)$, $\varphi(x_0) = v(r)$ and $\varphi \geq v$, which implies $\Delta\varphi \geq 1/2$ and so (4.138) holds.

There is a constant $\varepsilon_0 > 0$, depending on R_0, the dimension d and $\|v\|_{L^\infty}$, such that the function $\phi \in C^\infty(\mathbb{R})$, which solves

$$\phi''(r)+\frac{d-1}{r}\phi'(r) = \frac{1}{3}, \qquad \phi(R_0) = \phi(R_0+\varepsilon_0) = 2\|v\|_{L^\infty}, \tag{4.139}$$

changes sign on the interval $(R_0, R_0 + \varepsilon_0)$. We set

$$t_0 = \sup\left\{t : \{M \geq \phi + t\} \neq \emptyset\right\} > 0.$$

Since M is upper semi-continuous, there is some $r \in (R_0, R_0 + \varepsilon_0)$ such that $M(r) = \phi(r) + t_0$ and $M \leq \phi + t_0$, which is a contradiction with (4.138). □

In order to prove the boundedness of the local subsolutions for functionals of the form $E_f - E_1$, we will need the notion of $(\Delta-\mu)$-harmonic function.

Definition 4.62. Let μ be a capacitary measure on $\mathbb{R}^d$ such that $w_\mu \in L^1(\mathbb{R}^d)$ and let $B_R \subset \mathbb{R}^d$ be a given ball. For every $u \in H^1_\mu$ we will denote with h_u the solution of the problem

$$\min\left\{\int_{B_r} |\nabla v|^2\,dx + \int_{B_R} v^2\,d\mu\,:\, v \in H^1_\mu,\ u - v \in H^1_0(B_R)\right\}. \tag{4.140}$$

We will refer to h_u as the $(\Delta - \mu)$-harmonic function on B_R with boundary data u on ∂B_R.

Remark 4.63. *Properties of the* $(\Delta - \mu)$*-harmonic functions.*

- (*Uniqueness*). By the strict convexity of the functional in (4.140), we have that the problem (4.140) has a unique minimizer, *i.e.* h_u is uniquely determined;
- (*First variation*). Calculating the first variation of the functional from (4.140), we have

$$\int_{\mathbb{R}^d} \nabla h_u \cdot \nabla\psi\,dx + \int_{\mathbb{R}^d} h_u\psi\,d\mu = 0, \qquad \forall\psi \in H^1_\mu \cap H^1_0(B_R), \tag{4.141}$$

  and conversely, if the function $h_u \in H^1_\mu$ satisfies (4.141), then it minimizes (4.140);
- (*Comparison principle*). If $u, w \in H^1_\mu$ are two functions such that $w \geq u$ on ∂B_R, then $h_u \leq h_w$. Indeed, using $h_u \vee h_w \in H^1_\mu$ and

$h_w \wedge h_u \in H^1_\mu$ to test the minimality of h_w and h_u, respectively, we get

$$\int_{\{h_u>h_w\}} |\nabla h_u|^2\,dx + \int_{\{h_u>h_w\}} h_u^2\,d\mu = \int_{\{h_u>h_w\}} |\nabla h_w|^2\,dx + \int_{\{h_u>h_w\}} h_w^2\,d\mu,$$

which implies that $h_w \wedge h_u$ is also minimizer of (4.140) and so, $h_w \wedge h_u = h_u$.

Lemma 4.64. *Suppose that μ is a capacitary measure such that $w_\mu \in L^1(\mathbb{R}^d)$ and let the quasi-open set $\Omega \subset \mathbb{R}^d$ be a local subsolution for the functional $E_{\mu,f}(\Omega) - E_\mu(\Omega)$, where f is a bounded measurable function vanishing at infinity,* i.e. $\lim_{R\to+\infty} \|f\|_{L^\infty(B_R^c)} = 0$. *Then Ω is bounded.*

Proof. Without loss of generality, we may suppose that $\mu \geq I_\Omega$. Let, for generic quasi-open set $\omega \subset \mathbb{R}^d$, $R_\omega : L^\infty(\mathbb{R}^d) \to L^1(\mathbb{R}^d)$ be the resolvent operator associating to a function $f \in L^\infty(\mathbb{R}^d)$ the solution $w_{\mu,f,\omega}$. The subminimality of Ω with respect to $\omega \subset \Omega$

$$E_{\mu,f}(\Omega) - E_\mu(\Omega) \leq E_{\mu,f}(\omega) - E_\mu(\omega),$$

can be stated in terms of R_Ω and R_ω as follows:

$$\int_{\mathbb{R}^d} \big(R_\Omega(1) - f R_\Omega(f)\big)\,dx \leq \int_{\mathbb{R}^d} \big(R_\omega(1) - f R_\omega(f)\big)\,dx. \tag{4.142}$$

Moreover, by considering $f/2$ instead of f, we can suppose that the above inequality is strict, when $\omega \neq \Omega$.

We now show that choosing $\omega = \Omega \cap B_R$, for some R large enough, we can obtain equality in (4.142). Indeed, we have

$$\begin{aligned}
0 &\geq \int_{\mathbb{R}^d} \Big(\big(R_\Omega(1) - R_\omega(1)\big) - f\big(R_\Omega(f) - R_\omega(f)\big)\Big)\,dx\\
&\geq \int_{\mathbb{R}^d} \Big(\big(R_\Omega(1) - R_\omega(1)\big) - \big(R_\Omega(\|f\|_{L^\infty} f) - R_\omega(\|f\|_{L^\infty} f)\big)\Big)\,dx\\
&= \int_{B_R} \Big(\big(R_\Omega(1) - R_\omega(1)\big) - \big(R_\Omega(\|f\|_{L^\infty} f) - R_\omega(\|f\|_{L^\infty} f)\big)\Big)\,dx\\
&\quad + \int_{B_R^c} \big(R_\Omega(1) - R_\Omega(\|f\|_{L^\infty} f)\big)\,dx\\
&\geq \int_{B_R} \Big(\big(R_\Omega(1) - R_\omega(1)\big) - \big(R_\Omega(\|f\|_{L^\infty} f) - R_\omega(\|f\|_{L^\infty} f)\big)\Big)\,dx,
\end{aligned}$$

where the last inequality holds for $R > 0$ large enough and is due to Lemma 4.61. We now set for simplicity $w, u \in H^1_\mu$ to be respectively the solutions of the equations

$$-\Delta w + \mu w = 1 \quad \text{in} \quad H^1_\mu \quad \text{and} \quad -\Delta u + \mu u = \|f\|_{L^\infty} f \quad \text{in} \quad H^1_\mu.$$

Thus, the functions

$$h_w = R_\Omega(1) - R_\omega(1) \in H^1_\mu \quad \text{and} \quad h_u = R_\Omega(\|f\|_{L^\infty} f) - R_\omega(\|f\|_{L^\infty} f),$$

are $(\Delta - \mu)$-harmonic on the ball B_R. By the comparison principle, since $w \geq u$ on ∂B_R, we have that $h_w \geq h_u$ in B_R. Thus, for R large enough and $\omega = \Omega \cap B_R$, we have an equality in (4.142) which gives that $\Omega = \Omega \cap B_R$ and so Ω is bounded. □

Corollary 4.65. *Suppose that μ is a capacitary measure such that $w_\mu \in L^1(\mathbb{R}^d)$ and let $\Omega \subset \mathbb{R}^d$ be a quasi-open set, local subsolution for $\mathcal{F}$ as in* (4.128) *with respect to μ. If $F : \mathbb{R}^k \to \mathbb{R}$ is locally Lipschitz, then Ω is a bounded set.*

Proof. In view of Theorem 4.60 and Lemma 4.64, we have only to note that $w_\mu(x) \to 0$ as $|x| \to +\infty$. This fact was proved in [22] (see also [15] for a more precise account on the decay of w_μ) and we reproduce here the argument for the sake of completeness. Suppose, by absurd that there is some $\delta > 0$ and a sequence $x_n \in \mathbb{R}^d$ such that $|x_n| \to \infty$ and $w_\mu(x_n) \geq \delta$. Up to extracting a subsequence, we can suppose that $|x_n - x_m| \geq 2\delta$, for each pair of indices $n \neq m$. Since the function $w_\mu(x) - \dfrac{\delta^2 - |x - x_n|^2}{2d}$ is subharmonic, we have that

$$w_\mu(x_n) - \frac{\delta^2}{2d} \leq \fint_{B_\delta(x_n)} w_\mu \, dx,$$

and so, considering $\delta \leq 1$, we obtain

$$\frac{\delta}{2}|B_\delta| \leq \int_{B_\delta(x_n)} w_\mu \, dx, \ \forall n \in \mathbb{N},$$

which is a contradiction with the integrability of w_μ. □

Chapter 5
Shape supersolutions and quasi-minimizers

5.1. Introduction and motivation

In this chapter we consider measurable sets $\Omega \subset \mathbb{R}^d$, which are optimal for some given shape functional $\mathcal{F}$, with respect to *external* perturbations, *i.e.*

$$\mathcal{F}(\Omega) \leq \mathcal{F}(\Omega'), \quad \text{for every measurable set} \quad \Omega' \supset \Omega. \tag{5.1}$$

As in the previous chapter, we will try to recover some information on the set Ω starting from (5.1).

We start by a few examples which will help us establish some intuition on what to expect from the subsolutions of the energy and spectral functionals. To deal with these examples, we consider the following classical Lemma due to Alt and Caffarelli.

Lemma 5.1. *Suppose that $\mathcal{D} \subset \mathbb{R}^d$ is a given open set and that $u \in H_0^1(\mathcal{D})$ is a non-negative function such that*

$$\int_{\mathcal{D}} |\nabla u|^2\,dx + m|\{u > 0\}| \leq \int_{\mathcal{D}} |\nabla v|^2\,dx + m|\{v > 0\}|, \qquad \forall v \in H_0^1(\mathcal{D}), \quad v \geq u, \tag{5.2}$$

for some $m > 0$. Then the set $\Omega = \{u > 0\}$ is open. Moreover, if there is a function $f \in L^\infty(\mathcal{D})$ such that

$$-\Delta u = f \quad \text{in} \quad \Omega, \qquad u \in H_0^1(\Omega),$$

then u is locally Lipschitz continuous in $\mathcal{D}$.

Proof. Let $B_r(x_0) \subset \mathcal{D}$ be a given ball. Without loss of generality we can suppose that $x_0 = 0$. Let $v \in H^1(B_r)$ solve the problem

$$\min\left\{\int_{\mathbb{R}^d} |\nabla v|^2\,dx \,:\, v \in H^1(B_r),\ v \geq u \text{ in } B_r,\ v = u \text{ on } \partial B_r\right\}.$$

Setting $\widetilde{u} = \mathbb{1}_{B_r} v + \mathbb{1}_{B_r^c} u \in H_0^1(\mathcal{D})$ and using (5.2), we have

$$m|\{u>0\} \cup B_r| - m|\{u>0\}| \geq \int_{\mathbb{R}^d} |\nabla u|^2 \, dx - \int_{\mathbb{R}^d} |\nabla \widetilde{u}|^2 \, dx$$

$$\geq \frac{c_d}{r^2} |\{u=0\} \cap B_r| \left(\fint_{\partial B_r} u \, d\mathcal{H}^{d-1} \right)^2, \tag{5.3}$$

where c_d is a dimensional constant and the last inequality is due to (4.105) from Lemma 4.43. Thus, we have that $|B_r \cap \{u=0\}| > 0$ implies

$$\fint_{\partial B_r} u \, d\mathcal{H}^{d-1} \leq m C_d r, \tag{5.4}$$

and so, after integration

$$\fint_{B_r} u \, dx \leq m C_d r, \tag{5.5}$$

where C_d is a dimensional constant. We now recall that for quasi-every $x_0 \in \mathbb{R}^d$, we have

$$u(x_0) = \lim_{r \to 0^+} \fint_{B_r(x_0)} u \, dx. \tag{5.6}$$

Setting $u = 0$ on the set, where (5.6) does not hold, we have that for each $x_0 \in \{u > 0\}$ (5.6) holds. Now if $u(x_0) > 0$, then for some $r > 0$ small enough (5.5) does not hold and so $|B_r(x_0) \cap \{u = 0\}| = 0$. Now for $v \in H^1(B_r(x_0))$ as above, we have

$$0 = \int_{B_r(x_0)} |\nabla u|^2 \, dx - \int_{B_r(x_0)} |\nabla v|^2 \, dx = \int_{B_r(x_0)} |\nabla (u - v)|^2 \, dx,$$

and so $u \equiv v$ on $B_r(x_0)$. Since v is superharmonic, we obtain that $u > 0$ on $B_r(x_0)$ which gives that Ω is open.

We now set $\mathcal{D}_R := \{x \in \mathcal{D} : \operatorname{dist}(x, \partial\mathcal{D}) > R\}$. For fixed $R > 0$, we prove that $|\nabla u| \in L^\infty(\mathcal{D}_R)$. Suppose that $x_0 \in \mathcal{D}_R \cap \Omega$. If $\operatorname{dist}(x_0, \partial\Omega) > R/4$, then by the gradient estimate (see Lemma 5.9), we have

$$|\nabla u(x_0)| \leq C_d(1 + R^2)\|f\|_{L^\infty} + \frac{C_d}{R^{d+1}} \int_{B_R(x_0)} u \, dx.$$

If $\operatorname{dist}(x_0, \partial\Omega) < R/4$, then let $r = \operatorname{dist}(x_0, \partial\Omega) = |x_0 - y|$, for some

$y \in \partial\Omega$. Again by the gradient estimate

$$\begin{aligned}|\nabla u(x_0)| &\leq C_d(1+r^2)\|f\|_{L^\infty} + \frac{C_d}{r^{d+1}}\int_{B_r(x_0)} u\,dx \\ &\leq C_d(1+r^2)\|f\|_{L^\infty} + \frac{C_d}{r^{d+1}}\int_{B_{2r}(y)} u\,dx \\ &\leq C_d(1+r^2)\|f\|_{L^\infty} + C_d m,\end{aligned}$$

which concludes the proof. □

Remark 5.2. We note that if $\mathcal{D} = \mathbb{R}^d$, then we have that u is Lipschitz continuous on the whole $\mathbb{R}^d$.

We start with an example where this notion plays a fundamental role. For $f \in L^p(\mathbb{R}^d)$, we recall the notation

$$J_f(u) = \frac{1}{2}\int_{\mathbb{R}^d} |\nabla u|^2\,dx - \int_{\mathbb{R}^d} uf\,dx, \tag{5.7}$$

for the functional $J_f : H^1(\mathbb{R}^d) \cap L^{p'}(\mathbb{R}^d) \to \mathbb{R}$. If $p \in [2, +\infty]$ and $|\Omega| < +\infty$, we define the energy $E_f(\Omega)$ as

$$E_f(\Omega) = \min_{u \in H_0^1(\Omega)} J_f(u) = -\frac{1}{2}\int_{\mathbb{R}^d} f w_{f,\Omega}\,dx, \tag{5.8}$$

where $w_{f,\Omega}$ is the solution of

$$-\Delta w_{f,\Omega} = f \quad \text{in} \quad \Omega, \qquad w_{f,\Omega} \in H_0^1(\Omega),$$

which, in the case $f \equiv 1$, we denote with w_Ω.

Proposition 5.3. *Suppose that $\mathcal{D} \subset \mathbb{R}^d$ is a given open set and that the quasi-open set $\Omega \subset \mathbb{R}^d$ is a solution of the problem*

$$\min\Big\{E_f(\widetilde{\Omega}) + |\widetilde{\Omega}| : \Omega \subset \widetilde{\Omega} \subset \mathcal{D},\ \widetilde{\Omega}\ \textit{quasi-open}\Big\}, \tag{5.9}$$

where $f \in L^\infty(\mathbb{R}^d) \cap L^2(\mathbb{R}^d)$ is a given nonnegative function. Then Ω is an open set and the function $w_{f,\Omega}$ is locally Lipschitz continuous on $\mathcal{D}$.

Proof. We set for simplicity that $w := w_{f,\Omega}$ and we will prove that w satisfies the conditions of Lemma 5.1. Let $v \in H_0^1(\mathcal{D})$ be such that $v \geq$

w. Then, we have

$$\begin{aligned}\frac{1}{2}\int_{\mathcal{D}}|\nabla w|^2\,dx-\int_{\mathcal{D}}wf\,dx+|\{w>0\}|&=E_f(\Omega)+|\Omega|\\ &\le E_f(\{v>0\})+|\{v>0\}|\\ &\le\frac{1}{2}\int_{\mathcal{D}}|\nabla v|^2\,dx-\int_{\mathcal{D}}vf\,dx+|\{v>0\}|\\ &\le\frac{1}{2}\int_{\mathcal{D}}|\nabla v|^2\,dx-\int_{\mathcal{D}}wf\,dx+|\{v>0\}|,\end{aligned}$$

which finally gives (5.2). □

Proposition 5.4. *Suppose that $\mathcal{D}\subset\mathbb{R}^d$ is a given open set and that the quasi-open set $\Omega\subset\mathbb{R}^d$ is a solution of the problem*

$$\min\Big\{\lambda_1(\widetilde{\Omega})+|\widetilde{\Omega}|\,:\,\Omega\subset\widetilde{\Omega}\subset\mathcal{D},\ \widetilde{\Omega}\ \textit{quasi-open}\Big\}.\tag{5.10}$$

Then Ω is an open set and the first eigenfunction $u\in H_0^1(\Omega)$ is locally Lipschitz continuous on $\mathcal{D}$.

Proof. We suppose that u is non-negative and normalized in L^2. We note that we have $\Omega=\{u>0\}$. Let $v\in H_0^1(\mathcal{D})$ be such that $v\ge u$. Then, we have

$$\begin{aligned}\int_{\mathcal{D}}|\nabla u|^2\,dx+|\{u>0\}|=\lambda_1(\Omega)+|\Omega|&\le\lambda_1(\{v>0\})+|\{v>0\}|\\ &\le\frac{\int_{\mathcal{D}}|\nabla v|^2\,dx}{\int_{\mathcal{D}}v^2\,dx}+|\{v>0\}|\\ &\le\int_{\mathcal{D}}|\nabla v|^2\,dx+|\{v>0\}|,\end{aligned}$$

which gives (5.2). □

Remark 5.5. We note that in the propositions 5.3 and 5.4, we used only the optimality of Ω with respect to perturbations of the form $\widetilde{\Omega}=\Omega\cup B_r(x_0)$. Thus, the same result holds for quasi-open sets Ω, which are supersolutions for $E_f(\Omega)+|\Omega|$ and are such that $\{w_{f,\Omega}>0\}=\Omega$. We also note that this last equality, which is trivial if Ω is open, might need special attention if Ω is only quasi-open. In fact on quasi-open sets the strong maximum principle is known to hold only for functions f uniformly bounded from below by a positive constant on Ω.

Remark 5.6. We note that in the proofs of Proposition 5.3 and Proposition 5.4 we used the following two facts:

- The functionals $E_f + |\cdot|$ and $\lambda_1 + |\cdot|$ are *energy functional*, *i.e.* they can be written as minima of functionals on $H_0^1(\mathcal{D})$. For example, the optimal set Ω is given by $\Omega = \{w \neq 0\}$, where w solves the variational problem

$$\min\left\{\frac{1}{2}\int_{\mathbb{R}^d}|\nabla w|^2\,dx - \int_{\mathbb{R}^d} wf\,dx + |\{w \neq 0\}| \,:\, w \in H^1(\mathbb{R}^d)\right\}. \tag{5.11}$$

  Thus, we can restrict our attention to the functional space $H_0^1(\mathcal{D})$ instead to the family of quasi-open sets. We note also that this is not a property that all functionals have. The Dirichlet eigenvalues, for example, are defined through a min-max procedure, involving a whole k-dimensional subspace of $H^1(\mathbb{R}^d)$. This fact considerably complicates the analysis and will be one of the central arguments of this chapter.
- The second fact that was fundamental for our argument was the positivity of the state functions w and u. In fact, we were not able to reproduce Lemma 4.43 in the case when u changes sign. This obstacle was overcome by Briançon, Hayouni and Pierre in [17]. We will report their proof in Section 5.3 in the framework of quasi-minimizers.

In what follows we obtain the results from Propositions 5.3 and 5.4 for various functionals of spectral or energy type with penalizations with measure or perimeter. Of main interest will be the case when $\mathcal{D} = \mathbb{R}^d$, in which we expect the state functions to be globally Lipschitz.

5.2. Preliminary results

In this section we threat some preliminary results, which are crucial in the study of the regularity of the supersolutions. The results from Subsection 5.2.1 are mainly from [17], while the gradient estimate is classical and we report it here for convenience of the reader.

5.2.1. Pointwise definition of the solutions of PDEs on quasi-open sets

Let $f \in L^2(\mathbb{R}^d)$ and let Ω be a quasi-open set of finite measure. Consider the solution u of the equation

$$-\Delta u = f \quad \text{in} \quad \Omega, \qquad u \in H_0^1(\Omega). \tag{5.12}$$

Then the positive and the negative part $u_+ = \max\{u, 0\}$ and $u_- = \max\{-u, 0\}$ are solutions respectively of the equations

$$\begin{aligned} -\Delta u_+ &= f \quad \text{in} \quad \{u > 0\}, \qquad u_+ \in H_0^1\big(\{u > 0\}\big),\\ -\Delta u_- &= -f \quad \text{in} \quad \{u < 0\}, \qquad u_- \in H_0^1\big(\{u < 0\}\big). \end{aligned} \tag{5.13}$$

Thus, by Lemma 3.66 the operators

$$\Delta u_+ + f : H^1(\mathbb{R}^d) \to \mathbb{R} \qquad \text{and} \qquad \Delta u_- - f : H^1(\mathbb{R}^d) \to \mathbb{R},$$

are positive and correspond to a Radon capacitary measures, which we denote with

$$\mu_1 := \Delta u_+ + f \qquad \text{and} \qquad \mu_2 := \Delta u_- - f.$$

Moreover, if $f \in L^p(\mathbb{R}^d)$ for some $p \in (d/2, +\infty]$, then:

1. By Lemma 3.51, $u \in L^\infty(\mathbb{R}^d)$ and

$$\|u\|_{L^\infty} \le \frac{C_d}{2/d - 1/p}\|f\|_{L^p}|\Omega|^{2/d-1/p}.$$

2. By Theorem 3.68, every point $x \in \mathbb{R}^d$ is a Lebesgue point for u_+, u_- and u.

$$u_+(x) = \lim_{r\to 0} \fint_{\partial B_r(x)} u_+\, d\mathcal{H}^{d-1} \text{ and } u_-(x) = \lim_{r\to 0} \fint_{\partial B_r(x)} u_-\, d\mathcal{H}^{d-1}.$$

5.2.2. Gradient estimate for Sobolev functions with L^∞ Laplacian

Lemma 5.7. *Suppose that u is a bounded harmonic function on the ball $B_r \subset \mathbb{R}^d$. Then, its gradient in the ball $B_{r/2}$ can be estimated as follows:*

$$\|\nabla u\|_{L^\infty(B_{r/2})} \le \frac{2d}{r}\|u\|_{L^\infty(B_r)}.$$

Proof. Let us set $u_i := \dfrac{\partial u}{\partial x_i}$. Then u_i is harmonic in B_r and so the mean value property holds for any $x \in B_{r/2}$:

$$u_i(x) = \fint_{B_{r/2}(x)} u_i(y)\, dy = \frac{2^d}{\omega_d r^d}\int_{\partial B_{r/2}(x)} u\nu_i\, d\mathcal{H}^{d-1} \le \frac{2d}{r}\|u\|_{L^\infty(B_r)}.$$

□

Lemma 5.8 (see [75, Chapter 9]). *Consider the function $\Gamma : \mathbb{R}^d \times \mathbb{R}^d \to \mathbb{R}$ defined as*

$$\Gamma(x,y) = \begin{cases} \dfrac{1}{2\pi}\log|x-y|, & \text{if } d = 2,\\[2mm] \dfrac{1}{d(2-d)\omega_d}|x-y|^{2-d}, & \text{if } d > 2.\end{cases}$$

If $f \in L^\infty(B_r)$, then the function $u : B_r \to \mathbb{R}$, defined as

$$u(x) := \int_{B_r} \Gamma(x, y) f(y)\, dy.$$

has the following properties:

(a) *$u \in H^2(B_r)$ and $\Delta u = f$ almost everywhere in B_r,*
(b) *$u \in C^{1,\alpha}$, for any $\alpha \in (0, 1)$,*
(c) *$\|u\|_{L^\infty(B_r)} \le C_0 r\, \|f\|_{L^\infty(B_r)}$,*
(d) *$\|\nabla u\|_{L^\infty(B_r)} \le C_1\, \|f\|_{L^\infty(B_r)}$, where C_0 and C_1 are constants depending only on the dimension d.*

Lemma 5.9. *Suppose that $u \in H^1(B_r)$ satisfies the equation*[1]

$$-\Delta u = f \quad in \quad B_r,$$

for some function $f \in L^\infty(B_r)$. Then we can estimate the gradient in the ball $B_{r/2}$ as follows:

$$\|\nabla u\|_{L^\infty(B_{r/2})} \le C_d \|f\|_{L^\infty(B_r)} + \frac{2d}{r}\|u\|_{L^\infty(B_r)}.$$

Proof. Let u_N be the Newton potential from Lemma 5.8 and let $u_h = u - u_N$. Then u_h is harmonic in B_r and we have

$$\begin{aligned}
\|\nabla u\|_{L^\infty(B_{r/2})} &\le \|\nabla u_N\|_{L^\infty(B_{r/2})} + \|\nabla u_h\|_{L^\infty(B_{r/2})}\\
&\le C_1\|f\|_{L^\infty(B_r)} + \frac{2d}{r}\|u_h\|_{L^\infty(B_r)}\\
&\le C_1\|f\|_{L^\infty(B_r)} + \frac{2d}{r}\|u\|_{L^\infty(B_r)} + \frac{2d}{r}\|u_N\|_{L^\infty(B_r)}\\
&\le (C_1 + 2dC_0)\|f\|_{L^\infty(B_r)} + \frac{2d}{r}\|u\|_{L^\infty(B_r)},
\end{aligned}$$

where C_0 and C_1 are the constants from Lemma 5.8. □

Corollary 5.10. *Suppose that $\Omega \subset \mathbb{R}^d$ is an open set and suppose that $u \in H^1_0(\Omega)$ is a non-negative function satisfying*

$$-\Delta u = f \quad in \quad \Omega, \qquad u \in H^1_0(\Omega),$$

where $f \in L^\infty(\mathbb{R}^d)$. Suppose that there are constants $C > 0$ and $r_0 > 0$ such that

$$⨍_{B_r(x_0)} u\, dx \le Cr, \qquad \forall x_0 \in \partial\Omega,\ \forall 0 < r \le r_0.$$

[1] Note that no boundary values are imposed.

Then u is Lipschitz continuous on $\mathbb{R}^d$. In particular, on the set

$$\Omega_r := \big\{x \in \Omega : \, dist(x, \partial\Omega)\big\} < r_0/4,$$

we have the estimate

$$\|\nabla u\|_{L^\infty(\Omega_r)} \le C_d\Big((1+r_0^2)\|f\|_{L^\infty} + C\Big).$$

Proof. We will prove that $|\nabla u| \in L^\infty(\Omega)$. We first note that for every $x_0 \in \mathbb{R}^d$ and every $r > 0$, we have

$$\|u\|_{L^\infty(B_r(x_0))} \le \frac{r^2}{2d}\|f\|_{L^\infty} + \frac{1}{|B_r|}\int_{B_{2r}(x_0)} u\,dx. \tag{5.14}$$

Indeed, since $\Delta u + \|f\|_{L^\infty} \ge 0$ on $\mathbb{R}^d$, we have that the function

$$x \mapsto u(x) - \|f\|_{L^\infty}\frac{r^2 - |x - x_1|^2}{2d},$$

is sub-harmonic for every $x_1 \in B_r(x_0)$, and so

$$u(x_1) \le \frac{r^2}{2d}\|f\|_{L^\infty} + \frac{1}{|B_r|}\int_{B_r(x_1)} u\,dx \le \frac{r^2}{2d}\|f\|_{L^\infty} + \frac{1}{|B_r|}\int_{B_{2r}(x_0)} u\,dx.$$

Suppose now that $x_0 \in \Omega$. If $\mathrm{dist}(x_0, \partial\Omega) > r_0/4$, then by Lemma 5.9 we have

$$\begin{aligned}
|\nabla u|(x_0) &\le C_d\Big(\|f\|_{L^\infty} + r_0^{-1}\|u\|_{L^\infty(B_{r_0/8}(x_0))}\Big)\\
&\le C_d\Big((1+r_0^2)\|f\|_{L^\infty} + r_0^{-1-d}\int_{B_{r_0/4}(x_0)} u\,dx\Big)\\
&\le C_d\Big((1+r_0^2)\|f\|_{L^\infty} + r_0^{-1-d/2}\|u\|_{L^2}\Big).
\end{aligned}$$

If $r := \mathrm{dist}(x_0, \partial\Omega) \le r_0/4$, we set $y \in \partial\Omega$ to be such that $|y - x_0| = r$ and thus we have

$$\begin{aligned}
|\nabla u|(x_0) &\le C_d\Big(\|f\|_{L^\infty} + r^{-1}\|u\|_{L^\infty(B_{r/4}(x_0))}\Big)\\
&\le C_d\Big((1+r_0^2)\|f\|_{L^\infty} + r^{-d-1}\int_{B_{r/2}(x_0)} u\,dx\Big)\\
&\le C_d\Big((1+r_0^2)\|f\|_{L^\infty} + r^{-d-1}\int_{B_r(y)} u\,dx\Big)\\
&\le C_d\Big((1+r_0^2)\|f\|_{L^\infty} + C\Big).
\end{aligned}$$

□

5.2.3. Monotonicity formula

In this last preliminary subsection we restate the Caffarelli-Jerison-Kënig monotonicity formula in the case $-\Delta u = f$.

Theorem 5.11. *Let $\Omega \subset \mathbb{R}^d$ be a quasi-open set of finite measure, $f \in L^\infty(\Omega)$ and $u \in H^1(B_1)$ be the solution in Ω of the equation*

$$-\Delta u = f \quad in \quad \Omega, \qquad u \in H_0^1(\Omega). \tag{5.15}$$

Setting $u^+ = \sup\{u, 0\}$ and $u^- = \sup\{-u, 0\}$, there is a dimensional constant C_d such that for each $0 < r \leq 1/2$

$$\begin{aligned}&\left(\frac{1}{r^2}\int_{B_r}\frac{|\nabla u^+(x)|^2}{|x|^{d-2}}\,dx\right)\left(\frac{1}{r^2}\int_{B_r}\frac{|\nabla u^-(x)|^2}{|x|^{d-2}}\,dx\right)\\ &\leq C_d\left(\|f\|_{L^\infty}^2 + \int_\Omega u^2\,dx\right) \leq C_m,\end{aligned} \tag{5.16}$$

where $C_m = C_d\|f\|_{L^\infty}^2\left(1 + |\Omega|^{\frac{d+4}{d}}\right)$.

Proof. We apply Theorem 4.30 to

$$u_1 := \|f\|_{L^\infty}^{-1}u^+ \qquad \text{and} \qquad u_2 := \|f\|_{L^\infty}^{-1}u^-,$$

and substituting in (4.56) we obtain the first inequality in (5.16). The second one follows, using the equation (5.15):

$$\begin{aligned}\|u\|_{L^2}^2 &\leq C_d|\Omega|^{2/d}\|\nabla u\|_{L^2}^2 = C_d|\Omega|^{2/d}\int_\Omega fu\,dx\\ &\leq C_d|\Omega|^{2/d+1/2}\|f\|_{L^\infty}\|u\|_{L^2}.\end{aligned} \tag{5.17}$$

□

5.3. Lipschitz continuity of energy quasi-minimizers

Consider a function

$$f \in L^p(\mathbb{R}^d), \quad \text{where} \quad p \in \begin{cases}(1, +\infty], & \text{if} \quad d = 2,\\ \left[\dfrac{2d}{d+2}, +\infty\right], & \text{if} \quad d \geq 3,\end{cases} \tag{5.18}$$

and the functional

$$\begin{aligned}&J_f : H^1(\mathbb{R}^d) \cap L^{p'}(\mathbb{R}^d) \to \mathbb{R},\\ &J_f(u) := \frac{1}{2}\int_{\mathbb{R}^d}|\nabla u|^2\,dx - \int_{\mathbb{R}^d} uf\,dx,\end{aligned} \tag{5.19}$$

where $p' = \frac{p}{p-1}$.

The classical elliptic regularity theory studies the properties of the minimizers of J_f in the Sobolev space $H^1_0(\Omega)$, where Ω is a given fixed open set, usually bounded and regular. In this section we will study the regularity properties of the functions that minimize J_f in the whole space $H^1(\mathbb{R}^d)\cap L^{p'}(\mathbb{R}^d)$, up to a volume term Cr^d. In analogy with the situation arising in the theory of functions of bounded variation and the Mumford-Shah functional (see for instance [5] and [67]) we call these functions *quasi-minimizers*. Most of the theory in this section was exposed in [92] and also in [28], where it was applied to the problem of the regularity of the optimal sets for the kth Dirichlet Eigenvalues. Precisely we have the following definition.

Definition 5.12. We say that u is a **quasi-minimizer** for the functional J_f if there are positive constants $r_0 > 0$ and $C_0 > 0$ such that

$$J_f(u) \le J_f(u+\varphi) + C_0 r^d, \quad \text{for every} \quad r \in (0, r_0),\ x_0 \in \mathbb{R}^d \quad \text{and} \quad \varphi \in H^1_0(B_r(x_0)). \tag{5.20}$$

Remark 5.13. We note that the restriction $r < r_0$ can be removed from the quasi-minimality condition (5.20), up to changing the constant C_0 with $\widetilde{C}_0 = C_0 + J_f(u) r_0^{-d}$. Nevertheless, when we consider sequences of quasi-minimizers with possibly different constants r_0 and C_0 it is more convenient to work with the pair (r_0, C_0), since it is possible that one of the two constants (in our case r_0) degenerates, while the other remains controllable.

We also introduce the following more general notion of an α-quasi-minimizer.

Definition 5.14. Let $\alpha \in (d-1, d]$ be fixed and $f \in L^p(\mathbb{R}^d)$ and p be as in (5.18). We say that u is an **α-quasi-minimizer** for the functional J_f, defined in (5.19), if there are positive constants $r_0 > 0$ and $C_0 > 0$ such that

$$J_f(u) \le J_f(u+\varphi) + C_0 r^\alpha, \quad \text{for every} \quad r \in (0, r_0),\ x_0 \in \mathbb{R}^d \quad \text{and} \quad \varphi \in H^1_0(B_r(x_0)). \tag{5.21}$$

Remark 5.15. From now on the term quasi-minimizer will refer to the case $\alpha = d$.

Remark 5.16. We note that since $r_0 < +\infty$ and $\varphi \in H^1_0(B_{r_0}(x_0))$, then $\varphi \in L^{p'}(\mathbb{R}^d)$ and so the function $v := u + \varphi$ is a possible test function for the quasi-optimality of u in the domain of the functional J_f.

Remark 5.17. The function $u \in H^1(\mathbb{R}^d) \cap L^{p'}(\mathbb{R}^d)$ is an α-quasi-minimizer for the functional J_f, if and only if,

$$|\langle \Delta u + f, \varphi \rangle| \leq \frac{1}{2} \int_{\mathbb{R}^d} |\nabla \varphi|^2 \, dx + C_0 r^\alpha,$$
$$\forall r \in (0, r_0), \quad \forall x_0 \in \mathbb{R}^d, \quad \forall \varphi \in H^1_0(B_r(x_0)),$$

where with $\Delta u + f : H^1(\mathbb{R}^d) \cap L^{p'}(\mathbb{R}^d) \to \mathbb{R}$ we denote the functional

$$\langle \Delta u + f, \varphi \rangle := \int_{\mathbb{R}^d} \big(- \nabla u \cdot \nabla \varphi + f \varphi \big) \, dx, \quad \forall \varphi \in H^1(\mathbb{R}^d) \cap L^{p'}(\mathbb{R}^d).$$

In the following two elementary lemmas we give two more equivalent ways to state the quasi-minimality of u.

Lemma 5.18. *Let $\alpha \in (d-1, d]$ be given and $f \in L^p(\mathbb{R}^d)$ and p be as in (5.18). Then, for $u \in H^1(\mathbb{R}^d) \cap L^{p'}(\mathbb{R}^d)$, the following conditions are equivalent:*

(i) *u is an α-quasi-minimizer,* i.e. *there are constants $r_0 > 0$ and $C_0 > 0$ such that u satisfies*

$$|\langle \Delta u + f, \varphi \rangle| \leq \frac{1}{2} \int_{\mathbb{R}^d} |\nabla \varphi|^2 \, dx + C_0 r^\alpha, \quad \text{for every } \forall r \in (0, r_0),\ x \in \mathbb{R}^d \text{ and } \varphi \in H^1_0(B_r(x)); \tag{5.22}$$

(ii) *There are constants $r_1 > 0$, $C_1 > 0$ and $\delta_1 \in (0, +\infty]$ such that u satisfies the condition*

$$|\langle \Delta u + f, \varphi \rangle| \leq \frac{1}{2} \int_{\mathbb{R}^d} |\nabla \varphi|^2 \, dx + C_1 r^\alpha, \quad \text{for every } r \in (0, r_1),\ x \in \mathbb{R}^d \text{ and } \varphi \in H^1_0(B_r(x)) \text{ s.t. } \|\nabla \varphi\|_{L^2} \leq \delta_1; \tag{5.23}$$

(iii) *There are constants $r_2 > 0$ and $C_2 > 0$ such that u satisfies*

$$|\langle \Delta u + f, \psi \rangle| \leq C_2 r^{\alpha/2} \|\nabla \psi\|_{L^2}, \quad \text{for every } \forall r \in (0, r_2),\ x \in \mathbb{R}^d \text{ and } \psi \in H^1_0(B_r(x)). \tag{5.24}$$

Moreover,

1. *If u satisfies (ii) with r_1, C_1 and δ_1, then it satisfies* (iii) *with* $r_2 = \min\left\{r_1, \delta_1^{2/\alpha}\right\}$ *and* $C_2 = \frac{1}{2} + C_1$.
2. *If u satisfies (iii) with r_2 and C_2, then it satisfies (i) with $r_0 = r_2$ and* $C_0 = \frac{1}{2}C_2^2$.
3. *If u satisfies* (ii) *with r_1 and C_1, then it satisfies* (i) *with* $r_0 = \min\left\{r_1, \delta_1^{2/\alpha}\right\}$ *and* $C_0 = \frac{1}{2}\left(\frac{1}{2} + C_1\right)^2$.

Proof. We first prove the implication (ii)⇒(iii) and claim (1). Without loss of generality we set $x = 0$. We define

$$r_2 := \min\left\{r_1, \delta_1^{2/\alpha}\right\} \qquad \text{and} \qquad C_2 := \frac{1}{2} + C_1.$$

For given $r \in (0, r_2)$ and $\psi \in H_0^1(B_r)$ we consider the function $\varphi = r^{\alpha/2}\|\nabla\psi\|_{L^2}^{-1}\psi$.
By the choice of φ and r_2, we have

$$\varphi \in H_0^1(B_r) \qquad \text{and} \qquad \|\nabla\varphi\|_{L^2} = r^{\alpha/2} \le r_2^{\alpha/2} \le \delta_1.$$

Now testing (5.23) with φ we get

$$\frac{r^{\alpha/2}}{\|\nabla\psi\|_{L^2}}|\langle\Delta u + f, \psi\rangle| \le \frac{1}{2}r^\alpha\|\nabla\psi\|_{L^2}^{-2}\int_{\mathbb{R}^d}|\nabla\psi|^2\,dx + C_1r^\alpha = C_2r^\alpha,$$

that is, we proved (5.58) and also claim (1).

We now prove the implication (iii)⇒(i) and claim (2). Indeed, for every $\psi \in H_0^1(B_r(x))$, it is sufficient to use (iii) and the mean geometric-mean quadratic inequality obtaining

$$|\langle\Delta u + f, \psi\rangle| \le C_2r^{d/2}\|\nabla\psi\|_{L^2} \le \frac{1}{2}\int_{\mathbb{R}^d}|\nabla\psi|^2\,dx + \frac{C_2^2}{2}r^d.$$

The last claim (3) is a consequence of (1) and (2). □

In the particular case when $f \equiv 0$, the functional J_0 reduces simply to the Dirichlet integral

$$J_0 : H^1(\mathbb{R}^d) \to \mathbb{R}, \qquad J_0(u) = \int_{\mathbb{R}^d}|\nabla u|^2\,dx.$$

Under an integrability condition on a generic function f, the analysis of the quasi-minimizers of J_f can be reduced to the study of the quasi-minimizers for the Dirichlet integral J_0, which may significantly simplify the analysis. Indeed, we have the following result.

Lemma 5.19. *Suppose that $f \in L^p(\mathbb{R}^d)$ for $p \in [d, +\infty]$. Then every quasi-minimizer $u \in H^1(\mathbb{R}^d) \cap L^{p'}(\mathbb{R}^d)$ for the functional J_f, satisfying (5.58) with constants $C_2 > 0$ and $r_2 > 0$ (and $\alpha = d$) is also quasi-minimizer for J_0 satisfying*

$$|\langle \Delta u, \psi \rangle| \leq C_3 r^{d/2} \|\nabla \psi\|_{L^2}, \textit{for every } r \in (0, r_2),\ x_0 \in \mathbb{R}^d \quad \textit{and } \psi \in H_0^1(B_r(x_0)), \tag{5.25}$$

where $C_3 = C_2 + C_d \|f\|_{L^p} r_2^{1-d/p}$ and C_d is a dimensional constant.

Proof. Without loss of generality we fix $x_0 = 0$. Let $r < r_2$ and $\psi \in H_0^1(B_r)$. Then we have

$$\begin{aligned}\int_{\mathbb{R}^2} f\psi\, dx &\leq \|\psi\|_{L^2} \|f \mathbb{1}_{B_r}\|_{L^2} \leq \lambda_1(B_1)^{-1/2} r \|\nabla \psi\|_{L^2} \|f \mathbb{1}_{B_r}\|_{L^2} \\ &\leq \lambda_1(B_1)^{-1/2} r \|\nabla \psi\|_{L^2} \|f\|_{L^p} \|\mathbb{1}_{B_r}\|_{L^q} \\ &\leq \left(\lambda_1(B_1)^{-1/2} \|f\|_{L^p} r_2^{1-d/p}\right) r^{d/2} \|\nabla \psi\|_{L^2},\end{aligned}$$

where $\dfrac{1}{q} = \dfrac{1}{2} - \dfrac{1}{p}$. □

In what follows we study the regularity properties of the quasi-minimizers. The two main results of this section concern the solutions of u of elliptic equations of the form

$$-\Delta u = f \quad \text{in} \quad \Omega, \qquad u \in H_0^1(\Omega),$$

which are quasi-minimizers for the Dirichket integral J_0. Our main results are:

- the Lipschitz continuity of the quasi-minimizers;
- in the case $f = \lambda u$, the Lipschitz constant of u does not depend on the local geometry of the domain Ω, but only on the eigenvalue λ and the measure $|\Omega|$.

Lemma 5.20. *Suppose that $u \in H^1(\mathbb{R}^d)$ satisfies the following conditions:*

(a) *The positive and negative parts u^+ and u^- of u are such that $\Delta u^+ + 1 \geq 0$ and $\Delta u^- + 1 \geq 0$ as functionals on $H^1(\mathbb{R}^d) \cap L^1(\mathbb{R}^d)$.*
(b) *u is an α-quasi-minimizer of the Dirichlet integral J_0, for some $\alpha \in (d-1, d]$.*

Then the function $u : \mathbb{R}^d \to \mathbb{R}$ is continuous.

Proof. Consider a sequence $x_n \in \mathbb{R}^d$ converging to some $x_\infty \in \mathbb{R}^d$. Without loss of generality we suppose $x_\infty = 0$ and we set $\delta_n := |x_n|$ and $\xi_n := \delta_n^{-1} x_n$.

We will prove in a series of claims that $u(x_n)$ converges to $u(0)$. We consider the blow-up sequence

$$u_n \in H^1(\mathbb{R}^d), \qquad u_n(x) := u(\delta_n x).$$

- *The blow-up sequence is uniformly bounded.*
  By condition (a) we have that u is bounded, precisely for every $x_0 \in \mathbb{R}^d$ and every $R > 0$ the subharmonicity of the function $x \mapsto u(x) + \frac{|x-x_0|^2}{2d}$ gives the estimate

$$|u(x_0)| \le \fint_{B_R(x_0)} |u|\,dx + \frac{R^2}{d} \le \left(\fint_{B_R} |u|^2\,dx\right)^{1/2} + \frac{R^2}{d}$$
$$\le \frac{\|u\|_{L^2(\mathbb{R}^d)}}{(\omega_d R)^{d/2}} + \frac{R^2}{d},$$

  taking the minimum in $R \in (0, +\infty)$, we get

$$\|u\|_{L^\infty} \le C_d \|u\|_{L^2}^{\frac{4}{d+4}}, \tag{5.26}$$

  where C_d is a dimensional constant. Since $\|u_n\|_{L^\infty} = \|u\|_{L^\infty}$, for every $n \in \mathbb{N}$, the same inequality holds for u_n and so, u_n is a uniformly bounded sequence in $L^\infty(\mathbb{R}^d)$.
- *On a fixed ball $B_R \subset \mathbb{R}^d$ the blow-up sequence is asymptotically close in $H^1(B_R)$ to a sequence of harmonic functions with the same boundary values on ∂B_R.*
  For all $R \ge 1$ and $n \in \mathbb{N}$, we consider the harmonic function $v_{R,n}$ defined by

$$\Delta v_{R,n} = 0 \quad \text{in} \quad B_{R\delta_n}, \qquad v_{R,n} = u \quad \text{on} \quad \partial B_{R\delta_n},$$

  and its rescaling $v_n(x) := v_{R,n}(\delta_n x)$ satisfying

$$\Delta v_n = 0 \quad \text{in} \quad B_R, \qquad v_n = u_n \quad \text{on} \quad \partial B_R.$$

  For $n \in \mathbb{N}$ large enough we can test the quasi-minimality of u with

$\varphi = u - v_{R,n} \in H_0^1(B_{R\delta_n})$, obtaining

$$\begin{aligned}\int_{B_R} |\nabla(u_n - v_n)|^2\,dx &= \delta_n^{2-d}\int_{B_{R\delta_n}} |\nabla(u - v_{R,n})|^2\,dx\\ &= \delta_n^{2-d}\int_{B_{R\delta_n}} \nabla u\cdot\nabla(u - v_{R,n})\,dx\\ &\le C\delta_n^{2-d}\left(\int_{B_{R\delta_n}} |\nabla(u - v_{R,n})|^2\,dx\right)^{1/2}(R\delta_n)^{\alpha/2}\\ &\le CR^{d/2}\delta_n^{\beta/2}\left(\int_{B_R} |\nabla(u_n - v_n)|^2\,dx\right)^{1/2},\end{aligned}$$

where $\beta := 2 - d + \alpha > 1$. Thus, for n large enough, we have

$$\int_{B_R} |\nabla(u_n - v_n)|^2\,dx \le C^2R^\alpha\delta_n^\beta,$$

where C is the constant from (5.25).

- *Up to a subsequence, u_n converges in $H^1(B_{R/2})$ to a harmonic function $u_R \in H^1(B_{R/2})$.*
  By the previous point, it is sufficient to prove the claim for the sequence v_n. Since $v_n = u_n$ on ∂B_R, we have the following equiboundedness estimate:

$$\|v_n\|_{L^\infty(B_R)} \le C_d\|u\|_{L^2}^{\frac{4}{d+4}}.$$

On the other hand, the gradient estimate (Lemma 5.7) gives

$$\begin{aligned}\|\nabla v_n\|_{L^\infty(B_{3R/4})} &\le \frac{4d}{R}\|v_n\|_{L^\infty(B_R)},\\ \|\nabla(\partial_{x_i} v_n)\|_{L^\infty(B_{R/2})} &\le \frac{4d}{R}\|\partial_{x_i} v_n\|_{L^\infty(B_{3R/4})}\\ &\le \left(\frac{4d}{R}\right)^2\|v_n\|_{L^\infty(B_R)},\quad \forall i = 1,\dots,d.\end{aligned}$$

Thus on $\overline{B_{R/2}}$ the sequences v_n and ∇v_n are equi-bounded and equicontinuous. By the Ascoli-Arzela Theorem v_n converges in C^1 norm, up to a subsequence, to a function $u_R \in H^1(B_{R/2})$. Moreover, being v_R a weak limit of v_n in $H^1(B_{R/2})$, it is harmonic.

- *The limit u_R is constant on $B_{R/2}$. More precisely, $u_R \equiv u(0)$.*
  Repeating the argument from the previous point for every $R \in \mathbb{N}$ and taking a diagonal sequence, we have that up to a subsequence

u_n converges in $H^1(B_{R/2})$, for every $R > 0$. Moreover, for every $S > R$, we have that $v_S \equiv v_R$ on $B_{R/2}$. Thus, there is a harmonic function $v \in H^1_{\text{loc}}(\mathbb{R}^d)$ such that $v \equiv v_R$ on $B_{R/2}$, for every $R > 0$. By construction v is bounded and so, it is a constant. On the other hand, being 0 a Lebesgue point for u, we have

$$\fint_{B_1} u_n \, dx = \fint_{B_{\delta_n}} u \, dx \xrightarrow[n\to\infty]{} u(0),$$

and so $v \equiv u(0)$ on $\mathbb{R}^d$.

- *If $u(0) \geq 0$, then $u_n^- \to 0$ uniformly on $B_{R/2}$, for every $R > 0$.* Consider the function $\widetilde{u}_n \in H^1(B_R)$ satisfying

$$-\Delta\widetilde{u}_n = \delta_n^2 \quad \text{in} \quad B_R, \qquad \widetilde{u}_n = u_n^- \quad \text{on} \quad \partial B_R.$$

By the Poisson formula we have the following expression for $\widetilde{u}_n$:

$$\widetilde{u}_n(x) = \frac{R^2 - |x|^2}{d\omega_d R} \int_{\partial B_R} \frac{u_n^-(y)}{|x-y|^d} \, d\mathcal{H}^{d-1}(y) + \frac{\delta_n^2\,(R^2 - |x|^2)}{2d}, \quad \text{for every } x \in B_R,$$

and so, on $B_{R/2}$ we have

$$\|\widetilde{u}_n\|_{L^\infty(B_{R/2})} \leq 2^d \fint_{\partial B_R} u_n^- \, d\mathcal{H}^{d-1} + \delta_n^2 R^2 \xrightarrow[n\to\infty]{} 0,$$

where the right-hand side converges to zero, since u_n^- converges to zero strongly in $H^1(B_R)$. Now the claim follows since by (a) and the maximum principle $u_n^- \leq \widetilde{u}_n$ on B_R.

- *If $u(0) \geq 0$, then $u(0) \leq \liminf_{n\to\infty} u(x_n)$.*
For $0 < s < 1$ small enough consider the test function $\phi_s \in C_c^\infty(B_{2s}(x_n))$ such that

$$0 \leq \phi_s \leq 1 \text{ on } \mathbb{R}^d, \quad \phi_s \equiv 1 \text{ on } B_s(x_n) \quad \text{and} \quad \|\nabla\phi_s\|_{L^\infty} \leq \frac{C_d}{s},$$

for a dimensional constant C_d. By the quasi-optimality of u, there is a constant $C > 0$ such that we have

$$|\langle\Delta u, \phi_s\rangle| \leq C s^{\alpha/2} \|\nabla\phi_s\|_{L^2} \leq C s^{\alpha-1}.$$

Since $\mu_1 := \Delta u^+ + 1$ and $\mu_2 := \Delta u^- + 1$ are positive measures, we have

$$\begin{aligned} \mu_1(B_s(x_n)) \leq \langle\mu_1, \phi_s\rangle &= \langle\mu_1 - \mu_2, \phi_s\rangle + \langle\mu_2, \phi_s\rangle \\ &\leq C s^{\alpha-1} + \mu_2(B_{2s}(x_n)), \end{aligned}$$

and for every $s \leq 1$,

$$\Delta u^+(B_s(x_n)) \leq (C + \omega_d)s^{\alpha-1} + \Delta u^-(B_{2s}(x_n)). \tag{5.27}$$

Multiplying both sides of (5.27) by s^{1-d} and integrating for $s \in (0, \delta_n)$, we obtain

$$\fint_{\partial B_{\delta_n}(x_n)} u^+ \, d\mathcal{H}^{d-1} - u^+(x_n) \leq \frac{1}{2} \fint_{\partial B_{2\delta_n}(x_n)} u^- \, d\mathcal{H}^{d-1} + C\delta_n^{\alpha-d+1},$$

and for the rescaling $u_n(x) = u(\delta_n x)$, we get

$$\fint_{\partial B_1} u_n^+(\xi_n + \cdot)\, d\mathcal{H}^{d-1} - u_n^+(\xi_n) \leq \frac{1}{2} \fint_{\partial B_2} u_n^-(\xi_n + \cdot)\, d\mathcal{H}^{d-1} + C\delta_n^{\alpha-d+1}.$$

Up to a subsequence, we may assume that $\xi_n \to \xi_\infty$ in $\mathbb{R}^d$. Thus $u_n(\xi_n + \cdot)$ converges to the constant $u(0)$ strongly in $H^1(B_1)$. Together with the uniform convergence of u_n^- to zero, we get

$$u(0) \leq \liminf_{n\to\infty} u^+(x_n). \tag{5.28}$$

- *If $u(0) \geq 0$, then $u(0) = \lim_{n\to\infty} u(x_n)$.*
  Indeed, by (a) we have the upper semi-continuity inequality

$$u^+(0) \geq \limsup_{n\to\infty} u^+(x_n),$$

which together with (5.28) gives

$$u(0) = \lim_{n\to\infty} u^+(x_n).$$

Now since u_n^- converges uniformly to zero we have that $u^-(x_n)$ converges to zero and so

$$u(0) = \lim_{n\to\infty} u(x_n),$$

which concludes the proof. □

Lemma 5.21. *Suppose that $u \in H^1(\mathbb{R}^d)$ satisfies the following conditions:*

(a) *The positive and negative parts u^+ and u^- of u are such that $\Delta u^+ + 1 \geq 0$ and $\Delta u^- + 1 \geq 0$ as functionals on $H^1(\mathbb{R}^d) \cap L^1(\mathbb{R}^d)$.*
(b) *u is an α-quasi-minimizer of the Dirichlet integral J_0, for some $\alpha \in (d-1, d]$.*

Then we have the inequality

$$|\Delta|u||(B_r(x_0)) \le C\, r^{\beta}, \quad \textit{for every} \quad r < r_0, \tag{5.29}$$

where $\beta = \dfrac{\alpha+d-2}{2}$ *and the constant* C *is given by*

$$C = C_d\left(C_2 + 1 + r_0^{1+\frac{d-\alpha}{2}} + \|u\|_{L^2}^{\frac{2}{d+4}} r_0^{\frac{d-\alpha}{2}}\right), \tag{5.30}$$

and $r_0 = r_2/4$, *where* $r_2 > 0$ *and* $C_2 > 0$ *are the quasi-minimality constants from* (5.58) *corresponding to the case* $f \equiv 0$.

Proof. Without loss of generality we can suppose $x_0 = 0$. Again, we divide the proof in a series of claims.

For every $r > 0$, we consider the functions $v_r^+ \in H^1(B_r)$, $v_r^- \in H^1(B_r)$ and $v_r := v_r^+ - v_r^-$, where

$$\Delta v_r^{\pm} = 0 \quad \text{in} \quad B_r, \qquad v_r^{\pm} = u^{\pm} \quad \text{on} \quad \partial B_r.$$

- *For every* $r > 0$ *we have the bound*

$$\left(\int_{B_r} |\nabla(u^+ - v_r^+)|^2\,dx\right)^{1/2} \left(\int_{B_r} |\nabla(u^- - v_r^-)|^2\,dx\right)^{1/2} \le C_d r^d \left(1 + \|u\|_{L^2}^4\right). \tag{5.31}$$

Since the functions $v_r^{\pm}$ minimize the Dirichlet integral we have

$$\int_{B_r} |\nabla v_r^{\pm}|^2\,dx \le \int_{B_r} |\nabla u^{\pm}|^2\,dx,$$

and so, we obtain

$$\int_{B_r} |\nabla(u^{\pm} - v_r^{\pm})|^2\,dx = \int_{B_r} \nabla u^{\pm}\cdot\nabla(u^{\pm} - v_r^{\pm})\,dx \le 2\int_{B_r} |\nabla u^{\pm}|^2\,dx.$$

By the monotonicity formula (5.11) applied to u^+ and u^-, we get

$$\left(\fint_{B_r} |\nabla(u^+ - v_r^+)|^2\,dx\right)\left(\fint_{B_r} |\nabla(u^- - v_r^-)|^2\,dx\right) \le 4\left(\fint_{B_r} |\nabla u^+|^2\,dx\right)\left(\fint_{B_r} |\nabla u^-|^2\,dx\right) \le C_d\left(1 + \|u\|_{L^2}^4\right),$$

for a dimensional constant $C_d > 0$, which gives (5.31).

- *If r_2 and C_2 are the α-quasi-minimality constants from* (5.58) *corresponding to the case $f \equiv 0$, then for every $r < r_2$ we have*

$$\int_{B_r} |\nabla(u^+ - v_r^+)|^2\,dx + \int_{B_r} |\nabla(u^- - v_r^-)|^2\,dx \\ \le \left(C_2^2 + C_d(1 + \|u\|_{L^2}^4)r_0^{d-\alpha}\right)r^\alpha. \tag{5.32}$$

By the α-quasi-minimality of u, for every $r \le r_2$ we have

$$\int_{B_r} |\nabla(u - v_r)|^2\,dx = \int_{B_r} \nabla u \cdot \nabla(u - v_r)\,dx \\ \le C_2 r^{\alpha/2}\left(\int_{B_r} |\nabla(u - v_r)|^2\,dx\right)^{1/2},$$

and thus we obtain

$$\int_{B_r} |\nabla(u - v_r)|^2\,dx \le C_2^2 r^\alpha, \qquad \forall r \le r_2. \tag{5.33}$$

Now using the inequality

$$\int_{B_r} |\nabla(u^+ - v_r^+)|^2\,dx + \int_{B_r} |\nabla(u^- - v_r^-)|^2\,dx \\ \le \int_{B_r} |\nabla(u - v_r)|^2\,dx \\ + 2\left(\int_{B_r} |\nabla(u^+ - v_r^+)|^2\right)^{1/2}\left(\int_{B_r} |\nabla(u^- - v_r^-)|^2\right)^{1/2},$$

together with (5.31) and (5.33), we obtain (5.32).

- *For every $r > 0$ we have*

$$\left|\int_{B_r} (v_r^+ - u^+)\,d\mu_1\right| + \left|\int_{B_r} (v_r^- - u^-)\,d\mu_2\right| \\ \le \left(C_2^2 + C_d r_0^{d-\alpha} + C_d\|u\|_{L^2}^{\frac{16}{d+4}} r_0^{d-\alpha}\right)r^\alpha, \tag{5.34}$$

where we denote with μ_1 and μ_2 the positive measures

$$\mu_1 := \Delta u^+ + 1 \qquad \textit{and} \qquad \mu_2 := \Delta u^- + 1.$$

Indeed, by the definition of μ_1 and v_r^+, we have

$$\left|\int_{B_r} (v_r^+ - u^+)\,d\mu_1\right| = \left|\int_{B_r} \left(-\nabla(v_r^+ - u^+)\cdot\nabla u^+ + (v_r^+ - u^+)\right)dx\right| \\ \le \int_{B_r} \left(|\nabla(u^+ - v_r^+)|^2 + |v_r^+ - u^+|\right)dx \\ \le \left(C_2^2 + C_d r_0^{d-\alpha} + C_d\|u\|_{L^\infty}^4 r_0^{d-\alpha}\right)r^\alpha,$$

which together with (5.26) gives (5.34).

- *Setting* $\phi_r(x) := \dfrac{(r^2 - |x|^2)^+}{2d}$, *for every* $r > 0$, *we have*

$$\int_{\mathbb{R}^d} \phi_r \, d\mu_1 \le C_d\big(\|u^+\|_{L^\infty} + r_0^2\big) r^d. \tag{5.35}$$

Indeed, using the definition $\mu_1 = \Delta u^+ + 1$ and the equation $-\Delta\phi_r = 1$ on B_r, we have

$$\begin{aligned}
\int_{\mathbb{R}^d} \phi_r^2 \, d\mu_1 &= \int_{\mathbb{R}^d} \big(-2\phi_r \nabla u^+ \cdot \nabla\phi_r + \phi_r^2 \big) \, dx \\
&= \int_{\mathbb{R}^d} \big(-2\nabla(u^+\phi_r) \cdot \nabla\phi_r + 2u^+|\nabla\phi_r|^2 + \phi_r^2 \big) \, dx \\
&= \int_{\mathbb{R}^d} \big(-2u^+\phi_r + 2u^+|\nabla\phi_r|^2 + \phi_r^2 \big) \, dx \\
&= \int_{B_r} \left(2\left(\frac{|x|^2}{d^2} - \frac{r^2 - |x|^2}{2d} \right) u^+ + \phi_r^2 \right) dx \\
&\le C_d\big(\|u^+\|_{L^\infty} + r^2\big) r^{d+2}.
\end{aligned}$$

Applying the above inequality to ϕ_{2r} and using the fact that $\phi_{2r} \ge 3r^2/(2d)$ on B_r, we get

$$\int_{\mathbb{R}^d} \phi_{2r}^2 \, dx \ge \int_{\mathbb{R}^d} \phi_{2r}\phi_r \, dx \ge \frac{3r^2}{2d} \int_{\mathbb{R}^d} \phi_r \, dx,$$

which gives (5.35).

- *Conclusion.*
Let $r > 0$ be fixed and ϕ_r be as above. Setting $U := u^+ - v_r^+ - \phi_r$, we have

$$\Delta U = \mu_1 \quad \text{on} \quad B_r, \qquad U \in H_0^1(B_r).$$

Thus $U \le 0$ on B_r and for every $z \in B_{r/4}$ we have

$$\begin{aligned}
\frac{1}{d\omega_d} \int_0^{3r/4} s^{1-d} \mu_1(B_s(z)) \, ds &= \frac{1}{d\omega_d} \int_0^{3r/4} s^{1-d} \Delta U(B_s(z)) \, ds \\
&= \fint_{\partial B_{3r/4}(z)} U \, d\mathcal{H}^{d-1} - U(z) \\
&\le v_r^+(z) - u^+(z) + \phi_r(z).
\end{aligned}$$

Integrating both sides on $B_{r/4}$ with respect to the measure $d\mu_1(z)$, we obtain

$$\begin{aligned}
Cr^\alpha &\geq \int_{B_{r/4}} \left(v_r^+(z) - u^+(z) + \phi_r(z)\right) d\mu_1(z) \\
&\geq \frac{1}{d\omega_d} \int_{B_{r/4}} d\mu_1(z) \int_0^{3r/4} s^{1-d} \mu_1(B_s(z))\, ds \\
&\geq \frac{1}{d\omega_d} \int_{B_{r/4}} d\mu_1(z) \int_{r/2}^{3r/4} s^{1-d} \mu_1(B_s(z))\, ds \\
&\geq \frac{1}{d\omega_d} \int_{B_{r/4}} d\mu_1(z) \int_{r/2}^{3r/4} s^{1-d} \mu_1(B_{r/4})\, ds \\
&\geq C_d r^{2-d} \left[\mu_1(B_{r/4})\right]^2,
\end{aligned}$$

where $C = \left(C_2^2 + C_d r_0^{d-\alpha} + C_d \|u\|_{L^2}^{\frac{16}{d+4}} r_0^{d-\alpha}\right)$. Now since the analogous inequality holds for μ_2, we have

$$\begin{aligned}
|\Delta|u||(B_r) &\leq |\Delta u^+|(B_r) + |\Delta u^-|(B_r) \\
&\leq \mu_1(B_r) + |B_r| + \mu_2(B_r) + |B_r| \leq C r^{\frac{\alpha+d-2}{2}}.
\end{aligned}$$

□

Lemma 5.22. *Suppose that $u \in H^1(\mathbb{R}^d)$ satisfies the following conditions:*

(a) *The positive and negative parts u^+ and u^- of u are such that $\Delta u^+ + 1 \geq 0$ and $\Delta u^- + 1 \geq 0$ as functionals on $H^1(\mathbb{R}^d) \cap L^1(\mathbb{R}^d)$.*
(b) *u is an α-quasi-minimizer of the Dirichlet integral J_0, for some $\alpha \in (d-1, d]$.*

Then for every $x_0 \in \mathbb{R}^d$ such that $u(x_0) = 0$ we have

$$\|u\|_{L^\infty(B_r(x_0))} \leq C r^{1-\frac{d-\alpha}{2}}, \quad \textit{for every} \quad r < r_2/8, \tag{5.36}$$

where

$$C = C_d \left(C_2 + 1 + r_0^{1+\frac{d-\alpha}{2}} + \|u\|_{L^2}^{\frac{2}{d+4}} r_0^{\frac{d-\alpha}{2}}\right), \tag{5.37}$$

and r_2 and C_2 are the α-quasi-minimality constants from (5.58) *with $f \equiv 0$.*

Proof. Without loss of generality we assume $x_0 = 0$. By condition (a) the function $v(x) = |u|(x) - \frac{(2r)^2-|x|^2}{d}$ is subharmonic on B_r and $v \equiv |u|$ on ∂B_{2r}. Thus, by the Poisson formula we have

$$\begin{aligned}
\|u\|_{L^\infty(B_r)} &\le C_d \fint_{\partial B_{2r}} |u|\, d\mathcal{H}^{d-1} + \frac{4r^2}{d}\\
&\le C_d \int_0^{2r} s^{1-d}|\Delta |u||(B_s)\, ds + \frac{4r^2}{d}\\
&\le C_d \int_0^{2r} C\, s^{1-d+\beta}\, ds + \frac{4r^2}{d}\\
&\le C_d \left(C + r_0^{d-\beta}\right) r^{2-d+\beta},
\end{aligned}$$

where $\beta = \frac{d+\alpha-2}{2}$. □

Theorem 5.23. *Suppose that $u \in H^1(\mathbb{R}^d)$ satisfies the following conditions:*

(a) *u has support of finite measure: $|\{u \neq 0\}| < +\infty$ and there is a function $f \in L^\infty(\mathbb{R}^d)$ such that u satisfies the equation*

$$-\Delta u = f \quad \text{in} \quad \{u \neq 0\}, \qquad u \in H^1_0(\{u \neq 0\});$$

(b) *u is a quasi-minimizer for the Dirichlet integral J_0.*

Then u is Lipschitz continuous on $\mathbb{R}^d$ and the Lipschitz constant depends on the dimension d, the norm $\|f\|_{L^\infty}$, the measure $|\{u \neq 0\}|$ and the α-minimality constant C_2 and r_2 from (5.58).

Proof. We first note that the function $\widetilde{u} = \|f\|^{-1}_{L^\infty} u$ satisfies the conditions of Lemma 5.20 and that $\widetilde{u}$ satisfies (5.58) with $\widetilde{C}_2 = \|f\|^{-1}_{L^\infty} C_2$. Thus $\widetilde{u}$ is continuous and the set $\Omega := \{u \neq 0\}$ is open.

For every $r > 0$, denote with $\Omega_r \subset \Omega$ the neighbourhood of the boundary $\partial\Omega$ in Ω

$$\Omega_r := \Big\{x \in \Omega : \operatorname{dist}(x, \partial\Omega) < r\Big\}.$$

We now set $r_0 = r_2/16$ and we consider two cases:

- Suppose that $x_0 \in \Omega_{r_0}$. Let $y_0 \in \partial\Omega$ be such that $R := |x_0 - y_0| = \operatorname{dist}(x_0, \partial\Omega)$. We use the gradient estimate (Lemma 5.9) on the ball $B_R(x_0)$:

$$\begin{aligned}
|\nabla \widetilde{u}(x_0)| &\le C_d + \frac{2d}{R}\|\widetilde{u}\|_{L^\infty(B_R(x_0))}\\
&\le C_d + \frac{2d}{R}\|\widetilde{u}\|_{L^\infty(B_{2R}(y_0))} \le C_d + 2d\widetilde{C},
\end{aligned} \tag{5.38}$$

  where $\widetilde{C}$ is the constant from Lemma 5.22 with $\widetilde{C}_2$ in pace of C_2.

- Let $x_0 \in \Omega \setminus \Omega_{r_0}$. Again by the gradient estimate we have

$$|\nabla \widetilde{u}(x_0)| \leq C_d + \frac{4d}{r_0}\|\widetilde{u}\|_{L^\infty} \leq C_d\left(1 + r_0^{-1}\|\widetilde{u}\|_{L^2}^{\frac{4}{d+4}}\right). \tag{5.39}$$

By (5.38) and (5.39), we obtain that $\widetilde{u}$ is Lipschitz and

$$\|\nabla \widetilde{u}\|_{L^\infty} \leq C_d\left(1 + r_2^{-1}\|\widetilde{u}\|_{L^2}^{\frac{4}{d+4}} + \widetilde{C}\right). \tag{5.40}$$

□

Remark 5.24. We note that the Lipschitz constant of u depends on the range of the radii $r \in (0, r_2)$, for which the quasi-minimality of u holds. In particular the right-hand side of (5.40) explodes as $r_2 \to 0$.

Theorem 5.25. *Suppose that $u \in H^1(\mathbb{R}^d)$ satisfies the following conditions:*

(a) *u is normalized in $L^2(\mathbb{R}^d)$ and its support has finite measure:*

$$\int_{\mathbb{R}^d} u^2\,dx = 1 \qquad \text{and} \qquad |\{u \neq 0\}| < +\infty;$$

(b) *there is a positive real number $\lambda > 0$ such that u is an eigenfunction on its support,* i.e. *u satisfies the equation*

$$-\Delta u = \lambda u \quad \text{in} \quad \{u \neq 0\}, \qquad u \in H_0^1(\{u \neq 0\});$$

(c) *u is a quasi-minimizer for the Dirichlet integral J_0.*

Then u is Lipschitz continuous on $\mathbb{R}^d$ and

$$\|\nabla u\|_{L^\infty} \leq C_d(1 + C_2)\left(1 + \lambda^{\frac{d+8}{4}}\right)\left(1 + |\{u \neq 0\}|^{1/d}\right), \tag{5.41}$$

where C_d is a dimensional constant and C_2 is the quasi-minimality constant from (5.58) *with $\alpha = d$ and $f \equiv 0$.*

Proof. We first notice that u is bounded. More precisely, Proposition 3.83 provides us with the estimate

$$\|u\|_{L^\infty} \leq C_d \lambda^{d/4}. \tag{5.42}$$

By Theorem 5.23 with $f = \lambda u$, we already have that u is Lipschitz continuous and so, it remains to estimate the Lipschitz constant of u.

Let $\Omega = \{u \neq 0\}$ and let $r_0 = r_2/16$. We denote with Ω_{r_0} the set

$$\Omega_{r_0} := \Big\{x \in \Omega : \operatorname{dist}(x, \partial\Omega) < r_0\Big\}.$$

For every $x_0 \in \Omega_{r_0}$ we consider the projection $y_0 \in \partial\Omega$ such that $R := \operatorname{dist}(x_0, \partial\Omega) = |x_0 - y_0|$. As in Theorem 5.23 we use the gradient estimate (Lemma 5.9)

$$\begin{aligned} |\nabla u(x_0)| &\leq C_d \lambda \|u\|_{L^\infty} + \frac{2d}{R}\|u\|_{L^\infty(B_R(x_0))} \\ &\leq C_d \lambda^{\frac{d+4}{4}} + \frac{2d}{R}\|u\|_{L^\infty(B_{2R}(y_0))} \\ &\leq C_d \lambda^{\frac{d+4}{4}} + C_d\left(C_2 + \lambda^{\frac{d+4}{4}}(1+r_0) + \lambda^{\frac{d+2}{d}}\right), \end{aligned} \tag{5.43}$$

where the last inequality is due to Lemma 5.22 applied to $\big(C_d\lambda^{\frac{d+4}{4}}\big)^{-1}u$.

Consider the function $P \in C^\infty(\Omega)$ defined by

$$P := |\nabla u|^2 + \lambda u^2 - 2\lambda^2 \|u\|^2_{L^\infty} w,$$

where w is the solution of the equation

$$-\Delta w = 1 \quad \text{in} \quad \Omega, \qquad w \in H^1_0(\Omega).$$

Calculating the Laplacian of P in Ω we have

$$\begin{aligned} \Delta P = &\left(2[\operatorname{Hess}(u)]^2 - 2\lambda|\nabla u|^2\right) \\ &+ \left(2\lambda|\nabla u|^2 - 2\lambda^2 u^2\right) + 2\lambda^2\|u\|^2_{L^\infty} \geq 0 \quad \text{in} \quad \Omega, \end{aligned}$$

where we used the notation

$$[\operatorname{Hess}(u)]^2 := \sum_{i,j=1}^{d} \left|\frac{\partial^2 u}{\partial x_i \partial x_j}\right|^2.$$

Since P is subharmonic in Ω the maximum is achieved in a neighbourhood of the boundary $\partial\Omega$, i.e

$$\sup_{x\in\Omega} P(x) = \sup_{x\in\Omega_{r_0}} P(x).$$

Let now $x \in \Omega$. Then there is $x_0 \in \Omega_{r_0}$ such that

$$\begin{aligned} |\nabla u(x)| &\leq P(x) - \lambda u(x)^2 + 2\lambda^2\|u\|^2_{L^\infty} w(x) \\ &\leq P(x) + 2\lambda^2 \|u\|^2_{L^\infty}\|w\|_{L^\infty} \\ &\leq P(x_0) + 2\lambda^2 \|u\|^2_{L^\infty}\|w\|_{L^\infty} \\ &\leq |\nabla u(x_0)|^2 + \lambda\|u\|^2_{L^\infty} + 2\lambda^2\|u\|^2_{L^\infty}\|w\|_{L^\infty} \\ &\leq C_d\left(C_2 + \lambda^{\frac{d+4}{4}}(1+r_0) + \lambda^{\frac{d+2}{d}}\right)^2 + C_d\lambda^{\frac{d+6}{2}} + C_d\lambda^{\frac{d+8}{2}}|\Omega|^{\frac{2}{d}}, \end{aligned}$$

where we used (5.43), (5.42) and the inequality $\|w\|_{L^\infty} \le C_d|\Omega|^{2/d}$. Now by choosing $r_0 \le 1$ and algebra we obtain (5.41). □

5.4. Shape quasi-minimizers for Dirichlet eigenvalues

In this section we discuss the regularity of the eigenfunctions on sets which are minimal with respect to a given (spectral) shape functional.

Let $\mathcal{A}$ be the family of all Lebesgue measurable sets in $\mathbb{R}^d$ of finite measure. endowed with the equivalence relation $\Omega \sim \tilde{\Omega}$ if $|\Omega \Delta \tilde{\Omega}| = 0$.

Definition 5.26. Let $\mathcal{F} : \mathcal{A} \to \mathbb{R}$ be a given functional. We say that the measurable set $\Omega \in \mathcal{A}$ is a shape quasi-minimizer for the functional $\mathcal{F}$, if there exist constants $\Lambda > 0$ and $r_0 > 0$ such that for each ball $B_r(x) \subset \mathbb{R}^d$ with radius less than r_0 we have

$$\mathcal{F}(\Omega) \le \mathcal{F}(\widetilde{\Omega}) + \Lambda|B_r|, \text{ for every } \widetilde{\Omega} \in \mathcal{A} \text{ such that } \Omega\Delta\widetilde{\Omega} \subset B_r(x).$$

Remark 5.27. If the functional $\mathcal{F}$ is non increasing with respect to inclusions, then Ω is a shape quasi-minimizer, if and only if,

$$\mathcal{F}(\Omega) \le \mathcal{F}\big(\Omega \cup B_r(x)\big) + \Lambda|B_r|, \quad \text{for every} \quad x \in \mathbb{R}^d \quad \text{and} \quad r < r_0.$$

The shape quasi-minimality of a set Ω with respect to a functional $\mathcal{F}$ can usually be translated into a quasi-minimality condition on the state function (or functions) of $\mathcal{F}$ on Ω. In the following example we discuss the case when $\mathcal{F}$ is simply the Dirichlet Energy.

Example 5.28. Suppose that $\Omega \in \mathcal{A}$ is a shape quasi-minimizer for the Dirichlet Energy

$$\widetilde{E}_1(\Omega) := \min\left\{J_1(u) : u \in \widetilde{H}^1_0(\Omega)\right\},$$

where

$$J_1(u) := \frac{1}{2}\int_{\mathbb{R}^d} |\nabla u|^2\,dx - \int_{\mathbb{R}^d} u\,dx.$$

Then, for every $\widetilde{\Omega}$ such that $\widetilde{\Omega}\Delta\Omega \subset B_r(x)$, we have

$$J_1(w_\Omega) = \widetilde{E}_1(\Omega) \le \widetilde{E}_1(\widetilde{\Omega}) + \Lambda|B_r| \le J_1(w_\Omega + \varphi) + \Lambda|B_r|,$$

for every $\varphi \in H^1_0(B_r)$, where the energy function w_Ω is the solution of

$$-\Delta w_\Omega = 1 \quad \text{in} \quad \Omega, \qquad w_\Omega \in \widetilde{H}^1_0(\Omega).$$

Thus the function w_Ω is a quasi-minimizer for the functional J_1 and so, by Theorem 5.23, the it is Lipschitz continuous on $\mathbb{R}^d$.

In what follows we will study the case when the cost functional $\mathcal{F}$ depends on the spectrum of the Dirichlet Laplacian. This analysis in this case is more involved since the eigenfunctions are not defined through a single state function but are variationally characterized by a min-max procedure involving an entire linear subspace of the Sobolev space.

We recall that there are two ways to define the Dirichlet eigenvalues on a set $\Omega \subset \mathbb{R}^d$ of finite measure. The results in this section are valid in both cases. Precisely the kth eigenvalue of the Dirichlet Laplacian on Ω is defined as:

$$\lambda_k(\Omega) = \min_{S_k} \max_{u \in S_k \setminus \{0\}} \frac{\int_{\mathbb{R}^d} |\nabla u|^2\, dx}{\int_{\mathbb{R}^d} u^2\, dx},$$

where the minimum is defined over one of the following classes:

- all k-dimensional subspaces S_k of the Sobolev-like space $\widetilde{H}^1_0(\Omega)$, which we recall is defined as

$$\widetilde{H}^1_0(\Omega) = \Big\{ u \in H^1(\mathbb{R}^d) : \ |\{u \neq 0 \setminus \Omega\}| = 0 \Big\};$$

- all k-dimensional subspaces S_k of the Sobolev space $H^1_0(\Omega)$, which we recall is defined as

$$H^1_0(\Omega) = \Big\{ u \in H^1(\mathbb{R}^d) : \ \mathrm{cap}(\{u \neq 0\} \setminus \Omega) \Big\}.$$

In the lemma below, we shall assume that Ω is a generic set of finite measure and $l \geq 1$ is such that

$$\lambda_k(\Omega) = \cdots = \lambda_{k-l+1}(\Omega) > \lambda_{k-l}(\Omega). \tag{5.44}$$

Let $u_{k-l+1}, \ldots, u_k$ be l normalized orthogonal eigenfunctions corresponding to k-th eigenfunction of the Dirichlet Laplacian on Ω.

The following notation is used: given a vector $\alpha = (\alpha_{k-l+1}, ..., \alpha_k) \in \mathbb{R}^l$ and functions $u_{k_l+1}, \ldots, u_k \in H^1(\mathbb{R}^d)$, we denote with $\boldsymbol{u}_\alpha \in H^1(\mathbb{R}^d)$ the linear combination

$$\boldsymbol{u}_\alpha := \alpha_{k-l+1} u_{k-l+1} + ... + \alpha_k u_k. \tag{5.45}$$

Lemma 5.29. *Let $\Omega \subset \mathbb{R}^d$ be a set of finite measure and $l \geq 1$ is such that* (5.44) *holds. Then for every $\varepsilon > 0$ there is a constant $r_0 > 0$ such that:*

- *for every $r \in (0, r_0)$ and every $x_0 \in \mathbb{R}^d$,*
- *for every l-uple of functions $v_{k-l+1}, \ldots, v_k \in H^1_0(B_r(x_0))$ with $\|\nabla v_j\|_{L^2} \leq 1$, for $j = k - l + 1, \ldots, k$,*

there is a unit vector $(\alpha_{k-l+1}, \dots, \alpha_k) \in \mathbb{R}^l$ *such that*

$$\lambda_k(\Omega \cup B_r(x_0)) \le \frac{\int_{\mathbb{R}^d} |\nabla(\boldsymbol{u}_\alpha + \boldsymbol{v}_\alpha)|^2\,dx + \varepsilon \int_{\mathbb{R}^d} |\nabla \boldsymbol{v}_\alpha|^2\,dx}{\int_{\mathbb{R}^d} |\boldsymbol{u}_\alpha + \boldsymbol{v}_\alpha|^2\,dx - \varepsilon \int_{\mathbb{R}^d} |\nabla \boldsymbol{v}_\alpha|^2\,dx}, \tag{5.46}$$

where $u_{k-l+1}, \dots, u_k$ *are* l *orthonormal eigenfunctions corresponding to the multiple eigenvalue* $\lambda_k(\Omega)$ *and the linear combinations* $\boldsymbol{u}_\alpha$ *and* $\boldsymbol{v}_\alpha$ *are defined as in* (5.45).

Proof. Without loss of generality, we can suppose $x_0 = 0$. Let $\varepsilon > 0$ be fixed. For sake of simplicity we will choose from the start $r_0 < 1$.

By the definition of the kth Dirichlet eigenvalue, we have that

$$\lambda_k(\Omega \cup B_r) \le \max\left\{\frac{\int |\nabla u|^2 dx}{\int u^2 dx} : u \in \operatorname{span}\langle u_1, \dots, u_{k-l}, (u_{k-l+1} + v_{k-l+1}), \dots, (u_k + v_k)\rangle\right\}.$$

The maximum is attained for a linear combination

$$\alpha_1 u_1 + \dots + \alpha_{k-l} u_{k-l} + \alpha_{k-l+1}(u_{k-l+1} + v_{k-l+1}) + \dots + \alpha_k(u_k + v_k). \tag{5.47}$$

- *For* r_0 *small enough, the vector* $(\alpha_{k-l+1}, \dots, \alpha_k) \in \mathbb{R}^l$ *is non-zero.* Indeed, suppose that $r_0 > 0$ is such that

$$\lambda_{k-l}(\Omega) < \lambda_k(\Omega \cup B_{r_0}(x_0)), \quad \text{for every} \quad x_0 \in \mathbb{R}^d.$$

The existence of such an r_0 can be proved by contradiction, since for every sequecne $x_n \in \mathbb{R}^d$ the condition $r_n \to 0$ implies that $\Omega \cup B_{r_n}(x_n)$ γ-converges to Ω. For every $0 < r \le r_0$ we have

$$\lambda_{k-l}(\Omega) < \lambda_k(\Omega \cup B_{r_0}(x_0)) \le \lambda_k(\Omega \cup B_r(x_0)).$$

Now if, by absurd $\alpha_{k-l+1} = \dots = \alpha_k = 0$, then $\alpha_1^2 + \dots + \alpha_{k-l}^2 \neq 0$ and

$$\frac{\int_{\mathbb{R}^d} |\nabla(\alpha_1 u_1 + \dots + \alpha_{k-l} u_{k-l})|^2\,dx}{\int_{\mathbb{R}^d} (\alpha_1 u_1 + \dots + \alpha_{k-l} u_{k-l})^2\,dx} = \frac{1}{\alpha_1^2 + \dots + \alpha_{k-l}^2} \sum_{j=1}^{k-l} \alpha_j^2 \int_{\mathbb{R}^d} |\nabla u_j|^2\,dx \le \lambda_{k-l}(\Omega),$$

which proves the claim.

We now can suppose that the vector $(\alpha_{k-l+1}, \dots, \alpha_k)$ is unitary, *i.e.* we have

$$\int_{\mathbb{R}^d} \boldsymbol{u}_\alpha^2\,dx = \alpha_{k-l+1}^2 + \dots + \alpha_k^2 = 1.$$

- *If* $\alpha_1^2+\dots+\alpha_{k-l}^2>0$, *then*

$$\lambda_k(\Omega\cup B_r)\le\sup_{t\in\mathbb{R}}\frac{\lambda_{k-l}(\Omega)t^2+2t\int_{B_r}\nabla u\cdot\nabla \boldsymbol{v}_\alpha\,dx+\int_{\mathbb{R}^d}|\nabla(\boldsymbol{u}_\alpha+\boldsymbol{v}_\alpha)|^2\,dx}{t^2+2t\int_{B_r}u\boldsymbol{v}_\alpha\,dx+\int_{\mathbb{R}^d}|\boldsymbol{u}_\alpha+\boldsymbol{v}_\alpha|^2\,dx},\tag{5.48}$$

where

$$u:=\frac{1}{\sqrt{\alpha_1^2+\ldots+\alpha_{k-l}^2}}(\alpha_1u_1+\ldots+\alpha_{k-l}u_{k-l}).$$

Indeed the function $u\in\widetilde{H}_0^1(\Omega)$, defined as above, satisfies

$$\int_{\mathbb{R}^d}u^2\,dx=1\qquad\text{and}\qquad\int_{\mathbb{R}^d}|\nabla u|^2\,dx\le\lambda_{k-l}(\Omega).$$

Consequently, we have

$$\begin{aligned}\lambda_k(\Omega\cup B_r)&\le\sup_{t\in\mathbb{R}}\frac{\int_{\mathbb{R}^d}|\nabla(\boldsymbol{u}_\alpha+\boldsymbol{v}_\alpha+tu)|^2\,dx}{\int_{\mathbb{R}^d}|\boldsymbol{u}_\alpha+\boldsymbol{v}_\alpha+tu|^2\,dx}\\&=\sup_{t\in\mathbb{R}}\frac{t^2\int_{\mathbb{R}^d}|\nabla u|^2dx+2t\int_{\mathbb{R}^d}\nabla u\cdot\nabla(\boldsymbol{u}_\alpha+\boldsymbol{v}_\alpha)dx+\int|\nabla(\boldsymbol{u}_\alpha+\boldsymbol{v}_\alpha)|^2dx}{t^2\int_{\mathbb{R}^d}u^2\,dx+2t\int_{\mathbb{R}^d}u\,(\boldsymbol{u}_\alpha+\boldsymbol{v}_\alpha)\,dx+\int|\boldsymbol{u}_\alpha+\boldsymbol{v}_\alpha|^2\,dx}\\&\le\sup_{t\in\mathbb{R}}\frac{t^2\lambda_{k-l}(\Omega)+2t\int_{B_r}\nabla u\cdot\nabla\boldsymbol{v}_\alpha\,dx+\int_{\mathbb{R}^d}|\nabla(\boldsymbol{u}_\alpha+\boldsymbol{v}_\alpha)|^2dx}{t^2+2t\int_{B_r}u\boldsymbol{v}_\alpha\,dx+\int_{\mathbb{R}^d}|\boldsymbol{u}_\alpha+\boldsymbol{v}_\alpha|^2\,dx}.\end{aligned}$$

For simplicity, from now on we will use the notation $\lambda_j:=\lambda_j(\Omega)$, for every j. Moreover, we define the modulus of continuity

$$\theta(r):=\max_{x_0\in\mathbb{R}^d}\max_{j=1,\dots,k}\int_{B_r(x_0)}|\nabla u_j|^2\,dx,$$

and we note that θ is an increasing function in r, vanishing in zero.

- *Using the notation*

$$A:=\int_{\mathbb{R}^d}|\nabla(\boldsymbol{u}_\alpha+\boldsymbol{v}_\alpha)|^2\,dx\qquad\text{and}\qquad B:=\int_{\mathbb{R}^d}|\boldsymbol{u}_\alpha+\boldsymbol{v}_\alpha|^2\,dx,$$

we have the inequalities

$$\lambda_k-2\sqrt{\theta(r_0)}\le A\le\lambda_k+2\sqrt{\theta(r_0)}+1,\tag{5.49}$$

$$1-C_dr_0\le B\le 1+C_dr_0,\tag{5.50}$$

for a dimensional constant C_d.

Indeed, (5.49) follows since $\int_{\mathbb{R}^d} |\nabla v_\alpha|^2\,dx \le 1$ and

$$\left|\int_{\mathbb{R}^d} \nabla u_\alpha \cdot \nabla v_\alpha\,dx\right| \le \left(\int_{B_{r_0}} |\nabla u_\alpha|^2\,dx\right)^{1/2} \left(\int_{\mathbb{R}^d} |\nabla v_\alpha|^2\,dx\right)^{1/2} \le \sqrt{\theta(r_0)}.$$

For (5.50) we notice that since $v_\alpha \in H_0^1(B_{r_0})$ we have

$$\int_{\mathbb{R}^d} v_\alpha^2\,dx \le C_d r_0^2 \int_{\mathbb{R}^d} |\nabla v_\alpha|^2\,dx \le C_d r_0^2,$$

and

$$\left|\int_{\mathbb{R}^d} u_\alpha v_\alpha\,dx\right| \le \left(\int_{\mathbb{R}^d} u_\alpha^2\,dx\right)^{1/2} \left(\int_{\mathbb{R}^d} v_\alpha^2\,dx\right)^{1/2} \le C_d r_0.$$

- *Using the notation*

$$a := \int_{\mathbb{R}^d} \nabla u \cdot \nabla v_\alpha\,dx \qquad \text{and} \qquad b := \int_{\mathbb{R}^d} u v_\alpha\,dx,$$

and applying the Cauchy-Schwartz inequality we have the estimates

$$|a| \le \left(\int_{B_{r_0}} |\nabla u|^2 dx\right)^{1/2} \left(\int_{B_r} |\nabla v_\alpha|^2 dx\right)^{1/2} \le \sqrt{\theta(r_0)}\|\nabla v_\alpha\|_{L^2}, \quad (5.51)$$

$$|b| \le \left(\int_{B_r} u^2\,dx\right)^{1/2} \left(\int_{B_r} v_\alpha^2\,dx\right)^{1/2} \le C_d r_0 \|\nabla v_\alpha\|_{L^2}. \quad (5.52)$$

Consider the rational function $F : \mathbb{R} \to \mathbb{R}$ defined as

$$F(t) := \frac{t^2\lambda_{k-l} + 2at + A}{t^2 + 2bt + B}.$$

- *For r_0 small enough the maximum of F is attained on $\mathbb{R}$ and is one of the solutions of the equation*

$$t^2(\lambda_{k-l} b - a) + t(\lambda_{k-l} B - A) + (aB - bA) = 0.$$

Indeed, we have that $\lim_{t\to+\infty} F(t) = \lim_{t\to-\infty} F(t) = \lambda_{k-l}$. On the other hand, for small enough r_0 we have

$$F(0) = \frac{A}{B} \ge \frac{\lambda_k - 2\sqrt{\theta(r_0)}}{1 + C_d r_0} > \lambda_{k-l}.$$

Now the claim follows by computing the derivative of F.

- *For every $\varepsilon > 0$ there is $r_0 > 0$ such that for every $r < r_0$ and every $v_\alpha \in H^1_0(B_r)$ with $\|\nabla v_\alpha\|_{L^2} \le 1$, the maximum of F is achieved on the interval $(-\varepsilon\|\nabla v_\alpha\|_{L^2}, \varepsilon\|\nabla v_\alpha\|_{L^2})$.*
  We consider two cases:

  Case 1. Suppose that $\lambda_{k-l}b - a \neq 0$.
  Then, we have that $\max_{t\in\mathbb{R}} F(t) = \max\{F(t_1), F(t_2)\}$, where t_1 and t_2 are given by

$$t_{1,2} = \frac{A - \lambda_{k-l}B \pm \sqrt{(A - \lambda_{k-l}B)^2 - 4(\lambda_{k-l}b - a)(aB - bA)}}{2(\lambda_{k-l}b - a)}$$

$$= \frac{A - \lambda_{k-l}B}{2(\lambda_{k-l}b - a)}\left(1 \pm \sqrt{1 - \frac{4(\lambda_{k-l}b - a)(aB - bA)}{(A - \lambda_{k-l}.B)^2}}\right)$$

By (5.49), (5.50), (5.51) and (5.52), we can choose r_0 small enough, in order to have

$$\left|\frac{4(\lambda_{k-l}b - a)(aB - bA)}{(A - \lambda_{k-l}B)^2}\right| < \frac{1}{2}.$$

Since the function $x \mapsto \sqrt{1-x}$ is bounded and 1-Lipschitz on the interval $(-\frac{1}{2}, \frac{1}{2})$, we have the estimate

$$\begin{aligned}
|t_1| &= \left|\frac{A - \lambda_{k-l}B}{2(\lambda_{k-l}b - a)}\left(1 - \sqrt{1 - \frac{4(\lambda_{k-l}b - a)(aB - bA)}{(A - \lambda_{k-l}B)^2}}\right)\right| \\
&\le \left|\frac{A - \lambda_{k-l}B}{2(\lambda_{k-l}b - a)}\right| \cdot \left|\frac{4(\lambda_{k-l}b - a)(aB - bA)}{(A - \lambda_{k-l}B)^2}\right| \\
&= 2\left|\frac{aB - bA}{A - \lambda_{k-l}B}\right| \le 2\frac{|a|B + |b|A}{A - \lambda_{k-l}B} \qquad (5.53) \\
&\le 2\frac{\sqrt{\theta(r_0)}(1 + C_d r_0) + C_d r_0}{\lambda_k - 2\sqrt{\theta(r_0)} - \lambda_{k-l}(1 + C_d r_0)}\left(\int_{\mathbb{R}^d} |\nabla v_\alpha|^2 dx\right)^{1/2} \\
&\le \varepsilon\|\nabla v_\alpha\|_{L^2},
\end{aligned}$$

for r_0 small enough, where the last inequality is obtained using (5.49), (5.50), (5.51) and (5.52). On the other hand, for t_2, we have

$$\frac{1}{2}\left|\frac{A - \lambda_{k-l}B}{\lambda_{k-l}b - a}\right| \le |t_2| \le 2\left|\frac{A - \lambda_{k-l}B}{\lambda_{k-l}b - a}\right|. \qquad (5.54)$$

Note that if we choose r_0 such that $|t_1| < |t_2|$, then the maximum cannot be attained in t_2. In fact, $(\lambda_{k-l}b - a)t_2 > 0$ and so, in t_2, the derivative F' changes sign from negative to positive, if $t_2 > 0$ and from negative to positive, if $t_2 < 0$, which proves that the maximum is attained in t_1.

Case 2. Suppose that $\lambda_{k-l}b - a = 0$.
Then the maximum of F is achieved in

$$t = \frac{aB - bA}{\lambda_{k-l}B - A}.$$

Using the inequalities (5.49), (5.50), (5.51) and (5.52), we get

$$|t| \le \frac{\sqrt{\theta(r_0)}(1 + C_d r_0) + C_d r_0}{\lambda_k - 2\sqrt{\theta(r_0)} - \lambda_{k-l}(1 + C_d r_0)} \left(\int_{\mathbb{R}^d} |\nabla \boldsymbol{v}_\alpha|^2 dx \right)^{1/2}$$
$$\le \varepsilon \|\nabla \boldsymbol{v}_\alpha\|_{L^2},$$

which concludes the proof of the claim.

- *Conclusion.* Choosing r_0 such that the maximum of F is achieved on the interval $(-\varepsilon\|\nabla \boldsymbol{v}_\alpha\|_{L^2}, \varepsilon\|\nabla \boldsymbol{v}_\alpha\|_{L^2})$, we have

$$\lambda_k(\Omega \cup B_r) \le \max_{t\in\mathbb{R}} F(t) \le \frac{\lambda_{k-l}\,(\varepsilon\|\nabla \boldsymbol{v}_\alpha\|_{L^2})^2 + 2|a|\,(\varepsilon\|\nabla \boldsymbol{v}_\alpha\|_{L^2}) + A}{B - 2|b|\,(\varepsilon\|\nabla \boldsymbol{v}_\alpha\|_{L^2})}$$

$$\le \frac{\lambda_{k-l}\varepsilon^2\|\nabla \boldsymbol{v}_\alpha\|_{L^2}^2 + 2\varepsilon\sqrt{\theta(r_0)}\|\nabla \boldsymbol{v}_\alpha\|_{L^2}^2 + A}{B - 2C_d r_0 \varepsilon\|\nabla \boldsymbol{v}_\alpha\|_{L^2}^2},$$

which, by choosing r_0 small enough, concludes the proof. □

What we will really use is the following corollary, which we state as a separate lemma.

Lemma 5.30. *Let $\Omega \subset \mathbb{R}^d$ be a set of finite measure such that $\lambda_k(\Omega) > \lambda_{k-1}(\Omega)$. Then for every $\varepsilon > 0$ there is a constant $r_0 > 0$ such that:*

- *for every $r \in (0, r_0)$ and every $x_0 \in \mathbb{R}^d$,*
- *for every functions $v \in H_0^1(B_r(x_0))$ such that $\|\nabla v\|_{L^2} \le 1$,*

we have the estimate

$$\lambda_k(\Omega \cup B_r(x)) \le \frac{\int_{\mathbb{R}^d} |\nabla(u_k + v)|^2\,dx + \varepsilon \int_{\mathbb{R}^d} |\nabla v|^2\,dx}{\int_{\mathbb{R}^d} |u_k + v|^2\,dx - \varepsilon \int_{\mathbb{R}^d} |\nabla v|^2\,dx}, \qquad (5.55)$$

where u_k is a normalized eigenfunction on Ω, corresponding to the eigenvalue $\lambda_k(\Omega)$.

Remark 5.31. We note that in Lemma 5.30 we do not assume that $\lambda_k(\Omega)$ is simple, but only that $\lambda_k(\Omega) > \lambda_{k-1}(\Omega)$.

Remark 5.32. The constant r_0 depends on the set Ω and, in particular, on the gap $\lambda_k(\Omega) - \lambda_{k-1}(\Omega)$. In fact, if the gap vanishes, so does r_0.

Lemma 5.33. *Let $\Omega \subset \mathbb{R}^d$ be a shape quasi-minimizer for the kth Dirichlet eigenvalue λ_k such that $\lambda_k(\Omega) > \lambda_{k-1}(\Omega)$. Then every eigenfunction u_k, corresponding to the eigenvalue $\lambda_k(\Omega)$ and normalized in L^2, is Lipschitz continuous on $\mathbb{R}^d$ and*

$$\|\nabla u_k\|_{L^\infty} \le C_d(1+\Lambda)\Big(1+\lambda_k(\Omega)^{d+1}\Big)\Big(1+|\Omega|^{1/d}\Big), \tag{5.56}$$

where C_d is a dimensional constant and Λ is the shape quasi-minimality constant of Ω.

Proof. Let $u_k \in \widetilde{H}^1_0(\Omega)$ be a normalized eigenfunction corresponding to λ_k. By the shape quasi-minimality of Ω, we have

$$\lambda_k(\Omega) \le \lambda_k(\Omega \cup B_r(x)) + \Lambda|B_r|. \tag{5.57}$$

By choosing a radius $r_0 > 0$ small enough, we can apply the estimate (5.55) for any $v \in H^1_0(B_r)$ such that $\|\nabla v\|_{L^2} \le 1$, obtaining

$$\begin{aligned}
\lambda_k(\Omega) &\le \lambda_k(\Omega \cup B_r(x)) + \Lambda|B_r| \\
&\le \frac{\int_{\mathbb{R}^d} |\nabla(u_k+v)|^2\,dx + \varepsilon\int_{\mathbb{R}^d}|\nabla v|^2\,dx}{\int_{\mathbb{R}^d}|u_k+v|^2\,dx - \varepsilon\int_{\mathbb{R}^d}|\nabla v|^2\,dx} + \Lambda|B_r| \\
&\le \frac{\lambda_k(\Omega) + 2\int_{\mathbb{R}^d}\nabla u_k\cdot\nabla v\,dx + (1+\varepsilon)\int_{\mathbb{R}^d}|\nabla v|^2\,dx}{1+2\int_{\mathbb{R}^d}u_k v\,dx - \varepsilon\int_{\mathbb{R}^d}|\nabla v|^2\,dx} + \Lambda|B_r| \\
&\le \frac{\lambda_k(\Omega) + 2\int_{\mathbb{R}^d}\nabla u_k\cdot\nabla v\,dx + (1+\varepsilon)\int_{\mathbb{R}^d}|\nabla v|^2\,dx}{1-2\|u_k\|_{L^\infty}|B_r|^{1/2}C_d r_0\|\nabla v\|_{L^2} - \varepsilon\int_{\mathbb{R}^d}|\nabla v|^2\,dx} + \Lambda|B_r| \\
&\le \frac{\lambda_k(\Omega) + 2\int_{\mathbb{R}^d}\nabla u_k\cdot\nabla v\,dx + (1+\varepsilon)\int_{\mathbb{R}^d}|\nabla v|^2\,dx}{1-\varepsilon|B_r| - 2\varepsilon\int_{\mathbb{R}^d}|\nabla v|^2\,dx} + \Lambda|B_r|,
\end{aligned}$$

where for the last inequality we again choose r_0 small enough, depending also on the norm $\|u_k\|_{L^\infty}$. Taking the common denominator of the both sides we get

$$\begin{aligned}
-2\int_{\mathbb{R}^d}\nabla u_k\cdot\nabla v\,dx &\le \Big(\Lambda|B_r|-\lambda_k(\Omega)\Big)\Big(1-\varepsilon|B_r|-2\varepsilon\int_{\mathbb{R}^d}|\nabla v|^2\,dx\Big)\\
&\quad+\lambda_k(\Omega)+(1+\varepsilon)\int_{\mathbb{R}^d}|\nabla v|^2\,dx\\
&\le \big(\Lambda+\varepsilon\lambda_k(\Omega)\big)|B_r|\\
&\quad+\big(1+\varepsilon+2\varepsilon\lambda_k(\Omega)\big)\int_{\mathbb{R}^d}|\nabla v|^2\,dx\\
&\le 2\Lambda|B_r|+2\int_{\mathbb{R}^d}|\nabla v|^2\,dx,
\end{aligned}$$

for r_0 small enough. Since the above inequality holds for every $v\in H_0^1(B_r)$ with $\|\nabla v\|_{L^2}\le 1$, by Lemma 5.18 u_k satisfies the quasi-minimality condition

$$\begin{aligned}
&|\langle \Delta u_k,\psi\rangle|\le C_d(\Lambda+1)r^{d/2}\|\nabla\psi\|_{L^2},\\
&\text{for every } \forall r\in(0,r_0),\ x\in\mathbb{R}^d \text{ and } \psi\in H_0^1(B_r(x)),
\end{aligned}\tag{5.58}$$

where C_d is a dimensional constant and $r_0>0$ depends on Ω. Now the claim follows by Theorem 5.25. □

Remark 5.34. The exponent $d+1$ of $\lambda_k(\Omega)$ in (5.56) is just a rough estimate (for $d\ge 2$) of the exponent of λ in (5.41). We are only interested in the fact that this exponent is a dimensional constant.

5.5. Shape supersolutions of spectral functionals

In the previous section we studied the regularity of the state functions of the simplest spectral functional $\mathcal{F}(\Omega)=\lambda_k(\Omega)$ for domains Ω which satisfy some quasi-minimality condition as shapes in $\mathbb{R}^d$. In this setting we were able to give only partial regularity result for the eigenfunctions, under the additional condition $\lambda_k(\Omega)>\lambda_{k-1}(\Omega)$. In this section we will investigate another extremality condition on Ω, which is stronger than the shape quasi-minimality. As we will see this time we will obtain the Lipschitz regularity of the state functions without additional assumptions on the extremal domain Ω.

As in the previous section we will denote with $\mathcal{A}$ the family of measurable sets in $\mathbb{R}^d$.

Definition 5.35. We say that the measurable set $\Omega\subset\mathbb{R}^d$ is a shape supersolution for the functional $\mathcal{F}:\mathcal{A}\to\mathbb{R}$ if

$$\mathcal{F}(\Omega)\le\mathcal{F}(\widetilde{\Omega}),\qquad \text{for all measurable sets}\quad \widetilde{\Omega}\supset\Omega. \tag{5.59}$$

We now list some of the main properties of the shape supersolutions as well as some of the basic manipulations that we can do with the functional $\mathcal{F}$ without violating the superminimality property of a set Ω.

- Suppose that Ω is a shape supersolution for the functional $\mathcal{F}+\Lambda|\cdot|:\mathcal{A}\to\mathbb{R}$. Then the minimality of Ω can be expressed as

$$\mathcal{F}(\Omega)\le\mathcal{F}(\widetilde{\Omega})+\Lambda|\widetilde{\Omega}\setminus\Omega|,\qquad\text{for all measurable sets}\quad\forall\Omega\supset\Omega.$$

- If the functional $\mathcal{F}:\mathcal{A}\to\mathbb{R}$ is non increasing with respect to the inclusion, we have, by Remark 5.27, that every shape supersolution for $\mathcal{F}+\Lambda|\cdot|$ is also a shape quasi-minimizer.
- Suppose that Ω is a shape supersolution for the functional $\mathcal{F}:\mathcal{A}\to\mathbb{R}$ and that the functional $\mathcal{G}:\mathcal{A}\to\mathbb{R}$ is **increasing** with respect to the set inclusion, then Ω is a shape supersolution for $\mathcal{F}+\mathcal{G}$. This property will be used mainly in the following situations
  - *Adding measure*. Suppose that Ω is a shape supersolution for the functional $\mathcal{F}+\Lambda|\cdot|:\mathcal{A}\to\mathbb{R}$ and that $\Lambda'>\Lambda$. Then, taking $G=(\Lambda'-\Lambda)|\Omega|$, we get that Ω is a shape supersolution also for $\mathcal{F}+\Lambda'|\cdot|$.
  - *Deleting spectral terms*. Suppose that Ω is a shape supersolution for the functional $\mathcal{F}+\lambda_k$. Then, taking $G=-\lambda_k$, we get that Ω is a shape supersolution also for $\mathcal{F}$.

By adding a positive measure term to the functional $\mathcal{F}$ one can assure that the inequalities in (5.59) are strict. We state this property in a separate Remark since it will be used several times in crucial moments.

Remark 5.36. Suppose that $\mathcal{F}:\mathcal{A}\to\mathbb{R}$ is a given functional and that the measurable set $\Omega^*\subset\mathbb{R}^d$ is a shape supersolution for $\mathcal{F}+\Lambda|\cdot|$. Then for every $\Lambda'>\Lambda$ the set Ω^* is **the unique** solution of the shape optimization problem

$$\min\Big\{\mathcal{F}(\Omega)+\Lambda'|\Omega|\;:\;\Omega\subset\mathbb{R}^d\text{ Lebesgue measurable},\quad\Omega\supset\Omega^*\Big\}.$$

In Lemma 5.33 we showed that the kth eigenfunctions of the the shape quasi-minimizers for λ_k are Lipschitz continuous under the assumption $\lambda_k(\Omega)>\lambda_{k-1}(\Omega)$. In the next Theorem, we show that for shape supersolutions the later assumption can be dropped.

Throughout this section the kth Dirichlet eigenvalue on a set $\Omega\subset\mathbb{R}^d$ of finite measure will be defined as

$$\lambda_k(\Omega):=\min_{S_k}\max_{u\in S_k\setminus\{0\}}\frac{\int_{\mathbb{R}^d}|\nabla u|^2\,dx}{\int_{\mathbb{R}^d}u^2\,dx},$$

where the minimum is over all k-dimensional subspaces S_k of the Sobolev-like space $\widetilde{H}_0^1(\Omega)$.

Theorem 5.37. *Let $\Omega^* \subset \mathbb{R}^d$ be a bounded shape supersolution for λ_k with constant Λ. Then there is an eigenfunction $u_k \in \widetilde{H}_0^1(\Omega^*)$, normalized in L^2 and corresponding to the eigenvalue $\lambda_k(\Omega^*)$, which is Lipschitz continuous on $\mathbb{R}^d$.*

Proof. We first note that if $\lambda_k(\Omega^*) > \lambda_{k-1}(\Omega^*)$, then the claim follows by Lemma 5.33. Suppose now that $\lambda_k(\Omega^*) = \lambda_{k-1}(\Omega^*)$. For every $\varepsilon \in (0,1)$ consider the problem

$$\min\Big\{(1-\varepsilon)\lambda_k(\Omega) + \varepsilon\lambda_{k-1}(\Omega) + 2\Lambda|\Omega| \,:\, \Omega \supset \Omega^*\Big\}. \tag{5.60}$$

We consider the following two cases:

(i) Suppose that there is a sequence $\varepsilon_n \to 0$ and a sequence Ω_{ε_n} of corresponding minimizers for (5.60) such that $\lambda_k(\Omega_{\varepsilon_n}) > \lambda_{k-1}(\Omega_{\varepsilon_n})$. For each $n \in \mathbb{N}$, Ω_{ε_n} is a shape supersolution for λ_k with constant $2(1-\varepsilon_n)^{-1}\Lambda$ and so, by Lemma 5.33, we have that for each $n \in \mathbb{N}$ the normalized eigenfunctions $u_k^n \in \widetilde{H}_0^1(\Omega_{\varepsilon_n})$, corresponding to $\lambda_k(\Omega_{\varepsilon_n})$, are Lipschitz continuous on $\mathbb{R}^d$. We will prove that the Lipschitz constant is uniform and then we will pass to the limit. We first prove that Ω_{ε_n} γ-converges to Ω^* as $n \to \infty$. Indeed, by [25, Proposition 5.12], Ω_{ε_n} are all contained in some ball B_R with R big enough. Thus, there is a weak-γ-convergent subsequence of Ω_{ε_n} and let $\widetilde{\Omega}$ be its limit. Then $\widetilde{\Omega}$ is a solution of the problem

$$\min\Big\{\lambda_k(\Omega) + 2\Lambda|\Omega| \,:\, \Omega \supset \Omega^*\Big\}. \tag{5.61}$$

On the other hand, by Remark 5.36 we have that Ω^* is the unique solution of (5.61) and so, $\widetilde{\Omega} = \Omega^*$. Since the weak γ-limit Ω^* satisfies $\Omega^* \subset \Omega_{\varepsilon_n}$ for every $n \in \mathbb{N}$, then Ω_{ε_n} γ-converges to Ω^*. By the metrizability of the γ-convergence, we have that Ω^* is the γ-limit of Ω_{ε_n} as $n \to \infty$. As a consequence, we have that $\lambda_k(\Omega_{\varepsilon_n}) \to \lambda_k(\Omega^*)$ and by (5.56) we have that the sequence u_k^n is uniformly Lipschitz.

Then, we can suppose that, up to a subsequence $u_k^n \to u$ uniformly and weakly in $H_0^1(B_R)$, for some $u \in H_0^1(B_R)$, Lipschitz continuous on $\mathbb{R}^d$. By the weak convergence of u_k^n, we have that for each $v \in H_0^1(\Omega^*)$

$$\begin{aligned}\int_{\mathbb{R}^d} \nabla u \cdot \nabla v\,dx &= \lim_{n\to\infty}\int_{\mathbb{R}^d} \nabla u_k^n \cdot \nabla v\,dx = \lim_{n\to\infty}\lambda_k(\Omega_{\varepsilon_n})\int_{\mathbb{R}^d} u_k^n v\,dx\\ &= \lambda_k(\Omega^*)\int_{\mathbb{R}^d} uv\,dx.\end{aligned}$$

By the γ-convergence of Ω_{ε_n}, we have that $u \in H_0^1(\Omega^*)$ and so u is a k-th eigenfunction of the Dirichlet Laplacian on Ω^*.

(ii) Suppose that there is some $\varepsilon_0 \in (0,1)$ such that Ω_{ε_0} is a solution of (5.60) and $\lambda_k(\Omega_{\varepsilon_0}) = \lambda_{k-1}(\Omega_{\varepsilon_0})$. Then, Ω_{ε_0} is also a solution of (5.61) and, by Remark 5.36, $\Omega_{\varepsilon_0} = \Omega^*$. Thus we obtain that Ω^* is a shape supersolution for λ_{k-1} with constant $2\varepsilon_0^{-1}\Lambda$. If we have

$$\lambda_k(\Omega^*) = \lambda_{k-1}(\Omega^*) > \lambda_{k-2}(\Omega^*),$$

then, we can apply Lemma 5.33 obtaining that each eigenfunction corresponding to $\lambda_{k-1}(\Omega^*)$ is Lipschitz continuous on $\mathbb{R}^d$. On the other hand, if

$$\lambda_k(\Omega^*) = \lambda_{k-1}(\Omega^*) = \lambda_{k-2}(\Omega^*),$$

we consider, for each $\varepsilon \in (0,1)$, the problem

$$\min\Big\{(1-\varepsilon_0)\lambda_k(\Omega) + \varepsilon_0\left[(1-\varepsilon)\lambda_{k-1}(\Omega) + \varepsilon\lambda_{k-2}(\Omega)\right] + 3\Lambda|\Omega| \,:\, \Omega \supset \Omega^*\Big\}. \tag{5.62}$$

One of the following two situations may occur:

(a) There is a sequence $\varepsilon_n \to 0$ and a corresponding sequence Ω_{ε_n} of minimizers of (5.62) such that

$$\lambda_{k-1}(\Omega_{\varepsilon_n}) > \lambda_{k-2}(\Omega_{\varepsilon_n}).$$

(b) There is some $\varepsilon_1 \in (0,1)$ and Ω_{ε_1}, solution of (5.62), such that

$$\lambda_{k-1}(\Omega_{\varepsilon_1}) = \lambda_{k-2}(\Omega_{\varepsilon_1}).$$

If the case (a) occurs, then since Ω_{ε_n} is a shape quasi-minimizer for λ_{k-1}, by Lemma 5.33 we obtain the Lipschitz continuity of the eigenfunctions u_{k-1}^n, corresponding to λ_{k-1} on Ω_{ε_n}. Repeating the argument from (i), we obtain that Ω_{ε_n} γ-converges to Ω^* and that the sequence of eigenfunctions $u_{k-1}^n \in H_0^1(\Omega_{\varepsilon_n})$ uniformly converges to an eigenfunctions $u_{k-1} \in H_0^1(\Omega^*)$, corresponding to $\lambda_k(\Omega^*) = \lambda_{k-1}(\Omega^*)$. Since the Lipschitz constants of u_{k-1}^n are uniform, we have the conclusion.
If the case (b) occurs, then reasoning as in the case (ii), we have

that $\Omega_{\varepsilon_1} = \Omega^*$. Indeed, we have

$$\begin{aligned}
&(1-\varepsilon_0)\lambda_k(\Omega_{\varepsilon_1}) + \varepsilon_0\lambda_{k-1}(\Omega_{\varepsilon_1}) + 3\Lambda|\Omega_{\varepsilon_1}| \\
&= (1-\varepsilon_0)\lambda_k(\Omega_{\varepsilon_1}) + \varepsilon_0\left[(1-\varepsilon_1)\lambda_{k-1}(\Omega_{\varepsilon_1}) + \varepsilon_1\lambda_{k-2}(\Omega_{\varepsilon_1})\right] \\
&\quad + 3\Lambda|\Omega_{\varepsilon_1}| \\
&\le (1-\varepsilon_0)\lambda_k(\Omega^*) + \varepsilon_0\left[(1-\varepsilon_1)\lambda_{k-1}(\Omega^*) + \varepsilon_1\lambda_{k-2}(\Omega^*)\right] \\
&\quad + 3\Lambda|\Omega^*| \\
&= (1-\varepsilon_0)\lambda_k(\Omega^*) + \varepsilon_0\lambda_{k-1}(\Omega^*) + 3\Lambda|\Omega^*|.
\end{aligned} \tag{5.63}$$

On the other hand, we supposed that Ω^* is a solution of (5.60) with $\varepsilon = \varepsilon_0$ and so, it is the unique minimizer of the problem

$$\min\Big\{(1-\varepsilon_0)\lambda_k(\Omega) + \varepsilon_0\lambda_{k-1}(\Omega) + 3\Lambda|\Omega| :\ \Omega \supset \Omega^*\Big\}. \tag{5.64}$$

Thus, we have $\Omega^* = \Omega_{\varepsilon_1}$. We proceed considering, for any $\varepsilon \in (0,1)$, the problem

$$\begin{aligned}
\min\Big\{&(1-\varepsilon_0)\lambda_k(\Omega) + \varepsilon_0(1-\varepsilon_1)\lambda_{k-1}(\Omega) \\
&+\varepsilon_0\varepsilon_1\left[(1-\varepsilon)\lambda_{k-2}(\Omega) + \varepsilon\lambda_{k-3}(\Omega)\right] \\
&+4\Lambda|\Omega| :\ \Omega \supset \Omega^*\Big\},
\end{aligned} \tag{5.65}$$

and repeat the procedure described above. We note that this procedure stops after at most k iterations. Indeed, if Ω^* is a supersolution for λ_1 and $\lambda_k(\Omega^*) = \cdots = \lambda_1(\Omega^*)$, then we obtain the result applying Lemma 5.33 to λ_1. □

As a consequence, we obtain the following result on the optimal set for the kth Dirichlet eigenvalue.

Corollary 5.38. *Let $\Omega \subset \mathbb{R}^d$ be a solution of the problem*

$$\min\Big\{\lambda_k(\Omega) :\ \Omega \subset \mathbb{R}^d,\ \Omega \text{ measurable},\ |\Omega| = 1\Big\}.$$

Then there exists an eigenfunction $u_k \in \widetilde{H}^1_0(\Omega)$, corresponding to the eigenvalue $\lambda_k(\Omega)$, which is Lipschitz continuous on $\mathbb{R}^d$.

Remark 5.39. We note that Theorem 5.37 can be used to obtain information for the supersolutions of a general functional F. Indeed, let F be a functional defined on the family of sets of finite measure and suppose that there exist non-negative real numbers c_k, $k \in \mathbb{N}$, such that for each couple of sets $\Omega \subset \widetilde{\Omega} \subset \mathbb{R}^d$ of finite measure we have

$$c_k\big(\lambda_k(\Omega) - \lambda_k(\widetilde{\Omega})\big) \le F(\Omega) - F(\widetilde{\Omega}).$$

If Ω is a shape supersolution for $F+\Lambda|\cdot|$, then for any k such that $c_k > 0$, there is an eigenfunction $u_k \in H_0^1(\Omega)$, normalized in L^2 and corresponding to $\lambda_k(\Omega)$, which is Lipschitz continuous on $\mathbb{R}^d$. It is enough to note that, whenever $c_k > 0$, we have

$$\lambda_k(\Omega) - \lambda_k(\widetilde{\Omega}) \le c_k^{-1}\big(F(\Omega) - F(\widetilde{\Omega})\big) \le c_k^{-1}\Lambda|\widetilde{\Omega} \setminus \Omega|.$$

The conclusion follows by Theorem 5.37.

In order to prove a regularity result which involves all the eigenfunction corresponding to the eigenvalues that appear in a bi-Lipschitz functional of the form $F\big(\lambda_{k_1}(\Omega), \dots, \lambda_{k_p}(\Omega)\big)$, we need the following preliminary result.

Lemma 5.40. *Let $\Omega^* \subset \mathbb{R}^d$ be a supersolution for the functional $\lambda_k + \lambda_{k+1} + \cdots + \lambda_{k+p}$ with constant $\Lambda > 0$. Then there are L^2-orthonormal eigenfunctions $u_k, \dots, u_{k+p} \in \widetilde{H}_0^1(\Omega^*)$, corresponding to the eigenvalues $\lambda_k(\Omega^*), \dots, \lambda_{k+p}(\Omega^*)$, which are Lipschitz continuous on $\mathbb{R}^d$.*

Proof. We prove the lemma in two steps.

Step 1. Suppose that $\lambda_k(\Omega^*) > \lambda_{k-1}(\Omega^*)$. We first note that, by Lemma 5.33, if $j \in \{k, k+1, \dots, k+p\}$ is such that $\lambda_j(\Omega^*) > \lambda_{j-1}(\Omega^*)$, then any eigenfunction, corresponding to the eigenvalue $\lambda_j(\Omega^*)$, is Lipschitz continuous on $\mathbb{R}^d$. Let us now divide the eigenvalues $\lambda_k(\Omega^*), \dots, \lambda_{k+p}(\Omega^*)$ into clusters of equal consecutive eigenvalues. There exists $k = k_1 < k_2 < \cdots < k_s \le k + p$ such that

$$\begin{aligned}
\lambda_{k-1}(\Omega^*) < \lambda_{k_1}(\Omega^*) &= \cdots = \lambda_{k_2-1}(\Omega^*) \\
< \lambda_{k_2}(\Omega^*) &= \cdots = \lambda_{k_3-1}(\Omega^*) \\
&\dots \\
< \lambda_{k_s}(\Omega^*) &= \cdots = \lambda_{k+p}(\Omega^*).
\end{aligned}$$

Then, by the above observation, the eigenspaces corresponding to $\lambda_{k_1}(\Omega^*), \lambda_{k_2}(\Omega^*), \dots, \lambda_{k+p}(\Omega^*)$ consist on Lipschitz continuous functions. In particular, there exists a sequence of consecutive eigenfunctions $u_k, \dots, u_{k+p}$ satisfying the claim of the lemma.

Step 2. Suppose now that $\lambda_k(\Omega^*) = \lambda_{k-1}(\Omega^*)$. For each $\varepsilon \in (0, 1)$ we consider the problem

$$\min\Big\{\sum_{j=1}^{p} \lambda_{k+j}(\Omega)+(1-\varepsilon)\lambda_k(\Omega)+\varepsilon\lambda_{k-1}(\Omega)+2\Lambda|\Omega| \,:\, \Omega^* \subset \Omega \subset \mathbb{R}^d\Big\}. \tag{5.66}$$

As in Theorem 5.37, we have that at least one of the following cases occur:

(i) There is a sequence $\varepsilon_n \to 0$ and a corresponding sequence Ω_{ε_n} of minimizers of (5.66) such that, for each $n \in \mathbb{N}$,

$$\lambda_k(\Omega_{\varepsilon_n}) > \lambda_{k-1}(\Omega_{\varepsilon_n}).$$

(ii) There is some $\varepsilon_0 \in (0,1)$ for which there is Ω_{ε_0} a solution of (5.66) such that

$$\lambda_k(\Omega_{\varepsilon_0}) = \lambda_{k-1}(\Omega_{\varepsilon_0}).$$

In the first case Ω_{ε_n} is a supersolution to the functional $\lambda_k + \dots + \lambda_{k+p}$ with constant $\Lambda/(1-\varepsilon_n)$. Thus, by *Step* 1, there are orthonormal eigenfunctions $u_k^n, \dots, u_{k+p}^n \in H_0^1(\Omega_{\varepsilon_n})$, which are Lipschitz continuous on $\mathbb{R}^d$. Using the same approximation argument from Theorem 5.37, we obtain the claim. In the second case, reasoning again as in Theorem 5.37, we have that $\Omega_{\varepsilon_0} = \Omega^*$ and we have to consider two more cases. If $\lambda_{k-1}(\Omega^*) > \lambda_{k-2}(\Omega^*)$, we have the claim by *Step* 1. If $\lambda_{k-1}(\Omega^*) = \lambda_{k-2}(\Omega^*)$, then we consider the problem

$$\min\left\{\sum_{j=1}^{p}\lambda_{k+j}(\Omega) + (1-\varepsilon_0)\lambda_k(\Omega) + \varepsilon_0\left[(1-\varepsilon)\lambda_{k-1}(\Omega) + \varepsilon\lambda_{k-2}(\Omega)\right]\right.$$
$$\left. + 3\Lambda|\Omega| \;:\; \Omega^* \subset \Omega \subset \mathbb{R}^d\right\},$$

and proceed by repeating the argument above, until we obtain the claim or until we have a functional involving λ_1, in which case we apply one more time the result from *Step* 1. □

Theorem 5.41. *Let $F : \mathbb{R}^p \to \mathbb{R}$ be a bi-Lipschitz, increasing function in each variable and let $0 < k_1 < k_2 < \dots < k_p$ be natural numbers. Then for every bounded shape supersolution Ω^* of the functional*

$$\Omega \mapsto F\big(\lambda_{k_1}(\Omega), \dots, \lambda_{k_p}(\Omega)\big),$$

there exists a sequence of orthonormal eigenfunctions $u_{k_1}, \dots, u_{k_p}$, corresponding to the eigenvalues $\lambda_{k_j}(\Omega^*)$, $j = 1, \dots, p$, which are Lipschitz continuous on $\mathbb{R}^d$. Moreover,

- *if for some k_j we have $\lambda_{k_j}(\Omega^*) > \lambda_{k_j-1}(\Omega^*)$, then the full eigenspace corresponding to $\lambda_{k_j}(\Omega^*)$ consists only on Lipschitz functions;*
- *if $\lambda_{k_j}(\Omega^*) = \lambda_{k_{j-1}}(\Omega^*)$, then there exist at least $k_j - k_{j-1} + 1$ orthonormal Lipschitz eigenfunctions corresponding to $\lambda_{k_j}(\Omega^*)$.*

Proof. Let $c_1, \dots, c_p \in \mathbb{R}^+$ be strictly positive real numbers such that for each $x = (x_j), y = (y_j) \in \mathbb{R}^p$, such that $x_j \geq y_j, \forall j \in \{1, \dots, p\}$, we have

$$F(x) - F(y) \geq c_1(x_1 - y_1) + \cdots + c_p(x_p - y_p).$$

We note that if Ω^* is a supersolution of $F(\lambda_{k_1}, \dots, \lambda_{k_p})$, then Ω^* is also a supersolution for the functional

$$\widetilde{F} = \left[\min_{j\in\{1,\dots,p\}} c_j\right] \left(\lambda_{k_1} + \cdots + \lambda_{k_p}\right),$$

and, since $\min_{j\in\{1,\dots,p\}} c_j > 0$, we can assume $\min_{j\in\{1,\dots,p\}} c_j = 1$.

Reasoning as in Lemma 5.40, we divide the family $(\lambda_{k_1}(\Omega^*), \dots$ $\dots, \lambda_{k_p}(\Omega^*))$ into clusters of equal eigenvalues with consecutive indexes. There exist $1 \leq i_1 < i_2 \cdots < i_s \leq p - 1$ such that

$$\begin{aligned}\lambda_{k_1}(\Omega^*) = \cdots = \lambda_{k_{i_1}}(\Omega^*) &< \lambda_{k_{(i_1+1)}}(\Omega^*) = \cdots = \lambda_{k_{i_2}}(\Omega^*) \\ &< \lambda_{k_{(i_2+1)}}(\Omega^*) = \cdots = \lambda_{k_{i_3}}(\Omega^*) \\ &\dots \\ &< \lambda_{k_{(i_s+1)}}(\Omega^*) = \cdots = \lambda_{k_p}(\Omega^*).\end{aligned}$$

Since the eigenspaces, corresponding to different clusters, are orthogonal to each other, it is enough to prove the claim for the functionals defined as the sum of the eigenvalues in each cluster. In other words, it is sufficient to restrict our attention only to the case when Ω^* is a supersolution for the functional $F(\lambda_{k_1}, \dots, \lambda_{k_p}) = \sum_{j=1}^p \lambda_{k_j}$ and is such that

$$\lambda_{k_1}(\Omega^*) = \cdots = \lambda_{k_p}(\Omega^*). \tag{5.67}$$

Moreover, in this case Ω^* is also a supersolution (with possibly different constant Λ) for the sum of consecutive eigenvalues $\sum_{k=k_1}^{k_p} \lambda_k$. Indeed, it is enough to consider the functional

$$\widetilde{F}(\Omega) = \frac{1}{2}\sum_{j=1}^{p} \lambda_{k_j}(\Omega) + \theta \sum_{k=k_1}^{k_p} \lambda_k(\Omega),$$

for a suitable value of θ, *e.g.* $\theta = \dfrac{1}{2(k_p - k_1 + 1)}$. The conclusion then follows by Lemma 5.40. □

5.6. Measurable sets of positive curvature

In this section we study the properties of the shape supersolutions for functionals $\mathcal{F} : \mathcal{B}(\mathbb{R}^d) \to \mathbb{R}$ of the form

$$\mathcal{F}(\Omega) = \mathcal{G}(\Omega) + P(\Omega) + \alpha|\Omega|,$$

where

- $\mathcal{G} : \mathcal{B}(\mathbb{R}^d) \to \mathbb{R}$ is a decreasing (with respect to the set inclusion) functional on the family $\mathcal{B}(\mathbb{R}^d)$ of Borel measurable sets in $\mathbb{R}^d$;
- P is the perimeter in sense of De Giorgi;
- $\alpha \in \mathbb{R}$ is a given constant.

The results from this section involve only local arguments. Thus, we will prove then for sets which are shape supersolutions only locally. Most of the results are contained in the papers [59] and [58].

Definition 5.42. Let $\mathcal{F} : \mathcal{B}(\mathbb{R}^d) \to [0, +\infty]$ be a functional on the family of Borel sets $\mathcal{B}(\mathbb{R}^d)$ on $\mathbb{R}^d$ and let $\Omega \in \mathcal{B}(\mathbb{R}^d)$ be such that $\mathcal{F}(\Omega) < +\infty$. We say that:

- Ω is a **supersolution for $\mathcal{F}$ in the set $\mathcal{D} \subset \mathbb{R}^d$**, if

$$\mathcal{F}(\Omega) \leq \mathcal{F}(\widetilde{\Omega}), \text{ for every Borel set } \widetilde{\Omega} \supset \Omega \text{ such that } \widetilde{\Omega} \setminus \Omega \subset \mathcal{D}.$$

- Ω is a **local supersolution for $\mathcal{F}$**, if there is a constant $r_0 > 0$ such that Ω is a supersolution for $\mathcal{F}$ in $B_{r_0}(x)$ for every ball $B_{r_0}(x) \subset \mathbb{R}^d$.

The following simple Remark will play a crucial role in the study of spectral optimization problems with perimeter constraint.

Remark 5.43. Suppose that $\Omega \subset \mathbb{R}^d$ is a (local) subsolution for the functional $\mathcal{F} = \mathcal{G}+P+\alpha|\cdot|$, where $\mathcal{G} : \mathcal{B}(\mathbb{R}^d) \to \mathbb{R}$ is decreasing with respect to the set inclusion. Then Ω is a (local) supersolution also for $P + \alpha|\cdot|$. For every $\widetilde{\Omega} \supset \Omega$, by the monotonicity of $\mathcal{G}$ and the super-optimality of Ω, we have

$$\mathcal{G}(\Omega) + P(\Omega) + \alpha|\Omega| \leq \mathcal{G}(\widetilde{\Omega}) + P(\widetilde{\Omega}) + \alpha|\widetilde{\Omega}| \leq \mathcal{G}(\Omega) + P(\widetilde{\Omega}) + \alpha|\widetilde{\Omega}|,$$

which proves that

$$P(\Omega) + \alpha|\Omega| \leq P(\widetilde{\Omega}) + \alpha|\widetilde{\Omega}|, \quad \text{for every} \quad \widetilde{\Omega} \supset \Omega. \tag{5.68}$$

In particular, (5.68) holds for the sets Ω, which are supersolutions for functionals $\mathcal{F}$ of the form

$$\mathcal{F}(\Omega) = F\big(\lambda_1(\Omega), \dots, \lambda_k(\Omega)\big) + P(\Omega) + \alpha|\Omega|,$$

where $F : \mathbb{R}^k \to \mathbb{R}$ is a function on $\mathbb{R}^k$ increasing in each variable.

When we deal with shape optimization problems in a box $\mathcal{D}$, *a priori* we can only consider perturbations of a set $\Omega \subset \mathcal{D}$, which remain inside the box. The following lemma allows us to eliminate this restriction and work with the minimizers as if they are solutions of the problem in the free case $\mathcal{D} = \mathbb{R}^d$.

Lemma 5.44. *Let $\Omega \subset \mathcal{D}$ be two Borel sets in $\mathbb{R}^d$ and let $\mathcal{F} = P + \alpha|\cdot|$, where $\alpha \geq 0$. If $\mathcal{D}$ is a (local) supersolution for $\mathcal{F}$ and Ω is a (local) supersolution for $\mathcal{F}$ in $\mathcal{D}$, then Ω is a (local) supersolution for $\mathcal{F}$ in $\mathbb{R}^d$.*

Proof. Let $\widetilde{\Omega} \supset \Omega$. Since $\mathcal{D}$ is a supersolution, we get

$$\begin{aligned} &P(\widetilde{\Omega}; \mathcal{D}^c) + P(\mathcal{D}; \widetilde{\Omega}^c) + \alpha|\widetilde{\Omega} \cup \mathcal{D}| \\ &= \mathcal{F}(\widetilde{\Omega} \cup \mathcal{D}) \geq \mathcal{F}(\mathcal{D}) = P(\mathcal{D}; \widetilde{\Omega}) + P(\mathcal{D}; \widetilde{\Omega}^c) + \alpha|\mathcal{D}|, \end{aligned}$$

which gives

$$P(\mathcal{D}; \widetilde{\Omega}) \leq P(\widetilde{\Omega}; \mathcal{D}^c) + \alpha|\widetilde{\Omega} \setminus \mathcal{D}|. \tag{5.69}$$

On the other hand, we can test the super-optimality of Ω with $\widetilde{\Omega} \cap \mathcal{D}$ and then use (5.69) obtaining

$$\begin{aligned} \mathcal{F}(\Omega) \leq \mathcal{F}(\widetilde{\Omega} \cap \mathcal{D}) &= P(\widetilde{\Omega}; \mathcal{D}) + P(\mathcal{D}; \widetilde{\Omega}) + \alpha|\widetilde{\Omega} \cap \mathcal{D}| \\ &\leq P(\widetilde{\Omega}; \mathcal{D}) + P(\widetilde{\Omega}; \mathcal{D}^c) + \alpha|\widetilde{\Omega} \setminus \mathcal{D}| + \alpha|\widetilde{\Omega} \cap \mathcal{D}| \\ &= P(\widetilde{\Omega}) + \alpha|\widetilde{\Omega}| = \mathcal{F}(\widetilde{\Omega}). \end{aligned}$$

For the case of local supersolutions, it is enough to consider $\widetilde{\Omega}$ such that $\widetilde{\Omega} \setminus \mathcal{D} \subset \widetilde{\Omega} \setminus \Omega \subset B_r(x)$ and then use the same argument as above. □

5.6.1. Sets satisfying exterior density estimate

In this subsection we show that the local shape supersolutions for the functional $P + \alpha|\Omega|$ satisfy an exterior density estimate and we deduce some preliminary results based only on this property.

The following lemma is the first step in the analysis of the supersolutions for $P + \alpha|\cdot|$ and shows that they are in fact open sets. The result is classical (see, for instance, [67], [80, Theorem 16.14]) and so we only sketch the proof.

Lemma 5.45. *Let $\Omega \subset \mathbb{R}^d$ be a local supersolution for the functional $\mathcal{F} = P + \alpha|\cdot|$ with $\alpha \geq 0$. Then there exists a positive constant $\bar{c}$, depending only on the dimension d, such that for every $x \in \mathbb{R}^d$, one of the following situations occurs:*

(a) *there is $r > 0$ such that $|B_r(x) \cap \Omega^c| = 0$;*

(b) *there is* $r_1 > 0$, *depending on the dimension,* α *and the constant* r_0 *from Definition 5.59, such that*

$$|B_r(x) \cap \Omega^c| \geq \bar{c}|B_r|, \quad \textit{for every} \quad r \in (0, r_1).$$

Proof. Let $x \in \mathbb{R}^d$. Suppose that there is no $r > 0$ such that $B_r(x) \subset \Omega$. We will prove that (b) holds. Testing the super-optimality of Ω with the set $\widetilde{\Omega} := \Omega \cup B_r(x)$, for $r \leq r_0$, we get that for almost every $r \in (0, r_0)$,

$$P(\Omega, B_r(x)) \leq \mathcal{H}^{d-1}(\partial B_r(x) \cap \Omega^c) + \alpha|B_r(x) \cap \Omega^c|.$$

Applying the isoperimetric inequality to the set $B_r(x) \setminus \Omega$, we obtain

$$\begin{aligned} |B_r(x)\setminus\Omega|^{1-1/d} &\leq C_d\left(P(\Omega, B_r(x)) + \mathcal{H}^{d-1}(\partial B_r(x) \cap \Omega^c)\right) \\ &\leq 2C_d\mathcal{H}^{d-1}(\partial B_r(x) \cap \Omega^c) + C_d\alpha|B_r(x) \cap \Omega^c| \\ &\leq 2C_d\mathcal{H}^{d-1}(\partial B_r(x) \cap \Omega^c) \\ &\quad + \frac{1}{2}|B_r(x) \cap \Omega^c|^{1-1/d}, \end{aligned} \tag{5.70}$$

for $r \in (0, r_1)$, where $r_1 = \min\left\{r_0, (\omega_d^{1/d} C_d\alpha)^{-1}\right\}$, and $C_d > 0$ is a dimensional constant (in the case $\alpha = 0$, we set $r_1 = r_0$). Consider the function $\phi(r) := |B_r(x) \setminus \Omega|$. Note that $\phi(0) = 0$ and $\phi'(r) = \mathcal{H}^{d-1}(\partial B_r(x) \cap \Omega)$ and so, taking $\bar{c} = d(4C_d)^{-1}$ the estimate (5.70) gives

$$\bar{c} \leq \frac{d}{dr}\left(\phi(r)^{1/d}\right),$$

which after integration gives (b). □

Definition 5.46. If $\Omega \subset \mathbb{R}^d$ is a set if finite Lebesgue measure and if there is a constant $\bar{c} > 0$ such that for every point $x \in \mathbb{R}^d$ one of the conditions (a) and (b), from Lemma 5.45, holds, then we say that Ω satisfies an **exterior density estimate**.

In what follows we will denote with w_Ω the solution of the problem

$$-\Delta w_\Omega = 1 \quad \text{in} \quad \widetilde{H}_0^1(\Omega), \qquad w_\Omega \in \widetilde{H}_0^1(\Omega).$$

We first note that a classical argument provides the continuity of w_Ω on the sets with exterior density.

Proposition 5.47. *Let* $\Omega \subset \mathbb{R}^d$ *be a set of finite Lebesgue measure satisfying an exterior density estimate. Then there are positive constants* C *and* β *such that, for every* $x_0 \in \mathbb{R}^d$ *with the property*

$$|B_r(x_0) \cap \Omega^c| > 0, \quad \textit{for every} \quad r > 0,$$

we have

$$\|w_\Omega\|_{L^\infty(B_r(x_0))} \le r^\beta \|w_\Omega\|_{L^\infty(\mathbb{R}^d)}, \quad \textit{for every} \quad 0 < r < r_1. \tag{5.71}$$

In particular, if Ω is a perimeter supersolution, then the above conclusion holds.

Proof. Let $x_0 \in \mathbb{R}^d$ be such that that $|B_r(x_0) \cap \Omega^c| > 0$, for every $r > 0$. Without loss of generality we can suppose that $x_0 = 0$. Setting $w := w_\Omega$, we have that $\Delta w + 1 \ge 0$ in distributional sense on $\mathbb{R}^d$. Thus, on each ball $B_r(y)$ the function

$$u(x) := w(x) - \frac{r^2 - |x-y|^2}{2d},$$

is subharmonic and we have the mean value property

$$w(y) \le \frac{r^2}{2d} + \fint_{B_r(y)} w(x)\,dx. \tag{5.72}$$

Let us define $r_n = 4^{-n}$. For any $y \in B_{r_{n+1}}$, equation (5.72) implies

$$\begin{aligned} w(y) &\le \frac{r_n^2}{4d} + \fint_{B_{2r_{n+1}}(y)} w(x)\,dx \\ &\le \frac{r_n^2}{4d} + \frac{|\Omega \cap B_{2r_{n+1}}(y)|}{|B_{2r_{n+1}}(y)|} \|w\|_{L^\infty(B_{2r_{n+1}}(y))} \\ &\le \frac{r_n^2}{4d} + \left(1 - \frac{|\Omega^c \cap B_{r_{n+1}}|}{|B_{2r_{n+1}}|}\right) \|w\|_{L^\infty(B_{r_n}(0))} \\ &\le \frac{4^{-2n}}{4d} + \left(1 - 2^{-d}\bar{c}\right) \|w\|_{L^\infty(B_{r_n})}, \end{aligned} \tag{5.73}$$

where in the third inequality we have used the inclusion $B_{r_{n+1}} \subset B_{2r_{n+1}}(y)$ for every $y \in B_{r_{n+1}}$. Hence setting

$$a_n = \|w\|_{L^\infty(B_{r_n})},$$

we have

$$a_{n+1} \le \frac{8^{-n}}{4d} + (1 - 2^{-d}\bar{c})a_n,$$

which easily implies $a_n \le C a_0 4^{-n\beta}$ for some constants β and C depending only on $\bar{c}$. This gives (5.71). $\square$

Proposition 5.48. *Let $\Omega \subset \mathbb{R}^d$ be a set of finite Lebesgue measure satisfying an external density estimate. Then the set of points of density* 1,

$$\Omega_1 := \left\{ x \in \mathbb{R}^d : \exists \lim_{r \to 0} \frac{|\Omega \cap B_r(x)|}{|B_r(x)|} = 1 \right\},$$

is open and $\widetilde{H}_0^1(\Omega) = H_0^1(\Omega_1)$. In particular, if Ω is a local supersolution solution for the functional $\mathcal{F} = P + \alpha|\cdot|$, then Ω_1 is open and $\widetilde{H}_0^1(\Omega) = H_0^1(\Omega_1)$.

Proof. Thanks to Lemma 5.45, Ω_1 is an open set. It remains to prove the equality between the Sobolev spaces. We first recall that we have the equality

$$\widetilde{H}_0^1(\Omega) = H_0^1(\{w_\Omega > 0\}).$$

We now prove that $\Omega_1 = \{w_\Omega > 0\}$ up to a set of zero capacity. Consider a ball $B \subset \Omega_1$. By the weak maximum principle, $w_B \leq w_\Omega$ and so

$$\Omega_1 \subset \{w_\Omega > 0\}.$$

In order to prove the other inclusion, we recall that for every $x_0 \in \mathbb{R}^d$ we have

$$w_\Omega(x_0) = \lim_{r \to 0} \fint_{B_r(x_0)} w_\Omega \, dx,$$

By Proposition 5.47, $w_\Omega = 0$ on $\mathbb{R}^d \setminus \Omega_1$ which gives the converse inclusion. □

Proposition 5.49. *Let $\Omega \subset \mathbb{R}^d$ satisfy an exterior density estimate. Then $w_\Omega : \mathbb{R}^d \to \mathbb{R}$ is Hölder continuous and*

$$|w_\Omega(x) - w_\Omega(y)| \leq C|x - y|^\beta, \tag{5.74}$$

where β is the constant from Proposition 5.47.

Proof. Thanks to Proposition (5.48), up to a set of capacity zero, we can assume that Ω_1 is open and that w_Ω is the classical solution, with Dirichlet boundary conditions, of $-\Delta w_\Omega = 1$ in Ω_1. Consider two distinct points $x, y \in \mathbb{R}^d$. In case both x and y belong to Ω_1^c, the estimate (5.74) is trivial. Let us assume that $x \in \Omega_1$ and let $x_0 \in \partial\Omega_1$ be such that

$$|x - x_0| = \mathrm{dist}(x, \partial\Omega_1).$$

We distinguish two cases:

- Suppose that $y \in \mathbb{R}^d$ is such that

$$2|x - y| \geq \operatorname{dist}(x, \partial\Omega_1).$$

Hence $x, y \in B_{4|x-y|}(x_0)$ and by Proposition 5.47, we have that

$$w_\Omega(x) \leq C|x - y|^\beta \quad \text{and} \quad w_\Omega(y) \leq C|x - y|^\beta.$$

Thus we obtain

$$|w_\Omega(x) - w_\Omega(y)| \leq 2C|x - y|^\beta. \tag{5.75}$$

- Assume that $y \in \mathbb{R}^d$ is such that

$$2|x - y| \leq \operatorname{dist}(x, \partial\Omega_1).$$

Applying Lemma 5.9 to w_Ω in $B_{\operatorname{dist}(x,\partial\Omega_1)}(x) \subset \Omega_1$ we obtain

$$\begin{aligned} \|\nabla w_\Omega\|_{L^\infty(B_{\operatorname{dist}(x,\partial\Omega_1)/2}(x))} &\leq \frac{C_d \|w\|_{L^\infty(B_{\operatorname{dist}(x,\partial\Omega_1)}(x))}}{\operatorname{dist}(x, \partial\Omega_1)} \\ &\leq C_d \operatorname{dist}(x, \partial\Omega_1)^{\beta - 1}, \end{aligned} \tag{5.76}$$

which, since $\beta < 1$, together with our assumption and the mean value formula implies

$$|w_\Omega(x) - w_\Omega(y)| \leq C_d \operatorname{dist}(x, \partial\Omega_1)^{\beta-1}|x - y| \leq |x - y|^\beta. \qquad \square$$

5.6.2. Mean curvature bounds in viscosity sense

Let $\Omega \subset \mathbb{R}^d$ be an open set with smooth boundary. In a neighbourhood of a given boundary point $x \in \partial\Omega$ we can characterize (up to a coordinate change) Ω as

$$\Omega = \{(x_1, \dots, x_d) \in \mathbb{R}^d : \phi(x_1, \dots, x_{d-1}) > x_d\},$$

for a smooth function $\phi : \mathbb{R}^{d-1} \to \mathbb{R}$. Thus the mean curvature $H_\Omega(x)$ of Ω (with respect to the exterior normal) in a neighbourhood of x is given by

$$H_\Omega = \operatorname{div}\left(\frac{\nabla\phi}{\sqrt{1 + |\nabla\phi|^2}}\right) = \sum_{i=1}^{d-1} \frac{\partial_{ii}\phi}{\sqrt{1 + |\nabla\phi|^2}} - \sum_{i,j=1}^{d-1} \frac{\phi_i\phi_j\phi_{ij}}{(1 + |\nabla\phi|^2)^{3/2}},$$

and choosing the coordinates $x_1, \dots, x_d$ of x such that $|\nabla\phi|(x) = 0$ we get

$$H_\Omega(x) = \sum_{i=1}^{d-1} \phi_{ii}(x) =: \Delta_{d-1}\phi(x).$$

Definition 5.50. For an open set $\Omega \subset \mathbb{R}^d$ and $c \in \mathbb{R}$ we say that **the mean curvature of $\partial\Omega$ is bounded from below by c in viscosity sense** ($H_\Omega \geq c$), if for every open set $\omega \subset \Omega$ with smooth boundary and every point $x \in \partial\Omega \cap \partial\omega$ we have that $H_\omega(x) \geq c$.

Proposition 5.51. *Let $\Omega \subset \mathbb{R}^d$ be an open set of finite measure. If Ω is a local supersolution for the functional $P + \alpha|\cdot|$, then $H_\Omega \geq -\alpha$ in viscosity sense.*

Proof. Let $\omega \subset \Omega$ be an open set with smooth boundary and let $x_0 \in \partial\omega\cap\partial\Omega$. We can suppose that $x_0 = 0$ and that ω is locally a supergraph of a smooth function $\phi : \mathbb{R}^{d-1} \to \mathbb{R}$ such that $\phi(0) = |\nabla\phi(0)| = 0$. We can now suppose that $\{0\} = \partial\omega\cap\partial\Omega$, up to replace ω by a smooth set $\widetilde{\omega} \subset \omega$, which is locally a supergraph of the function $\widetilde{\phi}(x) = \phi(x) + |x|^4$. We consider now the family of sets $\omega_\varepsilon = -\varepsilon e_d + \widetilde{\omega}$, where $e_d = (0, \dots, 0, 1)$. By the choice of $\widetilde{\omega}$, for every $r > 0$ one can find $\varepsilon_0 > 0$ such that

$$\omega_\varepsilon \setminus \Omega \subset \omega_\varepsilon \setminus \omega \subset B_r, \text{ for every } 0 < \varepsilon < \varepsilon_0,$$

and so one can use the sets $\Omega_\varepsilon = \omega_\varepsilon \cup \Omega$ to test the local superminimality of Ω. Let $d_\varepsilon : \omega_\varepsilon \to \mathbb{R}$ be the distance function

$$d_\varepsilon(x) = \operatorname{dist}(x, \partial\omega_\varepsilon).$$

For small enough ε we have that d_ε is smooth in $\omega_\varepsilon \cap B_r$ up to the boundary $\partial\omega_\varepsilon$. By [67, Appendix B], we have that $H_\omega(0) = H_{\omega_\varepsilon}(-\varepsilon e_d) = \Delta d_\varepsilon(-\varepsilon e_d)$. if, by absurd $H_\omega(0) < -\alpha$, then for ε small enough we can suppose that $\Delta d_\varepsilon < -\alpha$ in $\omega_\varepsilon \cap B_r$ and so, denoting with ν_Ω the exterior normal to a set of finite perimeter Ω, we have

$$\begin{aligned}
-\alpha|\omega_\varepsilon \setminus \Omega| &> \int_{\omega_\varepsilon\setminus\Omega} \Delta d_\varepsilon(x)\,dx \\
&= \int_{\Omega\cap\partial\omega_\varepsilon} \nabla d_\varepsilon \cdot \nu_{\omega_\varepsilon}\,d\mathcal{H}^{d-1} - \int_{\omega_\varepsilon\cap\partial\Omega} \nabla d_\varepsilon \cdot \nu_\Omega\,d\mathcal{H}^{d-1} \\
&\geq P(\omega_\varepsilon; \Omega) - P(\Omega; \omega_\varepsilon),
\end{aligned}$$

which implies

$$P(\Omega) + \alpha|\Omega| > P(\Omega \cup \omega_\varepsilon) + \alpha|\Omega \cup \omega_\varepsilon|,$$

thus contradicting the local superoptimality of Ω. $\square$

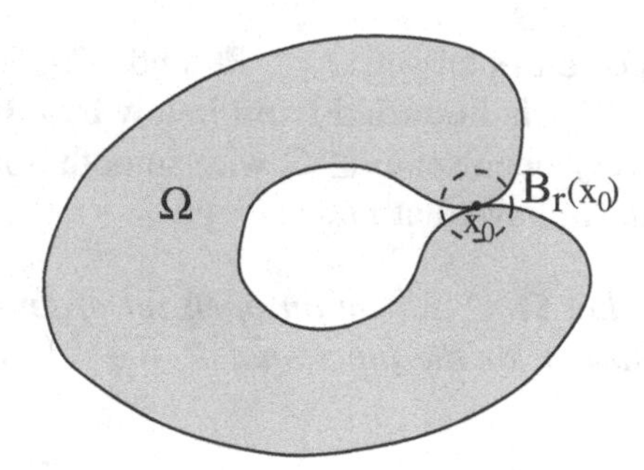

Figure 5.1. Ω has mean curvature bounded from below in viscosity sense, but is not a local supersolution for $P + \alpha|\cdot|$.

Remark 5.52. The converse implication is in general false. Indeed, the set Ω on Figure 5.1 has mean curvature bounded from below in viscosity sense. On the other hand it is not a supersolution for $P + \alpha|\cdot|$ since adding a ball $B_r(x_0)$ in the boundary point $x_0 \in \partial\Omega$ is an operation that decreases the perimeter by a linear term $\big(P(\Omega) - P(\Omega \cap B_r)\big) \sim r$.

The following Lemma is a generalization of [59, Lemma 5.3] and was proved in [58]. We prove that a set Ω, which has a bounded from below mean curvature in viscosity sense, has a distance function function $dist(x, \Omega^c)$ which is super harmonic in Ω in viscosity sense (see [39] for a nice account of theory of viscosity solutions). In case $\partial\Omega$ is smooth this easily implies that the mean curvature of $\partial\Omega$, computed with respect to the exterior normal, is positive (see for instance [66, Section 14.6]). A similar observation already appeared in [40], in the study of the regularity of minimal surfaces, and in [74, 82], in the study of free boundary type problems.

We recall that a continuous function $f : \Omega \to \mathbb{R}$ satisfies the inequality $\Delta f \leq \alpha$ in viscosity sense on Ω, if and only if,

$$\Delta\varphi(x_0) \leq \alpha, \quad \textit{for every } x_0 \in \Omega \textit{ and } \varphi \in C_c^\infty(\Omega) \textit{ such that } x_0 \textit{ is a local minimum for } f - \varphi.$$

Lemma 5.53. *Suppose that Ω is an open set such that $H_\Omega \geq -\alpha$ in viscosity sense. Then the distance function $d_\Omega(x) = dist(x, \partial\Omega)$ satisfies $\Delta d_\Omega \leq \alpha$ in viscosity sense.*

Proof. Suppose that $\varphi \in C_c^\infty(\Omega)$ is such that $\varphi \leq d_\Omega$ and suppose that $x_0 \in \Omega$ is such that $\varphi(x_0) = d_\Omega(x_0)$. In what follows we set $t = \varphi(x_0)$, $\Omega_t = \{\varphi > t\} \subset \{d_\Omega > t\}$ and $n = \frac{x_0 - y_0}{|x_0 - y_0|}$, where $y_0 \in \partial\mathcal{D}$ is chosen such that $|x_0 - y_0| = t$ (see Figure 5.2). We first prove that $\nabla\varphi(x_0) = n$.

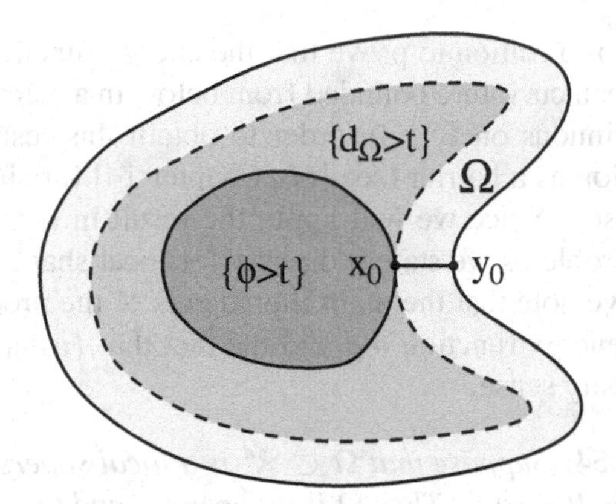

Figure 5.2. Testing the viscosity bound $H_\Omega \geq -\alpha$ with the set $\{\varphi > t\}$.

Indeed, on one hand the Lipschitz continuity of d_Ω gives

$$\varphi(x) - \varphi(x_0) \leq d_\Omega(x) - d_\Omega(x_0) \leq |x - x_0|,$$

and so $|\nabla\varphi|(x_0) \leq 1$. On the other hand, we have

$$\varphi(x_0) - \varphi(x_0 + \varepsilon n) \geq d_\Omega(x_0) - d_\Omega(x_0 + \varepsilon n) = \varepsilon,$$

which gives $|\nabla\varphi|(x_0) \geq \frac{\partial\varphi}{\partial n}(x_0) = 1$.

We now notice that φ is concave in the direction of n. Indeed

$$\begin{aligned}\frac{\partial^2\varphi}{\partial n^2}(x_0) &= \lim_{\varepsilon\to 0^+} \frac{\varphi(x_0 + \varepsilon n) + \varphi(x_0 - \varepsilon n) - 2\varphi(x_0)}{\varepsilon^2}\\ &\leq \lim_{\varepsilon\to 0^+} \frac{d_\Omega(x_0 + \varepsilon n) + d_\Omega(x_0 - \varepsilon n) - 2d_\Omega(x_0)}{\varepsilon^2}\\ &\leq \lim_{\varepsilon\to 0^+} \frac{(t + \varepsilon) + (t - \varepsilon) - 2t}{\varepsilon^2} = 0.\end{aligned}$$

Since $|\nabla\varphi|(x_0) = 1$, the level set Ω_t has smooth boundary in a neighbourhood of x_0 and $n = -\nu_{\Omega_t}(x_0)$ is the interior normal at $x_0 \in \partial\Omega_t$. Then we have

$$\Delta\varphi(x_0) = \frac{\partial^2\varphi}{\partial n^2}(x_0) - \frac{\partial\varphi}{\partial n}(x_0) H_{\Omega_t}(x_0) \leq -H_{\Omega_t}(x_0).$$

On the other hand setting $\omega = nt + \Omega_t$, we have $\omega \subset nt + \{d_\Omega > t\} \subset \Omega$, $y_0 \in \partial\omega$ and $H_\omega(y_0) = H_{\Omega_t}(x_0) \geq -\alpha$, which gives $\Delta\varphi(x_0) \leq \alpha$ and concludes the proof. □

We are now in position to prove that the energy functions on the sets, which have mean curvature bounded from below in a viscosity sense, are Lipschitz continuous on $\mathbb{R}^d$. In order to obtain this result, we use the distance function as a barrier (see [66, Chapter 14] for similar proofs in the smooth case). Since we will apply the result in the study of shape optimization problems we state it directly for local shape supersolutions $\Omega \subset \mathbb{R}^d$ and we note that the main ingredients of the proof are the continuity of the energy function w_Ω and the fact that H_Ω is bounded from below in viscosity sense.

Proposition 5.54. *Suppose that $\Omega \subset \mathbb{R}^d$ is a local supersolution for the functional $\mathcal{F} = P + \alpha|\cdot|$. Then Ω is an open set and the energy function w_Ω is Lipschitz continuous on $\mathbb{R}^d$ with a constant depending only on α, the dimension d and the measure $|\Omega|$.*

Proof. We set for simplicity $w = w_\Omega$. Consider the function

$$h(t) = Ct - bt^{1+\theta}$$

with derivatives $h'(t) = C - b(1+\theta)t^\theta$ and $h''(t) = -b\theta(1+\theta)t^{\theta-1}$.

In what follows we will show that we can choose the positive positive constants C, b and $\theta \in (0, 1]$ in such a way that the following inequality holds:

$$w_\Omega(x) \leq h(d_\Omega(x)) = C d_\Omega(x) - b\, d_\Omega(x)^{1+\theta}, \qquad \forall x \in \Omega. \tag{5.77}$$

We first note that on the interval $[0, (C/2b)^{1/\theta}]$, the derivative h' is positive and thus

$$h : \left[0, \left(\frac{C}{2b}\right)^{1/\theta}\right] \to \left[0, \frac{C}{2}\left(\frac{C}{2b}\right)^{1/\theta}\right],$$

is a diffeomorphism. If C, b and θ are such that

$$\frac{C}{2}\left(\frac{C}{2b}\right)^{1/\theta} \geq \frac{|\Omega|^{2/d}}{2d\omega_d^{2/d}} \geq \|w_\Omega\|_{L^\infty} \quad \text{and} \quad \left(\frac{C}{2b}\right)^{1/\theta} \geq \frac{|\Omega|^{1/d}}{\omega_d^{1/d}} \geq \max_\Omega d_\Omega, \tag{5.78}$$

then $h(d_\Omega)$ and $h^{-1}(w)$ are well defined, positive and have the same regularity as d_Ω and w. Suppose that there is $\varepsilon > 0$ such that the function $w_\varepsilon := (w - \varepsilon)^+$ satisfies

$$w_\varepsilon \leq h(d_\Omega) \text{ in } \Omega \text{ and } w_\varepsilon(x_0) = h(d_\Omega(x_0)), \text{ for some } \quad x_0 \in \Omega. \tag{5.79}$$

Then considering the function $u_\varepsilon = h^{-1}(w_\varepsilon)$, we get

$$u_\varepsilon \le d_\Omega \quad \text{in} \quad \Omega \qquad \text{and} \qquad u_\varepsilon(x_0) = d_\Omega(x_0).$$

By Lemma 5.53 we have $\Delta u_\varepsilon(x_0) \le \alpha$ and $|\nabla u_\varepsilon|^2(x_0) = 1$ and so

$$\begin{aligned} -1 &= \Delta w(x_0) \\ &= h''(u_\varepsilon(x_0))|\nabla u_\varepsilon|^2(x_0) + h'(u_\varepsilon(x_0))\Delta u_\varepsilon(x_0) \\ &\le -b\theta(1+\theta)u_\varepsilon(x_0)^{\theta-1} + h'(u_\varepsilon(x_0))\alpha \\ &\le -b\theta(1+\theta)u_\varepsilon(x_0)^{\theta-1} + (C - b(1+\theta)u_\varepsilon(x_0)^\theta)\alpha \\ &< -b\theta u_\varepsilon(x_0)^{\theta-1} + C\alpha. \end{aligned} \tag{5.80}$$

We now consider two cases.

- If $\alpha \le 0$, then it is sufficient to take

$$\theta = 1, \qquad b = 1 \qquad \text{and} \qquad C \ge \frac{|\Omega|^{1/d}}{\omega_d^{1/d}},$$

in order to have that there is no $\varepsilon > 0$ satisfying (5.79). Thus, we obtain

$$w_\Omega(x) \le \frac{|\Omega|^{1/d}}{\omega_d^{1/d}} d_\Omega(x) - d_\Omega^2(x), \qquad \forall x \in \overline{\Omega}. \tag{5.81}$$

- If $\alpha > 0$, then we choose θ, b and C as

$$\theta = \frac{1}{2}, \qquad \left(\frac{C}{2b}\right)^2 = \frac{C^{1/4}}{\alpha \vee 2} \qquad \text{and} \qquad \frac{C^{1/4}}{\alpha \vee 2} \ge \left(\frac{|\Omega|^{1/d}}{\omega_d^{1/d}}\right) \vee 16.$$

Then the conditions (5.78) are satisfied. Moreover, we have

$$\|w\|_{L^\infty} \le \frac{1}{2d}\left(\frac{|\Omega|^{1/d}}{\omega_d^{1/d}}\right)^2 \le \frac{1}{2}\frac{C^{1/2}}{(\alpha \vee 2)^2}, \tag{5.82}$$

and so

$$\|h^{-1}(w)\|_{L^\infty} \le \frac{1}{64(\alpha \vee 2)C^{1/4}}. \tag{5.83}$$

Indeed, if (5.83) does not hold, then

$$\begin{aligned} \|w\|_{L^\infty} &> h\left(\frac{1}{64(\alpha \vee 2)C^{1/4}}\right) = \frac{1}{64(\alpha \vee 2)C^{1/4}}\left(C - \frac{b}{64(\alpha \vee 2)C^{1/4}}\right) \\ &\ge \frac{C}{64(\alpha \vee 2)C^{1/4}}\left(1 - \frac{1}{2}\frac{1}{(\alpha \vee 2)^{1/2}C^{3/8}}\right) \ge \frac{C^{3/4}}{128(\alpha \vee 2)}, \end{aligned}$$

which is a contradiction with (5.82). Thus, (5.83) holds and so

$$\begin{aligned}\frac{1+C\alpha}{\|h^{-1}(w)\|_{L^\infty}^{-1/2}b/2} &= \frac{2(1+C\alpha)}{b}\|h^{-1}(w)\|_{L^\infty}^{1/2} \\ &\le \frac{4C^{1/8}(1+C\alpha)}{C(\alpha\vee 2)^{1/2}}\left(\frac{1}{64(\alpha\vee 2)C^{1/4}}\right)^{1/2} \\ &\le \frac{1+C\alpha}{2(\alpha\vee 2)C} \le 1.\end{aligned} \tag{5.84}$$

If, by absurd, (5.79) holds for some $\varepsilon > 0$, then both (5.80) and (5.84) must hold, which is impossible.

Thus, we finally obtain that (5.77) holds for every $\alpha \in \mathbb{R}$. Now by Corollary 5.10 (or simply arguing as in the proof of Proposition 5.49) we conclude that w is Lipschitz. □

Corollary 5.55. *Suppose that the set Ω is a supersolution for the functional $\mathcal{F} + P$, where $\mathcal{F}$ is decreasing with respect to the set inclusion. Then all the Dirichlet eigenfunctions on Ω are Lipschitz continuous.*

Proof. Since $\mathcal{F}$ is a decreasing functional, we have that Ω is also a perimeter supersolution. By Proposition 5.54, we have that w_Ω is Lipschitz. Now since for each $k \in \mathbb{N}$, there is a constant c_k such that $\|u_k\|_{L^\infty} \le C_k$, we have that $|u_k| \le C_k\lambda_k(\Omega)w_\Omega$. Thus, $|u_k(x)| \le C\text{dist}(x, \partial\Omega)$ for some constant $C > 0$, and the conclusion follows by a standard argument as in Corollary 5.10. □

5.7. Subsolutions and supersolutions

We conclude this chapter with a discussion on the combination of the techniques relative to subsolutions and supersolutions. There are several indications that this combination is sufficient to establish the regularity of the boundary of Ω and not only of the state functions on Ω.

Example 5.56. Suppose that Ω is both a subsolution and a supersolution for the functional $E(\Omega) + h(\Omega)$, where $h(\Omega) = \displaystyle\int_\Omega Q^2\,dx$, the weight function $Q : \mathbb{R}^d \to \mathbb{R}$ is smooth and E is the Dirichlet Energy

$$E(\Omega) = \min\left\{\frac{1}{2}\int_{\mathbb{R}^d}|\nabla u|^2\,dx - \int_{\mathbb{R}^d} u\,dx \;:\; u \in H_0^1(\Omega)\right\}.$$

It was proved in [17], using the classical technique of Alt and Caffarelli from [1], that the boundary $\partial\Omega$ is $C^{1,\alpha}$, for $\alpha \in (0, 1)$.

We note that the regularity of the function Q plays a fundamental role in the proof of this result in [1]. If Q is only measurable function such that $0 < \varepsilon \leq Q \leq \varepsilon^{-1}$, then the regularity of the boundary $\partial\Omega$ (if any!) is not known. More precisely, we state here the following:

Conjecture 5.57. Suppose that $0 < m \leq M < +\infty$ are two constants and suppose that the set Ω is a subsolution for $E + m|\cdot|$ and supersolution for $E + M|\cdot|$. Then the boundary $\partial\Omega$ is locally a graph of a Lipschitz function.

In this section we prove an analogous result for measurable sets Ω, which are subsolutions for $\widetilde{E} + mP$ and supersolutions for $\widetilde{E} + MP$, where P is the De Giorgi perimeter and $\widetilde{E}$ is the Dirichlet Energy

$$\widetilde{E}(\Omega) = \min\left\{\frac{1}{2}\int_{\mathbb{R}^d} |\nabla u|^2\,dx - \int_{\mathbb{R}^d} u\,dx \,:\, u \in \widetilde{H}^1_0(\Omega)\right\}.$$

The presence of the perimeter in the functional allows us to use the classical regularity theory of the quasi-minimizers of the perimeter, which considerably facilitates our task of achieving some regularity for Ω.

Remark 5.58. Suppose that the measurable set Ω is a supersolution for $\widetilde{E} + MP$. Then, by Remark 5.43 Ω is a perimeter supersolution. Thus, we may restrict our attention to sets, which are subsolutions for $\widetilde{E} + mP$ and supersolutions for the perimeter.

Theorem 5.59. *Let $\Omega \subset \mathbb{R}^d$ be a set of finite Lebesgue measure and finite perimeter. If Ω is an energy subsolution and a perimeter supersolution, then Ω is a bounded open set and its boundary is $C^{1,\alpha}$ for every $\alpha \in (0, 1)$ outside a closed set of dimension $d - 8$.*

Proof. First notice that, by Lemma 4.58, Ω is bounded. Moreover, since Ω is a perimeter supersolution, we can apply Proposition 5.48 and Proposition 5.54, obtaining that Ω is an open set and the energy function $w := w_\Omega$ is Lipschitz.

We now divide the proof in two steps.

Step 1 ($C^{1,\alpha}$ regularity up to $\alpha < 1/2$). Let $x_0 \in \partial\Omega$ and let $B_r(x_0)$ be a ball of radius less than 1. By Lemma 3.126, for each $\widetilde{\Omega} \subset \Omega$, such that $\widetilde{\Omega}\Delta\Omega \subset B_r(x_0)$, the subminimality of Ω implies (for $r \leq 1$)

$$\begin{aligned} m\big(P(\Omega) - P(\widetilde{\Omega})\big) &\leq \int_{B_r(x_0)} w\,dx \\ &\quad + C_d\left(r + \frac{\|w\|_{L^\infty(B_{2r}(x_0))}}{r}\right)\int_{\partial B_r(x_0)} w\,d\mathcal{H}^{d-1} \qquad (5.85) \\ &\leq C_d\|w\|_{L^\infty(B_{2r}(x_0))} r^{d-1}, \end{aligned}$$

where C_d is a dimensional constant. Now since w is Lipschitz and vanishes on $\partial\Omega$, we have $\|w\|_{L^\infty(B_{2r}(x_0))} \leq Cr$, hence equation (5.85), implies

$$P\big(\Omega, B_r(x_0)\big) \leq P\big(\widetilde{\Omega}, B_r(x_0)\big) + Cr^d, \tag{5.86}$$

where C depends on the dimension d, the constant m and the Lipschitz constant of w (which, in turn, depends only on the data of the problem). Moreover, by the perimeter subminimality, equation (5.86) clearly holds true also for outer variations. Splitting every local variation $\widetilde{\Omega}$ of Ω in an outer and inner variations, we obtain

$$\begin{aligned} P(\Omega, B_r) - P(\widetilde{\Omega}, B_r) &= P(\Omega, B_r) \\ &\quad - \Big(P(\widetilde{\Omega} \cup \Omega, B_r) + P(\widetilde{\Omega} \cap \Omega, B_r) - P(\Omega, B_r)\Big) \\ &\leq P(\Omega, B_r) - P(\Omega \cap \widetilde{\Omega}, B_r) \\ &\leq Cr^d, \text{ for every } \widetilde{\Omega} \subset \mathbb{R}^d \text{ such that } \widetilde{\Omega} \Delta \Omega \subset B_r. \end{aligned}$$

Hence Ω is a *almost-minimizer* for the perimeter in the sense of [90, 91]. From this it follows that $\partial\Omega$ is a $C^{1,\alpha}$ manifold, outside a closed *singular set* Σ of dimension $(d-8)$, for every $\alpha \in (0, 1/2)$.

• *Step* 2. We want to improve the exponent of Hölder continuity of the normal of $\partial\Omega$ in the regular (*i.e.* non-singular) points of the boundary. For this notice that, for every regular point $x_0 \in \partial\Omega$, there exists a radius r such that $\partial\Omega$ can be represented by the graph of a C^1 function ϕ in $B_r(x_0)$, that is, up to a rotation of coordinates

$$\Omega \cap B_r(x_0) = \big\{x_d > \phi(x_1, \dots, x_{d-1})\big\} \cap B_r(x_0).$$

For every $T \in C^1_c(B_r(x_0); \mathbb{R}^d)$ such that $T \cdot \nu_\Omega < 0$ and t is sufficiently small, we consider the local variation

$$\Omega_t = \big(\mathrm{Id} + tT\big)(\Omega) \subset \Omega.$$

By the energy subminimality we obtain

$$m\big(P(\Omega) - P(\Omega_t)\big) \leq E(\Omega_t) - E(\Omega). \tag{5.87}$$

Since T is supported in B_r and $\partial\Omega \cap B_r$ is C^1, we can perform the same computations as in [72, Chapter 5], to obtain that

$$E(\Omega_t) - E(\Omega) = -t \int_{\partial\Omega \cap B_r} \left|\frac{\partial w_\Omega}{\partial \nu}\right|^2 T \cdot \nu_\Omega \, d\mathcal{H}^{d-1} + o(t). \tag{5.88}$$

Moreover, see for instance [80, Theorem 17.5],

$$P(\Omega_t) = P(\Omega) + t \int_{\partial\Omega \cap B_r} \mathrm{div}_{\partial\Omega} T \, d\mathcal{H}^{d-1} + o(t) \tag{5.89}$$

where $\mathrm{div}_{\partial\Omega} T$ is the *tangential divergence* of T. Plugging (5.88) and (5.89) in (5.87), a standard computation (see [80, Theorem 11.8]), gives (in the distributional sense)

$$\mathrm{div}\left(\frac{\nabla\phi}{\sqrt{1+|\nabla\phi|^2}}\right) \leq \frac{1}{m}\left|\frac{\partial w_\Omega}{\partial \nu}\right|^2 \leq C,$$

where the last inequality is due to the Lipschitz continuity of w_Ω. Moreover applying (5.89) to outer variations of Ω (*i.e.* to variations such that $T \cdot \nu_\Omega > 0$) we get

$$\mathrm{div}\left(\frac{\nabla\phi}{\sqrt{1+|\nabla\phi|^2}}\right) \geq 0.$$

In conclusion ϕ is a C^1 function satisfying

$$\mathrm{div}\left(\frac{\nabla\phi}{\sqrt{1+|\nabla\phi|^2}}\right) \in L^\infty,$$

and classical elliptic regularity gives $\phi \in C^{1,\alpha}$, for every $\alpha \in (0, 1)$. □

Chapter 6
Spectral optimization problems in $\mathbb{R}^d$

6.1. Optimal sets for the kth eigenvalue of the Dirichlet Laplacian

The aim of this section is to study the optimal sets for functionals depending on the eigenvalues of the Dirichlet Laplacian. A typical example is the model problem

$$\min\Big\{\lambda_k(\Omega)\ :\ \Omega\subset\mathbb{R}^d,\ \Omega \text{ quasi-open },\ |\Omega|=c\Big\},\tag{6.1}$$

where $c>0$ is a given constant. The existence of an optimal set for the problem (6.1) was proved recently by Bucur (see [20]) and by Mazzoleni and Pratelli (see [81])two completely different techniques.

In [81] the authors reason on the minimizing sequence, proving that by modifying each set in an appropriate way, one can find another minimizing sequence composed of uniformly bounded sets. At this point the classical Buttazzo-Dal Maso theorem (see Theorem 2.82) can be applied.

The argument in [20] is based on a concentration-compactness principle in combination with an induction on k. The boundedness of the optimal set is fundamental for this argument and is obtained using the notion of energy subsolutions. We note that this technique can easily be generalized and applied to other situations (optimization of potentials, capacitary measures, etc). The price to pay is the fact that some minor restrictions are needed on the spectral functional. More precisely, for the penalized version of the problem it is required that the spectral functional is Lipschitz with respect to the eigenvalues involved, while in [81] was shown in the case of domains this assumption can be dropped.

We note that by a simple rescaling argument (see Remark 6.3), the problem (6.1) is equivalent to

$$\min\Big\{\lambda_k(\Omega)+m|\Omega|\ :\ \Omega\subset\mathbb{R}^d,\ \Omega \text{ quasi-open}\Big\},\tag{6.2}$$

for some positive constant m, which we call *Largange multiplier*. For general spectral functionals of the form

$$\mathcal{F}(\Omega) = F\big(\lambda_{k_1}(\Omega), \dots, \lambda_{k_p}(\Omega)\big),$$

the Lagrange multiplier problem is easier to threat, due to the fact that any quasi-open set can be used to test (6.2). The connection between the optimization problem at fixed measure and the penalized one is, in general, a technically difficult question; further complications appear if we optimize under additional geometric constraints.

Our first result in this section concerns the existence of an optimal set for the problem (6.2). Our result is more general and concerns shape optimization problems of the form

$$\min\Big\{F\big(\lambda_{k_1}(\Omega), \dots, \lambda_{k_p}(\Omega)\big) + |\Omega| : \ \Omega \subset \mathcal{D}, \ \Omega \text{ quasi-open}\Big\}, \tag{6.3}$$

where $k_1, \dots, k_p \in \mathbb{N}$ and the function $F : \mathbb{R}^p \to \mathbb{R}$ satisfies some mild monotonicity and continuity assumptions. More precisely we work with functionals satisfying the following definition.

Definition 6.1. We will say that the function $F : \mathbb{R}^p \to \mathbb{R}$ is:

- *increasing*, if for each $x \geq y \in \mathbb{R}^p$, we have that $F(x) \geq F(y)$[1];
- *diverging at infinity*, if $\lim_{x \to +\infty} F(x) = +\infty$. More precisely, $\lim_{n\to\infty} F(x_n) \to \infty$, for every sequence $x_n = (x_n^1, \dots, x_n^p) \in \mathbb{R}^p$ such that $\lim_{n\to\infty} x_n^i = +\infty$, for every $i = 1, \dots, p$.
- *increasing with growth at least* $a > 0$, if F is increasing and the constant $a > 0$ is such that, for every $x \geq y$, we have

$$F(x) - F(y) \geq a|x - y|.$$

Theorem 6.2. *Consider the set* $\{k_1, \dots, k_p\} \subset \mathbb{N}$ *and let* $F : \mathbb{R}^k \to \mathbb{R}$ *be an increasing and locally Lipschitz function diverging at infinity. Then there exists a quasi-open set, solution of the problem* (6.3). *Moreover, under the above assumptions on* F, *every solution of* (6.3) *is a bounded set of finite perimeter.*

If, furthermore, the function F *is increasing with growth rate at least* $a > 0$, *then for every optimal set* Ω, *there are orthonormal and Lipschitz continuous eigenfunctions* $u_{k_1}, \dots, u_{k_p} \in H_0^1(\Omega)$, *corresponding to the eigenvalues* $\lambda_{k_1}(\Omega), \dots, \lambda_{k_p}(\Omega)$.

[1] We say that $x = (x_1, \dots, x_p) \geq y = (y_1, \dots, y_p)$, if. $x_j \geq y_j$ for every $j = 1, \dots, p$.

Proof. Let Ω_n be a minimizing sequence for (6.3) in $\mathbb{R}^d$. By the Buttazzo-Dal Maso Theorem 2.82, for every $n \in \mathbb{N}$, there is a solution Ω_n^* of the problem

$$\min\Big\{F\big(\lambda_1(\Omega),\dots,\lambda_k(\Omega)\big)+|\Omega|:\ \Omega\subset\Omega_n,\ \Omega \text{ quasi-open}\Big\}. \quad (6.4)$$

We now note that

- the sequence Ω_n^* is still a minimizing sequence for 6.3;
- each Ω_n^* is a subsolution for the functional $F(\lambda_1,\dots,\lambda_k)+|\cdot|$.

By Theorem 4.46 Ω_n^* is a subsolution for $E(\Omega)+m|\Omega|$, where the constants m and ε from Definition 4.10 depend only on f, d and $\lambda_k(\Omega_n^*)$. Thus, by Lemma 4.17, we can cover Ω_n^* by N balls of radius r, where N and r do not depend on $n \in \mathbb{N}$. We can now translate the different clusters of balls and the corresponding components of Ω_n^* obtaining sets $\widetilde{\Omega}_n^*$ with the same spectrum and measure as Ω_n^*, for which there is some $R>0$ such that $\operatorname{diam}(\widetilde{\Omega}_n^*)<R$, for some R not depending on $n \in \mathbb{N}$. After an appropriate translation we can suppose $\widetilde{\Omega}_n^* \subset B_R$. Applying the Buttazzo-Dal Maso Theorem, we obtain the existence of a solution Ω of (6.3).

For the boundedness and the finiteness of the perimeter of the optimal sets, we note that by Theorem 4.46 any optimal set is an energy subsolution and so, it is sufficient to apply Theorem 4.22.

The existence of Lipschitz continuous eigenfunctions follows by Theorem 5.41. □

We now consider the spectral optimization problems at fixed measure

$$\min\Big\{F\big(\lambda_{k_1}(\Omega),\dots,\lambda_{k_p}(\Omega)\big):\ \Omega\subset\mathbb{R}^d,\ \Omega \text{ quasi-open},\ |\Omega|=c\Big\}, \quad (6.5)$$

where the constant $c>0$, the function $F:\mathbb{R}^p\to\mathbb{R}$ and $k_1,\dots,k_p\in\mathbb{N}$ are given. Before we continue with our main existence and regularity result in this case, we make some considerations in the case when the functionals involved are homogeneous. The following Proposition 6.3 holds in the following very general setting, in which are given:

- a family $\mathcal{A}$ of subsets of the Euclidean space $\mathbb{R}^d$ such that if $\Omega\in\mathcal{A}$, then also $t\Omega\in\mathcal{A}$, for every $t>0$;
- a positive functional $\mathcal{G}:\mathcal{A}\to(0,+\infty)$, which is β-homogeneous for some $\beta\in\mathbb{R}$, $\beta\neq 0$, *i.e.*

$$\mathcal{G}(t\Omega)=t^\beta\mathcal{G}(\Omega),\qquad \forall t>0,\quad \forall\Omega\in\mathcal{A};$$

- a functional $\mathcal{F} : \mathcal{A} \to \mathbb{R}$ which is α-homogeneous for some $\alpha \in \mathbb{R}$, $\alpha \neq 0$, *i.e.*

$$\mathcal{F}(t\Omega) = t^{\alpha}\mathcal{F}(\Omega), \qquad \forall t > 0, \quad \forall \Omega \in \mathcal{A}.$$

Proposition 6.3. *Let the family of subsets $\mathcal{A}$ and the functionals $\mathcal{F}$ and $\mathcal{G}$ be as above. Then the set $\Omega^* \in \mathcal{A}$ is a solution of the problem*

$$\min\Big\{\mathcal{F}(\Omega) + \Lambda\mathcal{G}(\Omega) : \ \Omega \in \mathcal{A}\Big\}, \tag{6.6}$$

if and only if, Ω^* is a solution of

$$\min\Big\{\mathcal{F}(\Omega) : \ \Omega \in \mathcal{A}, \ \mathcal{G}(\Omega) = \mathcal{G}(\Omega^*)\Big\}, \tag{6.7}$$

and the real function $f : (0, +\infty) \to \mathbb{R}$, given by

$$f(t) = t^{\alpha}\mathcal{F}(\Omega^*) + t^{\beta}\Lambda\mathcal{G}(\Omega^*),$$

has minimum in $t = 1$.

Proof. If $\Omega*$ is a solution of (6.6), then the claim follows by the fact that one can choose the sets $t\Omega^*$ as competitors in (6.6), as well as the sets Ω such that $\mathcal{G}(\Omega) = \mathcal{G}(\Omega^*)$.

For the converse claim, suppose that $\Omega \in \mathcal{A}$ and $t > 0$ is such that $\mathcal{G}(\Omega) = \mathcal{G}(t\Omega^*)$. By the homogeneity of $\mathcal{F}$ and the fact that Ω^* solves (6.7), we have that $\mathcal{F}(t\Omega^*) \leq \mathcal{F}(\Omega)$. Together with the fact that $f(t)$ achieves its minimum in $t = 1$ we get

$$\mathcal{F}(\Omega^*) + \Lambda\mathcal{G}(\Omega^*) \leq \mathcal{F}(t\Omega^*) + \Lambda\mathcal{G}(t\Omega^*) \leq \mathcal{F}(\Omega) + \Lambda\mathcal{G}(\Omega),$$

which proves that Ω^* minimizes (6.6). $\square$

Proposition 6.4. *Let the family of sets $\mathcal{A}$ and the functionals $\mathcal{F}$ and $\mathcal{G}$ be as in Proposition 6.3.*

(1) *If $\Omega^* \in \mathcal{A}$ is a solution of* (6.6), *for some $\Lambda \in \mathbb{R}$, then the set $\big(c/\mathcal{G}(\Omega^*)\big)^{1/\beta}\Omega^*$ is a solution of the problem*

$$\min\Big\{\mathcal{F}(\Omega) : \ \Omega \in \mathcal{A}, \ \mathcal{G}(\Omega) = c\Big\}. \tag{6.8}$$

(2) *If $\mathcal{F}$ is a positive functional, $\alpha\beta < 0$ and Ω^* is a solution of* (6.8), *then Ω^* is a solution of* (6.6) *with Lagrange multiplier $\Lambda = -\frac{\alpha\mathcal{F}(\Omega^*)}{\beta\mathcal{G}(\Omega^*)}$.*

Proof. For the first claim (1), let $t = \left(c/\mathcal{G}(\Omega^*)\right)^{1/\beta}$ and Ω be such that $\mathcal{G}(\Omega) = c$. Then $\mathcal{G}(t^{-1}\Omega) = \mathcal{G}(\Omega^*)$ and, by the optimality of Ω^*,

$$\mathcal{F}(\Omega) = t^\alpha \mathcal{F}(t^{-1}\Omega) \geq t^\alpha \mathcal{F}(\Omega^*) = \mathcal{F}(t\Omega^*),$$

which gives (1). In order to prove (2), we note that $t = 1$ is the unique minimizer of $f(t) = t^\alpha \mathcal{F}(\Omega^*) + \Lambda t^\beta \mathcal{G}(\Omega^*)$ and then apply Proposition 6.3. □

Example 6.5. If $\mathcal{A}$ is the family of quasi-open sets in $\mathbb{R}^d$, $\mathcal{F}(\Omega) = \lambda_k(\Omega)$ and $\mathcal{G}(\Omega) = |\Omega|$, we have that $\alpha = -2$, $\beta = d$ and the two problems (6.8) and (6.6) correspond, respectively, to (6.1) and (6.2).

If the functional $\mathcal{F}$ is not homogeneous, the question is more involved and, in general, there is no Lagrange multiplier Λ, that allows to transform the problem (6.7) into (6.6). For functionals of the form $\mathcal{F} = F(\lambda_{k_1}, \dots, \lambda_{k_p})$, we have the following result, which allows to apply the results from Chapters 4 and 5.

Proposition 6.6. *Let $\mathcal{G}$ be a positive and β-homogeneous functional. Suppose that the function $F : \mathbb{R}^p \to \mathbb{R}$ is increasing, locally Lipschitz continuous and with growth at least $a > 0$. Then, for every solution Ω of the problem*

$$\min\left\{F\left(\lambda_{k_1}(\Omega), \dots, \lambda_{k_p}(\Omega)\right) : \Omega \subset \mathbb{R}^d,\ \mathcal{G}(\Omega) = 1\right\}, \tag{6.9}$$

there are constants m and M such that Ω is a local (with respect to the distance d_γ) subsolution for the functional

$$F\left(\lambda_{k_1}(\Omega), \dots, \lambda_{k_p}(\Omega)\right) + m\mathcal{G}(\Omega),$$

and supersolution for $\mathcal{G}$ and for the functional

$$F\left(\lambda_{k_1}(\Omega), \dots, \lambda_{k_p}(\Omega)\right) + M\mathcal{G}(\Omega).$$

Proof. We first prove that Ω is a subsolution. Indeed, suppose that $U \subset \Omega$ and let $t = (\mathcal{G}(\Omega)/\mathcal{G}(U))^{1/\beta}$. We note that $\mathcal{G}(tU) = G(\Omega)$ and so tU can be used to test the optimality of Ω. Suppose that $t \leq 1$, *i.e.* $\mathcal{G}(U) \geq \mathcal{G}(\Omega)$. Then the inequality

$$F\left(\lambda_{k_1}(\Omega), \dots, \lambda_{k_p}(\Omega)\right) + m\mathcal{G}(\Omega) \leq F\left(\lambda_{k_1}(U), \dots, \lambda_{k_p}(U)\right) + m\mathcal{G}(U),$$

trivially holds for any $m > 0$.

Suppose that $t > 1$, *i.e.* $\mathcal{G}(U) < \mathcal{G}(\Omega)$. By the optimality of Ω, properties $(f2)$, $(f3)$, the trivial scaling properties of the eigenvalues and of the perimeter and the monotonicty of eigenvalues with respect to set inclusion, we obtain

$$\begin{aligned}
0 &\le F\left(\lambda_{k_1}(tU), \ldots, \lambda_{k_p}(tU)\right) - F\left(\lambda_{k_1}(\Omega), \ldots, \lambda_{k_p}(\Omega)\right) \\
&= F\left(\lambda_{k_1}(tU), \ldots, \lambda_{k_p}(tU)\right) - F\left(\lambda_{k_1}(U), \ldots, \lambda_{k_p}(U)\right) \\
&\qquad + F\left(\lambda_{k_1}(U), \ldots, \lambda_{k_p}(U)\right) - F\left(\lambda_{k_1}(\Omega), \ldots, \lambda_{k_p}(\Omega)\right) \\
&\le a(t^{-2} - 1)\left|\left(\lambda_{k_1}(U), \ldots, \lambda_{k_p}(U)\right)\right| \\
&\qquad + F\left(\lambda_{k_1}(U), \ldots, \lambda_{k_p}(U)\right) - F\left(\lambda_{k_1}(\Omega), \ldots, \lambda_{k_p}(\Omega)\right) \\
&\le a\left(\mathcal{G}(\Omega)\right)^{-\frac{2}{\beta}}\left(\mathcal{G}(U)^{\frac{2}{\beta}} - \mathcal{G}(\Omega)^{\frac{2}{\beta}}\right)\left|\left(\lambda_{k_1}(U), \ldots, \lambda_{k_p}(U)\right)\right| \\
&\qquad + F\left(\lambda_{k_1}(U), \ldots, \lambda_{k_p}(U)\right) - F\left(\lambda_{k_1}(\Omega), \ldots, \lambda_{k_p}(\Omega)\right)
\end{aligned}$$

where L is the (local) Lipschitz constant of f and $a > 0$ is the lower on the growth of F. Using the concavity of the function $z \mapsto z^{\frac{2}{\beta}}$ if $\beta < 2$, or the fact that $\mathcal{G}(U) < \mathcal{G}(\Omega)$ if $\beta \ge 2$, we can bound

$$\mathcal{G}(U)^{\frac{2}{\beta}} - \mathcal{G}(\Omega)^{\frac{2}{\beta}} \le C(\Omega)\left(\mathcal{G}(U) - \mathcal{G}(\Omega)\right),$$

which concludes the first part of the proof.

Consider the set $\widetilde{\Omega} \supset \Omega$. We first note that $\mathcal{G}(\widetilde{\Omega}) \ge \mathcal{G}(\Omega)$. Indeed, if this is not the case, we have

$$t := \left(\mathcal{G}(\Omega)/\mathcal{G}(\widetilde{\Omega})\right)^{1/\beta} > 1,$$

snd so, for any $k \in \mathbb{N}$, we have

$$\lambda_k(t\widetilde{\Omega}) < \lambda_k(\widetilde{\Omega}) \le \lambda_k(\Omega).$$

On the other hand $\mathcal{G}(t\widetilde{\Omega}) = \mathcal{G}(\Omega)$ and so, by the optimality of Ω and the strict monotonicity of F, we have

$$\begin{aligned}
0 &\le f\left(\lambda_{k_1}(t\widetilde{\Omega}), \ldots, \lambda_{k_p}(t\widetilde{\Omega})\right) - f\left(\lambda_{k_1}(\Omega), \ldots, \lambda_{k_p}(\Omega)\right) \\
&< f\left(\lambda_{k_1}(\widetilde{\Omega}), \ldots, \lambda_{k_p}(\widetilde{\Omega})\right) - f\left(\lambda_{k_1}(\Omega), \ldots, \lambda_{k_p}(\Omega)\right) \le 0,
\end{aligned}$$

which is a contradiction and so, we have $G(\widetilde{\Omega}) \geq G(\Omega)$ and $t \leq 1$. We now reason as in the subsolution's case.

$$\begin{aligned}
0 &\leq F\big(\lambda_{k_1}(t\widetilde{\Omega}), \dots, \lambda_{k_p}(t\widetilde{\Omega})\big) - F\big(\lambda_{k_1}(\Omega), \dots, \lambda_{k_p}(\Omega)\big) \\
&= F\big(\lambda_{k_1}(t\widetilde{\Omega}), \dots, \lambda_{k_p}(t\widetilde{\Omega})\big) - F\big(\lambda_{k_1}(\widetilde{\Omega}), \dots, \lambda_{k_p}(\widetilde{\Omega})\big) \\
&\quad + F\big(\lambda_{k_1}(\widetilde{\Omega}), \dots, \lambda_{k_p}(\widetilde{\Omega})\big) - F\big(\lambda_{k_1}(\Omega), \dots, \lambda_{k_p}(\Omega)\big) \\
&\leq L(t^{-2} - 1)\,\big|\big(\lambda_{k_1}(\widetilde{\Omega}), \dots, \lambda_{k_p}(\widetilde{\Omega})\big)\big| \\
&\quad + F\big(\lambda_{k_1}(\widetilde{\Omega}), \dots, \lambda_{k_p}(\widetilde{\Omega})\big) - F\big(\lambda_{k_1}(\Omega), \dots, \lambda_{k_p}(\Omega)\big) \\
&\leq L(\mathcal{G}(\Omega))^{-\frac{2}{\beta}}\Big(\mathcal{G}(\widetilde{\Omega})^{\frac{2}{\beta}} - \mathcal{G}(\Omega)^{\frac{2}{\beta}}\Big)\big|\big(\lambda_{k_1}(\Omega), \dots, \lambda_{k_p}(\Omega)\big)\big| \\
&\quad + F\big(\lambda_{k_1}(\widetilde{\Omega}), \dots, \lambda_{k_p}(\widetilde{\Omega})\big) - F\big(\lambda_{k_1}(\Omega), \dots, \lambda_{k_p}(\Omega)\big),
\end{aligned}$$

where L is the Lipschitz constant of f. Now the conclusions follows estimating the difference $\mathcal{G}(\widetilde{\Omega})^{\frac{2}{\beta}} - \mathcal{G}(\Omega)^{\frac{2}{\beta}}$, as in the previous case. □

Remark 6.7. We note that the conclusions of Proposition 6.6 hold also if we substitute $\lambda_{k_1}, \dots, \lambda_{k_p}$ with any p-uple $\mathcal{F}_1, \dots, \mathcal{F}_p$ of functionals, which are positive, decreasing with respect to the inclusion and α-homogeneous, for some $\alpha < 0$.

We are now in position to prove an existence of optimal sets for problems with measure constraint.

Theorem 6.8. *Consider the set* $\{k_1, ..., k_p\} \subset \mathbb{N}$ *and suppose that the function* $f : \mathbb{R}^p \to \mathbb{R}$ *is increasing, locally Lipschitz continuous with growth at least* $a > 0$. *Then there exists a solution of the problem* (6.5). *Moreover, any solution* Ω *of* (6.5) *is a bounded set with finite perimeter and there are orthonormal Lipschitz continuous eigenfunctions* $u_{k_1}, ..., u_{k_p} \in H_0^1(\Omega)$, *corresponding to the eigenvalues* $\lambda_{k_1}(\Omega), ..., \lambda_{k_p}(\Omega)$.

Proof. We argue by induction on the number of variables p. If $p = 1$, then thanks to the monotonicity of f, any solution of (6.2) is also a solution of (6.5) and so we have the claim by Theorem 6.2 and Remark 6.3.

Consider now the functional

$$\mathcal{F}(\Omega) = F\big(\lambda_{k_1}(\Omega), \dots, \lambda_{k_p}(\Omega)\big),$$

and let Ω_n be a minimizing sequence for (6.5). We now apply the quasi-open version (see Remark 3.132) of Theorem 3.131 to the sequence Ω_n. Note that the vanishing (Theorem 3.131 (ii)) cannot occur since the sequence $\big(\lambda_{k_1}(\Omega_n), \dots, \lambda_{k_p}(\Omega_n)\big) \in \mathbb{R}^p$ remains bounded. On the other hand, by the translation invariance of λ_k, we can reduce the case Theorem 3.131 (i2) to (i1). Thus we have two possibilities for the sequence Ω_n: *compactness* (i1) and *dichotomy* (iii).

If the *compactness* occurs, then by (i1) and the continuity of f, we have

$$\lim_{n\to\infty} F\big(\lambda_{k_1}(\Omega_n),\dots,\lambda_{k_p}(\Omega_n)\big) = F\big(\lambda_{k_1}(\mu),\dots,\lambda_{k_p}(\mu)\big),$$

where the capacitary measure $\mu \in \mathcal{M}_{\text{cap}}^T(\mathbb{R}^d)$ is the γ-limit of I_{Ω_n}. Let $\Omega := \Omega_\mu$. Then $\mu \geq I_\Omega$ and by the monotonicity of λ_k and f, we have

$$F\big(\lambda_{k_1}(\mu),\dots,\lambda_{k_p}(\mu)\big) \leq F\big(\lambda_{k_1}(\Omega),\dots,\lambda_{k_p}(\Omega)\big).$$

Thus, it is sufficient to note that $|\Omega| \leq c$, which follows since Ω_n weak-γ-converges to Ω and so we can apply Lemma 2.48.

Suppose now that the *dichotomy* occurs. We may suppose that $\Omega_n = A_n \cup B_n$, where the Lebesgue measure of A_n and B_n is uniformly bounded from below and $\text{dist}(A_n, B_n) \to \infty$. Moreover, up to extracting a subsequence, we may suppose that there is some $1 \leq l < p$ and two sets of natural numbers

$$1 \leq \alpha_1 < \dots < \alpha_l \qquad \text{and} \qquad 1 \leq \beta_{l+1} < \dots < \beta_p,$$

such that for every $n \in \mathbb{N}$, we have that the following to sets of real numbers coincide:

$$\big\{\lambda_{\alpha_1}(A_n), ..., \lambda_{\alpha_l}(A_n), \lambda_{\beta_{l+1}}(B_n), ..., \lambda_{\beta_p}(B_n)\big\} = \big\{\lambda_{k_1}(\Omega_n), ..., \lambda_{k_p}(\Omega_n)\big\}.$$

Indeed, if all the eigenvalues of Ω_n are realized by, say, A_n arguing as in the proof of Theorem 6.51 we can construct a strictly better minimizing sequence. Moreover, without loss of generality we may assume that

$$\lambda_{\alpha_i}(A_n) = \lambda_{k_i}(\Omega_n),\ \forall i = 1, ..., l, \ \text{and}\ \lambda_{\beta_j}(B_n) = \lambda_{k_j}(\Omega_n),\ \forall j = l+1, ..., p.$$

We can also suppose that for every i and j, the following limits exist:

$$\lambda^*_{\alpha_i} := \lim_{n\to\infty} \lambda_{\alpha_i}(A_n) \qquad \text{and} \qquad \lambda^*_{\beta_j} := \lim_{n\to\infty} \lambda_{\beta_j}(B_n).$$

By scaling we also have that without loss of generality

$$|A_n| = c_\alpha \qquad \text{and} \qquad |B_n| = c_\beta,$$

where c_α and c_β are fixed positive constants.

Let $F_\alpha : \mathbb{R}^l \to \mathbb{R}$ be the restriction of F to the l-dimensional hyperplane

$$\Big\{(x_1,\dots,x_p) \in \mathbb{R}^p : x_j = \lambda^*_{\beta_j},\ j = l+1,\dots,p\Big\}.$$

Since $l < p$, by the inductive assumption, there is a solution A^* of the problem

$$\min\Big\{F_\alpha\big(\lambda_{\alpha_1}(A),\dots,\lambda_{\alpha_l}(A)\big) : A\subset\mathbb{R}^d,\ A \text{ quasi-open},\ |A|=c_\alpha\Big\},\quad (6.10)$$

and since F is locally Lipschitz, we have

$$\begin{aligned}
&\lim_{n\to\infty} F\big(\lambda_{\alpha_1}(A_n),\dots,\lambda_{\alpha_l}(A_n),\lambda_{\beta_{l+1}}(B_n),\dots,\lambda_{\beta_p}(B_n)\big)\\
&\qquad= \lim_{n\to\infty} F\big(\lambda_{\alpha_1}(A_n),\dots,\lambda_{\alpha_l}(A_n),\lambda^*_{\beta_{l+1}},\dots,\lambda^*_{\beta_p}\big)\\
&\qquad\geq f\big(\lambda_{\alpha_1}(A^*),\dots,\lambda_{\alpha_l}(A^*),\lambda^*_{\beta_{l+1}},\dots,\lambda^*_{\beta_p}\big)\\
&\qquad= \lim_{n\to\infty} F\big(\lambda_{\alpha_1}(A^*),\dots,\lambda_{\alpha_l}(A^*),\lambda_{\beta_{l+1}}(B_n),\dots,\lambda_{\beta_p}(B_n)\big),
\end{aligned}$$

and thus the minimum in (6.10) is smaller than the infimum in (6.5). Moreover, A^* is bounded and so, up to translating B_n, we may suppose that $\mathrm{dist}(A^*, B_n) > 0$, for all $n \in \mathbb{N}$. Thus, the sequence $A^* \cup B_n$ is minimizing for (6.5).

Let now $F_\beta : \mathbb{R}^{p-l} \to \mathbb{R}$ be the restriction of F to the $(p-l)$-dimensional hyperplane

$$\Big\{(x_1,\dots,x_p)\in\mathbb{R}^p : x_i = \lambda_{\alpha_i}(A^*),\ i=1,\dots,l\Big\},$$

and let B^* be a solution of the problem

$$\min\Big\{F_\beta\big(\lambda_{\beta_{l+1}}(B),\dots,\lambda_{\beta_p}(B)\big) : B\subset\mathbb{R}^d,\ B \text{ quasi-open},\ |B|=c_\beta\Big\}.\quad (6.11)$$

Clearly the minimum in (6.11) is smaller than the minimum in (6.10) and so than that in (6.5). On the other hand, since both A^* and B^* are bounded and the functionals we consider are translation invariant, we may suppose that $\mathrm{dist}(A^*, B^*) > 0$. Thus the set $\Omega^* := A^* \cup B^*$ is a solution of (6.5).

In order to prove the boundedness of a generic optimal set Ω and the finiteness of its perimeter, we first note that, by Proposition 6.6 with $\mathcal{G}(\Omega) = |\Omega|$, we have that that Ω is a subsolution for the functional $F\big(\lambda_{k_1},\dots,\lambda_{k_p}\big) + |\cdot|$. Thus, by Theorem 4.46, Ω is an energy subsolution an so the claim follows by Theorem 4.22. □

6.2. Spectral optimization problems in a box revisited

In Section 2.4, we proved the Buttazzo-Dal Maso Theorem (see Theorem 2.82), which concerns general decreasing and lower semi-continuous (with respect to the strong-γ-convergence) shape functionals. Here we discuss more deeply the case when the box is an open subset of $\mathbb{R}^d$, proving some additional properties of the optimal sets. We start by noting that the technique from the previous section can be used to easily show that the box $\mathcal{D} \subset \mathbb{R}^d$ need not be bounded or of finite measure in order to have an existence for the problem

$$\min\Big\{F\big(\lambda_{k_1}(\Omega),\dots,\lambda_{k_p}(\Omega)\big)+|\Omega|\ :\ \Omega\subset\mathcal{D},\ \Omega\text{ quasi-open}\Big\}. \tag{6.12}$$

Theorem 6.9. *Suppose that the function $F:\mathbb{R}^p\to\mathbb{R}$ is locally Lipschitz continuous and increasing. Suppose that the open set $\mathcal{D}\subset\mathbb{R}^d$ vanishes at infinity,* i.e. *is such that*

$$\lim_{n\to\infty}\sup_{x\in\mathbb{R}^d}|(\mathcal{D}\setminus B_n)\cap B_R(x)|=0,$$

for every $R>0$. Then there is a solution of (6.12). *Moreover, any solution of* (6.12) *is a bounded quasi-open set of finite perimeter.*

Proof. Consider a minimizing sequence Ω_n and let Ω_n^* be the solution of

$$\min\Big\{F\big(\lambda_1(\Omega),\dots,\lambda_k(\Omega)\big)+|\Omega|\ :\ \Omega\subset\Omega_n,\ \Omega\text{ quasi-open}\Big\}. \tag{6.13}$$

As in Theorem 6.2, we have that each Ω_n^* can be covered by N balls of radius r, where N and r do not depend on $n\in\mathbb{N}$. Let A_n be an open set of at most N balls of radius r such that $\Omega_n^*\subset A_n$. We can suppose that the number of connected components of A_n is constantly equal to $N_C\le N$. Moreover, each connected component A_n^j, for $j=1,\dots,N_C$ is such that $\operatorname{diam}(A_n^j)<R$, for some universal R not depending on n and j. Since Ω_n^* is minimizing, we can also suppose that for each $j=1,\dots,N_C$,

$$\liminf_{n\to\infty}|A_n^j\cap\Omega_n^*|>0.$$

Thus, by the condition (b), the sequence $\operatorname{dist}(0,A_n^j)$ remains bounded as $n\to\infty$. Thus, there is some $\widetilde{R}>0$ such that $\Omega_n^*\subset B_{\widetilde{R}}$ and so, we can apply the Buttazzo-Dal Maso Theorem 2.82, obtaining the existence of an optimal set. The boundedness and the finiteness of the perimeter are again due to Theorem 4.46 and Theorem 4.22. □

Remark 6.10. The problem at fixed measure also admits optimal sets

$$\min\Big\{F\big(\lambda_{k_1}(\Omega),\dots,\lambda_{k_p}(\Omega)\big)\ :\ \Omega\subset\mathcal{D},\ \Omega\text{ quasi-open},\ |\Omega|=c\Big\}, \tag{6.14}$$

when the box $\mathcal{D}$ has finite measure. Since the presence of the external constraint $\mathcal{D}$ can significantly complicate the passage from the problem at fixed measure (6.14) to the penalized problem (6.12). Below we provide an example for an optimal sets (at fixed measure), which is bounded and has infinite perimeter.

Example 6.11. Suppose that $\mathcal{D} = \mathcal{D}_1 \cup \mathcal{D}_2 \subset \mathbb{R}^d$, where

$$\mathcal{D}_1 = \Big\{(x, y) \in \mathbb{R}^d : x > 1, 0 \le y \le 1/x^2\Big\}, \tag{6.15}$$

and $\mathcal{D}_2 = \mathcal{D}_1 + (2, 0)$. Thus, the solution of the problem

$$\min\Big\{\lambda_1(\Omega) : \Omega \subset \mathcal{D},\ \Omega \text{ quasi-open}, |\Omega| = 1\Big\}, \tag{6.16}$$

is one of the sets $\mathcal{D}_1$ or $\mathcal{D}_2$, which are both unbounded with infinite perimeter. A more complicated counter-examples can be given also in the case when $\mathcal{D}$ is connected. In conclusion, we note that this example shows that the analogue of Proposition 6.6 in a box $\mathcal{D}$ is in general false, since the subsolutions for $\lambda_1 + m|\cdot|$ are necessarily bounded sets.

In the rest of this section, we aim to prove some regularity properties of the optimal quasi-sets for low eigenvalues. In particular, we prove that the problem

$$\min\big\{\lambda_k(\Omega) + m|\Omega| : \Omega \subset \mathcal{D},\ \Omega \text{ open}\big\}, \tag{6.17}$$

has solution in the cases $k = 1$ and $k = 2$, when $\mathcal{D}$ is an open set vanishing at infinity. We note that for $\mathcal{D} = \mathbb{R}^d$ this is trivial since the solutions are given, respectively, by a ball (for $k = 1$) and two equal balls (for $k = 2$).

It was first proved in [17] that if $\mathcal{D}$ is open, then every solution of the problem

$$\min\big\{\lambda_1(\Omega) + m|\Omega| : \Omega \subset \mathcal{D},\ \Omega \text{ quasi-open}\big\}, \tag{6.18}$$

is a bounded open set. The analogous problem for higher eigenvalues (even for λ_2) remained open for a long time, the reason being that the available regularity techniques were based on the classical approach by Alt and Caffarelli (see [1]) and can be applied for functionals of energy type.

As far as we know, the first result for higher eigenvalues, was obtained by Michel Pierre who claimed that if $\mathcal{D}$ is an open set of finite measure and Ω is a solution of

$$\min\big\{\lambda_2(\Omega) + m|\Omega| : \Omega \subset \mathcal{D},\ \Omega \text{ quasi-open}\big\}, \tag{6.19}$$

such that $\lambda_2(\Omega) > \lambda_1(\Omega)$, then Ω is (equivalent to) an open set. This, in fact, gives the existence of an open solution of (6.19), provided that the following conjecture holds:

Conjecture 6.12. Suppose that $\mathcal{D}$ is a connected bounded open set. Then any solution of (6.19) is given by two disjoint equal balls or is equivalent in measure to a set Ω such that $\lambda_2(\Omega) > \lambda_1(\Omega)$.[2]

In [29] a direct proof was given to the fact that every solution of (6.19) contains an open set, which is solution of the same problem. It was proved that, if u_2 is a sign-changing second eigenfunction on the optimal quasi-open set Ω, then the two quasi-open level sets $\{u_2 > 0\}$ and $\{u_2 < 0\}$ can be separated by two open sets, in which case regularity results for the problem (6.18) can be applied.

We start discussing the regularity of the optimal quasi-open set for the first eigenvalue of the Dirichlet Laplacian (originally proved in [17]).

Proposition 6.13. *Suppose that the quasi-open set Ω is a solution of the problem* (6.18)*, where $\mathcal{D}$ is an open set. Then Ω is open and the first eigenfunction $u \in H_0^1(\Omega)$ is locally Lipschitz continuous in $\mathcal{D}$. If, moreover, the external constraint $\mathcal{D}$ is such that its energy function $w_{\mathcal{D}}$ is Lipschitz continuous on $\mathbb{R}^d$, then u is also Lipschitz continuous on $\mathbb{R}^d$.*

Proof. We first note that the openness of Ω and the local Lipschitz continuity of u follow by Proposition 5.3. Moreover, as we saw in the proof of Lemma 5.1, there is a constant $C_d > 0$ such that, for every ball $B_r(x_0) \subset \mathcal{D}$, we have

$$\Big(|B_r(x_0) \setminus \Omega| > 0\Big) \quad \Rightarrow \quad \Big(\fint_{\partial B_r(x_0)} u \, d\mathcal{H}^{d-1} \leq mC_d r\Big). \tag{6.20}$$

Suppose now that $w := w_{\mathcal{D}}$ is Lipschitz continuous. Since $u \in L^\infty$, by the maximum principle, there is a constant C such that $u \leq Cw$. Let now $x_0 \in \partial\Omega$ and let $0 < r \leq r_0$. If we have that $B_r(x_0) \subset \mathcal{D}$, then (6.20) holds. If there is $y \in \partial\mathcal{D}$ such that $|x_0 - y| < r$, then $u \leq 2CLr$ on $\partial B_r(x_0)$, where L is the Lipscitz constant of w, and so (6.20) holds again with $2CL$ in place of mC_d. Now the conclusion follows by Corollary 5.10. □

[2] We note that if Ω is a solution of (6.19), then there are disjoint quasi-open sets $\Omega_1, \Omega_2 \subset \Omega$ such that $\Omega_1 \cup \Omega_2$ is also a solution of (6.19) (it is sufficient to take the level sets $\Omega_1 = \{u_2 > 0\}$ and $\Omega_2 = \{u_2 < 0\}$ of the second eigenfunction u_2 on Ω). Our conjecture is based on the supposition that we can add part of the common boundary of Ω_1 and Ω_2, thus obtaining a quasi-connected quasi-open set of the same measure.

Before we proceed, with the study of the problem (6.19), we need a regularity result for the optimal set for λ_1 for fixed measure. The main tool is the following Lemma due to Briançon, Hayouni and Pierre (see [17]).

Lemma 6.14. *Suppose that Ω is a solution of the problem*

$$\min\Big\{\lambda_1(\Omega)\,:\,\Omega\subset\mathcal{D},\ \Omega \textit{ quasi-open},\ |\Omega|=c\Big\},\tag{6.21}$$

where $c\le|\mathcal{D}|$ and $\mathcal{D}$ is a quasi-open set of finite measure. Then, there is some $m>0$ such that Ω is a supersolution for $\lambda_1+m|\cdot|$ in $\mathcal{D}$.

Proof. We will prove that there is some $m>0$ such that Ω is a solution of the problem

$$\min\Big\{\lambda_1(\Omega)+m(|\Omega|-c)^+\,:\,\Omega\subset\mathcal{D},\ \Omega\text{ quasi-open}\Big\}.\tag{6.22}$$

Suppose that Ω_m is a solution of (6.22). We have two case. If $|\Omega_m|\le c$, then we have

$$\begin{aligned}\lambda_1(\Omega_m)&=\lambda_1(\Omega_m)+m(|\Omega_m|-c)^+\le\lambda_1(\Omega)+m(|\Omega|-c)^+\\&=\lambda_1(\Omega)\le\lambda_1(\Omega_m),\end{aligned}$$

and so, all the inequalities are equalities, which gives the optimality of Ω. Suppose that $|\Omega_m|>c$ and let u be the first normalized eigenfunction on Ω_m. Then Ω_m is a local shape subsolution for $\lambda_1+m|\cdot|$ and so, by Theorem 4.48 and the following Remark 4.51, we have

$$\lambda_1(\Omega)\ge\lambda_1(\Omega_m)\ge c_d\sqrt{m}|\Omega_m|^{\frac{d-2}{2d}}\ge c_d\sqrt{m}c^{\frac{d-2}{2d}},$$

which is absurd for m large enough (at least for $d\ge2$, while the case $d=1$ is trivial). □

Corollary 6.15. *Suppose that Ω is a solution of* (6.21), *where $\mathcal{D}\subset\mathbb{R}^d$ is a connected open set of finite measure. Then Ω is an open set and the first eigenfunction u of Ω is locally Lipschitz continuous on $\mathcal{D}$. If, moreover, the energy function $w_{\mathcal{D}}$ is Lipschitz continuous on $\mathbb{R}^d$, then u is also Lipschitz continuous on $\mathbb{R}^d$.*

We are now in position to state our first result concerning the optimal set for λ_2.

Proposition 6.16. *Suppose that $\mathcal{D}\subset\mathbb{R}^d$ is an open set of finite measure and that Ω is a solution of the problem*

$$\min\Big\{\lambda_2(\Omega)+m|\Omega|\,:\,\Omega\subset\mathcal{D},\ \Omega \textit{ quasi-open}\Big\}.\tag{6.23}$$

Then there is an open set $\omega\subset\Omega$, which is also a solution of (6.23).

Proof. Let $u_2 \in H_0^1(\Omega)$ be the second normalized eigenfunction of the Dirichlet Laplacian on Ω. Note that we can assume that u_2 changes sign. Indeed, if $u_2 \geq 0$, then $\Omega = \{u_1 > 0\} \cup \{u_2 > 0\}$ and moreover, by the optimality of Ω, we have $\lambda_1(\{u_1 > 0\}) = \lambda_1(\{u_2 > 0\})$, and so $u_1 - u_2$ is a second eigenfunction which changes sign on Ω. Let now $\Omega_+ = \{u_2 > 0\}$ and $\Omega_- = \{u_2 < 0\}$. Since $\lambda_2(\Omega) = \lambda_2(\Omega_+ \cup \Omega_-)$, we have that $\Omega_+ \cup \Omega_-$ is also a solution of (6.23). Suppose that $\omega \subset \Omega_+$. Then

$$\begin{aligned}\lambda_1(\omega) + |\omega| + |\Omega_-| &= \lambda_2(\omega \cup \Omega_-) + |\omega \cup \Omega_-| \\ &\geq \lambda_2(\Omega_+ \cup \Omega_-) + |\Omega_+ \cup \Omega_-| \\ &= \lambda_1(\Omega_+) + |\Omega_+| + |\Omega_-|,\end{aligned}$$

and so, Ω_+ and, analogously, Ω_- are subsolutions for $\lambda_1 + |\cdot|$ and, as a consequence, energy subsolutions. By Proposition 4.40 there are open sets $\mathcal{D}_+$ and $\mathcal{D}_-$ in $\mathcal{D}$ such that $\Omega_+ \subset \mathcal{D}_+$, $\Omega_- \subset \mathcal{D}_-$, $\Omega_+ \cap \mathcal{D}_- = \emptyset$ and $\Omega_- \cap \mathcal{D}_+ = \emptyset$. We note that Ω_+ is contained in exactly one connected component of $\mathcal{D}_+$. Indeed, if this is not the case, we remove the parts of Ω_+ contained in the other connected components of $\mathcal{D}_+$, thus obtaining a set $\widetilde{\Omega}_+ \cup \Omega_-$ with the same second eigenvalue as $\Omega_+ \cup \Omega_-$ and lower measure. Thus Ω_+ is a solution of

$$\min\Big\{\lambda_1(\Omega) : \Omega \subset \mathcal{D}_+,\ \Omega \text{ quasi-open},\ |\Omega| = |\Omega_+|\Big\},$$

where $\mathcal{D}_+$ is a connected open set. By Corollary 6.15, we get that Ω_+ is open. Analogously, also Ω_- is open, which concludes the proof. □

6.3. Spectral optimization problems with internal constraint

In this section we consider problems of the form

$$\min\Big\{F\big(\lambda_{k_1}(\Omega), \dots, \lambda_{k_p}(\Omega)\big) + |\Omega| : \mathcal{D}^i \subset \Omega \subset \mathbb{R}^d, \Omega \text{ quasi-open}\Big\}, \quad (6.24)$$

where $\{k_1, \dots, k_p\} \subset \mathbb{N}$ and $\mathcal{D}^i \subset \mathbb{R}^d$ is a given quasi-open set[3], to which we usually refer to as *internal constraint*. Before we state our main results we need some preliminary results.

6.3.1. Some tools in the presence of internal constraint

The following is a generalization of the notion of a subsolution

[3] The index i stands for *internal*.

Definition 6.17. Given the quasi-open set A, we say that the quasi-open set Ω is a shape subsolution in A for the functional F if

$$\mathcal{F}(\Omega) \leq \mathcal{F}(\omega), \quad \forall \omega \subset \Omega, \ \omega \text{ quasi-open}, \ \Omega \Delta \omega \subset A. \tag{6.25}$$

We say that Ω is a local shape subsolution, if there is some $\varepsilon > 0$ such that (6.25) holds only for quasi-open sets ω such that $d_\gamma(I_\Omega, I_\omega) < \varepsilon$.

We will use this notion in the presence of internal constraint $\mathcal{D}^i$, taking $A = \mathbb{R}^d \setminus \mathcal{D}^i$. The following Theorems are analogous to (4.22) and Theorem 4.46, so we limit ourselves to state the precise results.

Theorem 6.18. *Suppose that the set Ω is a local shape subsolution in A for the functional $E(\Omega) + m|\Omega|$. Then there are constants $C > 0$ and $r_0 > 0$, depending only on m, d, ε and A, such that for every $0 < r < r_0$, the set $\Omega \cap A_r$ can be covered by Cr^{-d-1} balls of radius r, where $A_r = \{x \in \mathbb{R}^d : dist(x, A) > r\}$. Moreover the perimeter of Ω in A, $P(\Omega; A)$ is finite.*

Theorem 6.19. *Suppose that the set Ω is a shape subsolution in A for the functional*

$$\Omega \mapsto F\big(\lambda_1(\Omega), \ldots, \lambda_k(\Omega)\big) + |\Omega|,$$

where $F : \mathbb{R}^k \to \mathbb{R}$ is a locally Lipschitz function in $\mathbb{R}^k$. Then there are positive constants $m > 0$ and $\varepsilon > 0$, depending only on d, Ω and f, such that Ω is a local shape subsolution in A for the functional $E(\Omega) + m|\Omega|$, where ε is the constant from Definition 6.17.

A fundamental tool allowing to understand the behaviour of a minimizing sequence for (6.24) in $\mathbb{R}^d$ is the concentration-compactness principle for quasi-open sets. We state here the result in the presence of internal constraint.

Theorem 6.20. *Let Ω_n be a sequence of quasi-open sets of uniformly bounded measure, all containing a given non-empty quasi-open set $\mathcal{D}^i$. Then, there exists a subsequence, still denoted by Ω_n, such that one of the following situations occurs.*

(i) ***Compactness.*** *The sequence Ω_n γ-converges to a capacitary measure μ and R_{Ω_n} converges in the uniform operator topology of $L^2(\mathbb{R}^d)$ to R_μ. Moreover, we have that $\mathcal{D}^i \subset \Omega_\mu$.*

(ii) ***Dichotomy.*** *There exists a sequence of subsets $\tilde{\Omega}_n \subseteq \Omega_n$, such that:*

- $\|R_{\Omega_n} - R_{\tilde{\Omega}_n}\|_{\mathcal{L}(L^2(\mathbb{R}^d))} \to 0$;

- *$\tilde{\Omega}_n$ is a union of two disjoint quasi-open sets $\tilde{\Omega}_n = \Omega_n^+ \cup \Omega_n^-$;*
- *$d(\Omega_n^+, \Omega_n^-) \to \infty$;*
- *$\liminf_{n\to\infty} |\Omega_n^\pm| > 0$;*
- *$\limsup_{n\to\infty} |\Omega_n^+ \cap \mathcal{D}^i| = 0$ or $\limsup_{n\to\infty} |\Omega_n^- \cap \mathcal{D}^i| = 0$.*

Proof. Since Ω_n is a sequence of quasi-open sets of uniformly bounded measure we can apply the quasi-open version (see Remark 3.132) of Theorem 3.131. Thus it is sufficient to prove that the compactness at infinity (i2) and the vanishing (ii) cannot occur. Indeed, the vanishing cannot occur, since by the maximum principle we have $w_{\Omega_n} \geq w_{\mathcal{D}^i}$, for every $n \in \mathbb{N}$.

Suppose that we have that compactness at infinity, *i.e.* there is a divergent sequence x_n such that $w_{x_n+\Omega_n}$ converges in $L^1(\mathbb{R}^d)$ (and so, also in $L^2(\mathbb{R}^d)$). We note that the energy function solution $w_{\mathcal{D}^i+x_n}$ is just $w_{\mathcal{D}^i}$ translated by x_n. By the maximum principle, we have that $w_{\Omega_n+x_n} \geq w_{\mathcal{D}^i+x_n}$ and so

$$\int_{\mathbb{R}^d} w_{\mathcal{D}^i+x_n} w_{\Omega_n+x_n}\,dx \geq \int_{\mathbb{R}^d} w_{\mathcal{D}^i}^2\,dx > 0.$$

On the other hand, since $x_n \to \infty$, we have that $w_{\mathcal{D}^i+x_n} \rightharpoonup 0$ weakly in $L^2(\mathbb{R}^d)$. By the strong convergence of $w_{\Omega_n+x_n}$ in $L^2(\mathbb{R}^d)$ we have

$$\int_{\mathbb{R}^d} w_{\mathcal{D}^i+x_n} w_{\Omega_n+x_n}\,dx \to 0,$$

which is a contradiction.

It remains to check that the last claim from the dichotomy case. Indeed, since $d(\Omega_n^+, \Omega_n^-) \to \infty$, we have that one of the sequences of characteristic functions $\mathbb{1}_{\Omega_n^+}$ or $\mathbb{1}_{\Omega_n^-}$ has a subsequence, which converges weakly in $L^2(\mathbb{R}^d)$ to zero. Taking into account that $\mathbb{1}_{\mathcal{D}}^i \in L^2(\mathbb{R}^d)$, we have the claim. □

6.3.2. Existence of an optimal set

We start by a discussion of the case of bounded internal constraint $\mathcal{D}^i$, in which the existence can be obtained in the same manner as in Theorem 6.2.

Let $F : \mathbb{R}^p \to \mathbb{R}$ be a given increasing and locally Lipschitz function which diverges at infinity. Suppose that $\mathcal{D}^i$ is a bounded quasi-open set. Then the problem (6.24) has a solution. Indeed, suppose that Ω_n is a minimizing sequence for (6.24) and, for each $n \in \mathbb{N}$, consider the solution Ω_n^* of the problem

$$\min\Big\{F\big(\lambda_{k_1}(\Omega), \dots, \lambda_{k_p}(\Omega)\big)+|\Omega| : \mathcal{D}^i \subset \Omega \subset \Omega_n,\ \Omega \text{ quasi-open}\Big\}. \quad (6.26)$$

Then Ω_n^* is a subsolution for $F\big(\lambda_{k_1}(\Omega), \ldots, \lambda_{k_p}(\Omega)\big) + |\Omega|$ in B_R^c, where B_R is a ball containing $\mathcal{D}$. By Theorem 6.19, we have that each Ω_n^* is a local shape subsolution in B_R^c for $E(\Omega) + m|\Omega|$, for some universal constant m and so Theorem 6.18 applies. Reasoning as in Theorem 6.2, we can suppose that the sets Ω_n^* are all contained in a ball of sufficiently large radius $\widetilde{R} >> 0$. Applying the Buttazzo-Dal Maso Theorem, we obtain the existence of a solution of (6.24).

We note that this argument works only if the internal constraint $\mathcal{D}^i$ is bounded. The reason is that Theorem 6.18 gives only that we can choose Ω_n to be in the set $\mathcal{D}_R^i = \big\{x \,:\, \mathrm{dist}(x, \mathcal{D}^i) < R\big\}$, for some $R > 0$ large enough. But the set $\mathcal{D}_R^i$ has finite measure only if $\mathcal{D}^i$ is bounded. Thus, for the general case we will use an argument based on the concentration-compactness principle from Theorem 6.20.

In order to prove existence for general internal obstacles $\mathcal{D}^i$, we first consider the problem

$$\min\Big\{\lambda_k(\Omega) + m|\Omega| \,:\, \mathcal{D}^i \subset \Omega \subset \mathbb{R}^d,\ \Omega \text{ quasi-open}\Big\}, \tag{6.27}$$

where $k \in \mathbb{N}$, $m > 0$ and $\mathcal{D}^i \subset \mathbb{R}^d$ is a quasi-open sets. We have the following existence result.

Theorem 6.21. *Let $\mathcal{D}^i \subset \mathbb{R}^d$ be a quasi-open set of finite Lebesgue measure and suppose that the set $\mathbb{R}^d \setminus \overline{\mathcal{D}^i}$ contains a ball of radius R, where $R > 0$ is a constant depending on k, m and d. Then the problem (6.27) has a solution. Moreover, any solution Ω of (6.27) is such that $\Omega \subset (\mathcal{D}^i + B_{\widetilde{R}})$, where $\widetilde{R} > 0$ is a constant depending only $\mathcal{D}^i$, k and m. In particular, if $\mathcal{D}^i$ is bounded the optimal sets are also bounded. Finally, there is an eigenfunction $u_k \in H_0^1(\Omega)$, corresponding to the eigenvalue $\lambda_k(\Omega)$, which is Lipschitz continuous on $\mathbb{R}^d$.*

Proof. We note that in the case $\mathcal{D}^i = \emptyset$ the claim follows by Theorem 6.2. Thus we suppose $0 < |\mathcal{D}^i| < \infty$. We also note that if an optimal set exists, then Theorem 6.18 and Theorem 6.19 give the last claim.

Let Ω_n be a minimizing sequence for (6.27). We apply to Ω_n the concentration-compactness principle 6.20. If the compactness occurs, then we obtain the existence immediately. Thus, we only need to check what happens in the dichotomy case.

We first prove that (b) holds, then the dichotomy is impossible and so we have the existence. In fact, if the dichotomy occurs and Ω_n^+ and Ω_n^- are as in Theore 6.20, then we can suppose that $\mathrm{dist}(0, \Omega_n^-) \to \infty$. But then (b) implies that $\lambda_1(\Omega_n^-) \to \infty$ and so, for n large enough

$$\lambda_k(\Omega_n^+ \cup \Omega_n^-) = \lambda_k(\Omega_n^+) \le \lambda_k(\Omega_n^+ \cup \mathcal{D}^i),$$

which is absurd, since $\liminf_{n\to\infty} |\Omega_n| < \liminf |\Omega_n^+ \cup \mathcal{D}|$.

Suppose now that (a) holds and that we have dichotomy. We also suppose that

$$\lim_{n\to\infty} |\Omega_n^-| = c_- > 0 \qquad \text{and} \qquad \lim_{n\to\infty} |\Omega_n^- \cap \mathcal{D}^i| = 0.$$

Since Ω_n is a minimizing sequence, we can assume:

- $\lambda_k(\Omega_n^+) > \lambda_k(\Omega_n^+ \cup \Omega_n^-)$, since otherwise we would have

$$\begin{aligned}\liminf_{n\to\infty} \lambda_k(\Omega_n^+) + m|\Omega_n^+ \cup \mathcal{D}^i| &\le \liminf_{n\to\infty} \lambda_k(\Omega_n) + m|\Omega_n^+ \cup \mathcal{D}^i| \\ &\le \liminf_{n\to\infty} \lambda_k(\Omega_n) + m|\Omega_n| - mc_-,\end{aligned}$$

  which is a contradiction;
- $\lambda_k(\Omega_n^-) > \lambda_k(\Omega_n^+ \cup \Omega_n^-)$, since otherwise we would have that the disjoint union $\Omega^* \cup \mathcal{D}^i$ is optimal for (6.27), where Ω^* is the optimal set for λ_k with measure constraint c_- placed in such a way that $\Omega^* \cap \mathcal{D}^i = \emptyset$. In the case $k = 1$, this is a contradiction with the minimality. In fact in this case Ω^* is a ball of measure c_- which does not intersect $\mathcal{D}^i$. Taking a ball B of slightly larger measure intersecting $\mathcal{D}^i$, we obtain a better competitor for (6.27).

Thus, we obtained that for $k = 1$ the dichotomy does not appear and so we have the first step of the induction.

For $k > 1$, we can assume that there is some $1 \le l \le k-1$ such that

$$\lambda_k(\Omega_n^+ \cup \Omega_n^-) = \max\Big\{\lambda_{k-l}(\Omega_n^+),\, \lambda_l(\Omega_n^-)\Big\}.$$

Let $(\Omega_n^+)^*$ be the solution of

$$\min\Big\{\lambda_{k-l}(\Omega) + m|\Omega| \,:\, \mathcal{D}^i \subset \Omega \subset \Omega_n^+,\ \Omega \text{ quasi-open}\Big\},$$

and let Ω_-^* be a solution of

$$\min\Big\{\lambda_l(\Omega) \,:\, \Omega \subset \mathbb{R}^d,\ \Omega \text{ quasi-open},\ |\Omega| = c_-\Big\}.$$

By Theorem 6.19 and Theorem 6.18, we have that all $(\Omega_n^+)^*$ can be covered by a finite number of balls of sufficiently small radius. We now translate the connected components of this cover in $\mathbb{R}^d \setminus \overline{\mathcal{D}^i}$, obtaining a set $\widetilde{\Omega}_n^+$ which has the same measure and spectrum as $(\Omega_n^+)^*$ and is contained in $\mathcal{D}^i + B_R$ for some R not depending on n. We now can choose

Ω_-^* in such a way to not intersect any of the sets $\widetilde{\Omega}_n^+$. We claim that the sequence $\widetilde{\Omega}_n^+ \cup \Omega_-^*$ is still minimizing for (6.27). Indeed, we have

$$\begin{aligned}
&\lim_{n\to\infty} \lambda_k(\Omega_n) + m|\Omega_n| \\
&= \lim_{n\to\infty} \lambda_k(\Omega_n^+ \cup \Omega_n^-) + m|\Omega_n^+ \cup \mathcal{D}^i| + m|\Omega_n^-| \\
&= \lim_{n\to\infty} \max\left\{\lambda_{k-l}(\Omega_n^+), \lambda_l(\Omega_n^-)\right\} + m|\Omega_n^+ \cup \mathcal{D}^i| + m|\Omega_n^-| \\
&= \lim_{n\to\infty} \max\left\{\lambda_{k-l}(\Omega_n^+) + m|\Omega_n^+ \cup \mathcal{D}^i|\,,\ \lambda_l(\Omega_n^-) + m|\Omega_n^+ \cup \mathcal{D}^i|\right\} + mc_- \\
&\geq \lim_{n\to\infty} \max\left\{\lambda_{k-l}(\widetilde{\Omega}_n^+) + m|\widetilde{\Omega}_n^+|\,,\ \lambda_l(\Omega_-^*) + m|\widetilde{\Omega}_n^+|\right\} + mc_- \\
&= \lim_{n\to\infty} \max\left\{\lambda_{k-l}(\widetilde{\Omega}_n^+), \lambda_l(\Omega_-^*)\right\} + m|\widetilde{\Omega}_n^+ \cup \Omega_-^*|.
\end{aligned}$$

We now again apply the concentration compactness principle, this time to the sequence Ω_n^+. If Ω_n^+ γ-converges to a capacitary measure μ, then the set $\Omega_\mu \cup \Omega_-^*$ is a solution of (6.27). If we are in the dichotomy case of Theorem 6.20, then we reapply the above argument to the sequence $\widetilde{\Omega}_n^+$, obtaining a minimizing sequence of sets composed of optimal sets for some λ_l in $\mathbb{R}^d$ and a sequence of sets containing $\mathcal{D}^i$ laying at finite distance from the internal constraint $\mathcal{D}^i$. We note that this procedure stops since, as we saw above, the dichotomy in the case $k = 1$ is impossible for minimizing sequences.

The existence of Lipschitz continuous eigenfunction follows by Theorem 5.37. □

We are now in position to state our main result.

Theorem 6.22. *Let $\mathcal{D}^i \subset \mathbb{R}^d$ be quasi-open sets such that $\mathcal{D}^i$ has finite Lebesgue measure and the set $\mathbb{R}^d \setminus \overline{\mathcal{D}^i}$ contains a ball of radius R, where $R > 0$ is a constant depending on k, m and d. Then for every increasing and locally Lipschitz function $F : \mathbb{R}^k \to \mathbb{R}$, the problem* (6.24) *has a solution.*

Any solution Ω of (6.24) *is such that $\Omega \subset (\mathcal{D}^i + B_{\widetilde{R}})$, where $\widetilde{R} > 0$ is a constant depending only $\mathcal{D}^i$, f and m. Moreover, if F has growth bounded from below*[4]*, then there are orthonormal eigenfunctions $u_{k_1}, \ldots$ $\ldots, u_{k_p}$, corresponding to the eigenvalues $\lambda_{k_1}(\Omega), \ldots, \lambda_{k_p}(\Omega)$, which are Lipschitz continuous on $\mathbb{R}^d$.*

[4] Recall that a function $F : \mathbb{R}^p \to \mathbb{R}$ has growth bounded from below, if there is a constant $a > 0$ such that for each $x \geq y \in \mathbb{R}^p$, we have $F(x) - F(y) \geq a|x - y|$.

Proof. The proof follows by induction on the number of variables of F, exactly as in Theorem 6.8, the first step of the induction being proved in Theorem 6.22. The Lipschitz regularity of the eigenfunctions follows by Theorem 5.41. □

Using the same argument we can deal with the fixed measure version of the above results. As we saw in the case of external constraint, the presence of the geometric obstacle makes the passage from the problem at fixed measure to the penalized problem quite complicated. Thus, proving the boundedness of the optimal set, which was one of the fundamental steps in Theorem 6.22 and Theorem 6.8, becomes a difficult and in some cases impossible task. Thus, the existence result for the problem

$$\min\Big\{F\big(\lambda_{k_1}(\Omega),\dots,\lambda_{k_p}(\Omega)\big) : \mathcal{D}^i\subset\Omega\subset\mathbb{R}^d,\ \Omega \text{ quasi-open},\ |\Omega|=c\Big\}, \tag{6.28}$$

relies on the following result.

Proposition 6.23. *Suppose that the internal constraint $\mathcal{D}^i$ satisfies*[5]

$$\limsup_{t\to1^+}\frac{|\mathcal{D}^i\setminus t\mathcal{D}^i|}{t-1}<\infty. \tag{6.29}$$

Suppose that the function $F:\mathbb{R}^p\to\mathbb{R}$ is locally Lipschitz and that there is $a>0$ such that

$$F(x)-F(y)\ge a|x-y|,\ \forall y\ge x\in\mathbb{R}^p.$$

Then every solution of the problem (6.28) *is a shape subsolution for the functional $F\big(\lambda_{k_1},\dots,\lambda_{k_p}\big)+m|\cdot|$, for some $m>0$, depending on a, $\mathcal{D}^i$ and the dimension d.*

Proof. Let Ω be a solution of (6.28). Suppose by contradiction, that for each $\varepsilon>0$, there is some quasi-open set Ω_ε such that $\mathcal{D}^i\subset\Omega_\varepsilon\subset\Omega$,

$$F\big(\lambda_{k_1}(\Omega_\varepsilon),\dots,\lambda_{k_p}(\Omega_\varepsilon)\big)+\varepsilon|\Omega_\varepsilon|<F\big(\lambda_{k_1}(\Omega),\dots,\lambda_{k_p}(\Omega)\big)+\varepsilon|\Omega|, \tag{6.30}$$

and note that by the optimality of Ω we necessarily have $|\Omega\setminus\Omega_\varepsilon|>0$.

By the compactness of the inclusion $H_0^1(\Omega)\subset L^2(\Omega)$, we can suppose, up to a subsequence that Ω_ε γ-converges to some capacitary measure μ, whose regular set Ω_μ is such that

$$|\Omega_\mu|\le\liminf_{\varepsilon\to0}|\Omega_\varepsilon|,$$

[5] This condition is for instance satisfied if $\mathcal{D}^i$ is bounded and Lipschitz, or if $\mathcal{D}^i$ is starshaped.

$$\lambda_k(\Omega_\mu) \le \lambda_k(\mu) = \lim_{\varepsilon\to 0} \lambda_k(\Omega_\varepsilon), \quad \forall k \in \mathbb{N}.$$

Thus, by (6.30) we have that

$$\lambda_{k_1}(\Omega_\mu) = \lambda_{k_1}(\Omega)\ , \ \dots\ , \ \lambda_{k_p}(\Omega_\mu) = \lambda_{k_p}(\Omega).$$

Note that $|\Omega_\mu| = |\Omega| = \lim_{\varepsilon\to 0} |\Omega_\varepsilon|$. Indeed, if this is not the case, then the set $t\Omega_\mu \cup \mathcal{D}^i$, for some $t > 1$ such that $|t\Omega_\varepsilon \cup \mathcal{D}^i| = |\Omega|$, is a better competitor than Ω in (6.28).

Let $\Omega'_\varepsilon = t_\varepsilon \Omega_\varepsilon \cup \mathcal{D}^i$, where t_ε is such that $|\Omega'_\varepsilon| = c$. Then, we have that

$$\begin{aligned}
&F\big(\lambda_{k_1}(\Omega_\varepsilon), \dots, \lambda_{k_p}(\Omega_\varepsilon)\big) + \varepsilon|\Omega_\varepsilon| \\
&< F\big(\lambda_{k_1}(\Omega), \dots, \lambda_{k_p}(\Omega)\big) + \varepsilon|\Omega| \\
&\le F\big(\lambda_{k_1}(\Omega'_\varepsilon), \dots, \lambda_{k_p}(\Omega'_\varepsilon)\big) + \varepsilon|\Omega'_\varepsilon| \\
&\le F\big(\lambda_{k_1}(t_\varepsilon\Omega_\varepsilon), \dots, \lambda_{k_p}(t_\varepsilon\Omega_\varepsilon)\big) + \varepsilon|t_\varepsilon\Omega_\varepsilon \cup \mathcal{D}^i| \\
&\le F\big(t_\varepsilon^{-2}\lambda_{k_1}(\Omega_\varepsilon), \dots, t_\varepsilon^{-2}\lambda_{k_p}(\Omega_\varepsilon)\big) + \varepsilon\left(|t_\varepsilon\Omega_\varepsilon| + |\mathcal{D}^i \setminus t_\varepsilon\Omega_\varepsilon|\right) \\
&\le F\big(t_\varepsilon^{-2}\lambda_{k_1}(\Omega_\varepsilon), \dots, t_\varepsilon^{-2}\lambda_{k_p}(\Omega_\varepsilon)\big) + \varepsilon\left(|t_\varepsilon\Omega_\varepsilon| + |\mathcal{D}^i \setminus t_\varepsilon\mathcal{D}^i|\right),
\end{aligned}$$

and so

$$a\,\frac{t_\varepsilon^2 - 1}{t_\varepsilon^2}\left|\big(\lambda_{k_1}(\Omega_\varepsilon), \dots, \lambda_{k_p}(\Omega_\varepsilon)\big)\right| \le \varepsilon\left((t_\varepsilon^d - 1)|\Omega_\varepsilon| + |\mathcal{D}^i \setminus t_\varepsilon\mathcal{D}^i|\right). \tag{6.31}$$

Passing to the limit as $\varepsilon \to 0$ we have $t_\varepsilon \to 1^+$ and so, by (6.29), there is some constant C such that for ε small enough

$$\left|\big(\lambda_{k_1}(\Omega_\varepsilon), \dots, \lambda_{k_p}(\Omega_\varepsilon)\big)\right| \le \varepsilon C.$$

Passing to the limit as $\varepsilon \to 0$, we have a contradiction. □

As a consequence of this result and the argument from Theorem 6.21 and Theorem 6.8, we have the following:

Theorem 6.24. *Suppose that the function $F:\mathbb{R}^p \to \mathbb{R}$ is locally Lipschitz, diverges at infinity and that there is some $a > 0$ such that*

$$F(x) - F(y) \ge a|x - y|, \ \forall y \ge x \in \mathbb{R}^p.$$

Suppose that $\mathcal{D}^i \subset \mathbb{R}^d$ is a quasi-open set such that $\mathbb{R}^d \setminus \overline{\mathcal{D}^i}$ contains a ball of sufficiently large radius and we have

$$\limsup_{t\to 1+} \frac{|\mathcal{D}^i \setminus t\mathcal{D}^i|}{t-1} < \infty.$$

Then the problem (6.28) *has a solution. Moreover, any solution Ω of* (6.28) *is such that $\Omega \subset \mathcal{D}^i + B_{\widetilde{R}}$, where $\widetilde{R} > 0$ is a constant depending only $\mathcal{D}^i$, f and c.*

6.3.3. Existence of open optimal sets for low eigenvalues

In this subsection we prove that the problem

$$\min\left\{\lambda_k(\Omega) + m|\Omega| \,:\, \mathcal{D}^i \subset \Omega \subset \mathbb{R}^d,\ \Omega \text{ open}\right\}, \tag{6.32}$$

admits open solutions for $k = 1, 2$. The case $k = 1$ was treated in [25] by the classical Alt-Caffarelli technique, where was proved that any optimal set is necessarily open. An analogous result for $k = 2$ was, as far as we know, the first complete result concerning the openness of an optimal set for higher eigenvalues. Our approach was inspired by the Pierre's claim for the optimal sets in a box and that the internal obstacle $\mathcal{D}^i$ can be used to glue together the two level sets $\{u_2 < 0\}$ and $\{u_2 > 0\}$ of the second eigenfunction $u_2 \in H_0^1(\Omega)$, thus proving that the optimal set Ω must be (quasi-)connected and so, $\lambda_2(\Omega) > \lambda_1(\Omega)$.

We start discussing the regularity of the optimal quasi-open set for the first eigenvalue of the Dirichlet Laplacian.

Proposition 6.25. *Suppose that the quasi-open set Ω is a solution of the problem*

$$\min\left\{\lambda_1(\Omega) + m|\Omega| \,:\, \mathcal{D}^i \subset \Omega \subset \mathbb{R}^d,\ \Omega \textit{ open}\right\}, \tag{6.33}$$

where $\mathcal{D}^i$ is an open set of finite measure. Then Ω is open and the first eigenfunction $u \in H_0^1(\Omega)$ is Lipschitz continuous on $\mathbb{R}^d$.

Proof. We first note that by Theorem 6.13[6], there is a Lipschitz continuous first eigenfunction $u_1 \in H_0^1(\Omega)$. Then $\Omega = \{u_1 > 0\} \cup \mathcal{D}^i$, which is an open set. □

[6] Alternatively, one may use Proposition 5.3.

Proposition 6.26. *Suppose that Ω is a solution of the problem*

$$\min\Big\{\lambda_2(\Omega)+|\Omega| : \mathcal{D}^i\subset\Omega\subset\mathbb{R}^d,\ \Omega \textit{ quasi-open}\Big\}, \tag{6.34}$$

where $\mathcal{D}^i$ is a connected open set. Then there is an open set $\omega\subset\Omega$, which is also a solution of (6.34).

Proof. Let $u_2\in H_0^1(\Omega)$ be the second normalized eigenfunction of the Dirichlet Laplacian on Ω. Suppose first that u_2 changes sign and consider the set $\omega=\{u_2\neq 0\}\cup\mathcal{D}^i$. If $\lambda_2(\omega)>\lambda_1(\Omega)$, then by Lemma 5.33 we have that u_2 is Lipschitz and so, ω is open. If $\lambda_2(\Omega)=\lambda_1(\Omega)$, then u_2 is also the first eigenfunction on ω and so both u_2^+ and u_2^- are first eigenfunctions. Thus, if $\{u_2>0\}\cap\mathcal{D}^i\neq\emptyset$, by the strong maximum principle on the connected open set $\mathcal{D}^i$, we have that $\mathcal{D}^i\subset\{u_2>0\}$ and by the optimality of ω, $\{u_2<0\}$ is a ball. Thus, we have that

$$\lambda_1(\{u_2>0\})=\lambda_1(\{u_2<0\})=C_d|\{u_2<0\}|^{-d/2},$$

and so, we have that $\{u_2>0\}$ is the solution of

$$\min\Big\{\lambda_1(\Omega)+C_d\lambda_1(\Omega)^{-d/2}+|\Omega| : \mathcal{D}_i\subset\Omega\Big\}.$$

Consider the function $f(t)=t+C_dt^{-d/2}$ and note that its minimum is achieved for $t=\lambda_1(B)$, where B is the ball minimizing $\lambda_1+|\cdot|$ in $\mathbb{R}^d$. If $\{u_2>0\}$ is not a ball, then we have that $f'(\lambda_1(\{u_2>0\}))>0$ and so $\{u_2>0\}$ is a local supersolution for $\lambda_1+m|\cdot|$, for some $m>0$. Thus, applying again Lemma 5.1 as in Proposition 5.4, we have the claim in the case when u_2 changes sign. If $u_2>0$ the argument is the same as in the disconnected case $\lambda_2(\Omega)=\lambda_1(\Omega)$. □

6.3.4. On the convexity of the optimal set for λ_1

Suppose that $\mathcal{D}^i\subset\mathcal{D}\subset\mathbb{R}^d$ are given (quasi-)open sets and let Ω be a solution of

$$\min\Big\{\lambda_1(\Omega) : \mathcal{D}^i\subset\Omega\subset\mathcal{D},\ \Omega \text{ quasi-open},\ |\Omega|=c\Big\}. \tag{6.35}$$

It is natural to ask if some of the qualitative properties of the obstacles $\mathcal{D}^i$ and $\mathcal{D}$ are transferred to the optimal set Ω. The boundedness for example is such a property, *i.e.* if $\mathcal{D}^i$ is bounded, then so is Ω. A long-standing conjecture concerns the convexity of the optimal set.

Conjecture 6.27. Suppose that Ω is a solution of

$$\min\Big\{\lambda_1(\Omega)\,:\,\Omega\subset\mathcal{D},\ \Omega \text{ quasi-open},\ |\Omega|=c\Big\},$$

where the external constraint $\mathcal{D}^e$ is a bounded convex open set. Then Ω is convex.

Here we give a negative answer to the analogous question for a convex internal constraint. More precisely, we prove that a solution Ω of the optimization problem

$$\min\Big\{\lambda_1(\Omega)\,:\,\mathcal{D}^i\subset\Omega\subset\mathbb{R}^2,\ \Omega \text{ quasi-open},\ |\Omega|=c\Big\}, \tag{6.36}$$

might not be convex, even if the constraint $\mathcal{D}^i$ is convex.
Consider the sequence of internal constraints $\mathcal{D}^i_n$, where $\mathcal{D}^i_n=(-\frac{1}{n},\frac{1}{n})\times(-1,1)$ and consider the sequence of optimal sets Ω_n for the problem (6.36) with internal constraint $\mathcal{D}^i_n$.

Proposition 6.28. *For every $c<4/\pi$, there is $N>0$ such that Ω_n is not convex for all $n\geq N$.*

Proof. We begin with some observations on the optimal sets.

1. By a Steiner symmetrization argument, all the sets Ω_n are Steiner symmetric with respect to the axes x and y (in consequence, they are also star-shaped sets).
2. For n large enough, we consider the set $\Omega'_n=D_n\cup B^*(c-\frac{4}{n})$, where for any constant $a>0$, $B^*(a)$ denotes the ball with center in 0 and measure a. By the optimality of Ω_n, we have

$$\lambda_1(\Omega_n)\leq\lambda_1(\Omega'_n)\leq\lambda_1\big(B^*(c-\frac{4}{n})\big).$$

By Theorem 6.20, Ω_n has a γ-converging subsequence, still denoted by Ω_n. Let Ω be the γ-limit of this subsequence. Then

- $\lambda_1(\Omega)\leq\liminf_{n\to\infty}\lambda_1(\Omega_n)\leq\liminf_{n\to\infty}\lambda_1\big(B^*(c-\frac{4}{n})\big)=\lambda_1\big(B^*(c)\big);$
- $|\Omega|\leq\liminf_{n\to\infty}|\Omega_n|=c.$

Using the fact that the ball is the unique minimizer of λ_1 under a measure constraint, we obtain $\Omega=B^*(c)$. Consider now the two small balls B', of center $(0, \sqrt{\frac{c}{\pi}} - \varepsilon)$ and radius ε, and B'', of center $(0, -\sqrt{\frac{c}{\pi}} + \varepsilon)$ and radius ε. Then we have

$$\Omega_n \cap B' \xrightarrow[n\to\infty]{\gamma} \Omega \cap B' = B' \qquad \text{and} \qquad \Omega_n \cap B'' \xrightarrow[n\to\infty]{\gamma} \Omega \cap B'' = B''.$$

Then there is some n large enough such that both sets $B' \cap \Omega_n$ and $B'' \cap \Omega_n$ are non-empty, and Ω_n cannot be convex (see Figure 6.1).

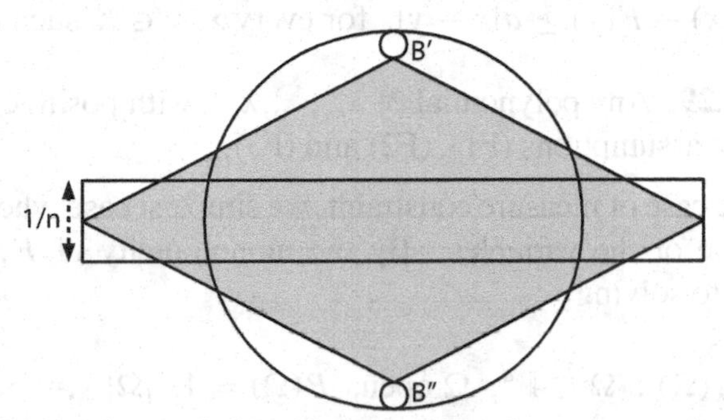

Figure 6.1. Convex internal obstacle does not imply convex optimal set.

In fact, if by contradiction Ω_n was convex, then we should have that the rhombus R with vertices $(-1, 0)$, $(0, -\sqrt{\frac{c}{\pi}} + \varepsilon)$, $(1, 0)$ and $(0, \sqrt{\frac{c}{\pi}} - \varepsilon)$ is contained in Ω_n. But

$$|R| = 2(\sqrt{\frac{c}{\pi}} - \varepsilon) > c,$$

for ε small enough and $c \leq 4/\pi$, which is in contradiction with the measure constraint. □

6.4. Optimal sets for spectral functionals with perimeter constraint

In this section we study the existence and regularity of optimal sets for spectral functionals under a perimeter constraint in $\mathbb{R}^d$. In particular we study the shape optimization problem

$$\min\left\{F\big(\lambda_{k_1}(\Omega), \dots, \lambda_{k_p}(\Omega)\big) : \Omega \subset \mathbb{R}^d, \Omega \text{ open}, P(\Omega) = 1, |\Omega| < \infty\right\}, \tag{6.37}$$

where the function $F : \mathbb{R}^p \to \mathbb{R}$ is such that:

(F1) F is locally Lipschitz continuous;
(F2) F is diverging at infinity, *i.e.* $\lim_{x \to +\infty} F(x) = +\infty$ in sense of Definition 6.1;

(F3) F is increasing with growth at least $a > 0$, in sense of Definition 6.1, on every compact set $K \subset \mathbb{R}^p \setminus \{0\}$, *i.e.* for any $x = (x_1, \dots, x_p) \in \mathbb{R}^p$ and $y = (y_1, \dots, y_p) \in \mathbb{R}^p$ such that $x \geq y$, *i.e.* satisfying $x_j \geq y_j$, for every $j = 1, \dots, p$, we have $F(x) \geq F(y)$. Moreover for every compact set $K \subset \mathbb{R}^d \setminus \{0\}$, there exists a constant $a > 0$ such that

$$F(x) - F(y) \geq a|x - y|, \text{ for every } x, y \in K \text{ such that } x \geq y.$$

Remark 6.29. Any polynomial of $\lambda_{k_1}, \dots, \lambda_{k_p}$, with positive coefficients, satisfies the assumptions (F1), (F2) and (F3).

As in the case of measure constraint, we simplest case when F depends only on one of the variables. By the monotonicity of F, this case is equivalent to solving

$$\min\Big\{\lambda_k(\Omega) : \ \Omega \subset \mathbb{R}^d, \ \Omega \text{ open}, \ P(\Omega) = 1, \ |\Omega| < +\infty\Big\}, \tag{6.38}$$

which, by Remark 6.3, is equivalent to

$$\min\Big\{\lambda_k(\Omega) + mP(\Omega) : \ \Omega \subset \mathbb{R}^d, \ \Omega \text{ open}, \ |\Omega| < +\infty\Big\}, \tag{6.39}$$

for some constant $m > 0$. In this case, we have the following result.

Theorem 6.30. *The shape optimization problem* (6.39) *has a solution. Moreover, any optimal set Ω is bounded and connected. The boundary $\partial\Omega$ is $C^{1,\alpha}$, for every $\alpha \in (0, 1)$, outside a closed set of Hausdorff dimension at most $d - 8$.*

Proof. We prove this theorem in four steps.

Step 1 (*Existence of generalized solution*). We claim that, for any $k \in \mathbb{N}$ and $m > 0$, there exists a solution of the problem

$$\min\Big\{\widetilde{\lambda}_k(\Omega) + mP(\Omega) : \ \Omega \subset \mathbb{R}^d, \ \Omega \text{ measurable}, \ |\Omega| < \infty\Big\}. \tag{6.40}$$

Let Ω_n be a minimizing sequence for (6.40). By the concentration-compactness principle (Theorem 3.131), we have two possibilities for the minimizing sequence: *compactness* and *dichotomy*.

Suppose that the compactness occurs. Since Ω_n is minimizing, there is a constant $C > 0$ such that $P(\Omega_n) \leq C$. Thus we may suppose that $\mathbb{1}_{\Omega_n}$ converges to $\mathbb{1}_\Omega$ in $L^1_{\text{loc}}(\mathbb{R}^d)$ and since $\mathbb{1}_{\Omega_n}$ is concentrated, we have that the convergence takes place in $L^1(\mathbb{R}^d)$ and $P(\Omega) \leq \liminf_{n\to\infty} P(\Omega_n)$.

On the other hand, the sequence of measures $|\Omega_n|$ is also bounded and so the sequence of energy functions w_n, solutions of

$$-\Delta w_n = 1 \quad \text{in} \quad \Omega_n, \qquad w_n \in \widetilde{H}^1_0(\Omega_n),$$

is bounded in $L^\infty(\mathbb{R}^d)$. The sequence $\widetilde{I}_{\Omega_n}$ converges to a capacitary measure μ in $\mathbb{R}^d$, *i.e.* $w_n \to w_\mu$ in $L^1(\mathbb{R}^d)$, where w_μ is the energy function of μ. Since $w_n \le C\mathbb{1}_{\Omega_n}$, for dome universal $C > 0$, we obtain that $w_\mu \le C\mathbb{1}_\Omega$.Thus $\Omega_\mu := \{w_\mu > 0\} \subset \Omega$ and so, $\mu \ge \widetilde{I}_\Omega$, which in turn gives

$$\widetilde{\lambda}_k(\Omega) \le \lambda_k(\mu) = \lim_{n\to\infty} \widetilde{\lambda}_k(\Omega_n).,$$

and so, if the compactness occurs, then Ω is a solution of (6.40).

Suppose now that the dichotomy occurs. Then we may suppose that $\Omega_n = \Omega_n^+ \cup \Omega_n^-$, where $\text{dist}(\Omega_n^+, \Omega_n^-) \ge n$ and

$$P(\Omega_n) = P(\Omega_n^+) + P(\Omega_n^-), \qquad \widetilde{\lambda}_k(\Omega_n) = \max\left\{\widetilde{\lambda}_l(\Omega_n^+), \widetilde{\lambda}_k(\Omega_n^-)\right\},$$

where $l \in \{0, \dots, k\}$ is fixed. Since Ω_n is minimizing, we may suppose $l \in \{1, \dots k-1\}$. In particular, if $k = 1$, then the dichotomy cannot occur.

We now prove the existence of a solution of (6.40) reasoning by induction. if $k = 1$, then the existence holds since for every minimizing sequence, the compactness case of Theorem 3.131 necessarily occurs. Suppose now that the existence holds for $1, \dots, k-1$ and let Ω_n be a minimizing sequence for the functional $\lambda_k + mP$. If the compactness occurs for Ω_n, then the existence holds immediately. If we are in the dichotomy case, then we consider the solutions Ω_+ and Ω_- of the problems

$$\min\left\{\widetilde{\lambda}_l(\Omega) : \Omega \subset \mathbb{R}^d,\ \Omega \text{ measurable},\ |\Omega| < \infty,\ P(\Omega) = \lim_{n\to\infty} P(\Omega_n^+)\right\},$$

$$\min\left\{\widetilde{\lambda}_{k-l}(\Omega) : \Omega \subset \mathbb{R}^d, \Omega \text{ measurable},\ |\Omega| < \infty,\ P(\Omega) = \lim_{n\to\infty} P(\Omega_n^-)\right\},$$

which admit solutions by the inductive assumption and Remark 6.3. We now note that

$$\widetilde{\lambda}_l(\Omega_+) \le \liminf_{n\to\infty} \widetilde{\lambda}_l(\Omega_n^+) \qquad \text{and} \qquad \widetilde{\lambda}_{k-l}(\Omega_-) \le \liminf_{n\to\infty} \widetilde{\lambda}_{k-l}(\Omega_n^-),$$

and since we can suppose that Ω_+ and Ω_- are disjoint and distant sets, we have

$$\begin{aligned}\widetilde{\lambda}_k(\Omega_+ \cup \Omega_-) &\le \max\left\{\widetilde{\lambda}_l(\Omega_+), \widetilde{\lambda}_{k-l}(\Omega_-)\right\} \\ &\le \liminf_{n\to\infty} \max\left\{\widetilde{\lambda}_l(\Omega_n^+), \widetilde{\lambda}_{k-l}(\Omega_n^-)\right\} = \liminf_{n\to\infty} \widetilde{\lambda}_k(\Omega_n),\end{aligned}$$

which gives that the disjoint union $\Omega_+ \cup \Omega_-$ is a solution of (6.40).

Step 2 (*Existence of open solution*). Let Ω be a solution of (6.40). Then Ω is a supersolution for $\widetilde{\lambda}_k + mP$ and, since $\widetilde{\lambda}_k$ is decreasing with respect to the inclusion, Ω is a supersolution for the perimeter. Now by Proposition 5.48 we have that Ω is an open set and $H_0^1(\Omega) = \widetilde{H}_0^1(\Omega)$. In particular, by the variational definition of the Dirichlet eigenvalues, we have $\widetilde{\lambda}_k(\Omega) = \lambda_k(\Omega)$. Let now $U \subset \mathbb{R}^d$ be any open set. Then

$$\begin{aligned}\lambda_k(\Omega) + mP(\Omega) &= \widetilde{\lambda}_k(\Omega) + mP(\Omega)\\ &\le \widetilde{\lambda}_k(U) + mP(U)\\ &\le \lambda_k(U) + mP(U),\end{aligned}$$

which, by the arbitrariness of U proves that Ω is a solution of (6.39). Moreover, we proved that there is a solution of (6.39) which is also a solution of (6.40) and so, any solution of (6.39) which is also a solution of (6.40).

Step 3 (*Boundedness and regularity*). Let Ω be a solution of (6.39) (and thus, of (6.40)). Then Ω is a perimeter supersolution and, by the results from Section 4.6, it is also a subsolution for the functional $\widetilde{E} + \widetilde{m}P$, for some $\widetilde{m} > 0$. By Theorem 5.59, this implies that Ω is a bounded open set with $C^{1,\alpha}$ boundary, for every $\alpha < 1$.

Step 4 (*Connectedness of the optimal set*). We first prove the result in dimension $d \le 7$, in which case the singular set of the boundary $\partial\Omega$ is empty. We first note that, since Ω is a solution of (6.39), it has a finite number (at most k) of connected components. Suppose, by contradiction, that there are at least two connected components of Ω. If we take one of them and translate it until it touches one of the others, then we obtain a set $\widetilde{\Omega}$ which is still a solution of (6.42). Using the regularity of the contact point for the two connected components, it is easy to construct an outer variation of $\widetilde{\Omega}$ which decreases the perimeter (see Figure 6.2). In fact, assuming that the contact point is the origin, up to a rotation of the coordinate axes, we can find a small cylinder C_r and two $C^{1,\alpha}$ functions g_1 and g_2 such that

$$g_1(0) = g_2(0) = |\nabla g_1(0)| = |\nabla g_2(0)| = 0, \tag{6.41}$$

and

$$\widetilde{\Omega}^c \cap C_r = \big\{g_1(x_1, \dots, x_{d-1}) \le x_d \le g_2(x_1, \dots, x_{d-1})\big\} \cap C_r.$$

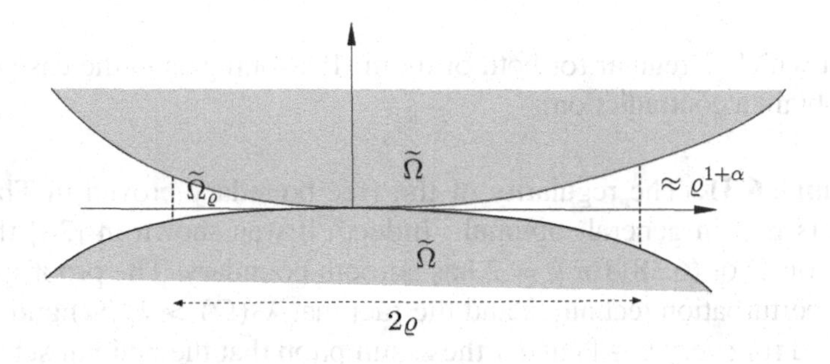

Figure 6.2. The variation from *Step* 4 of the proof of Theorem 6.30.

Now, for $\varrho < r$, consider the set $\widetilde{\Omega}_\varrho := \widetilde{\Omega} \cup C_\varrho \supset \widetilde{\Omega}$. It is easy see that, thanks to (6.41) and the $C^{1,\alpha}$ regularity of g_1 and g_2,

$$P(\widetilde{\Omega}_\varrho) - P(\widetilde{\Omega}) \leq C_\alpha \varrho^{d-1+\alpha} - C_d \varrho^{d-1} < 0,$$

for ϱ small enough, which contradicts the minimality of $\widetilde{\Omega}$ [7].

We now consider the case $d \geq 8$. In this case the singular set may be non-empty and so, in order to perform the operation described above, we need to be sure that the contact point is not singular.

Suppose, by contradiction, that the optimal set Ω is disconnected, *i.e.* there exist two non-empty open sets $A, B \subset \Omega$ such that $A \cup B = \Omega$ and $A \cap B = \emptyset$. We have

$$\partial A \cup \partial B \subset \partial\Omega = \partial^M \Omega,$$

where the last inequality follows by classical density estimates. By Federer's criterion [80, Theorem 16.2], A and B have finite perimeter. Arguing as in [3, Theorem 2, Section 4], we deduce that $P(\Omega) = P(A) + P(B)$.

Since both A and B are bounded, there is some $x_0 \in \mathbb{R}^d$ such that $\text{dist}(A, x_0 + B) > 0$. Then the set $\Omega' = A \cup (x_0 + B)$ is also a solution of (6.42). Let $x \in \partial A$ and $y \in \partial(x_0 + B)$ be such that $|x - y| = \text{dist}(A, x_0 + B)$. Since the ball with center $(x + y)/2$ and radius $|x - y|/2$ does not intersect Ω', we have that in both x and y, Ω' satisfies the exterior ball condition. Hence both x and y are regular points [8].

Consider now the set $\Omega'' = (-x + A) \cup (-y + x_0 + B)$. It is a solution of (6.42) and has at least two connected components, which meet in a

[7] Another way to conclude is to notice that for $\widetilde{\Omega}$ the origin is not a regular point, a contradiction with Theorem 5.59.

[8] This can be easily seen, since any tangent cone at these points is contained in an half-space and hence it has to coincide with it, see [88, Theorem 36.5]

point which is regular for both of them. Reasoning as in the case $d \leq 7$, we obtain a contradiction. □

Remark 6.31. The regularity of the free boundary proved in Theorem 6.30 is not, in general, optimal. Indeed, it was shown in [24] that the solution Ω of (6.38) for $k = 2$ has smooth boundary. The proof is based on a perturbation technique and the fact that $\lambda_2(\Omega) > \lambda_1(\Omega)$ and can be applied for every $k \in \mathbb{N}$ under the assumption that the optimal set is such that $\lambda_k(\Omega) > \lambda_{k-1}(\Omega)$. On the other hand it is expected (due to some numerical computations) that the optimal set Ω for λ_3 in $\mathbb{R}^2$ is a ball and, in particular, $\lambda_3(\Omega) = \lambda_2(\Omega)$.

We are now in position to state the following more general result

Theorem 6.32. *Suppose that $F : \mathbb{R}^p \to \mathbb{R}$ satisfies the assumptions $(F1)$, $(F2)$ and $(F3)$. Then the shape optimization problem*

$$\min\Big\{F\big(\lambda_{k_1}(\Omega),\dots,\lambda_{k_p}(\Omega)\big) : \Omega\subset\mathbb{R}^d,\ \Omega \text{ open},\ P(\Omega)=1,\ |\Omega|<+\infty\Big\}, \tag{6.42}$$

has a solution. Moreover, any optimal set Ω is bounded and connected and its boundary $\partial\Omega$ is $C^{1,\alpha}$, for every $\alpha \in (0,1)$, outside a closed set of Hausdorff dimension at most $d-8$.

Proof. We first consider the problem

$$\min\Big\{F\big(\widetilde{\lambda}_{k_1}(\Omega),\dots,\widetilde{\lambda}_{k_p}(\Omega)\big) : \Omega\subset\mathbb{R}^d,\ \Omega \text{ measurable},\ P(\Omega)=1, |\Omega|<+\infty\Big\}. \tag{6.43}$$

By Proposition 6.6 with $\mathcal{G} = P$, we have that any solution Ω of (6.43) is a subsolution for $F\big(\widetilde{\lambda}_{k_1}(\Omega),\dots,\widetilde{\lambda}_{k_p}(\Omega)\big) + mP(\Omega)$ and asupersolution for $F\big(\widetilde{\lambda}_{k_1}(\Omega),\dots,\widetilde{\lambda}_{k_p}(\Omega)\big) + MP(\Omega)$ for some $m, M > 0$. Thus, by Theorem 4.57, Ω is a supersolution for $\widetilde{E} + \widetilde{m}P$, for some $\widetilde{m} > 0$ and, by Remark 5.43, Ω is a perimeter supersolution. Thus, by Theorem 5.59 Ω is a bounded open set with $C^{1,\alpha}$, outside a set of dimension at most $d-8$, for every $\alpha \in (0,1)$. Moreover, since Ω is a perimeter supersolution, we have $H^1_0(\Omega) = \widetilde{H}^1_0(\Omega)$ and so, by the same argument as in Theorem 6.30, Ω is a solution of (6.42) and every solution of 6.42 is also a solution of (6.43).

The existence of a solution of (6.43) follows by induction on the number of variables p, using the same argument as in Theorem 6.8.

In conclusion, the connectedness of the optimal set follows as in *Step* 4 of the proof of Theorem 6.30. □

6.5. Optimal potentials for Schrödinger operators

In this section we consider optimization problems concerning potentials in place of sets, *i.e.* we consider variational problems of the form

$$\min\Big\{\mathcal{F}(V)\ :\ V\in\mathcal{V}\Big\},\tag{6.44}$$

where $\mathcal{V}$ is an admissible class of nonnegative Borel functions on the open set $\Omega\subset\mathbb{R}^d$ and $\mathcal{F}$ is a cost functional on the family of capacitary measures $\mathcal{M}^+_{\text{cap}}(\Omega)$. This problem was extensively studied far a great variety of cost functionals $\mathcal{F}$ and admissible sets $\mathcal{V}$. We refer to [71, Chapter 8] for a extensive survey on the known results (before [34] and [26]).

The admissible classes we study in this section are determined by a function $\Psi:[0,+\infty]\to[0,+\infty]$

$$\mathcal{V}=\Big\{V:\Omega\to[0,+\infty]: V \text{ Lebesgue measurable},\ \int_\Omega\Psi(V)\,dx\le 1\Big\}.$$

The cost functional $\mathcal{F}$ is typically given through the solution of some partial differential equation involving the operator $-\Delta+V$ on Ω as, for example, the functional

$$\mathcal{F}(V)=F\big(\lambda_1(V),\dots,\lambda_k(V)\big)+\int_\Omega V^p\,dx,$$

where $\lambda_k(V):=\lambda_k(V dx+I_\Omega)$ and $p\in\mathbb{R}$.

6.5.1. Optimal potentials in bounded domain

In this subsection we consider the case when Ω is a bounded open set. Our first result concerns constraints of the form $\Phi(x)=x^p$, for some $p\ge 1$. More precisely, we have the following result:

Theorem 6.33. *Suppose that $\Omega\subset\mathbb{R}^d$ is a bounded open set. Let $\mathcal{F}: L^1_+(\Omega)\to\mathbb{R}$ be a functional, lower semicontinuous with respect to the γ-convergence, and let $\mathcal{V}$ be a weakly $L^1(\Omega)$ compact set. Then the problem*

$$\min\Big\{\mathcal{F}(V)\ :\ V\in\mathcal{V}\Big\},\tag{6.45}$$

admits a solution.

Proof. Let (V_n) be a minimizing sequence in $\mathcal{V}$. By the compactness assumption on $\mathcal{V}$, we may assume that V_n tends weakly $L^1(\Omega)$ to some

$V \in \mathcal{V}$. By Proposition 3.118, we have that V_n γ-converges to V and so, by the semicontinuity of F,

$$\mathcal{F}(V) \leq \liminf_{n\to\infty} \mathcal{F}(V_n),$$

which gives the conclusion. □

Corollary 6.34. *Let $F : \mathbb{R}^k \to \mathbb{R}$ be a lower semi-continuous function. let Ω be a a given quasi-open set of finite measure and let $p \geq 1$ be a given real numbers. Then, there exists a solution of the problem*

$$\min\Big\{F\big(\lambda_1(V), \dots, \lambda_k(V)\big) + \int_\Omega V^p\,dx \ :\ V : \Omega \to [0, +\infty]\ \textit{measurable}\Big\}, \tag{6.46}$$

admits a solution.

Proof. It is sufficient to note that both functionals $F(\lambda_1, \dots, \lambda_k)$ and $V \mapsto \int_\Omega V^p\,dx$ are lower semi-continuous with respect to the γ-convergence. Indeed, for the second one, it is sufficient to note that, by Theorem 3.119 on the bounded sets of positive functions in L^p the γ-convergence and the weak convergence in L^p are equivalent. □

Remark 6.35. It is more appropriate to refer to the problem (6.46) as to a *maximization* problem. In fact, in the typical case when the function f is increasing, the solution of (6.46) is the potential constantly equal to zero on Ω. In order to have non-trivial solutions one has to choose f to be a decreasing function on $\mathbb{R}^k$.

We now turn our attention to the case when Φ is a decreasing function. In this case it is natural to expect that the problem (6.44) has a non-trivial solution for increasing functions f. Before we state our main existence result in this case, we will need two preliminary Lemmas. The first one (Lemma 6.36) is a classical result who can also be found in [31] and [5]. The second one (Lemma 6.37) is a classical semi-continuity result, which can be found in [31]. We report here the proofs for the sake of completeness

Lemma 6.36. *Consider an open set $\Omega \subset \mathbb{R}^d$ and a σ-finite Borel measure ν on Ω. Let $\{\phi_n\}_{n\in\mathbb{N}}$ be a sequence of positive Borel functions on $\mathbb{R}$ and let $\phi = \sup_n \phi_n$. Then, we have that*

$$\int_\Omega \phi\,d\nu = \sup\Big\{\sum_{i\in I} \int_{A_i} \phi_i\,d\nu\Big\},$$

where the supremum is over all finite subsets $I \subset \mathbb{N}$ and over all families $\{A_i\}_{i\in I}$ of disjoint open sets with compact closure in Ω.

Proof. By the monotone convergence theorem, it is enough to prove that for each $k \in \mathbb{N}$, we have

$$\int_\Omega \sup_{1\le i\le k} \phi_i \, d\nu = \sup\left\{\sum_{i=1}^k \int_{A_i} \phi_i \, d\nu\right\}.$$

Let $B_i = \{\phi_i = \sup_{1\le i\le k} \phi_i\}$ and $C_i = B_i \setminus \cup_{j<i} B_j$. Then $C_1, \dots, C_k$ are disjoint Borel subsets of Ω and

$$\int_\Omega \sup_{1\le i\le k} \phi_i \, d\nu = \sum_{i=1}^k \int_{C_i} \phi_i \, d\nu.$$

Approximating each C_i with compact sets K_{ij}, from inside, and then aproximating each compact set K_{ij} with open sets A_{ijl} such that $\{A_{ijl}\}_{1\le i\le k}$ is a family of disjoint sets, we have the claim. $\square$

Lemma 6.37. *Let $1 < p, q < \infty$ and let $u_n \in L^p(\Omega)$ and $v_n \in L^q(\Omega)$ be two sequences of positive functions on the open set $\Omega \subset \mathbb{R}^d$ such that u_n converges strongly in L^p to $u \in L^p(\Omega)$ and v_n converges weakly in L^q to $v \in L^q(\Omega)$. Suppose that $H : [0.+\infty] \to [0,+\infty]$ is a convex function. Then we have*

$$\int_\Omega u H(v)\, dx \le \liminf_{n\to\infty} \int_\Omega u_n H(v_n)\, dx.$$

Proof. Let us first prove the claim for $H(x) = x$. Indeed, if $q' > p$, then for each $t \ge 0$, $u_n \wedge t$ converges strongly $L^{q'}$ to $u \wedge t$ and so, we have that

$$\int_\Omega v(u\wedge t)\, dx = \lim_{n\to\infty} \int_\Omega v_n(u_n \wedge t)\, dx \le \liminf_{n\to\infty} \int_\Omega v_n u_n \, dx, \quad (6.47)$$

and we obtain the thesis passing to the limit as $t \to \infty$. If $q' \le p$, then for each $R > 0$, we have that $\mathbb{1}_{B_R} u_n$ converges strongly in $L^{q'}$ to $1_{B_R} u$ and so

$$\int_\Omega v \mathbb{1}_{B_R} u \, dx = \lim_{n\to\infty} \int_\Omega v_n \mathbb{1}_{B_R} u_n \, dx \le \liminf_{n\to\infty} \int_\Omega v_n u_n \, dx, \quad (6.48)$$

and we obtain the claim passing to the limit as $R \to \infty$.

We now prove the Lemma for generic function H. Let $a_n, b_n \in \mathbb{R}$ be such that for each $x \in \mathbb{N}$

$$H(x) = \sup_{n\in\mathbb{N}} \{a_n x + b_n\},$$

and let $A_1, \dots, A_k$ be disjoint open subsets of Ω. On each A_j consider a function $\phi_j \in C_c^\infty(A_k)$ such that $0 \le \phi_j \le 1$. Then, we have that $a, b \in \mathbb{R}$

$$\begin{aligned}
\sum_{j=1}^k \int_\Omega (av+b)^+\phi_j u\,dx &\le \liminf_{n\to\infty} \sum_{j=1}^k \int_\Omega (av_n+b)^+\phi_j u_n\,dx \\
&\le \liminf_{n\to\infty} \sum_{j=1}^k \int_\Omega H(v_n)\phi_j u_n\,dx \\
&\le \liminf_{n\to\infty} \int_\Omega H(v_n) u_n\,dx.
\end{aligned} \tag{6.49}$$

Taking the supremum over all $\phi_j \in C_c^\infty(A_j)$ such that $0 \le \phi_j \le 1$, we obtain that

$$\sum_{j=1}^k \int_{A_j} (av+b)^+ u\,dx \le \liminf_{n\to\infty} \int_\Omega H(v_n)u_n\,dx. \tag{6.50}$$

Now the claim follows by Lemma 6.36. □

The following existence result was proved in [34].

Theorem 6.38 (Buttazzo-Dal Maso Theorem for potentials). *Let $\Omega \subset \mathbb{R}^d$ be a bounded open set and $\Psi : [0, +\infty] \to [0, +\infty]$ a strictly decreasing function such that there exists $\varepsilon > 0$ for which the function $x \mapsto \Psi^{-1}(x^{1+\varepsilon})$, defined on $[0, +\infty]$, is convex. Then, for any functional $\mathcal{F} : \mathcal{M}_{\mathrm{cap}}(\Omega) \to \mathbb{R}$, which is increasing and lower semi-continuous with respect to the γ-convergence, the problem* (6.44) *has a solution.*

Proof. Let $V_n \in \mathcal{A}(\Omega)$ be a minimizing sequence for problem (6.44). Then, $v_n := \big(\Psi(V_n)\big)^{1/(1+\varepsilon)}$ is a bounded sequence in $L^{1+\varepsilon}(\Omega)$ and so, up to a subsequence, we have that v_n converges weakly in $L^{1+\varepsilon}$ to some $v \in L^{1+\varepsilon}(\Omega)$. We will prove that $V := \Psi^{-1}(v^{1+\varepsilon})$ is a solution of (6.44). Clearly $V \in \mathcal{A}(\Omega)$ and so it remains to prove that $\mathcal{F}(V) \le \liminf_n \mathcal{F}(V_n)$. By the compactness of the γ-convergence in a bounded domain, we can suppose that, up to a subsequence, V_n γ-converges to a capacitary measure $\mu \in \mathcal{M}_{\mathrm{cap}}(\Omega)$. We claim that the following inequalities hold true:

$$\mathcal{F}(V) \le \mathcal{F}(\nu) \le \liminf_{n\to\infty} \mathcal{F}(V_n). \tag{6.51}$$

In fact, the second inequality in (6.51) is the lower semi-continuity of F with respect to the γ-convergence, while the first needs a more careful examination. By the definition of γ-convergence, we have that for any $u \in H_0^1(\Omega)$, there is a sequence $u_n \in H_0^1(\Omega)$ which converges to u in $L^2(\Omega)$ and is such that

$$\begin{aligned}\int_\Omega |\nabla u|^2 dx + \int_\Omega u^2 d\mu &= \lim_{n\to\infty} \int_\Omega |\nabla u_n|^2 dx + \int_\Omega u_n^2 V_n \, dx \\ &= \lim_{n\to\infty} \int_\Omega |\nabla u_n|^2 dx + \int_\Omega u_n^2 \Psi^{-1}(v_n^{1+\varepsilon})\, dx \\ &\geq \int_\Omega |\nabla u|^2 dx + \int_\Omega u^2 \Psi^{-1}(v^{1+\varepsilon})\, dx \\ &= \int_\Omega |\nabla u|^2 dx + \int_\Omega u^2 V \, dx,\end{aligned} \tag{6.52}$$

where the inequality in (6.52) is due to the strong-weak lower semi-continuity result from Lemma 6.37. Thus, for any $u \in H_0^1(\Omega)$, we have that

$$\int_\Omega u^2 \, d\mu \geq \int_\Omega u^2 V \, dx,$$

and so, $V \leq \mu$. Since $\mathcal{F}$ is increasing, we obtain the first inequality in (6.51) and so the conclusion. □

Remark 6.39. The condition on the admissible set in Theorem 6.38 is satisfied by the following functions:

1. $\Psi(x) = x^{-p}$, for any $p > 0$;
2. $\Psi(x) = e^{-\alpha x}$, for $\alpha > 0$.

Indeed, if $\Psi(x) = x^{-p}$, then

$$\Psi^{-1}(x^{1+\varepsilon}) = x^{-(1+\varepsilon)/p},$$

is convex for any $\varepsilon > 0$. If $\Psi(x) = e^{-\alpha x}$, then the function

$$\Psi^{-1}(x^{1+\varepsilon}) = -\frac{1+\varepsilon}{\alpha} \log x,$$

is convex, also for any $\varepsilon > 0$.

Remark 6.40. In particular, Theorem 6.38 provides an existence result for the following problem

$$\min \left\{ \lambda_k(V) : \; V : \Omega \to [0, +\infty] \text{ measurable}, \; \int_\Omega V^{-p} \, dx = 1 \right\}, \tag{6.53}$$

where $k \in \mathbb{N}$, $p > 0$ and Ω is a bounded open set.

6.5.2. Optimal potentials in $\mathbb{R}^d$

In this subsection we consider optimization problems for spectral funcionals in $\mathbb{R}^d$. In particular, we consider the problem

$$\min\Big\{\lambda_k(V)\,:\,V:\mathbb{R}^d\to[0,+\infty]\text{ measurable},\ \int_{\mathbb{R}^d}V^{-p}dx=1\Big\}. \tag{6.54}$$

We note that the cost functional $\lambda_k(V)$ and the constraint $\int_{\mathbb{R}^d}V^{-p}\,dx$ have the following rescaling properties:

Remark 6.41 (Scaling). Suppose that u_k is the kth eigenfunction. Then we have

$$-\Delta u_k+Vu_k=\lambda_k u_k,$$

and rescaling the eigenfunction u_k, we have

$$-\Delta\big(u_k(x/t)\big)+V_t u_k(x/t)=t^{-2}\lambda_k u_k(x/t),$$

where

$$V_t(x):=t^{-2}V(x/t). \tag{6.55}$$

Repeating the same argument for every eigenfunction, we have that

$$\lambda_k(V_t)=t^{-2}\lambda_k(V). \tag{6.56}$$

On the other hand, we have

$$\int_{\mathbb{R}^d}V_t^{-p}\,dx=\int_{\mathbb{R}^d}t^{2p}V(x/t)^{-p}\,dx=t^{2p+d}\int_{\mathbb{R}^d}V^{-p}\,dx. \tag{6.57}$$

Now as in the case of eigenvalues on sets, we have

Remark 6.42 (Existence of a Lagrange multiplier). The potential $\widetilde{V}:\mathbb{R}^d\to[0,+\infty]$ is a solution of

$$\min\Big\{\lambda_k(V)+m\int_{\mathbb{R}^d}V^{-p}\,dx\,:\ V:\mathbb{R}^d\to[0,+\infty]\text{ measurable}\Big\}, \tag{6.58}$$

if and only if, for every $t>0$, we have that $\widetilde{V}_t$, defined as in (6.55), is a solution of

$$\min\Big\{\lambda_k(V)\,:\,V:\mathbb{R}^d\to[0,+\infty]\text{ measurable},\ \int_{\mathbb{R}^d}V^{-p}dx=\int_{\mathbb{R}^d}\widetilde{V}_t^{-p}dx\Big\}, \tag{6.59}$$

and the function

$$f(t):=t^{-2}\lambda_k(\widetilde{V})+mt^{2p+d}\int_{\mathbb{R}^d}\widetilde{V}^{-p}\,dx,$$

achieves its minimum, on the interval $(0,+\infty)$, in the point $t=1$.

In the case $k = 1$, the existence holds for every $p > 0$ by a standard variational argument.

Proposition 6.43 (Faber-Krahn inequality for potentials). *For every $p > 0$ there is a solution V_p of the problem* (6.54) *with $k = 1$. Moreover, there is an optimal potential V_p given by*

$$V_p = \left(\int_{\mathbb{R}^d} |u_p|^{2p/(p+1)}\,dx\right)^{1/p} |u_p|^{-2/(1+p)}, \qquad (6.60)$$

where u_p is a radially decreasing minimizer of

$$\min\left\{\int_{\mathbb{R}^d} |\nabla u|^2 dx + \left(\int_{\mathbb{R}^d} |u|^{2p/(p+1)} dx\right)^{(p+1)/p} : u \in H^1(\mathbb{R}^d), \int_{\mathbb{R}^d} u^2 dx = 1\right\}. \qquad (6.61)$$

Moreover, u_p has a compact support, hence the set $\{V_p < +\infty\}$ is a ball of finite radius in $\mathbb{R}^d$.

Proof. Let us first show that the minimum in (6.61) is achieved. Let $u_n \in H^1(\mathbb{R}^d)$ be a minimizing sequence of positive functions normalized in L^2. Note that by the classical Pólya-Szegö inequality (see for example [78]) we may assume that each of these functions is radially decreasing in $\mathbb{R}^d$ and so we will use the identification $u_n = u_n(r)$. In order to prove that the minimum is achieved it is enough to show that the sequence u_n converges in $L^2(\mathbb{R}^d)$. Indeed, since u_n is a radially decreasing minimizing sequence, there exists $C > 0$ such that for each $r > 0$ we have

$$u_n(r)^{2p/(p+1)} \leq \frac{1}{|B_r|}\int_{B_r} u_n^{2p/(p+1)}\,dx \leq \frac{C}{r^d}.$$

Thus, for each $R > 0$, we obtain

$$\int_{B_R^c} u_n^2\,dx \leq C_1 \int_R^{+\infty} r^{-d(p+1)/p}\, r^{d-1}\,dr = C_2 R^{-1/p}, \qquad (6.62)$$

where C_1 and C_2 do not depend on n and R. Since the sequence u_n is bounded in $H^1(\mathbb{R}^d)$, it converges locally in $L^2(\mathbb{R}^d)$ and, by (6.62), this convergence is also strong in $L^2(\mathbb{R}^d)$. Thus, we obtain the existence of a radially symmetric and decreasing solution u_p of (6.61).

We now note that for any $u \in L^2(\mathbb{R}^d)$ and $V^{-p} \in L^1(\mathbb{R}^d)$, we have

$$\begin{aligned}\left(\int_{\mathbb{R}^d} |u|^{2p/(p+1)}\,dx\right)^{(p+1)/p} &\leq \int_{\mathbb{R}^d} u^2 V\,dx \left(\int_{\mathbb{R}^d} V^{-p}\,dx\right)^{1/p}\\ &= \int_{\mathbb{R}^d} u^2 V\,dx.\end{aligned}$$

Thus, for any $u \in H^1(\mathbb{R}^d)$, such that $\int_{\mathbb{R}^d} u^2\,dx = 1$, we have

$$\int_{\mathbb{R}^d} |\nabla u|^2\,dx + \left(\int_{\mathbb{R}^d} |u|^{2p/(p+1)}\,dx\right)^{(p+1)/p} \leq \int_{\mathbb{R}^d} |\nabla u|^2\,dx + \int_{\mathbb{R}^d} u^2 V\,dx,$$

which gives that the minimum in (6.61) is smaller than $\lambda_1(V)$, for any V such that $\int_{\mathbb{R}^d} V^{-p}\,dx$ and so, it is also smaller than the minimum in (6.54) for $k = 1$. We now note that, writing the Euler-Lagrange equation for u_p, which minimizes (6.61), we have that u_p is the first eigenfunction for the operator $-\Delta + V_p$ on $\mathbb{R}^d$. Thus, we obtain that V_p solves (6.54) for $k = 1$.

We now prove that the support of u_p is a ball of finite radius. By the radial symmetry of u_p we can write it in the form $u_p(x) = u_p(|x|) = u_p(r)$, where $r = |x|$. With this notation, u_p satisfies the equation:

$$-u_p'' - \frac{d-1}{r}u_p' + C_p u_p^s = \lambda u_p,$$

where $s = (p-1)/(p+1) < 1$ and $C_p > 0$ is a constant depending on p. After multiplication by u_p' and integration, we get

$$-u_p'(r) \geq \left(\frac{C_p}{s+1}u_p(r)^{s+1} - \frac{\lambda}{2}u_p(r)^2\right)^{1/2}.$$

Now, since u_p vanishes at infinity, we obtain for $r > 0$ large enough

$$-u_p'(r) \geq \left(\frac{C_p}{2(s+1)}u_p(r)^{s+1}\right)^{1/2}.$$

Integrating both sides of the above inequality, we conclude that u_p has a compact support. □

We now prove an existence result in the case $k = 2$. By Proposition 6.43, there exists optimal potential V_p, for λ_1, such that the set of finiteness $\{V_p < +\infty\}$ is a ball. Thus, we have a situation analogous to the Faber-Krahn inequality, which states that the minimum

$$\min\left\{\lambda_1(\Omega)\ :\ \Omega \subset \mathbb{R}^d,\ |\Omega| = c\right\}, \tag{6.63}$$

is achieved for the ball of measure c. We recall that, starting from (6.63), one may deduce, by a simple argument (see for instance [71]), the Krahn-Szegö inequality, which states that the minimum

$$\min\left\{\lambda_2(\Omega)\ :\ \Omega \subset \mathbb{R}^d,\ |\Omega| = c\right\}, \tag{6.64}$$

is achieved for a disjoint union of equal balls. In the case of potentials one can find two optimal potentials for λ_1 with disjoint sets of finiteness and then apply the argument from the proof of the Krahn-Szegö inequality.

Proposition 6.44 (Krahn-Szegö inequality for potentials). *There exists an optimal potential, solution of* (6.54) *for* $k = 2$. *Moreover, it can be chosen to be of the form* $\min\{V_1, V_2\}$, *where* V_1 *and* V_2 *are optimal potentials for* λ_1, *whose sets of finiteness* $\{V_1 < +\infty\}$ *and* $\{V_2 < +\infty\}$ *are disjoint balls and, moreover,* V_1 *is a translation of* V_2.

Proof. Given V_1 and V_2 as above, we prove that for every $V : \mathbb{R}^d \to [0, +\infty]$ with $\int_{\mathbb{R}^d} V^{-p}\,dx = 1$, we have

$$\lambda_2\big(\min\{V_1, V_2\}\big) \le \lambda_2(V).$$

Indeed, let u_2 be the second eigenfunction of $-\Delta+V$. We first suppose that u_2 changes sign on $\mathbb{R}^d$ and consider the functions $V_+ = \sup\left\{V, \widetilde{I}_{\{u_2\le 0\}}\right\}$ and $V_- = \sup\left\{V, \widetilde{I}_{\{u_2\ge 0\}}\right\}$[9]. We note that

$$1 \ge \int_{\mathbb{R}^d} V^{-p}\,dx \ge \int_{\mathbb{R}^d} V_+^{-p}\,dx + \int_{\mathbb{R}^d} V_-^{-p}\,dx.$$

Moreover, on the sets $\{u_2 > 0\}$ and $\{u_2 < 0\}$, the following equations are satisfied:

$$-\Delta u_2^+ + V_+u_2^+ = \lambda_2(V)u_2^+, \qquad -\Delta u_2^- + V_-u_2^- = \lambda_2(V)u_2^-,$$

and so, multiplying respectively by u_2^+ and u_2^-, we get

$$\lambda_2(V) \ge \lambda_1(V_+), \qquad \lambda_2(V) \ge \lambda_1(V_-), \tag{6.65}$$

where we have equalities, if and only if, u_2^+ and u_2^- are the first eigenfunctions corresponding to $\lambda_1(V_+)$ and $\lambda_1(V_-)$. Let now $\widetilde{V}_+$ and $\widetilde{V}_-$ be optimal potentials for λ_1 from Proposition 6.43, corresponding to the constraints

$$\int_{\mathbb{R}^d} \widetilde{V}_+^{-p}\,dx = \int_{\mathbb{R}^d} V_+^{-p}\,dx \qquad \text{and} \qquad \int_{\mathbb{R}^d} \widetilde{V}_-^{-p}\,dx = \int_{\mathbb{R}^d} V_-^{-p}\,dx.$$

[9] We recall that, for any measurable $A \subset \mathbb{R}^d$, we have

$$\widetilde{I}_A(x) = \begin{cases} +\infty, & x \in A, \\ 0, & x \notin A. \end{cases}$$

By Proposition 6.43, the sets of finiteness of $\widetilde{V}_+$ and $\widetilde{V}_-$ are compact, hence we may assume (up to translations) that they are also disjoint. By the monotonicity of λ_1, we have

$$\max\big\{\lambda_1(V_1), \lambda_1(V_2)\big\} \le \max\big\{\lambda_1(\widetilde{V}_+), \lambda_1(\widetilde{V}_-)\big\},$$

and so, we obtain

$$\begin{aligned}\lambda_2\big(\min\{V_1, V_2\}\big) &\le \max\big\{\lambda_1(\widetilde{V}_+), \lambda_1(\widetilde{V}_-)\big\}\\ &\le \max\big\{\lambda_1(V_+), \lambda_1(V_-)\big\} \le \lambda_2(V),\end{aligned}$$

as required.
If u_2 does not change sign, then we consider $V_+ = \sup\{V, \widetilde{I}_{\{u_2=0\}}\}$ and $V_- = \sup\{V, \widetilde{I}_{\{u_1=0\}}\}$, where u_1 is the first eigenfunction of $-\Delta + V$. Then the claim follows by the same argument as above. □

We now turn our attention to the general case $k > 2$.

Remark 6.45 (Compactness of the embedding $H^1_V \hookrightarrow L^1$). We first note that if $p \in (0, 1]$ and $\int_{\mathbb{R}^d} V^{-p}\,dx < +\infty$, then for every $R > 0$ the solution w_R of the equation

$$-\Delta w_R + V w_R = 1, \qquad w_R \in H^1(B_R) \cap L^2(V dx),$$

is such that

$$\begin{aligned}&\int_{\mathbb{R}^d} w_R\,dx\\ &\le \left(\int_{\mathbb{R}^d} w_R^{\frac{2p}{p+1}}\,dx\right)^{\frac{(1+p)(d+2)}{2(d+2p)}} \left(\int_{\mathbb{R}^d} w_R^{\frac{2d}{d-2}}\,dx\right)^{\frac{(d-2)(1-p)}{2(d+2p)}}\\ &\le \left(\left(\int_{\mathbb{R}^d} w_R^2 V\,dx\right)^{\frac{p}{1+p}} \left(\int_{\mathbb{R}^d} V^{-p}\,dx\right)^{\frac{1}{1+p}}\right)^{\frac{(1+p)(d+2)}{2(d+2p)}}\\ &\quad\times\left(\left(C_d \int_{\mathbb{R}^d} |\nabla w_R|^2\,dx\right)^{\frac{d}{d-2}}\right)^{\frac{(d-2)(1-p)}{2(d+2p)}}\\ &\le \left(\int_{\mathbb{R}^d} w_R^2 V\,dx\right)^{\frac{p(d+2)}{2(d+2p)}} \left(\int_{\mathbb{R}^d} V^{-p}\,dx\right)^{\frac{d+2}{2(d+2p)}}\\ &\quad\times\left(C_d \int_{\mathbb{R}^d} |\nabla w_R|^2\,dx\right)^{\frac{d(1-p)}{2(d+2p)}}\\ &\le C\left(\int_{\mathbb{R}^d} w_R\,dx\right)^{1/2},\end{aligned}$$

for some appropriate constant $C > 0$. Thus we have that the sequence w_R is uniformly bounded in $L^1(\mathbb{R}^d)$ and so the energy function $w_V = \sup_R w_R$ is in $L^1(\mathbb{R}^d)$, which in turn gives that the inclusion $H^1_V(\mathbb{R}^d) \hookrightarrow L^1(\mathbb{R}^d)$ is compact and, in particular, the spectrum of $-\Delta + V$ is discrete.

We now apply the results from Chapter 3 and Chapter 4 to obtain the existence of optimal potential in $\mathbb{R}^d$.

Theorem 6.46. *Suppose that $p \in (0, 1)$. Then, for every $k \in \mathbb{N}$, there is a solution of the problem* (6.54). *Moreover, any solution V of* (6.54) *is constantly equal to $+\infty$ outside a ball of finite radius.*

Proof. By Remark 6.42, every solution of (6.54) is a solution also of the penalized problem (6.58), for some appropriately chosen Lagrange multiplier $m > 0$. Thus, by Theorem 4.54 and Lemma 4.55, we have that if V is optimal for (6.58), then it is constantly $+\infty$ outside a ball of finite radius.

The proof of the existence part follows by induction on k. The first step $k = 1$ being proved in Proposition (6.43). We prove the claim for $k > 1$, provided that the existence holds for all $1, \dots, k-1$.

Let V_n be a minimizing sequence for (6.54). By Remark 6.45, we have that the sequence w_{V_n} is uniformly bounded in $L^1(\mathbb{R}^d)$ and so, by Theorem 3.130, we have two possibilities for the sequence of capacitary measures $V_n dx$: *compactness* and *dichotomy*.

If the compactness occurs, then there is a capacitary measure μ such that the sequence $V_n dx$ γ-converges to μ. By Proposition 3.111, we have that $\|\cdot\|_{H^1_{V_n}}$ Γ-converges in $L^2(\mathbb{R}^d)$ to $\|\cdot\|_{H^1_\mu}$. Now, by the same argument as in Theorem (6.38), we have that $V = \mu_a$, is a solution of (6.54).

If the dichotomy occurs, then we can suppose that $V_n = V_n^+ \vee V_n^-$, where

$$1/V_n = 1/V_n^+ + 1/V_n^-, \qquad \operatorname{dist}\big(\{V_n^+ < \infty\}, \{V_n^- < \infty\}\big) \to +\infty.$$

Since V_n is minimizing, there is $1 \le l \le k-1$ such that

$$\lambda_k(V_n) = \lambda_l(V_n^+) \ge \lambda_{k-l}(V_n^-).$$

Taking the solutions, V^+ and V^- respectively of

$$\min\Big\{\lambda_l(V) : V : \mathbb{R}^d \to [0, +\infty] \text{ measurable}, \int_{\mathbb{R}^d} V^{-p}\,dx = \lim_{n\to\infty} \int_{\mathbb{R}^d} V_n^+\,dx\Big\},$$

$$\min\Big\{\lambda_{k-l}(V) : V : \mathbb{R}^d \to [0, +\infty] \text{measurable}, \int_{\mathbb{R}^d} V^{-p}\,dx = \lim_{n\to\infty} \int_{\mathbb{R}^d} V_n^-\,dx\Big\},$$

in such a way that $\operatorname{dist}(\{V^+ < \infty\}, \{V^- < \infty\}) > 0$, we have that $V = V^+ \wedge V^-$ is a solution of (6.54). □

6.6. Optimal measures for spectral-torsion functionals

In this section we consider spectral optimization problems for operators depending on capacitary measures. The admissible class of measures is determined through the torsion energy

$$E(\mu) = \min\left\{\frac{1}{2}\int_{\mathbb{R}^d} |\nabla u|^2\,dx + \frac{1}{2}\int_{\mathbb{R}^d} u^2\,d\mu - \int_{\mathbb{R}^d} u\,dx \;:\; u \in L^1(\mathbb{R}^d) \cap H^1_\mu(\mathbb{R}^d)\right\},$$

while the spectrum corresponding to the measure μ is defined as

$$\lambda_k(\mu) = \min_{K \subset H^1_\mu} \max_{u \in K} \frac{\int_{\mathbb{R}^d} |\nabla u|^2\,dx + \int_{\mathbb{R}^d} u^2\,d\mu}{\int_{\mathbb{R}^d} u^2\,dx}, \tag{6.66}$$

where the minimum is over all k-dimensional spaces $K \subset H^1_\mu$. We recall that if the $E(\mu) < +\infty$, then the torsion energy function $w_\mu \in L^1(\mathbb{R}^d)$ ($\mu \in \mathcal{M}^T_{\text{cap}}(\mathbb{R}^d)$), we have that the embedding $H^1_\mu \subset L^1(\mathbb{R}^d)$ is compact and the spectrum of the operator $(-\Delta + \mu)$ is discrete and is given precisely by (6.66).

Fixed a capacitary measure ν on $\mathbb{R}^d$ such that $w_\nu \in L^1(\mathbb{R}^d)$, we will prove the existence of optimal capacitary measures for the problem

$$\min\left\{F\big(\lambda_1(\mu), \dots, \lambda_k(\mu)\big) : \mu \text{ capacitary measure},\; E(\mu) = c,\, \mu \geq \nu\right\}, \tag{6.67}$$

where $c \in [E(\nu), 0)$ and $F : \mathbb{R}^k \to \mathbb{R}$ is a given function. We note that the case $\nu = I_{\mathcal{D}}$, where $\mathcal{D} \subset \mathbb{R}^d$ is a bounded quasi-open set, corresponds to an optimization problem in the box $\mathcal{D}$.

Theorem 6.47. *Let ν be a capacitary measure of finite torsion on $\mathbb{R}^d$ and let $F : \mathbb{R}^k \to \mathbb{R}$ be a given lower semi-continuous function. Then, for any $c \in [E(\nu), 0)$, the optimization problem* (6.67) *has a solution.*

Proof. Consider a minimizing sequence μ_n for (6.67). By Corollary 3.117, we have that up to a subsequence μ_n γ-converges to some capacitary measure $\mu \in \mathcal{M}^T_{\text{cap}}(\mathbb{R}^d)$ such that $\mu \geq \nu$. Thus, we have

$$E(\mu) = -\frac{1}{2}\int_{\mathbb{R}^d} w_\mu\,dx = -\lim_{n\to\infty}\frac{1}{2}\int_{\mathbb{R}^d} w_{\mu_n}\,dx = \lim_{n\to\infty} E(\mu_n).$$

By the semi-continuity of F and of the spectrum λ_k, with respect to the γ-convergence, we have that

$$F\big(\lambda_1(\mu),\dots,\lambda_k(\mu)\big) \le \liminf_{n\to\infty} F\big(\lambda_1(\mu_n),\dots,\lambda_k(\mu_n)\big),$$

which concludes the proof. □

In $\mathbb{R}^d$ the existence of an optimal set is more involved due to the lack of the compactness provided by the box $\mathcal{D}$. In this case we consider the model problem

$$\min\Big\{\lambda_k(\mu)\,:\ \mu \text{ capacitary measure},\ E(\mu)=c\Big\}. \tag{6.68}$$

As in the case of potentials, we note that the functionals $\lambda_k(\mu)$ and $E(\mu)$ have the following rescaling properties:

Remark 6.48 (Scaling). Suppose that u_k is the kth eigenfunction of $(-\Delta+\mu)$. Then we have

$$-\Delta u_k + \mu u_k = \lambda_k(\mu)u_k,$$

and rescaling the eigenfunction u_k, we have

$$-\Delta\big(u_k(x/t)\big) + \mu_t u_k(x/t) = t^{-2}\lambda_k(\mu)u_k(x/t),$$

where $\mu_t := t^{d-2}\mu(\cdot/t)$, *i.e.* for every $\phi\in L^1(\mu)$, we have

$$\int_{\mathbb{R}^d}\phi(x/t)\,d\mu_t(x) := t^{d-2}\int_{\mathbb{R}^d}\phi\,d\mu. \tag{6.69}$$

Repeating the same argument for every eigenfunction, we have that

$$\lambda_k(\mu_t) = t^{-2}\lambda_k(\mu). \tag{6.70}$$

On the other hand, we have

$$-\Delta\big(w_\mu(x/t)\big) + t^{d-2}\mu(x/t)w_\mu(x/t) = t^{-2},$$

and so,

$$w_{\mu_t}(x) = t^2 w_\mu(x/t) \qquad \text{and} \qquad E(\mu_t) = t^{d+2}E(\mu). \tag{6.71}$$

As in the cases of optimization of domains and potentials, we have:

Remark 6.49 (Existence of a Lagrange multiplier). The capacitary measure $\widetilde{\mu} \in \mathcal{M}_{\text{cap}}^T(\mathbb{R}^d)$ is a solution of

$$\min\left\{\lambda_k(\mu) - mE(\mu) : \mu \in \mathcal{M}_{\text{cap}}^T(\mathbb{R}^d)\right\}, \tag{6.72}$$

if and only if, for every $t > 0$, the capacitary measure $\widetilde{\mu}_t$, defined as in (6.69), is a solution of

$$\min\left\{\lambda_k(\mu) : \mu \in \mathcal{M}_{\text{cap}}^T(\mathbb{R}^d),\ E(\mu) = E(\widetilde{\mu}_t)\right\}, \tag{6.73}$$

and the function

$$f(t) := t^{-2}\lambda_k(\widetilde{\mu}) - mt^{2+d}E(\widetilde{\mu}),$$

achieves its minimum, on the interval $(0, +\infty)$, for $t = 1$.

Theorem 6.50. *For every $k \in \mathbb{N}$ and $c < 0$, there is a solution of the problem* (6.68). *Moreover, for any solution μ of* (6.68), *there is a ball B_R such that $\mu \geq I_{B_R}$.*

Proof. Suppose first that μ is a solution of (6.68). By Remark 6.49, μ is also a solution of the problem (6.72), for some constant $m > 0$. Let Ω_μ be the set of finiteness of the capacitary measure μ. By the optimality of μ, we have that Ω_μ is a subsolution for the functional

$$\Omega \mapsto \lambda_k(\mu \vee I_\Omega) - mE(\mu \vee I_\Omega).$$

By Corollary 4.65, we have that Ω_μ is a bounded set and so there is a ball B_R such that $\mu \geq I_{B_R}$.

The proof of the existence part follows by induction on k. Suppose that $k = 1$ and let μ_n be a minimizing sequence for the problem

$$\min\left\{\lambda_1(\mu) - mE(\mu) : \mu \in \mathcal{M}_{\text{cap}}^T(\mathbb{R}^d)\right\}. \tag{6.74}$$

By the concentration-compactness principle (Theorem 3.130), we have two possibilities: compactness and dichotomy. If the compactness occurs, we have that, up to a subsequence, μ_n γ-converges to some $\mu \in \mathcal{M}_{\text{cap}}^T(\mathbb{R}^d)$. Thus, by the continuity of λ_1 and E, we have that μ is a solution of (6.74). We now show that the dichotomy cannot occur. Indeed, if we suppose that $\mu_n = \mu_n^+ \vee \mu_n^-$, where μ_n^+ and μ_n^- have distant sets of finiteness, then

$$\lambda_1(\mu_n) = \min\{\lambda_1(\mu_n^1), \lambda_1(\mu_n^+)\} \qquad \text{and} \qquad E(\mu_n) = E(\mu_n^+) + E(\mu_n^-).$$

Since, by Theorem 3.130

$$\liminf_{n\to\infty}\big(-E(\mu_n^+)\big)>0 \qquad \text{and} \qquad \liminf_{n\to\infty}\big(-E(\mu_n^-)\big)>0,$$

we obtain that one of the sequences μ_n^+ and μ_n^-, say μ_n^+ is such that

$$\liminf_{n\to\infty}\big(\lambda_1(\mu_n^+)-mE(\mu_n^+)\big)<\liminf_{n\to\infty}\big(\lambda_1(\mu_n)-mE(\mu_n)\big),$$

which is a contradiction and so, the compactness is the only possible case for μ_n.

We now prove the claim for $k>1$, provided that the existence holds for all $1,\dots,k-1$.

Let μ_n be a minimizing sequence for (6.54). The sequence w_{μ_n} is uniformly bounded in $L^1(\mathbb{R}^d)$ and so, by Theorem 3.130, we have two possibilities for the sequence of capacitary measures μ_n: *compactness* and *dichotomy*.

If the compactness occurs, then there is a capacitary measure μ such that the sequence μ_n γ-converges to μ, which by the continuity of λ_k and the energy E, is a solution of (6.68).

If the dichotomy occurs, then we can suppose that $\mu_n=\mu_n^+\wedge\mu_n^-$, where the sets of finiteness $\Omega_{\mu_n^+}$ and $\Omega_{\mu_n^-}$ are such that

$$\operatorname{dist}\big(\Omega_{\mu_n^+},\Omega_{\mu_n^-}\big)\to+\infty, \qquad E(\mu_n)=E(\mu_n^+)+E(\mu_n^-),$$

$$\lim_{n\to\infty}E(\mu_n^+)<0 \qquad \text{and} \qquad \lim_{n\to\infty}E(\mu_n^-)<0.$$

Since μ_n is a minimizing sequence, there is a constant $1\le l\le k-1$ such that

$$\lambda_k(\mu_n)=\lambda_l(\mu_n^+)\ge\lambda_{k-l}(\mu_n^-).$$

Taking the solutions, μ^+ and μ^- respectively of

$$\min\Big\{\lambda_l(\mu)\,:\,\mu\in\mathcal{M}_{\mathrm{cap}}^T(\mathbb{R}^d),\ E(\mu)=\lim_{n\to\infty}E(\mu_n^+)\Big\},$$

$$\min\Big\{\lambda_{k-l}(\mu)\,:\,\mu\in\mathcal{M}_{\mathrm{cap}}^T(\mathbb{R}^d),\ E(\mu)=\lim_{n\to\infty}E(\mu_n^-)\Big\},$$

in such a way that $\operatorname{dist}\big(\Omega_{\mu^+},\Omega_{\mu^-}\big)>0$, we have that $\mu=\mu^+\wedge\mu^-$ is a solution of (6.68). □

Remark 6.51. The Kohler-Jobin inequality (we refer to [14] and the references therein for more details on this isoperimetric inequality) states that the ball B minimizes the first eigenvalue $\lambda_1(\Omega)$ among all (open) sets Ω of fixed torsion $T(\Omega)=T(B)$. Since the family $\{I_\Omega\,:\,\Omega\subset\mathbb{R}^d \text{ open}\}\subset\mathcal{M}_{\mathrm{cap}}^T(\mathbb{R}^d)$ is dense in $\mathcal{M}_{\mathrm{cap}}^T(\mathbb{R}^d)$ (see [33]), we have that the measure I_B solves (6.68) for $k=1$.

6.7. Multiphase spectral optimization problems

Let $\mathcal{D} \subset \mathbb{R}^d$ be a quasi-open set of finite measure, let $p \in \mathbb{N}$ and let

$$k_1, \dots, k_p \in \mathbb{N} \qquad \text{and} \qquad m_1, \dots, m_p \in (0, +\infty),$$

be given numbers. We consider the problem

$$\min\Big\{\sum_{j=1}^{p} \big(\lambda_{k_j}(\Omega_j) + m_j|\Omega_j|\big) : \big(\Omega_1, \dots, \Omega_p\big) \text{ quasi-open partition of } \mathcal{D}\Big\}, \tag{6.75}$$

where we say that the p-uple of quasi-open sets $\big(\Omega_1, \dots, \Omega_p\big)$ is a quasi-open partition of Ω, if

$$\bigcup_{j=1}^{p} \Omega_j \subset \mathcal{D} \qquad \text{and} \qquad \Omega_i \cap \Omega_j = \emptyset, \text{ for } i \neq j \in \{1, \dots, p\}. \tag{6.76}$$

We say that the partition is open, if all the sets Ω_j are open.

Remark 6.52. We note that the existence of optimal partitions holds thanks to Theorem 2.84.

In this section we study the qualitative properties of the optimal partitions and we prove the existence of an open optimal partition in the case when the eigenvalues involved in (6.75) are only λ_1 and λ_2. The results we present here were obtained in [29]. We refer also to [12] for some numerical computations and further study of the qualitative properties of the optimal partitions. For the existence part we use the general result from Theorem 2.84, the openness and the other properties of the optimal partitions follow by the results on the interaction between the energy subsolutions and the regularty results from Section 6.3.3.

We start by a result on the multiphase optimization problems in their full generality, *i.e.* we consider the variational problem

$$\min\Big\{g\big(\mathcal{F}_1(\Omega_1), \dots, \mathcal{F}_p(\Omega_p)\big) + \sum_{i=1}^{p} m_i|\Omega_i| : \big(\Omega_1, \dots, \Omega_p\big) \text{ quasi-open partition of } \mathcal{D}\Big\}, \tag{6.77}$$

where

(P1) the function $g : \mathbb{R}^p \to \mathbb{R}$ is increasing in each variable and lower semi-continuous;

(P2) the functionals $\mathcal{F}_1, \dots, \mathcal{F}_p$ on the family of quasi-open sets are decreasing with respect to inclusions and continuous for the γ-convergence;

(P3) the multipliers $m_1, \dots m_p$ are given positive constants.

Definition 6.53. We say that the functional $\mathcal{F}$, defined on the family of quasi-open sets in $\mathbb{R}^d$, is locally γ-Lipschitz for subdomains (or simply γ-Lip), if for each quasi-open set $\Omega \subset \mathbb{R}^d$, there are constants $C > 0$ and $\varepsilon > 0$ such that

$$|\mathcal{F}(\widetilde{\Omega}) - \mathcal{F}(\Omega)| \leq C d_\gamma(\widetilde{\Omega}, \Omega),$$

for every quasi-open set $\widetilde{\Omega} \subset \Omega$, such that $d_\gamma(\widetilde{\Omega}, \Omega) \leq \varepsilon$.

Remark 6.54. Following Theorem 4.46, we have that the functional associated to the k-th eigenvalue of the Dirichlet Laplacian $\Omega \mapsto \lambda_k(\Omega)$ is γ-Lip, for every $k \in \mathbb{N}$.

Theorem 6.55. *Let $\mathcal{D} \subset \mathbb{R}^d$ be a quasi-open set of finite measure. Under the conditions* (P1), (P2) *and* (P3), *the problem* (6.77) *has a solution.*

Suppose that the function $g : \mathbb{R}^p \to \mathbb{R}$ is locally Lipschitz and that each of the functionals $\mathcal{F}_i$, $i = 1, \dots, p$ is γ-Lip. If the quasi-open partition $(\Omega_1, \dots, \Omega_p)$ is a solution of (6.77), *then every quasi-open set Ω_i, $i = 1, \dots, p$, is an energy subsolution. In particular, we have*

(i) *the quasi-open sets Ω_i are bounded and have finite perimeter;*
(ii) *there are no triple points,* i.e. *if i, j and k are three different numbers, then*

$$\partial^M \Omega_i \cap \partial^M \Omega_j \cap \partial^M \Omega_k = \emptyset.^{10}$$

(iii) *There are open sets $\mathcal{D}_1, \dots, \mathcal{D}_p \subset \mathbb{R}^d$ such that*

$$\Omega_i \subset \mathcal{D}_i,\ \forall i \qquad \text{and} \qquad \Omega_i \cap \mathcal{D}_j = \emptyset,\ \text{if } i \neq j.$$

Proof. The existence part follows by Theorem 2.84. We now prove that each Ω_i is an energy subsolution. We set for simplicity $i = 1$ and let $\widetilde{\Omega}_1 \subset \Omega_1$ be a quasi-open set such that $d_\gamma(\widetilde{\Omega}_1, \Omega_1) < \varepsilon$. We now use the partition $(\widetilde{\Omega}_1, \Omega_2, \dots, \Omega_p)$ to test the optimality of $(\Omega_1, \dots, \Omega_p)$. By the Lipschitz continuity of g, the γ-Lip condition on $\mathcal{F}_1, \dots, \mathcal{F}_h$ and the minimality of $(\Omega_1, \dots, \Omega_h)$, we have

$$\begin{aligned} m_1\big(|\Omega_1| - |\widetilde{\Omega}_1|\big) &\leq g\big(\mathcal{F}_1(\widetilde{\Omega}_1), \mathcal{F}_2(\Omega_2), \dots, \mathcal{F}_h(\Omega_h)\big) \\ &\quad - g\big(\mathcal{F}_1(\Omega_1), \mathcal{F}_2(\Omega_2), \dots, \mathcal{F}_h(\Omega_h)\big) \\ &\leq L\big(\mathcal{F}_1(\widetilde{\Omega}_1) - \mathcal{F}_1(\Omega_1)\big) \leq CL\, d_\gamma(\widetilde{\Omega}_1, \Omega_1) \\ &\leq CL\big(E(\widetilde{\Omega}_1) - E(\Omega_1)\big), \end{aligned}$$

where L is the Lipschitz constant of g and C the constant from Definition 6.53. Repeating the argument for Ω_i, we obtain that it is a local shape subsolution for the functional $E(\Omega) + (CL)^{-1} m_i |\Omega|$. The claims (i), (ii) and (iii) follow by Theorem 4.22, Proposition 4.40 and Theorem 4.44. □

[10] We recall that by $\partial^M \Omega$ we denote the measure theoretic boundary of Ω.

Remark 6.56. A consequence of the claim (iii) of Theorem 6.55, we have that each cell Ω_i of a given optimal partition $(\Omega_1, \dots, \Omega_p)$ is a solution of the problem

$$\min\left\{\mathcal{F}_i(\Omega) :\ \Omega \subset \mathcal{D}_i \cap \mathcal{D},\ \Omega \text{ quasi-open},\ |\Omega| = |\Omega_i|\right\}. \tag{6.78}$$

Theorem 6.57. *Let $\mathcal{D} \subset \mathbb{R}^d$ be a bounded open set. Then every partition $(\Omega_1, \dots, \Omega_p)$, optimal for* (6.75), *is composed of energy subsolutions satisfying the conditions* (i), (ii) *and* (iii) *of Theorem* 6.55. *Moreover, we have that*

(iv) *For every $i \in \{1, \dots, p\}$, there is an open set $\mathcal{D}_i \subset \mathcal{D}$ such that the set Ω_i is a solution of the problem*

$$\min\left\{\lambda_{k_i}(\Omega) + m_i|\Omega| :\ \Omega \subset \mathcal{D}_i \text{ quasi-open}\right\}. \tag{6.79}$$

(v) *If $k_i = 1$, then the set Ω_i is open and connected.*

(vi) *If $k_i = 2$, then there are non-empty disjoint connected open sets ω_i^+ and ω_i^-, which are subsolutions for the functional $\lambda_1 + m_i|\cdot|$ and are such that the set $\omega_i := \omega_i^+ \cup \omega_i^- \subset \Omega_i$ is also a solution* (6.79) *and the partition $(\Omega_1, \dots, \omega_i, \dots, \Omega_p)$, of* (6.75).

Proof. We first note that, by Theorem 4.46, we have that λ_k is γ-Lip and so, satisfies the hypotheses of Theorem 6.55.

In order to prove (iv), we set $i = 1$ and then we note that by Theorem 6.55 (iii), there is an open set $\mathcal{D}_1 \subset \mathcal{D}$ such that

$$\Omega_1 \subset \mathcal{D}_1 \qquad \text{and} \qquad \mathcal{D}_1 \cap \Omega_i = \emptyset, \text{ for } i \geq 2.$$

Thus, we can use any quasi-open set $\Omega \subset \mathcal{D}_1$ and the associated quasi-open partition $(\Omega, \Omega_2, \dots, \Omega_p)$ to test the optimality of $(\Omega_1, \dots, \Omega_p)$, which gives that Ω_1 solves (6.79).

Now (v) and (vi) are consequences of (iv) and Proposition 6.15 and Proposition 6.16 from Section 6.3.3. □

Remark 6.58. We note that if we know that, for a generic bounded open set $\mathcal{D} \subset \mathbb{R}^d$, the problem

$$\min\left\{\lambda_k(\Omega) + m|\Omega| :\ \Omega \subset \mathcal{D},\ \Omega \text{ quasi-open}\right\},$$

has an open solution, then also the multiphase problem (6.75) has an open solution.

Chapter 7
Appendix: Shape optimization problems for graphs

In the previous chapters we discussed a wide variety of spectral optimization problems. In particular, we have a theory, which can be successfully applied to study the existence of optimal sets in the very general context of metric measure spaces. The variables in this case were always subsets of a given ambient space, since most of the geometric and analytical objects can be viewed as subspaces of some bigger space, this is quite a reasonable assumption. The more restrictive assumption, and the one that provided enough structure to develop the theory, concerns the cost functionals. More precisely, to each subset Ω of the ambient space X we associate in a specific way a subspace $H(\Omega)$ of some prescribed functional space H on X. The cost functionals with respect to which we optimize are in fact of the form $F(\Omega) = \mathcal{F}(H(\Omega))$, where $\mathcal{F}$ is a functional on the subspaces of H.

If we have a functional F for which we cannot prescribe a functional space H and representation of the form above, then the question becomes more involved. This is the case for example with the problem

$$\min\Big\{\mu_k(\Omega)\ :\ \Omega \subset \mathbb{R}^d,\ \Omega \text{ open},\ |\Omega| = 1\Big\},$$

where $\mu_k(\Omega)$ is the kth eigenvalue of the Neumann Laplacian on Ω. A similar problem occurs when we consider the problem

$$\min\Big\{\lambda_k(M) : \dim(M) = m,\ M \text{ embedded in } \mathbb{R}^d,\ \partial M = \mathcal{D},\ \mathcal{H}^m(M) \le 1\Big\},$$

where $\mathcal{D} \subset \mathbb{R}^d$ is a given compact embedded manifold of dimension $m-1$ and the optimization is over all embedded manifolds $M \subset \mathbb{R}^d$ of dimension $2 \le m < d$, with respect to the kth Dirichlet eigenvalue on M. By $\mathcal{H}^m$, as usual, we denote the m-dimensional Hausdorff measure on $\mathbb{R}^d$. The one dimensional analogue of this problem can be stated as

$$\min\Big\{\lambda_k(C) : C \subset \mathbb{R}^d \text{ closed connected set},\ \mathcal{D} \subset C,\ \mathcal{H}^1(C) \le 1\Big\}, \quad (7.1)$$

where $\mathcal{D}$ is a given (finite) closed set and λ_k is defined through an appropriately chosen functional space on C of continuous functions vanishing on $\mathcal{D}$. In this Chapter we will concentrate our attention on (7.1) in the case $k = 1$ and in the case of the Dirichlet Energy $\mathcal{E}(C)$[1].

Our main result is an existence theorem for optimal metric graphs, where the cost functional is the extension of the energy functional defined above. In Section 7.3 we show some explicit examples of optimal metric graphs. The last section contains a discussion, on the possible extensions of our result to other similar problems, as well as some open questions.

7.1. Sobolev space and Dirichlet Energy of a rectifiable set

Let $\mathcal{C} \subset \mathbb{R}^d$ be a closed connected set of finite length, *i.e.* $\mathcal{H}^1(\mathcal{C}) < \infty$, where $\mathcal{H}^1$ denotes the one-dimensional Hausdorff measure. On the set $\mathcal{C}$ we consider the metric

$$d_{\mathcal{C}}(x, y) = \inf\left\{ \int_0^1 |\dot{\gamma}(t)|\, dt \ :\ \gamma : [0,1] \to \mathbb{R}^d \text{ Lipschitz}, \right.$$
$$\left. \gamma([0,1]) \subset \mathcal{C},\ \gamma(0) = x,\ \gamma(1) = y \right\},$$

which is finite since, by the First Rectifiability Theorem (see [6, Theorem 4.4.1]), there is at least one rectifiable curve in C connecting x to y. For any function $u : \mathcal{C} \to \mathbb{R}$, Lipschitz with respect to the distance d (we also use the term d-Lipschitz), we define the norm

$$\|u\|^2_{H^1(\mathcal{C})} = \int_{\mathcal{C}} |u(x)|^2\, d\mathcal{H}^1(x) + \int_{\mathcal{C}} |u'|(x)^2\, d\mathcal{H}^1(x),$$

where

$$|u'|(x) = \limsup_{y \to x} \frac{|u(y) - u(x)|}{d_{\mathcal{C}}(x, y)}.$$

The Sobolev space $H^1(\mathcal{C})$ is the closure of the d-Lipschitz functions on $\mathcal{C}$ with respect to the norm $\|\cdot\|_{H^1(\mathcal{C})}$.

Remark 7.1. The inclusion $H^1(\mathcal{C}) \subset C(\mathcal{C}; \mathbb{R})$ is compact, where $C(\mathcal{C}; \mathbb{R})$ indicates the space of real-valued functions on $\mathcal{C}$, continuous with respect to the metric d. In fact, for each $x, y \in \mathcal{C}$, there is a rectifiable curve

[1] The change of notation with respect to the previous chapters is due to the fact that the letter E is reserved for the number of edges of graph.

$\gamma : [0, d(x,y)] \to \mathcal{C}$ connecting x to y, which we may assume arc-length parametrized. Thus, for any $u \in H^1(\mathcal{C})$, we have that

$$\begin{aligned}|u(x) - u(y)| &\leq \int_0^{d(x,y)} \left| \frac{d}{dt} u(\gamma(t)) \right| dt \\ &\leq d(x,y)^{1/2} \left(\int_0^{d(x,y)} \left| \frac{d}{dt} u(\gamma(t)) \right|^2 dt \right)^{1/2} \\ &\leq d(x,y)^{1/2} \|u'\|_{L^2(\mathcal{C})},\end{aligned}$$

and so, u is $1/2$-Hölder continuous. On the other hand, for any $x \in \mathcal{C}$, we have that

$$\begin{aligned}\int_{\mathcal{C}} u(y) d\mathcal{H}^1(y) &\geq \int_{\mathcal{C}} \left(u(x) - d(x,y)^{1/2} \|u'\|_{L^2(\mathcal{C})} \right) d\mathcal{H}^1(y) \\ &\geq l u(x) - l^{3/2} \|u'\|_{L^2(\mathcal{C})},\end{aligned}$$

where $l = \mathcal{H}^1(\mathcal{C})$. Thus, we obtain the L^∞ bound

$$\|u\|_{L^\infty} \leq l^{-1/2} \|u\|_{L^2(\mathcal{C})} + l^{1/2} \|u'\|_{L^2(\mathcal{C})} \leq (l^{-1/2} + l^{1/2}) \|u\|_{H^1(\mathcal{C})}.$$

and so, by the Ascoli-Arzelá Theorem, we have that the inclusion is compact.

Remark 7.2. By the same argument as in Remark 7.1 above, we have that for any $u \in H^1(\mathcal{C})$, the $(1,2)$-Poincaré inequality holds, *i.e.*

$$\int_{\mathcal{C}} \left| u(x) - \frac{1}{l} \int_{\mathcal{C}} u \, d\mathcal{H}^1 \right| d\mathcal{H}^1(x) \leq l^{3/2} \left(\int_{\mathcal{C}} |u'|^2 d\mathcal{H}^1 \right)^{1/2}. \tag{7.2}$$

Moreover, if $u \in H^1(\mathcal{C})$ is such that $u(x) = 0$ for some point $x \in \mathcal{C}$, then we have the Poincaré inequality:

$$\|u\|_{L^2(\mathcal{C})} \leq l^{1/2} \|u\|_{L^\infty(\mathcal{C})} \leq l \|u'\|_{L^2(\mathcal{C})}. \tag{7.3}$$

Since $\mathcal{C}$ is supposed connected, by the Second Rectifiability Theorem (see [6, Theorem 4.4.8]) there exists a countable family of injective arc-length parametrized Lipschitz curves $\gamma_i : [0, l_i] \to \mathcal{C}$, $i \in \mathbb{N}$ and an $\mathcal{H}^1$-negligible set $N \subset \mathcal{C}$ such that

$$\mathcal{C} = N \cup \left(\bigcup_i Im(\gamma_i) \right),$$

where $Im(\gamma_i) = \gamma_i([0, l_i])$. By the chain rule (see Lemma 7.3 below) we have

$$\left|\frac{d}{dt}u(\gamma_i(t))\right| = |u'|(\gamma_i(t)), \qquad \forall i \in \mathbb{N}$$

and so, we obtain for the norm of $u \in H^1(\mathcal{C})$:

$$\|u\|^2_{H^1(\mathcal{C})} = \int_{\mathcal{C}} |u(x)|^2 \, d\mathcal{H}^1(x) + \sum_i \int_0^{l_i} \left|\frac{d}{dt}u(\gamma_i(t))\right|^2 dt. \tag{7.4}$$

Moreover, we have the inclusion

$$H^1(\mathcal{C}) \subset \oplus_{i\in\mathbb{N}} H^1([0, l_i]), \tag{7.5}$$

which gives the reflexivity of $H^1(\mathcal{C})$ and the lower semicontinuity of the $H^1(\mathcal{C})$ norm, with respect to the strong convergence in $L^2(\mathcal{C})$.

Lemma 7.3. *Let $\gamma : [0, l] \to \mathbb{R}^d$ be an injective arc-length parametrized Lipschitz curve with $\gamma([0, l]) \subset \mathcal{C}$. Then we have*

$$\left|\frac{d}{dt}u(\gamma(t))\right| = |u'|(\gamma(t)), \qquad \textit{for } \mathcal{L}^1\textit{-a.e. } t \in [0, l]. \tag{7.6}$$

Proof. Let $u : \mathcal{C} \to \mathbb{R}$ be a Lipschitz map with Lipschitz constant $Lip(u)$ with respect to the distance d. We prove that the chain rule (7.6) holds in all the points $t \in [0, l]$ which are Lebesgue points for $\left|\frac{d}{dt}u(\gamma(t))\right|$ and such that the point $\gamma(t)$ has density one, *i.e.*

$$\lim_{r\to 0} \frac{\mathcal{H}^1\big(C \cap B_r(\gamma(t))\big)}{2r} = 1, \tag{7.7}$$

(thus almost every points, see for istance [82]) where $B_r(x)$ indicates the ball of radius r in $\mathbb{R}^d$. Since, $\mathcal{H}^1$-almost all points $x \in \mathcal{C}$ have this property, we obtain the conclusion. Without loss of generality, we consider $t = 0$. Let us first prove that $|u'|(\gamma(0)) \geq \left|\frac{d}{dt}u(\gamma(0))\right|$. We have that

$$|u'|(\gamma(0)) \geq \limsup_{t\to 0} \frac{|u(\gamma(t)) - u(\gamma(0))|}{d(\gamma(t), \gamma(0))} = \left|\frac{d}{dt}u(\gamma(0))\right|,$$

since γ is arc-length parametrized. On the other hand, we have

$$\begin{aligned} |u'|(x) &= \limsup_{y\to x} \frac{|u(y)-u(x)|}{d(y,x)} \\ &= \lim_{n\to\infty} \frac{|u(y_n)-u(x)|}{d(y_n,x)} \\ &= \lim_{n\to\infty} \frac{|u(\gamma_n(r_n))-u(\gamma_n(0))|}{r_n} \\ &\le \lim_{n\to\infty} \frac{1}{r_n}\int_0^{r_n} \left|\frac{d}{dt}u(\gamma_n(t))\right| dt \end{aligned} \tag{7.8}$$

where $y_n \in C$ is a sequence of points which realizes the lim sup and $\gamma_n : [0, r_n] \to \mathbb{R}^d$ is a geodesic in C connecting x to y_n. Let $S_n = \{t : \gamma_n(t) = \gamma(t)\} \subset [0, r_n]$, then, we have

$$\begin{aligned} \int_0^{r_n} \left|\frac{d}{dt}u(\gamma_n(t))\right|^2 dt &\le \int_{S_n} \left|\frac{d}{dt}u(\gamma(t))\right|^2 dt + Lip(u)\,(r_n - |S_n|) \\ &\le \int_0^{r_n} \left|\frac{d}{dt}u(\gamma(t))\right|^2 dt \\ &\quad + Lip(u)\,(\mathcal{H}^1(B_{r_n}(\gamma(0))\cap C) - 2r_n), \end{aligned} \tag{7.9}$$

and so, since $\gamma(0)$ is of density 1, we conclude applying this estimate to (7.9). □

Given a set of points $\mathcal{D} = \{D_1, \ldots, D_k\} \subset \mathbb{R}^d$ we define the admissible class $\mathcal{A}(\mathcal{D}; l)$ as the family of all closed connected sets C containing $\mathcal{D}$ and of length $\mathcal{H}^1(C) = l$. For any $C \in \mathcal{A}(\mathcal{D}; l)$ we consider the space of Sobolev functions which satisfy a Dirichlet condition at the points D_i:

$$H_0^1(C; \mathcal{D}) = \{u \in H^1(C) : u(D_j) = 0,\ j = 1\ldots, k\},$$

which is well-defined by Remark 7.1. For the points D_i we use the term *Dirichlet points*. The *Dirichlet Energy* of the set C with respect to $D_1, \ldots, D_k$ is defined as

$$\mathcal{E}(C; \mathcal{D}) = \min\left\{J(u) \ : \ u \in H_0^1(C; \mathcal{D})\right\}, \tag{7.10}$$

where

$$J(u) = \frac{1}{2}\int_C |u'|(x)^2\, d\mathcal{H}^1(x) - \int_C u(x)\, d\mathcal{H}^1(x). \tag{7.11}$$

Remark 7.4. For any $\mathcal{C} \in \mathcal{A}(\mathcal{D}; l)$ there exists a unique minimizer of the functional $J : H^1_0(\mathcal{C}; \mathcal{D}) \to \mathbb{R}$. In fact, by Remark 7.1 we have that a minimizing sequence is bounded in H^1 and compact in L^2. The conclusion follows by the semicontinuity of the L^2 norm of the gradient, with respect to the strong L^2 convergence, which is an easy consequence of equation (7.4). The uniqueness follows by the strict convexity of the L^2 norm and the sub-additivity of the gradient $|u'|$. We call the minimizer of J the *energy function* of $\mathcal{C}$ with Dirichlet conditions in $D_1, \dots, D_k$.

Remark 7.5. Let $u \in H^1(\mathcal{C})$ and $v : \mathcal{C} \to \mathbb{R}$ be a positive Borel function. Applying the chain rule, as in (7.4), and the one dimensional co-area formula (see for instance [5]), we obtain a co-area formula for the functions $u \in H^1(\mathcal{C})$:

$$\begin{aligned} \int_{\mathcal{C}} v(x)|u'|(x)\, d\mathcal{H}^1(x) &= \sum_i \int_0^{l_i} \left|\frac{d}{dt}u(\gamma_i(t))\right| v(\gamma_i(t))\, dt \\ &= \sum_i \int_0^{+\infty} \Big(\sum_{u\circ\gamma_i(t)=\tau} v \circ \gamma_i(t) \Big)\, d\tau \qquad (7.12) \\ &= \int_0^{+\infty} \Big(\sum_{u(x)=\tau} v(x) \Big)\, d\tau. \end{aligned}$$

7.1.1. Optimization problem for the Dirichlet Energy on the class of connected sets

We study the following shape optimization problem:

$$\min\left\{\mathcal{E}(\mathcal{C}; \mathcal{D}) \ :\ \mathcal{C} \in \mathcal{A}(\mathcal{D}; l)\right\}, \qquad (7.13)$$

where $\mathcal{D} = \{D_1, \dots, D_k\}$ is a given set of points in $\mathbb{R}^d$ and l is a prescribed length.

Remark 7.6. When $k = 1$ problem (7.13) reads as

$$\mathcal{E} = \min\left\{\mathcal{E}(\mathcal{C}; D) \ :\ \mathcal{H}^1(C) = l,\ D \in \mathcal{C}\right\}, \qquad (7.14)$$

where $D \in \mathbb{R}^d$ and $l > 0$. In this case the solution is a line of length l starting from D (see Figure 7.1). A proof of this fact, in a slightly different context, can be found in [64] and we report it here for the sake of completeness.

Let $\mathcal{C} \in \mathcal{A}(D; l)$ be a generic connected set and let $w \in H^1_0(\mathcal{C}; D)$ be its energy function, *i.e.* the minimizer of J on $\mathcal{C}$. Let $v : [0, l] \to \mathbb{R}$ be such that $\mu_w(\tau) = \mu_v(\tau)$, where μ_w and μ_v are the distribution function of w and v respectively, defined by

$$\mu_w(\tau) = \mathcal{H}^1(w \le \tau) = \sum_i \mathcal{H}^1(w_i \le \tau), \qquad \mu_v(\tau) = \mathcal{H}^1(v \le \tau).$$

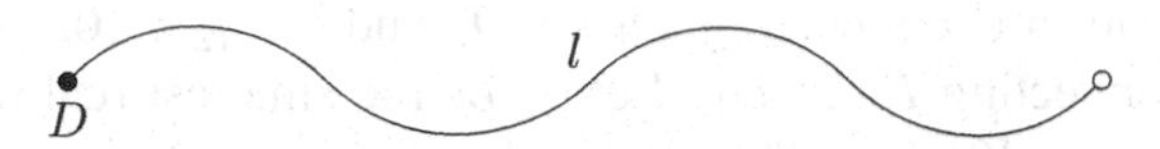

Figure 7.1. The optimal graph with only one Dirichlet point.

It is easy to see that, by the Cavalieri Formula, $\|v\|_{L^p([0,l])} = \|w\|_{L^p(C)}$, for each $p \geq 1$. By the co-area formula (7.12)

$$\begin{aligned}\int_C |w'|^2 \, d\mathcal{H}^1 &= \int_0^{+\infty} \Big(\sum_{w=\tau} |w'|\Big) \, d\tau \geq \int_0^{+\infty} \Big(\sum_{w=\tau} \frac{1}{|w'|}\Big)^{-1} d\tau \\ &= \int_0^{+\infty} \frac{d\tau}{\mu'_w(\tau)},\end{aligned} \tag{7.15}$$

where we used the Cauchy-Schwartz inequality and the identity

$$\mu_w(t) = \mathcal{H}^1(\{w \leq t\}) = \int_{w \leq t} \frac{|w'|}{|w'|} \, ds = \int_0^t \Big(\sum_{w=s} \frac{1}{|w'|}\Big) \, ds$$

which implies that $\mu'_w(t) = \sum_{w=t} \frac{1}{|w'|}$. The same argument applied to v gives:

$$\int_0^l |v'|^2 \, dx = \int_0^{+\infty} \Big(\sum_{v=\tau} |v'|\Big) \, d\tau = \int_0^{+\infty} \frac{d\tau}{\mu'_v(\tau)}. \tag{7.16}$$

Since $\mu_w = \mu_v$, the conclusion follows.

The following Theorem shows that it is enough to study the problem (7.13) on the class of finite graphs embedded in $\mathbb{R}^d$. Consider the subset $\mathcal{A}_N(\mathcal{D}; l) \subset \mathcal{A}(\mathcal{D}; l)$ of those sets C, for which there exists a finite family $\gamma_i : [0, l_i] \to \mathbb{R}$, $i = 1, \ldots, n$ with $n \leq N$, of injective rectifiable curves such that $\cup_i \gamma_i([0, l_i]) = C$ and $\gamma_i((0, l_i)) \cap \gamma_j((0, l_j)) = \emptyset$, for each $i \neq j$.

Theorem 7.7. *Consider the set of distinct points* $\mathcal{D} = \{D_1, \ldots, D_k\} \subset \mathbb{R}^d$ *and* $l > 0$. *We have that*

$$\inf\big\{\mathcal{E}(C; \mathcal{D}) \; : \; C \in \mathcal{A}(\mathcal{D}; l)\big\} = \inf\big\{\mathcal{E}(C; \mathcal{D}) \; : \; C \in \mathcal{A}_N(\mathcal{D}; l)\big\}, \tag{7.17}$$

where $N = 2k - 1$. *Moreover, if* C *is a solution of the problem* (7.13), *then there is also a solution* $\widetilde{C}$ *of the same problem such that* $\widetilde{C} \in \mathcal{A}_N(\mathcal{D}; l)$.

Proof. Consider a connected set $C \in \mathcal{A}(\mathcal{D}; l)$. We show that there is a set $\widetilde{C} \in \mathcal{A}_N(\mathcal{D}; l)$ such that $\mathcal{E}(\widetilde{C}; \mathcal{D}) \leq \mathcal{E}(C; \mathcal{D})$. Let $\eta_1 : [0, a_1] \to C$

be a geodesic in $\mathcal{C}$ connecting D_1 to D_2 and let $\eta_2 : [0, a] \to \mathcal{C}$ be a geodesic connecting D_3 to D_1. Let a_2 be the smallest real number such that $\eta_2(a_2) \in \eta_1([0, a_1])$. Then, consider the geodesic η_3 connecting D_4 to D_1 and the smallest real number a_3 such that $\eta_3(a_3) \in \eta_1([0, a_1]) \cup \eta_2([0, a_2])$. Repeating this operation, we obtain a family of geodesics η_i, $i = 1, \dots, k-1$ which intersect each other in a finite number of points. Each of these geodesics can be decomposed in several parts according to the intersection points with the other geodesics (see Figure 7.2).

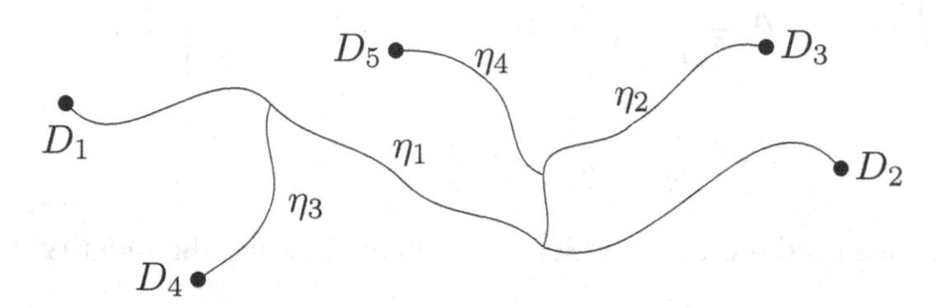

Figure 7.2. Construction of the set $\mathcal{C}'$.

So, we can consider a new family of geodesics (still denoted by η_i), $\eta_i : [0, l_i] \to \mathcal{C}, i = 1, \dots, n$, which does not intersect each other in internal points. Note that, by an induction argument on $k \geq 2$, we have $n \leq 2k-3$. Let $\mathcal{C}' = \cup_i \eta_i([0, l_i]) \subset \mathcal{C}$. By the Second Rectifiability Theorem (see [6, Theorem 4.4.8]), we have that

$$\mathcal{C} = \mathcal{C}' \cup E \cup \Gamma,$$

where $\mathcal{H}^1(E) = 0$ and $\Gamma = \left(\bigcup_{j=1}^{+\infty} \gamma_j\right)$, where $\gamma_j : [0, l_j] \to \mathcal{C}$ for $j \geq 1$ is a family of Lipschitz curves in C. Moreover, we can suppose that $\mathcal{H}^1(\Gamma \cap \mathcal{C}') = 0$. In fact, if $\mathcal{H}^1(Im(\gamma_j) \cap \mathcal{C}') \neq 0$ for some $j \in \mathbb{N}$, we consider the restriction of γ_j to (the closure of) each connected component of $\gamma_j^{-1}(\mathbb{R}^d \setminus \mathcal{C}')$.

Let $w \in H_0^1(\mathcal{C}; \mathcal{D})$ be the energy function on C and let $v : [0, \mathcal{H}^1(\Gamma)] \to \mathbb{R}$ be a monotone increasing function such that $|\{v \leq \tau\}| = \mathcal{H}^1(\{w \leq \tau\} \cap \Gamma)$. Reasoning as in Remark 7.6, we have that

$$\frac{1}{2}\int_0^{\mathcal{H}^1(\Gamma)} |v'|^2\, dx - \int_0^{\mathcal{H}^1(\Gamma)} v\, dx \leq \frac{1}{2}\int_\Gamma |w'|^2\, d\mathcal{H}^1 - \int_\Gamma w\, d\mathcal{H}^1. \quad (7.18)$$

Let $\sigma : [0, \mathcal{H}^1(\Gamma)] \to \mathbb{R}^d$ be an injective arc-length parametrized curve such that $Im(\sigma) \cap \mathcal{C}' = \sigma(0) = x'$, where $x' \in \mathcal{C}'$ is the point where $w_{|\mathcal{C}'}$ achieves its maximum. Let $\widetilde{\mathcal{C}} = \mathcal{C}' \cup Im(\sigma)$. Notice that $\widetilde{\mathcal{C}}$ connects the points $D_1, \dots, D_k$ and has length $\mathcal{H}^1(\widetilde{\mathcal{C}}) = \mathcal{H}^1(\mathcal{C}') + \mathcal{H}^1(Im(\sigma)) =$

$\mathcal{H}^1(\mathcal{C}') + \mathcal{H}^1(\Gamma) = l$. Moreover, we have

$$\mathcal{E}(\widetilde{\mathcal{C}}; \mathcal{D}) \leq J(\widetilde{w}) \leq J(w) = \mathcal{E}(\mathcal{C}; \mathcal{D}), \tag{7.19}$$

where $\widetilde{w}$ is defined by

$$\widetilde{w}(x) = \begin{cases} w(x), & \text{if } x \in \mathcal{C}', \\ v(t) + w(x') - v(0), & \text{if } x = \sigma(t). \end{cases} \tag{7.20}$$

We have then (7.19), *i.e.* the energy decreases. We conclude by noticing that the point x' where we attach σ to $\mathcal{C}'$ may be an internal point for η_i, *i.e.* a point such that $\eta_i^{-1}(x') \in (0, l_i)$. Thus, the set $\widetilde{\mathcal{C}}$ is composed of at most $2k - 1$ injective arc-length parametrized curves which does not intersect in internal points, *i.e.* $\widetilde{\mathcal{C}} \in \mathcal{A}_{2k-1}(\mathcal{D}; l)$. □

Remark 7.8. Theorem 7.7 above provides a nice class of admissible sets, where to search for a minimizer of the energy functional $\mathcal{E}$. Indeed, according to its proof, we may limit ourselves to consider only graphs $\mathcal{C}$ such that:

1. $\mathcal{C}$ is a tree, *i.e.* it does not contain any closed loop;
2. the Dirichlet points D_i are vertices of degree one (endpoints) for $\mathcal{C}$;
3. there are at most $k - 1$ other vertices; if a vertex has degree three or more, we call it Kirchhoff point;
4. there is at most one vertex of degree one for $\mathcal{C}$ which is not a Dirichlet point. In this vertex the energy function w satisfies Neumann boundary condition $w' = 0$ and so we call it Neumann point.

The previous properties are also necessary conditions for the optimality of the graph $\mathcal{C}$ (see Proposition 7.19 for more details).

As we show in Example 7.24, the problem (7.13) may not have a solution in the class of connected sets. It is worth noticing that the lack of existence only occurs for particular configurations of the Dirichlet points D_i and not because of some degeneracy of the cost functional $\mathcal{E}$. In fact, we are able to produce other examples in which an optimal graph exists (see Section 7.3).

7.2. Sobolev space and Dirichlet Energy of a metric graph

Let $V = \{V_1, \ldots, V_N\}$ be a finite set and let $E \subset \{e_{ij} = \{V_i, V_j\}\}$ be a set of pairs of elements of V. We define combinatorial graph (or just graph) a pair $\Gamma = (V, E)$. We say the set $V = V(\Gamma)$ is the set of vertices of Γ and the set $E = E(\Gamma)$ is the set of edges. We denote with $|E|$ and

$|V|$ the cardinalities of E and V and with $\deg(V_i)$ the degree of the vertex V_i, *i.e.* the number of edges incident to V_i.

A *path* in the graph Γ is a sequence $V_{\alpha_0}, \dots, V_{\alpha_n} \in V$ such that for each $k = 0, \dots, n-1$, we have that $\{V_{\alpha_k}, V_{\alpha_{k+1}}\} \in E$. With this notation, we say that the path connects V_{i_0} to V_{i_α}. The path is said to be *simple* if there are no repeated vertices in $V_{\alpha_0}, \dots, V_{\alpha_n}$. We say that the graph $\Gamma = (V, E)$ is connected, if for each pair of vertices $V_i, V_j \in V$ there is a path connecting them. We say that the connected graph Γ is a tree, if after removing any edge, the graph becomes not connected.

If we associate a non-negative length (or weight) to each edge, *i.e.* a map $l : E(\Gamma) \to [0, +\infty)$, then we say that the couple (Γ, l) determines a metric graph of length

$$l(\Gamma) := \sum_{i<j} l(e_{ij}).$$

A function $u : \Gamma \to \mathbb{R}^n$ on the metric graph Γ is a collection of functions $u_{ij} : [0, l_{ij}] \to \mathbb{R}$, for $1 \leq i \neq j \leq N$, such that:

1. $u_{ji}(x) = u_{ij}(l_{ij} - x)$, for each $1 \leq i \neq j \leq N$,
2. $u_{ij}(0) = u_{ik}(0)$, for all $\{i, j, k\} \subset \{1, \dots, N\}$,

where we used the notation $l_{ij} = l(e_{ij})$. A function $u : \Gamma \to \mathbb{R}$ is said continuous ($u \in C(\Gamma)$), if $u_{ij} \in C([0, l_{ij}])$, for all $i, j \in \{1, \dots, n\}$. We call $L^p(\Gamma)$ the space of p-summable functions ($p \in [1, +\infty)$), *i.e.* the functions $u = (u_{ij})_{ij}$ such that

$$\|u\|^p_{L^p(\Gamma)} := \frac{1}{2} \sum_{i,j} \|u_{ij}\|^p_{L^p(0,l_{ij})} < +\infty,$$

where $\|\cdot\|_{L^p(a,b)}$ denotes the usual L^p norm on the interval $[a, b]$. As usual, the space $L^2(\Gamma)$ has a Hilbert structure endowed by the scalar product:

$$\langle u, v\rangle_{L^2(\Gamma)} := \frac{1}{2} \sum_{i,j} \langle u_{ij}, v_{ij}\rangle_{L^2(0,l_{ij})}.$$

We define the Sobolev space $H^1(\Gamma)$ as:

$$H^1(\Gamma) = \Big\{u \in C(\Gamma) : u_{ij} \in H^1([0, l_{ij}]),\ \forall i, j \in \{1, \dots, n\}\Big\},$$

which is a Hilbert space with the norm

$$\begin{aligned} \|u\|^2_{H^1(\Gamma)} &= \frac{1}{2} \sum_{i,j} \|u_{ij}\|^2_{H^1([0,l_{ij}])} \\ &= \frac{1}{2} \sum_{i,j} \left(\int_0^{l_{ij}} |u_{ij}|^2\, dx + \int_0^{l_{ij}} |u'_{ij}|^2\, dx \right). \end{aligned}$$

Remark 7.9. Note that for $u \in H^1(\Gamma)$ the family of derivatives $(u'_{ij})_{1\le i\ne j\le N}$ is not a function on Γ, since $u'_{ij}(x) = \frac{\partial}{\partial x}u_{ji}(l_{ij}-x) = -u'_{ji}(l_{ij}-x)$. Thus, we work with the function $|u'| = \big(|u'_{ij}|\big)_{1\le i\ne j\le N} \in L^2(\Gamma)$.

Remark 7.10. The inclusions $H^1(\Gamma)\subset C(\Gamma)$ and $H^1(\Gamma)\subset L^2(\Gamma)$ are compact, since the corresponding inclusions, for each of the intervals $[0,l_{ij}]$, are compact. By the same argument, the H^1 norm is lower semi-continuous with respect to the strong L^2 convergence of the functions in $H^1(\Gamma)$.

For any subset $W = \{W_1,\dots,W_k\}$ of the set of vertices $V(\Gamma) = \{V_1,\dots,V_N\}$, we introduce the Sobolev space with *Dirichlet boundary conditions* on W:

$$H_0^1(\Gamma;W) = \big\{u\in H^1(\Gamma)\ :\ u(W_1)=\dots=u(W_k)=0\big\}.$$

Remark 7.11. Arguing as in Remark 7.1 we have that for each $u \in H_0^1(\Gamma;W)$ and, more generally, for each $u\in H^1(\Gamma)$ such that $u(V_\alpha)=0$ for some $\alpha = 1,\dots,N$, the Poincaré inequality

$$\|u\|_{L^2(\Gamma)} \le l^{1/2}\|u\|_{L^\infty} \le l\|u'\|_{L^2(\Gamma)}, \tag{7.21}$$

holds, where

$$\|u'\|^2_{L^2(\Gamma)} := \int_\Gamma |u'|^2\,dx := \sum_{i,j}\int_0^{l_{ij}} |u'_{ij}|^2\,dx.$$

On the metric graph Γ, we consider the Dirichlet Energy with respect to W:

$$\mathcal{E}(\Gamma;W) = \inf\big\{J(u)\ :\ u\in H_0^1(\Gamma;W)\big\}, \tag{7.22}$$

where the functional $J : H_0^1(\Gamma;W)\to\mathbb{R}$ is defined by

$$J(u) = \frac{1}{2}\int_\Gamma |u'|^2 dx - \int_\Gamma u\,dx. \tag{7.23}$$

Lemma 7.12. *Given a metric graph Γ of length l and Dirichlet points $\{W_1,\dots,W_k\}\subset V(\Gamma) = \{V_1,\dots,V_N\}$, there is a unique function $w = (w_{ij})_{1\le i\ne j\le N} \in H_0^1(\Gamma;W)$ which minimizes the functional J. Moreover, we have*

(i) *for each $1\le i\ne j\le N$ and each $t\in(0,l_{ij})$, $-w''_{ij} = 1$;*

(ii) *at every vertex $V_i \in V(\Gamma)$, which is not a Dirichlet point, w satisfies the Kirchhoff's law:*

$$\sum_j w'_{ij}(0) = 0,$$

where the sum is over all j for which the edge e_{ij} exists;

Furthermore, the conditions (i) *and* (ii) *uniquely determine w.*

Proof. The existence is a consequence of Remark 7.10 and the uniqueness is due to the strict convexity of the L^2 norm. For any $\varphi \in H_0^1(\Gamma; W)$, we have that 0 is a critical point for the function

$$\varepsilon \mapsto \frac{1}{2}\int_\Gamma |(w+\varepsilon\varphi)'|^2\,dx - \int_\Gamma (w+\varepsilon\varphi)\,dx.$$

Since φ is arbitrary, we obtain the first claim. The Kirchhoff's law at the vertex V_i follows by choosing φ supported in a "small neighborhood" of V_i. The last claim is due to the fact that if $u \in H_0^1(\Gamma; W)$ satisfies (i) and (ii), then it is an extremal for the convex functional J and so, $u = w$. $\square$

Remark 7.13. As in Remark 7.5 we have that the co-area formula holds for the functions $u \in H^1(\Gamma)$ and any positive Borel (on each edge) function $v : \Gamma \to \mathbb{R}$:

$$\begin{aligned}
\int_\Gamma v(x)|u'|(x)\,dx &= \sum_{1\le i<j\le N}\int_0^{l_{ij}} |u'_{ij}(x)|\,v(x)\,dx \\
&= \sum_{1\le i<j\le N}\int_0^{+\infty}\Big(\sum_{u_{ij}(x)=\tau} v(x)\Big)\,d\tau \qquad (7.24)\\
&= \int_0^{+\infty}\Big(\sum_{u(x)=\tau} v(x)\Big)\,d\tau.
\end{aligned}$$

7.2.1. Optimization problem for the Dirichlet Energy on the class of metric graphs

We say that the continuous function $\gamma = (\gamma_{ij})_{1\le i\ne j\le N} : \Gamma \to \mathbb{R}^d$ is an *immersion* of the metric graph Γ into $\mathbb{R}^d$, if for each $1 \le i \ne j \le N$ the function $\gamma_{ij} : [0, l_{ij}] \to \mathbb{R}^d$ is an injective arc-length parametrized curve. We say that $\gamma : \Gamma \to \mathbb{R}^d$ is an *embedding*, if it is an immersion which is also injective, *i.e.* for any $i \ne j$ and $i' \ne j'$, we have

1. $\gamma_{ij}((0, l_{ij})) \cap \gamma_{i'j'}([0, l_{i'j'}]) = \emptyset$,
2. $\gamma_{ij}(0) = \gamma_{i'j'}(0)$, if and only if, $i = i'$.

Remark 7.14. Suppose that Γ is a metric graph of length l and that $\gamma : \Gamma \to \mathbb{R}^d$ is an embedding. Then the set $\mathcal{C} := \gamma(\Gamma)$ is rectifiable of length $\mathcal{H}^1(\gamma(\Gamma)) = l$ and the spaces $H^1(\Gamma)$ and $H^1(\mathcal{C})$ are isometric as Hilbert spaces, where the isomorphism is given by the composition with the function γ.

Consider a finite set of distinct points $\mathcal{D} = \{D_1, \dots, D_k\} \subset \mathbb{R}^d$ and let $l \geq St(\mathcal{D})$, where $St(\mathcal{D})$ is the length of the Steiner set, the minimal among the ones connecting all the points D_i (see [6] for more details on the Steiner problem). Consider the shape optimization problem:

$$\min\Big\{\mathcal{E}(\Gamma;\mathcal{V}) : \Gamma \in CMG, l(\Gamma) = l, \\ \mathcal{V} \subset V(\Gamma), \exists \gamma : \Gamma \to \mathbb{R}^d \text{ immersion}, \gamma(\mathcal{V}) = \mathcal{D}\Big\}, \tag{7.25}$$

where CMG indicates the class of connected metric graphs. Note that since $l \geq St(\mathcal{D})$, there is a metric graph and an *embedding* $\gamma : \Gamma \to \mathbb{R}^d$ such that $\mathcal{D} \subset \gamma(V(\Gamma))$ and so the admissible set in the problem (7.25) is non-empty, as well as the admissible set in the problem

$$\min\Big\{\mathcal{E}(\Gamma;\mathcal{V}) \ : \ \Gamma \in CMG, \ l(\Gamma) = l, \\ \mathcal{V} \subset V(\Gamma), \ \exists \gamma : \Gamma \to \mathbb{R}^d \text{ embedding}, \gamma(\mathcal{V}) = \mathcal{D}\Big\}. \tag{7.26}$$

We will see in Theorem 7.18 that problem (7.25) admits a solution, while Example 7.24 shows that in general an optimal embedded graph for problem (7.26) may not exist.

Remark 7.15. By Remark 7.14 and by the fact that the functionals we consider are invariant with respect to the isometries of the Sobolev space, we have that the problems (7.13) and (7.26) are equivalent, *i.e.* if $\Gamma \in CMG$ and $\gamma : \Gamma \to \mathbb{R}^d$ is an embedding such that the pair (Γ, γ) is a solution of (7.26), then the set $\gamma(\Gamma)$ is a solution of the problem (7.13). On the other hand, if C is a solution of the problem (7.13), by Theorem 7.7, we can suppose that $C = \bigcup_{i=1}^N \gamma_i([0, l_i])$, where γ_i are injective arc-length parametrized curves, which does not intersect internally. Thus, we can construct a metric graph Γ with vertices the set of points $\{\gamma_i(0), \gamma_i(l_i)\}_{i=1}^N \subset \mathbb{R}^d$, and N edges of lengths l_i such that two vertices are connected by an edge, if and only if they are the endpoints of the same curve γ_i. The function $\gamma = (\gamma_i)_{i=1,\dots,N} : \Gamma \to \mathbb{R}^d$ is an embedding by construction and by Remark 7.14, we have $\mathcal{E}(C;\mathcal{D}) = \mathcal{E}(\Gamma;\mathcal{D})$.

Theorem 7.16. *Let $\mathcal{D} = \{D_1, \dots, D_k\} \subset \mathbb{R}^d$ be a finite set of points and let $l \geq St(\mathcal{D})$ be a positive real number. Suppose that Γ is a connected* *metric graph of length* l, $\mathcal{V} \subset V(\Gamma)$ *is a set of vertices of* Γ *and* $\gamma : \Gamma \to \mathbb{R}^d$ *is an immersion (embedding) such that* $\mathcal{D} = \gamma(\mathcal{V})$. *Then there exists a connected metric graph* $\widetilde{\Gamma}$ *of at most* $2k$ *vertices and* $2k-1$ *edges, a set* $\widetilde{\mathcal{V}} \subset V(\widetilde{\Gamma})$ *of vertices of* $\widetilde{\Gamma}$ *and an immersion (embedding)* $\widetilde{\gamma} : \widetilde{\Gamma} \to \mathbb{R}^d$ *such that* $\mathcal{D} = \widetilde{\gamma}(\widetilde{\mathcal{V}})$ *and*

$$\mathcal{E}(\widetilde{\Gamma}; \widetilde{\mathcal{V}}) \leq \mathcal{E}(\Gamma; \mathcal{V}). \tag{7.27}$$

Proof. We repeat the argument from Theorem 7.7. We first construct a connected metric graph Γ' such that $V(\Gamma') \subset V(\Gamma)$ and the edges of Γ' are appropriately chosen paths in Γ. The edges of Γ, which are not part of any of these paths, are symmetrized in a single edge, which we attach to Γ' in a point, where the restriction of w to Γ' achieves its maximum, where w is the energy function for Γ.

Suppose that $V_1, \dots, V_k \in \mathcal{V} \subset V(\Gamma)$ are such that $\gamma(V_i) = D_i$, $i = 1, \dots, k$. We start constructing Γ' by taking $\widetilde{\mathcal{V}} := \{V_1, \dots, V_k\} \subset V(\Gamma')$. Let $\sigma_1 = \{V_{i_0}, V_{i_1}, \dots, V_{i_s}\}$ be a path of different vertices (*i.e.* simple path) connecting $V_1 = V_{i_s}$ to $V_2 = V_{i_0}$ and let $\tilde{\sigma}_2 = \{V_{j_0}, V_{j_1}, \dots, V_{j_t}\}$ be a simple path connecting $V_1 = V_{j_t}$ to $V_3 = V_{j_0}$. Let $t' \in \{1, \dots, t\}$ be the smallest integer such that $V_{j_{t'}} \in \sigma_1$. Then we set $V_{j_{t'}} \in V(\Gamma')$ and $\sigma_2 = \{V_{j_0}, V_{j_1}, \dots, V_{j_{t'}}\}$. Consider a simple path $\tilde{\sigma}_3 = \{V_{m_0}, V_{m_1}, \dots, V_{m_r}\}$ connecting $V_1 = V_{m_r}$ to $V_3 = V_{m_0}$ and the smallest integer r' such that $V_{m_{r'}} \in \sigma_1 \cup \sigma_2$. We set $V_{m_{r'}} \in V(\Gamma')$ and $\sigma_3 = \{V_{m_0}, V_{m_1}, \dots, V_{m_{r'}}\}$. We continue the operation until each of the points $V_1, \dots, V_k$ is in some path σ_j. Thus we obtain the set of vertices $V(\Gamma')$. We define the edges of Γ' by saying that $\{V_i, V_{i'}\} \in E(\Gamma')$ if there is a simple path σ connecting V_i to $V_{i'}$ and which is contained in some path σ_j from the construction above; the length of the edge $\{V_i, V_{i'}\}$ is the sum of the lengths of the edges of Γ which are part of σ. We notice that $\Gamma' \in CMG$ is a tree with at most $2k-2$ vertices and $2k-2$ edges. Moreover, even if Γ' is not a subgraph of Γ ($E(\Gamma')$ may not be a subset of $E(\Gamma)$), we have the inclusion $H^1(\Gamma') \subset H^1(\Gamma)$.

Consider the set $E'' \subset E(\Gamma)$ composed of the edges of Γ which are not part of none of the paths σ_j from the construction above. We denote with l'' the sum of the lengths of the edges in E''. For any $e_{ij} \in E''$ we consider the restriction $w_{ij} : [0, l_{ij}] \to \mathbb{R}$ of the energy function w on e_{ij}. Let $v : [0, l''] \to \mathbb{R}$ be the monotone function defined by the equality $|\{v \geq \tau\}| = \sum_{e_{ij} \in E''} |\{w_{ij} \geq \tau\}|$. Using the co-area formula (7.24) and repeating the argument from Remark 7.14, we have that

$$\frac{1}{2}\int_0^{l''} |v'|^2 dx - \int_0^{l''} v(x)\, dx \leq \sum_{e_{ij} \in E''} \left(\frac{1}{2}\int_0^{l_{ij}} |w'_{ij}|^2 dx - \int_0^{l_{ij}} w_{ij}\, dx \right). \tag{7.28}$$

Let $\widetilde{\Gamma}$ be the graph obtained from Γ by creating a new vertex W_1 in the point, where the restriction $w_{|\Gamma'}$ achieves its maximum, and another vertex W_2, connected to W_1 by an edge of length l''. It is straightforward to check that $\widetilde{\Gamma}$ is a connected metric tree of length l and that there exists an immersion $\widetilde{\gamma} : \widetilde{\Gamma} \to \mathbb{R}^d$ such that $\mathcal{D} = \widetilde{\gamma}(\widetilde{\mathcal{V}})$. The inequality (7.27) follows since, by (7.28), $J(\widetilde{w}) \leq J(w)$, where $\widetilde{w}$ is defined as w on the edges $E(\Gamma') \subset E(\widetilde{\Gamma})$ and as v on the edge $\{W_1, W_2\}$. □

Before we prove our main existence result, we need a preliminary Lemma.

Lemma 7.17. *Let Γ be a connected metric tree and let $\mathcal{V} \subset V(\Gamma)$ be a set of Dirichlet vertices. Let $w \in H_0^1(\Gamma; \mathcal{V})$ be the energy function on Γ with Dirichlet conditions in $\mathcal{V}$,* i.e. *the function that realizes the minimum in the definition of $\mathcal{E}(\Gamma; \mathcal{V})$. Then, we have the bound $\|w'\|_{L^\infty} \leq l(\Gamma)$.*

Proof. Up to adding vertices in the points where $|w'| = 0$, we can suppose that on each edge $e_{ij} := \{V_i, V_j\} \in E(\Gamma)$ the function $w_{ij} : [0, l_{ij}] \to \mathbb{R}^+$ is monotone. Moreover, up to relabel the vertices of Γ we can suppose that if $e_{ij} \in V(\Gamma)$ and $i < j$, then $w(V_i) \leq w(V_j)$. Fix $V_i, V_{i'} \in V(\Gamma)$ such that $e_{ii'} \in E(\Gamma)$. Note that, since the derivative is monotone on each edge, it suffices to prove that $|w'_{ii'}(0)| \leq l(\Gamma)$. It is enough to consider the case $i < i'$, *i.e.* $w'_{ii'}(0) > 0$. We construct the graph $\widetilde{\Gamma}$ inductively, as follows (see Figure 7.3):

1. $V_i \in V(\widetilde{\Gamma})$;
2. if $V_j \in V(\widetilde{\Gamma})$ and $V_k \in V(\Gamma)$ are such that $e_{jk} \in E(\Gamma)$ and $j < k$, then $V_k \in V(\widetilde{\Gamma})$ and $e_{jk} \in E(\widetilde{\Gamma})$.

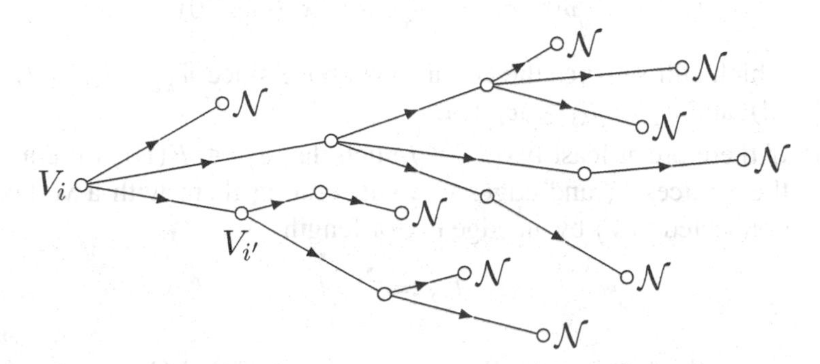

Figure 7.3. The graph $\widetilde{\Gamma}$; with the letter $\mathcal{N}$ we indicate the Neumann vertices.

The graph $\widetilde{\Gamma}$ constructed by the above procedure and the restriction $\widetilde{w} \in H^1(\widetilde{\Gamma})$ of w to $\widetilde{\Gamma}$ have the following properties:

(a) On each edge $e_{jk} \in E(\widetilde{\Gamma})$, the function $\widetilde{w}_{jk}$ is non-negative, monotone and $\widetilde{w}''_{jk} = -1$;
(b) $\widetilde{w}(V_j) > \widetilde{w}(V_k)$ whenever $e_{jk} \in E(\widetilde{\Gamma})$ and $j > k$;
(c) if $V_j \in V(\widetilde{\Gamma})$ and $j > i$, then there is exactly one $k < j$ such that $e_{kj} \in E(\widetilde{\Gamma})$;
(d) for j and k as in the previous point, we have that

$$0 \leq \widetilde{w}'_{kj}(l_{kj}) \leq \sum_s \widetilde{w}'_{js}(0),$$

where the sum on the right-hand side is over all $s > j$ such that $e_{sj} \in E(\widetilde{\Gamma})$. If there are not such s, we have that $\widetilde{w}'_{kj}(l_{kj}) = 0$.

The first three conditions follow by the construction of $\widetilde{\Gamma}$, while condition (d) is a consequence of the Kirchkoff's law for w.
We prove that for any graph $\widetilde{\Gamma}$ and any function $\widetilde{w} \in H^1(\widetilde{\Gamma})$, for which the conditions (a), (b), (c) and (d) are satisfied, we have that

$$\sum_j \widetilde{w}'_{ij}(0) \leq l(\widetilde{\Gamma}),$$

where the sum is over all $j \geq i$ and $e_{ij} \in E(\widetilde{\Gamma})$. It is enough to observe that each of the operations (i) and (ii) described below, produces a graph which still satisfies (a), (b), (c) and (d). Let $V_j \in V(\widetilde{\Gamma})$ be such that for each $s > j$ for which $e_{js} \in E(\widetilde{\Gamma})$, we have that $\widetilde{w}'_{js}(l_{js}) = 0$ and let $k < j$ be such that $e_{jk} \in E(\widetilde{\Gamma})$.

(i) If there is only one $s > j$ with $e_{js} \in E(\widetilde{\Gamma})$, then we erase the vertex V_j and the edges e_{kj} and e_{js} and add the edge e_{ks} of length $l_{ks} := l_{kj} + l_{js}$. On the new edge we define $\widetilde{w}_{ks} : [0, l_{sk}] \to \mathbb{R}^+$ as

$$\widetilde{w}_{ks}(x) = -\frac{x^2}{2} + l_{ks}\, x + \widetilde{w}_{kj}(0),$$

which still satisfies the conditions above since $\widetilde{w}'_{kj} - l_{kj} \leq l_{js}$, by (d), and $\widetilde{w}'_{ks} = l_{ks} \geq \widetilde{w}'_{kj}(0)$.
(ii) If there are at least two $s > j$ such that $e_{js} \in E(\widetilde{\Gamma})$, we erase all the vertices V_s and edges e_{js}, substituting them with a vertex V_S connected to V_j by an edge e_{jS} of length

$$l_{jS} := \sum_s l_{js},$$

where the sum is over all $s > j$ with $e_{js} \in E(\widetilde{\Gamma})$. On the new edge, we consider the function $\widetilde{w}_{jS}$ defined by

$$\widetilde{w}_{jS}(x) = -\frac{x^2}{2} + l_{jS}\, x + \widetilde{w}(V_j),$$

which still satisfies the conditions above since

$$\sum_{\{s:s>j\}} \widetilde{w}'_{js}(0) = \sum_{\{s:s>j\}} l_{js} = l_{jS} = \widetilde{w}'_{jS}(0).$$

We apply (i) and (ii) until we obtain a graph with vertices V_i, V_j and only one edge e_{ij} of length $l(\widetilde{\Gamma})$. The function we obtain on this graph is $-\frac{x^2}{2} + l(\widetilde{\Gamma})x$ with derivative in 0 equal to $l(\widetilde{\Gamma})$. Since, after applying (i) and (ii), the sum $\sum_{j>i} \widetilde{w}'_{ij}(0)$ does not decrease, we have the claim. □

Theorem 7.18. *Consider a set of distinct points $\mathcal{D} = \{D_1, \ldots, D_k\} \subset \mathbb{R}^d$ and a positive real number $l \geq St(\mathcal{D})$. Then there exists a connected metric graph Γ, a set of vertices $\mathcal{V} \subset V(\Gamma)$ and an immersion $\gamma : \Gamma \to \mathbb{R}^d$ which are solution of the problem* (7.25). *Moreover, Γ can be chosen to be a tree of at most $2k$ vertices and $2k - 1$ edges.*

Proof. Consider a minimizing sequence (Γ_n, γ_n) of connected metric graphs Γ_n and immersions $\gamma_n : \Gamma_n \to \mathbb{R}^d$. By Theorem 7.16, we can suppose that each Γ_n is a tree with at most $2k$ vertices and $2k - 1$ edges. Up to extracting a subsequence, we may assume that the metric graphs Γ_n are the same graph Γ but with different lengths l^n_{ij} of the edges e_{ij}. We can suppose that for each $e_{ij} \in E(\Gamma)$ $l^n_{ij} \to l_{ij}$ for some $l_{ij} \geq 0$ as $n \to \infty$. We construct the graph $\widetilde{\Gamma}$ from Γ identifying the vertices $V_i, V_j \in V(\Gamma)$ such that $l_{ij} = 0$. The graph $\widetilde{\Gamma}$ is a connected metric tree of length l and there is an immersion $\widetilde{\gamma} : \widetilde{\Gamma} \to \mathbb{R}^d$ such that $\mathcal{D} \subset \widetilde{\gamma}(\widetilde{\Gamma})$. In fact if $\{V_1, \ldots V_N\}$ are the vertices of Γ, up to extracting a subsequence, we can suppose that for each $i = 1, \ldots, N$ $\gamma_n(V_i) \to X_i \in \mathbb{R}^d$. We define $\widetilde{\gamma}(V_i) := X_i$ and $\gamma_{ij} : [0, l_{ij}] \to \mathbb{R}^d$ as any injective arc-length parametrized curve connecting X_i and X_j, which exists, since

$$l_{ij} = \lim l^n_{ij} \geq \lim |\gamma_n(V_i) - \gamma_n(V_j)| = |X_i - X_j|.$$

To prove the theorem, it is enough to check that

$$\mathcal{E}(\widetilde{\Gamma}; \mathcal{V}) = \lim_{n\to\infty} \mathcal{E}(\Gamma_n; \mathcal{V}).$$

Let $w^n = (w^n_{ij})_{ij}$ be the energy function on Γ_n. Up to a subsequence, we may suppose that for each $i = 1, \ldots, N$, $w^n(V_i) \to a_i \in \mathbb{R}$ as $n \to \infty$. Moreover, by Lemma 7.17, we have that if $l_{ij} = 0$, then $a_i = a_j$. On each of the edges $e_{ij} \in E(\widetilde{\Gamma})$, where $l_{ij} > 0$, we define the function $w_{ij} : [0, l_{ij}] \to \mathbb{R}$ as the parabola such that $w_{ij}(0) = a_i$, $w_{ij}(l_{ij}) = a_j$ and $w''_{ij} = -1$ on $(0, l_{ij})$. Then, we have

$$\frac{1}{2}\int_0^{l^n_{ij}} |(w^n_{ij})'|^2\,dx - \int_0^{l^n_{ij}} w^n_{ij}\,dx \xrightarrow[n\to\infty]{} \frac{1}{2}\int_0^{l_{ij}} |(w_{ij})'|^2\,dx - \int_0^{l_{ij}} w_{ij}\,dx,$$

and so, it is enough to prove that $\widetilde{w} = (w_{ij})_{ij}$ is the energy function on $\widetilde{\Gamma}$, *i.e.* (by Lemma 7.12) that the Kirchoff's law holds in each vertex of $\widetilde{\Gamma}$. This follows since for each $1 \le i \neq j \le N$ we have

1. $(w_{ij}^n)'(0) \to w_{ij}'(0)$, as $n \to \infty$, if $l_{ij} \neq 0$;
2. $|(w_{ij}^n)'(0) - (w_{ij}^n)'(l_{ij}^n)| \le l_{ij}^n \to 0$, as $n \to \infty$, if $l_{ij} = 0$.

The proof is then concluded. □

The proofs of Theorem 7.16 and Theorem 7.18 suggest that a solution $(\Gamma, \mathcal{V}, \gamma)$ of the problem (7.25) must satisfy some optimality conditions. We summarize this additional information in the following Proposition.

Proposition 7.19. *Consider a connected metric graph Γ, a set of vertices $\mathcal{V} \subset V(\Gamma)$ and an immersion $\gamma : \Gamma \to \mathbb{R}^d$ such that $(\Gamma, \mathcal{V}, \gamma)$ is a solution of the problem (7.25). Moreover, suppose that all the vertices of degree two are in the set $\mathcal{V}$. Then we have that:*

(i) *the graph Γ is a tree;*
(ii) *the set $\mathcal{V}$ has exactly k elements, where k is the number of Dirichlet points $\{D_1, \dots, D_k\}$;*
(iii) *there is at most one vertex $V_j \in V(\Gamma) \setminus \mathcal{V}$ of degree one;*
(iv) *if there is no vertex of degree one in $V(\Gamma) \setminus \mathcal{V}$, then the graph Γ has at most $2k - 2$ vertices and $2k - 3$ edges;*
(v) *if there is exactly one vertex of degree one in $V(\Gamma) \setminus \mathcal{V}$, then the graph Γ has at most $2k$ vertices and $2k - 1$ edges.*

Proof. We use the notation $V(\Gamma) = \{V_1, \dots, V_N\}$ for the vertices of Γ and e_{ij} for the edges $\{V_i, V_j\} \in E(\Gamma)$, whose lengths are denoted by l_{ij}. Moreover, we can suppose that for $j = 1, \dots, k$, we have $\gamma(V_j) = D_j$, where $D_1, \dots, D_k$ are the Dirichlet points from problem (7.25) and so, $\{V_1, \dots, V_k\} \subset \mathcal{V}$. Let $w = (w_{ij})_{ij}$ be the energy function on Γ with Dirichlet conditions in the points of $\mathcal{V}$.

(i) Suppose that we can remove an edge $e_{ij} \in E(\Gamma)$, such that the graph $\Gamma' = (V(\Gamma), E(\Gamma) \setminus e_{ij})$ is still connected. Since $w_{ij}'' = -1$ on $[0, l_{ij}]$ we have that at least one of the derivatives $w_{ij}'(0)$ and $w_{ij}'(l_{ij})$ is not zero. We can suppose that $w_{ij}'(l_{ij}) \neq 0$. Consider the new graph $\widetilde{\Gamma}$ to which we add a new vertex: $V(\widetilde{\Gamma}) = V(\Gamma) \cup V_0$, then erase the edge e_{ij} and create a new one $e_{i0} = \{V_i, V_0\}$, of the same length, connecting V_i to V_0: $E(\widetilde{\Gamma}) = \big(E(\Gamma) \setminus e_{ij}\big) \cup e_{i0}$. Let $\widetilde{w}$ be the energy function on $\tilde{\Gamma}$ with Dirichlet conditions in $\mathcal{V}$. When seen as a subspaces of $\oplus_{ij} H^1([0, l_{ij}])$, we have that $H_0^1(\Gamma; \mathcal{V}) \subset H_0^1(\widetilde{\Gamma}; \mathcal{V})$ and so $\mathcal{E}(\widetilde{\Gamma}; \mathcal{V}) \le \mathcal{E}(\Gamma; \mathcal{V})$, where the equality occurs, if and only if the energy functions w and $\widetilde{w}$ have the same components in $\oplus_{ij} H^1([0, l_{ij}])$. In particular, we must have that $w_{ij} = \widetilde{w}_{i0}$ on the interval $[0, l_{ij}]$, which is impossible since $w'_{ij}(l_{ij}) \neq 0$ and $\widetilde{w}'_{i0}(l_{ij}) = 0$.

(ii) Suppose that there is a vertex $V_j \in \mathcal{V}$ with $j > k$ and let $\widetilde{w}$ be the energy function on Γ with Dirichlet conditions in $\{V_1, \dots, V_k\}$. We have the inclusion $H_0^1(\Gamma; \mathcal{V}) \subset H_0^1(\Gamma; \{V_1, \dots, V_k\})$ and so, the inequality $J(\widetilde{w}) = \mathcal{E}(\Gamma; \{V_1, \dots, V_k\}) \leq \mathcal{E}(\Gamma; \mathcal{V}) = J(w)$, which becomes an equality if and only if $\widetilde{w} = w$, which is impossible. Indeed, if the equality holds, then in V_j, w satisfies both the Dirichlet condition and the Kirchoff's law. Since w is positive, for any edge e_{ji} we must have $w_{ji}(0) = 0$, $w'_{ji}(0) = 0$, $w''_{ji} = -1$ ad $w_{ji} \geq 0$ on $[0, l_{ji}]$, which is impossible.

(iii) Suppose that there are two vertices V_i and V_j of degree one, which are not in $\mathcal{V}$, *i.e.* $i, j > k$. Since Γ is connected, there are two edges, $e_{ii'}$ and $e_{jj'}$ starting from V_i and V_j respectively. Suppose that the energy function $w \in H_0^1(\Gamma; \{V_1, \dots, V_k\})$ is such that $w(V_i) \geq w(V_j)$. We define a new graph $\widetilde{\Gamma}$ by erasing the edge $e_{jj'}$ and creating the edge e_{ij} of length $l_{jj'}$. On the new edge e_{ij} we consider the function $w_{ij}(x) = w_{jj'}(x) + w(V_i) - w(V_j)$. The function $\widetilde{w}$ on $\widetilde{\Gamma}$ obtained by this construction is such that $J(\widetilde{w}) \leq J(w)$, which proves the conclusion.

The points (iv) and (v) follow by the construction in Theorem 7.16 and the previous claims (i), (ii) and (iii). □

Remark 7.20. Suppose that $V_j \in V(\Gamma) \setminus \mathcal{V}$ is a vertex of degree one and let V_i be the vertex such that $e_{ij} \in E(\Gamma)$. Then the energy function w with Dirichlet conditions in $\mathcal{V}$ satisfies $w'_{ji}(0) = 0$. In this case, we call V_j a Neumann vertex. By Proposition 7.19, an optimal graph has at most one Neumann vertex.

In some situations, we can use Theorem 7.16 to obtain an existence result for (7.13).

Proposition 7.21. *Suppose that D_1, D_2 and D_3 be three distinct, non colinear points in $\mathbb{R}^d$ and let $l > 0$ be a real number such that there exists a closed set of length l connecting D_1, D_2 and D_3. Then the problem* (7.13) *has a solution.*

Proof. Let the graph Γ be a solution of (7.25) and let $\gamma : \Gamma \to \mathbb{R}^d$ be an immersion of Γ such that $\gamma(V_j) = D_j$ for $j = 1, 2, 3$. Note that if the immersion γ is such that the set $\gamma(\Gamma) \subset \mathbb{R}^d$ is represented by the same graph Γ, then $\gamma(\Gamma)$ is a solution of (7.13) since we have

$$E(\Gamma; \{V_1, V_2, V_3\}) = E(C; D_1, D_2, D_3).$$

By Proposition 7.19, we can suppose that Γ is obtained by a tree Γ' with vertices V_1, V_2 and V_3 by attaching a new edge (with a new vertex in one of the extrema) to some vertex or edge of Γ'. Since we are free to choose the immersion of the new edge, we only need to show that we can choose γ in order to have that the set $\gamma(\Gamma')$ is represented by Γ'. On the other hand we have only two possibilities for Γ' and both of them can be seen as embedded graphs in $\mathbb{R}^d$ with vertices D_1, D_2 and D_3. □

7.3. Some examples of optimal metric graphs

In this section we show three examples. In the first one we deal with two Dirichlet points, the second concerns three aligned Dirichlet points and the third one deals with the case in which the Dirichlet points are vertices of an equilateral triangle. In the first and the third one we find the minimizer explicitly as an embedded graph, while in the second one we limit ourselves to prove that there is no embedded minimizer of the energy, *i.e.* the problem (7.26) does not admit a solution.

In the following example we use a symmetrization technique similar to the one from Remark 7.6.

Example 7.22. Let D_1 and D_2 be two distinct points in $\mathbb{R}^d$ and let $l \geq |D_1 - D_2|$ be a real number. Then the problem

$$\min\Big\{\mathcal{E}(\Gamma; \{V_1, V_2\}) : \Gamma \in CMG,\ l(\Gamma) = l,\ V_1, V_2 \in V(\Gamma),\\ \text{exists } \gamma : \Gamma \to \mathbb{R} \text{ immersion},\ \gamma(V_1) = D_1, \gamma(V_2) = D_2\Big\}. \tag{7.29}$$

has a solution (Γ, γ), where Γ is a metric graph with vertices $V(\Gamma) = \{V_1, V_2, V_3, V_4\}$ and edges $E(\Gamma) = \{e_{13} = \{V_1, V_3\}, e_{23} = \{V_2, V_3\}, e_{43} = \{V_4, V_3\}\}$ of lengths $l_{13} = l_{23} = \frac{1}{2}|D_1 - D_2|$ and $l_{34} = l - |D_1 - D_2|$, respectively. The map $\gamma : \Gamma \to \mathbb{R}^d$ is an embedding such that $\gamma(V_1) = D_1, \gamma(V_2) = D_2$ and $\gamma(V_3) = \frac{D_1+D_2}{2}$ (see Figure 7.4).

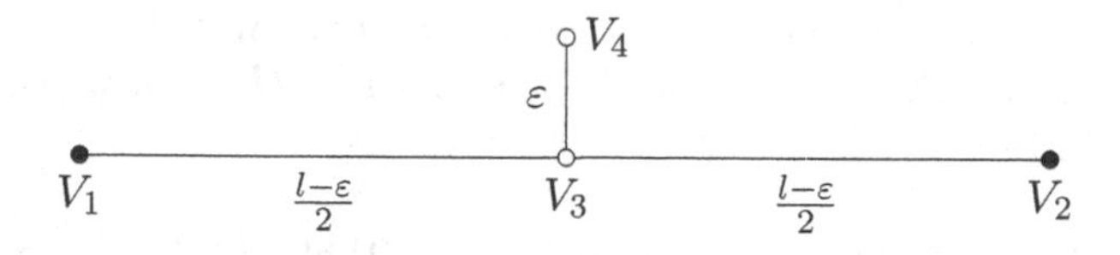

Figure 7.4. The optimal graph with two Dirichlet points.

To fix the notations, we suppose that $|D_1 - D_2| = l - \varepsilon$. Let $u = (u_{ij})_{ij}$ be the energy function of a generic metric graph Σ and immersion $\sigma : \Sigma \to \mathbb{R}^d$ with $D_1, D_2 \in \sigma(V(\Sigma))$. Let $M = \max\{u(x) : x \in \Sigma\} > 0$. We construct a candidate $v \in H_0^1(\Gamma; \{V_1, V_2\})$ such that $J(v) \leq J(u)$, which immediately gives the conclusion.

We define v by the following three *increasing* functions

$$v_{13} = v_{23} \in H^1([0, (l-\varepsilon)/2]), \qquad v_{34} \in H^1([0, \varepsilon]),$$

with boundary values

$$v_{13}(0) = v_{23}(0) = 0, \quad v_{13}((l-\varepsilon)/2) = v_{23}((l-\varepsilon)/2) = v_{34}(0) = m < M,$$

and level sets uniquely determined by the equality $\mu_u = \mu_v$, where μ_u and μ_v are the distribution functions of u and v respectively, defined by

$$\mu_u(t) = \mathcal{H}^1(\{u \leq t\}) = \sum_{e_{ij} \in E(\Sigma)} \mathcal{H}^1(\{u_{ij} \leq t\}),$$
$$\mu_v(t) = \mathcal{H}^1(\{v \leq t\}) = \sum_{j=1,2,4} \mathcal{H}^1(\{v_{j3} \leq t\}).$$

As in Remark 7.6 we have $\|v\|_{L^1(\Gamma)} = \|u\|_{L^1(C)}$ and

$$\begin{aligned} \int_\Sigma |u'|^2 \, dx &= \int_0^M \Big(\sum_{u=\tau} |u'| \Big) d\tau \\ &\geq \int_0^M n_u^2(\tau) \Big(\sum_{u=\tau} \frac{1}{|u'|(\tau)} \Big)^{-1} d\tau = \int_0^M \frac{n_u^2(\tau)}{\mu_u'(\tau)} \, d\tau \end{aligned} \tag{7.30}$$

where $n_u(\tau) = \mathcal{H}^0(\{u = \tau\})$. The same argument holds for v on the graph Γ but, this time, with the equality sign:

$$\int_\Gamma |v'|^2 dx = \int_0^M \Big(\sum_{v=\tau} |v'| \Big) d\tau = \int_0^M \frac{n_v^2(\tau)}{\mu_v'(\tau)} \, d\tau, \tag{7.31}$$

since $|v'|$ is constant on $\{v = \tau\}$, for every τ. Then, in view of (7.30) and (7.31), to conclude it is enough to prove that $n_u(\tau) \geq n_v(\tau)$ for almost every τ. To this aim we first notice that, by construction $n_v(\tau) = 1$ if $\tau \in [m, M]$ and $n_v(\tau) = 2$ if $\tau \in [0, m)$. Since n_u is decreasing and greater than 1 on $[0, M]$, we only need to prove that $n_u \geq 2$ on $[0, m]$. To see this, consider two vertices $W_1, W_2 \in V(\Sigma)$ such that $\sigma(W_1) = D_1$ and $\sigma(W_2) = D_2$. Let η be a simple path connecting W_1 to W_2 in Σ. Since σ is an immersion we know that the length $l(\eta)$ of η is at least $l - \varepsilon$.

By the continuity of u, we know that $n_u \geq 2$ on the interval $[0, \max_\eta u)$. Since $n_v = 1$ on $[m, M]$, we need to show that $\max_\eta u \geq m$. Otherwise, we would have

$$l(\eta) \leq |\{u \leq \max_\eta u\}| < |\{u \leq m\}| = |\{v \leq m\}| = |D_1 - D_2| \leq l(\eta),$$

which is impossible.

Remark 7.23. In the previous example the optimal metric graph Γ is such that for any (admissible) immersion $\gamma : \Gamma \to \mathbb{R}^d$, we have $|\gamma(V_1) - \gamma(V_3)| = l_{13}$ and $|\gamma(V_2) - \gamma(V_3)| = l_{23}$, *i.e.* the point $\gamma(V_3)$ is necessary the midpoint $\frac{D_1+D_2}{2}$, so we have a sort of *rigidity* of the graph Γ. More generally, we say that an edge e_{ij} is *rigid*, if for any admissible immersion $\gamma : \Gamma \to \mathbb{R}^d$, *i.e.* an immersion such that $\mathcal{D} = \gamma(\mathcal{V})$, we have $|\gamma(V_i) - \gamma(V_j)| = l_{ij}$, in other words the realization of the edge e_{ij} in $\mathbb{R}^d$ via any immersion γ is a segment. One may expect that in the optimal graph all the edges, except the one containing the Neumann vertex, are rigid. Unfortunately, we are able to prove only the weaker result that:

1. if the energy function w, of an optimal metric graph Γ, has a local maximum in the interior of an edge e_{ij}, then the edge is rigid; if the maximum is global, then Γ has no Neumann vertices;
2. if Γ contains a Neumann vertex V_j, then w achieves its maximum at it.

To prove the second claim, we just observe that if it is not the case, then we can use an argument similar to the one from point (iii) of Proposition 7.19, erasing the edge e_{ij} containing the Neumann vertex V_j and creating an edge of the same length that connects V_j to the point, where w achieves its maximum, which we may assume a vertex of Γ (possibly of degree two).

For the first claim, we apply a different construction which involves a symmetrization technique. In fact, if the edge e_{ij} is not rigid, then we can create a new metric graph of smaller energy, for which there is still an immersion which satisfies the conditions in problem (7.25). In this there are points $0 < a < b < l_{ij}$ such that $l_{ij} - (b - a) \geq |\gamma(V_i) - \gamma(V_j)|$ and $\min_{[a,b]} w_{ij} = w_{ij}(a) = w_{ij}(b) < \max_{[a,b]} w_{ij}$. Since the edge is not rigid, there is an immersion γ such that $|\gamma_{ij}(a) - \gamma_{ij}(b)| > |b - a|$. The problem (7.29) with $D_1 = \gamma_{ij}(a)$ and $D_2 = \gamma_{ij}(b)$ has as a solution the T-like graph described in Example 7.22. This shows, that the original graph could not be optimal, which is a contradiction.

Example 7.24. Consider the set of points $\mathcal{D} = \{D_1, D_2, D_3\} \subset \mathbb{R}^2$ with coordinates respectively $(-1, 0)$, $(1, 0)$ and $(n, 0)$, where n is a positive integer. Given $l = (n+2)$, we aim to show that for n large enough there is no solution of the optimization problem

$$\min\Big\{\mathcal{E}(\Gamma;\mathcal{V}) : \Gamma \in CMG,\ l(\Gamma) = l,\ \mathcal{V} \subset V(\Gamma),$$
$$\exists\gamma : \Gamma \to \mathbb{R} \text{ embedding},\ \mathcal{D} = \gamma(\mathcal{V})\Big\}.$$

In fact, we show that all the possible solutions of the problem

$$\min\Big\{\mathcal{E}(\Gamma;\mathcal{V}) : \Gamma \in CMG,\ l(\Gamma) = l,\ \mathcal{V} \subset V(\Gamma),$$
$$\exists\gamma : \Gamma \to \mathbb{R} \text{ immersion},\ \mathcal{D} = \gamma(\mathcal{V})\Big\}, \tag{7.32}$$

are metric graphs Γ for which there is no embedding $\gamma : \Gamma \to \mathbb{R}^2$ such that $\mathcal{D} \subset \gamma(V(\Gamma))$. Moreover, there is a sequence of embedded metric graphs which is a minimizing sequence for the problem (7.32).

More precisely, we show that the only possible solution of (7.32) is one of the following metric trees:

(i) Γ_1 with vertices $V(\Gamma_1) = \{V_1, V_2, V_3, V_4\}$ and edges $E(\Gamma_1) = (e_{14} = \{V_1, V_4\}, e_{24} = \{V_2, V_4\}, e_{34} = \{V_3, V_4\}$ of lengths $l_{14} = l_{24} = 1$ and $l_{34} = n$, respectively. The set of vertices in which the Dirichlet condition holds is $\mathcal{V}_1 = \{V_1, V_2, V_3\}$.

(ii) Γ_2 with vertices $V(\Gamma_2) = \{W_i\}_{i=1}^6$, and edges $E(\Gamma_2) = \{e_{14}, e_{24}, e_{35}, e_{45}, e_{56}\}$,where $e_{ij} = \{W_i, W_j\}$ for $1 \le i \ne j \le 6$ of lengths $l_{14} = 1 + \alpha, l_{24} = 1 - \alpha,\ l_{35} = n - \beta, l_{45} = \beta - \alpha, l_{56} = \alpha$, where $0 < \alpha < 1$ and $\alpha < \beta < n$. The set of vertices in which the Dirichlet condition holds is $\mathcal{V}_1 = \{V_1, V_2, V_3\}$. A possible immersion γ is described in Figure 7.5.

Figure 7.5. The two candidates for a solution of (7.32).

We start showing that if there is an optimal metric graph with no Neumann vertex, then it must be Γ_1. In fact, by Proposition 7.19, we know that the optimal metric graph is of the form Γ_1, but we have no information on the lengths of the edges, which we set as $l_i = l(e_{i4})$, for $i = 1, 2, 3$ (see Figure 7.6). We can calculate explicitly the minimizer of the energy functional and the energy itself in function of l_1, l_2 and l_3.

The minimizer of the energy $w : \Gamma \to \mathbb{R}$ is given by the functions $w_i : [0, l_i] \to \mathbb{R}$, where $i = 1, 2, 3$ and

$$w_i(x) = -\frac{x^2}{2} + a_i x.$$

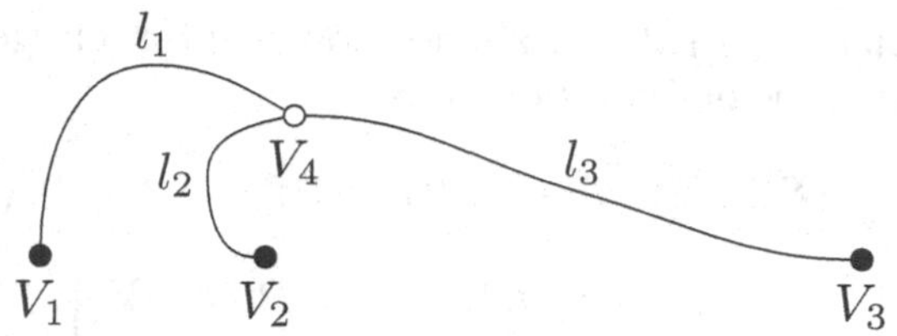

Figure 7.6. A metric tree with the same topology as Γ_1.

where

$$a_1 = \frac{l_1}{2} + \frac{l_2 l_3 (l_1 + l_2 + l_3)}{2(l_1 l_2 + l_2 l_3 + l_3 l_1)},$$

and a_2 and a_3 are defined by a cyclic permutation of the indices. As a consequence, we obtain that the derivative along the edge e_{14} in the vertex V_4 is given by

$$w_1'(l_1) = -l_1 + a_1 = -\frac{l_1}{2} + \frac{l_2 l_3 (l_1 + l_2 + l_3)}{2(l_1 l_2 + l_2 l_3 + l_3 l_1)}, \tag{7.33}$$

and integrating the energy function w on Γ, we obtain

$$\mathcal{E}(\Gamma; \{V_1, V_2, V_3\}) = -\frac{1}{12}(l_1^3 + l_2^3 + l_3^3) - \frac{(l_1 + l_2 + l_3)^2 l_1 l_2 l_3}{4(l_1 l_2 + l_2 l_3 + l_3 l_1)}.$$

Studying this function using Lagrange multipliers is somehow complicated due to the complexity of its domain. Thus we use a more geometric approach applying the symmetrization technique described in Remark 7.6 in order to select the possible candidates. We prove that if the graph is optimal, then all the edges must be rigid (this would force the graph to coincide with Γ_1). Suppose that the optimal graph Γ is not rigid, *i.e.* there is a non-rigid edge. Then, for $n > 4$, we have that $l_2 < l_1 < l_3$ and so, by (7.33), we obtain $w_3'(l_3) < w_1'(l_1) < w_2'(l_2)$. As a consequence of the Kirchoff's law we have $w_3'(l_3) < 0$ and $w_2'(l_2) > 0$ and so, w has a local maximum on the edge e_{34} and is increasing on e_{14}. By Remark 7.23, we obtain that the edge e_{34} is rigid.

We first prove that $w_1'(l_1) > 0$. In fact, if this is not the case, *i.e.* $w_1'(l_1) < 0$, by Remark 7.23, we have that the edges e_{14} is also rigid and so, $l_1 + l_3 = |D_1 - D_3| = n + 1$, *i.e.* $l_2 = 1$. Moreover, by (7.33), we have that $w_1'(l_1) < 0$, if and only if $l_1^2 > l_2 l_3 = l_3$. The last inequality does not hold for $n > 11$, since, by the triangle inequality, $l_2 + l_3 \geq |D_2 - D_3| = n - 1$, we have $l_1 \leq 3$. Thus, for n large enough, we have that w is increasing on the edge e_{14}.

We now prove that the edges e_{14} and e_{24} are rigid. In fact, suppose that e_{24} is not rigid. Let $a \in (0, l_1)$ and $b \in (0, l_2)$ be two points close to l_1 and l_2 respectively and such that $w_{14}(a) = w_{24}(b) < w(V_4)$ since w_{14} and w_{24} are strictly increasing. Consider the metric graph $\widetilde{\Gamma}$ whose vertices and edges are

$$V(\widetilde{\Gamma}) = \{V_1 = \widetilde{V}_1,\ V_2 = \widetilde{V}_2,\ V_3 = \widetilde{V}_3,\ V_4 = \widetilde{V}_4,\ \widetilde{V}_5,\ \widetilde{V}_6\},$$
$$E(\widetilde{\Gamma}) = \{e_{15},\ e_{25},\ e_{45},\ e_{34},\ e_{46}\},$$

where $e_{ij} = \{\widetilde{V}_i, \widetilde{V}_j\}$ and the lengths of the edges are respectively (see Figure 7.7)

$$\widetilde{l}_{15} = a,\ \widetilde{l}_{25} = b,\ \widetilde{l}_{45} = l_2 - b,\ \widetilde{l}_{34} = l_3,\ \widetilde{l}_{46} = l_1 - a.$$

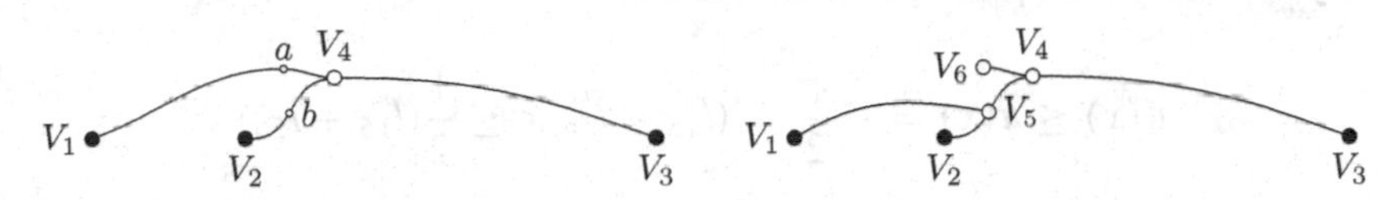

Figure 7.7. The graph Γ (on the left) and the modified one $\widetilde{\Gamma}$ (on the right).

The new metric graph is still a competitor in the problem (7.32) and there is a function $w \in H_0^1(\widetilde{\Gamma}; \{V_1, V_2, V_3\})$ such that $\mathcal{E}(\widetilde{\Gamma}; \{V_1, V_2, V_3\}) < J(\widetilde{w}) = J(w)$, which is a contradiction with the optimality of Γ. In fact, it is enough to define $\widetilde{w}$ as

$$\widetilde{w}_{15} = w_{14}|_{[0,a]},\ \widetilde{w}_{25} = w_{24}|_{[0,b]},\ \widetilde{w}_{54} = w_{24}|_{[b,l_2]},$$
$$\widetilde{w}_{34} = w_{34},\qquad \widetilde{w}_{64} = w_{14}|_{[a,l_1]},$$

and observe that $\widetilde{w}$ is not the energy function on the graph $\widetilde{\Gamma}$ since it does not satisfy the Neumann condition in $\widetilde{V}_6$. In the same way, if we suppose that w_{14} is not rigid, we obtain a contradiction, and so all the three edges must be rigid, *i.e.* $\Gamma = \Gamma_1$.

In a similar way we prove that a metric graph Γ with a Neumann vertex can be a solution of (7.32) only if it is of the same form as Γ_2. We proceed in two steps: first, we show that, for n large enough, the edge containing the Neumann vertex has a common vertex with the longest edge of the graph; then we can conclude reasoning analogously to the previous case. Let Γ be a metric graph with vertices $V(\Gamma) = \{V_i\}_{i=1}^6$, and edges $E(\Gamma) = \{e_{15}, e_{24}, e_{34}, e_{45}, e_{56}\}$, where $e_{ij} = \{V_i, V_j\}$ for $1 \le i \ne j \le 6$.

We prove that $w(V_6) \le \max_{e_{34}} w$, *i.e.* the graph Γ is not optimal, since, by Remark 7.23, the maximum of w must be achieved in the Neumann vertex V_6 (the case $E(\Gamma) = \{e_{14}, e_{25}, e_{34}, e_{45}, e_{56}\}$ is analogous). Let $w_{15} : [0, l_{15}] \to \mathbb{R}$, $w_{65} : [0, l_{65}] \to \mathbb{R}$ and $w_{34} : [0, l_{34}] \to \mathbb{R}$ be the restrictions of the energy function w of Γ to the edges e_{15}, e_{65} and e_{34} of lengths l_{15}, l_{65} and l_{34}, respectively. Let $u : [0, l_{15} + l_{56}] \to \mathbb{R}$ be defined as

$$u(x) = \begin{cases} w_{15}(x), & x \in [0, l_{15}], \\ w_{56}(x - l_{15}), & x \in [l_{15}.l_{15} + l_{56}]. \end{cases}$$

If the metric graph Γ is optimal, then the energy function on w_{54} on the edge e_{45} must be decreasing and so, by the Kirchhoff's law in the vertex V_5, we have that $w'_{15}(l_{15}) + w'_{65}(l_{65}) \leq 0$, *i.e.* the left derivative of u at l_{15} is less than the right one:

$$\partial_- u(l_{15}) = w'_{15}(l_{15}) \leq w'_{56}(0) = \partial_+ u(l_{15}).$$

By the maximum principle, we have that

$$u(x) \leq \widetilde{u}(x) = -\frac{x^2}{2} + (l_{15} + l_{56})x \leq \frac{1}{2}(l_{15} + l_{56})^2.$$

On the other hand, $w_{34}(x) \geq v(x) = -\frac{x^2}{2} + \frac{l_{34}}{2}x$, again by the maximum principle on the interval $[0, l_{34}]$. Thus we have that

$$\max_{x \in [0, l_{34}]} w_{34}(x) \geq \max_{x \in [0, l_{34}]} v(x) = \frac{1}{8} l_{34}^2 > \frac{1}{2}(l_{15} + l_{56})^2 \geq w(V_6),$$

for n large enough.

Repeating the same argument, one can show that the optimal metric graph Γ is not of the form $V(\Gamma) = (V_1, V_2, V_3, V_4, V_5)$, $E(\Gamma) = \{V_1, V_4\}, \{V_2, V_4\}, \{V_3, V_4\}, \{V_4, V_5\}$.

Thus, we obtained that the if the optimal graph has a Neumann vertex, then the corresponding edge must be attached to the longest edge. To prove that it is of the same form as Γ_2, there is one more case to exclude, namely: Γ with vertices, $V(\Gamma) = (V_1, V_2, V_3, V_4, V_5)$, $E(\Gamma) = \{\{V_1, V_2\}, \{V_2, V_4\}, \{V_3, V_4\}, \{V_4, V_5\}\}$ (see Figure 7.8). By Example 7.22, the only possible candidate of this form is the graph with lengths $l(\{V_1, V_2\}) = |D_1 - D_2| = 2$, $l(\{V_2, V_4\}) = \frac{n-1}{2}$, $l(\{V_3, V_4\}) = \frac{n-1}{2}$, $l(\{V_4, V_5\}) = 2$. In this case, we compare the energy of Γ and Γ_1, by an explicit calculation:

$$\begin{aligned} \mathcal{E}(\Gamma; \{V_1, V_2, V_3\}) &= -\frac{n^3 - 3n^2 + 6n}{24} > -\frac{n^2(n+1)^2}{12(2n+1)} \\ &= \mathcal{E}(\Gamma_1; \{V_1, V_2, V_3\}), \end{aligned}$$

for n large enough.

Before we pass to our last example, we need the following Lemma.

Figure 7.8. The graph Γ_1 (on the left) has lower energy than the graph Γ (on the right).

Lemma 7.25. *Let $w_a : [0,1] \to \mathbb{R}$ be given by $w_a(x) = -\frac{x^2}{2} + ax$, for some positive real number a. If $w_a(1) \le w_A(1) \le \max_{x\in[0,1]} w_a(x)$, then $J(w_A) \le J(w_a)$, where $J(w) = \frac{1}{2}\int_0^1 |w'|^2\,dx - \int_0^1 w\,dx$.*

Proof. It follows by performing the explicit calculations. □

Example 7.26. Let D_1, D_2 and D_3 be the vertices of an equilateral triangle of side 1 in $\mathbb{R}^2$, *i.e.*

$$D_1 = (-\frac{\sqrt{3}}{3}, 0),\ D_2 = (\frac{\sqrt{3}}{6}, -\frac{1}{2}),\ D_3 = (\frac{\sqrt{3}}{6}, \frac{1}{2}).$$

We study the problem (7.25) with $\mathcal{D} = \{D_1, D_2, D_3\}$ and $l > \sqrt{3}$. We show that the solutions may have different qualitative properties for different l and that there is always a symmetry breaking phenomena, *i.e.* the solutions does not have the same symmetries as the initial configuration $\mathcal{D}$. We first reduce our study to the following three candidates (see Figure 7.9):

1. The metric tree Γ_1, defined by with vertices $V(\Gamma) = \{V_1, V_2, V_3, V_4\}$ and edges $E(\Gamma) = \{e_{14}, e_{24}, e_{34}\}$, where $e_{ij} = \{V_i, V_j\}$ and the lengths of the edges are respectively $l_{24} = l_{34} = x$, $l_{14} = \frac{\sqrt{3}}{2} - \sqrt{x^2 - \frac{1}{4}}$, for some $x \in [1/2, 1/\sqrt{3}]$. Note that the length of Γ_1 is less than $1 + \sqrt{3}/2$, *i.e.* it is a possible solution only for $l \le 1 + \sqrt{3}/2$. The new vertex V_4 is of Kirchhoff type and there are no Neumann vertices.
2. The metric tree Γ_2 with vertices $V = (V_1, V_2, V_3, V_4, V_5)$ and $E(\Gamma) = \{e_{14}, e_{24}, e_{34}, e_{45}\}$, where $e_{ij} = \{V_i, V_j\}$ and the lengths of the edges $l_{14} = l_{24} = l_{34} = 1/\sqrt{3}$, $l_{45} = l - \sqrt{3}$, respectively. The new vertex V_4 is of Kirchhoff type and V_5 is a Neumann vertex.
3. The metric tree Γ_3 with vertices $V(\Gamma) = \{V_1, V_2, V_3, V_4, V_5, V_6\}$ and edges $E(\Gamma) = \{e_{15}, e_{24}, e_{34}, e_{45}, e_{56}\}$, where $e_{ij} = \{V_i, V_j\}$ and the lengths of the edges are $l_{24} = l_{34} = x$, $l_{15} = \frac{lx}{2(2l-3x)} + \frac{\sqrt{3}}{4} - \frac{1}{4}\sqrt{4x^2 - 1}$, $l_{45} = \frac{\sqrt{3}}{4} - \frac{lx}{2(2l-3x)} - \frac{1}{4}\sqrt{4x^2 - 1}$ and $l_{56} = l - 2x - \sqrt{3}/2 + \frac{1}{2}\sqrt{4x^2 - 1}$. The new vertices V_4 and V_5 are of Kirchhoff type and V_6 is a Neumann vertex.

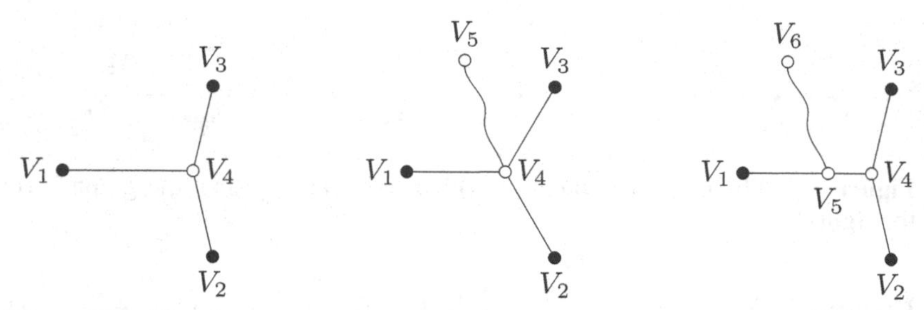

Figure 7.9. The three competing graphs.

Suppose that the metric graph Γ is optimal and has the same vertices and edges as Γ_1. Without loss of generality, we can suppose that the maximum of the energy function w on Γ is achieved on the edge e_{14}. If $l_{24} \neq l_{34}$, we consider the metric graph $\widetilde{\Gamma}$ with the same vertices and edges as Γ and lengths $\widetilde{l}_{14} = l_{14}, \widetilde{l}_{24} = \widetilde{l}_{34} = (l_{24}+l_{34})/2$. An immersion $\widetilde{\gamma} : \widetilde{\Gamma} \to \mathbb{R}^2$, such that $\widetilde{\gamma}(V_j) = D_j$, for $j = 1, 2, 3$ still exists and the energy decreases, *i.e.* $\mathcal{E}(\widetilde{\Gamma}; \{V_1, V_2, V_3\}) < \mathcal{E}(\Gamma; \{V_1, V_2, V_3\})$. In fact, let $v = \widetilde{w}_{24} = \widetilde{w}_{34} : [0, \frac{l_{24}+l_{34}}{2}] \to \mathbb{R}$ be an increasing function such that $2|\{v \geq \tau\}| = |\{w_{24} \geq \tau\}| + |\{w_{34} \geq \tau\}|$. By the classical Polya-Szegö inequality and by the fact that w_{24} and w_{34} have no constancy regions, we obtain that

$$J(\widetilde{w}_{24}) + J(\widetilde{w}_{34}) < J(w_{24}) + J(w_{34}),$$

and so it is enough to construct a function $\widetilde{w}_{14} : [0, l_{14}] \to \mathbb{R}$ such that $\widetilde{w}_{14}(l_{14}) = \widetilde{w}_{24} = \widetilde{w}_{34}$ and $J(\widetilde{w}_{14}) \leq J(w_{14})$. Consider a function such that $\widetilde{w}_{14}'' = -1$, $\widetilde{w}_{14}(0) = 0$ and $\widetilde{w}_{14}(l_{14}) = \widetilde{w}_{24}(l_24) = \widetilde{w}_{34}(l_{34})$. Since we have the inequality $w_{14}(l_{14}) \leq \widetilde{w}_{14}(l_{14}) \leq \max_{[0,l_{14}]} w_{14} = \max_\Gamma w$, we can apply Lemma 7.25 and so, $J(\widetilde{w}_{14}) \leq J(w_{14})$. Thus, we obtain that $l_{24} = l_{34}$ and that both the functions w_{24} and w_{34} are increasing (in particular, $l_{14} \geq l_{24} = l_{34}$). If the maximum of w is achieved in the interior of the edge e_{14} then, by Remark 7.23, the edge e_{14} must be rigid and so, all the edges must be rigid. Thus, Γ coincides with Γ_1 for some $x \in (\frac{1}{2}, \frac{1}{\sqrt{3}}]$. If the maximum of w is achieved in the vertex V_4, then applying one more time the above argument, we obtain $l_{14} = l_{24} = l_{34} = \frac{1}{\sqrt{3}}$, *i.e.* Γ is Γ_1 corresponding to $x = \frac{1}{\sqrt{3}}$.

Suppose that the metric graph Γ is optimal and that has the same vertices as Γ_2. If $w = (w_{ij})_{ij}$ is the energy function on Γ with Dirichlet conditions in $\{V_1, V_2, V_3\}$, we have that w_{14}, w_{24} an w_{34} are increasing on the edges e_{14}, e_{24} and e_{34}. As in the previous situation $\Gamma = \Gamma_1$, by a symmetrization argument, we have that $l_{14} = l_{24} = l_{34}$. Since any level set $\{w = \tau\}$ contains exactly 3 points, if $\tau < w(V_4)$, and 1 point, if $\tau \geq w(V_4)$, we can apply the same technique as in Example 7.22 to obtain that $l_{14} = l_{24} = l_{34} = \frac{1}{\sqrt{3}}$.

Suppose that the metric graph Γ is optimal and that has the same vertices and edges as Γ_3. Let w be the energy function on Γ with Dirichlet conditions in $\{V_1, V_2, V_3\}$. Since we assume Γ optimal, we have that w_{45} is increasing on the edge e_{45} and $w(V_5) \geq w_{ij}$, for any $\{i, j\} \neq \{5, 6\}$. Applying the symmetrization argument from the case $\Gamma = \Gamma_1$ and Lemma 7.25, we obtain that $l_{24} = l_{34} = x$ and that the functions $w_{24} = w_{34}$ are increasing on $[0, l_{24}]$. Let $a \in [0, l_{15}]$ be such that $w_{15}(a) = w(V_4)$. By a symmetrization argument, we have that necessarily $l_{15} - a = l_{45}$ an that $w_{45}(x) = w_{15}(x - a)$. Moreover, the edges e_{15} and e_{45} are rigid. Indeed, for any admissible immersion $\gamma = (\gamma_{ij})_{ij} : \Gamma \to \mathbb{R}^2$, we have that the graph $\widetilde{\Gamma}$ with vertices $V(\widetilde{\Gamma}) = \{\widetilde{V}_1, V_4, V_5, V_6\}$ and edges $E(\widetilde{\Gamma}) = \big\{\{\widetilde{V}_1, V_5\}, \{V_4, V_5\}, \{V_5, V_6\}\big\}$, is a solution for the problem (7.29) with $D_1 := \gamma_{15}(a)$ and $D_2 := \gamma(V_4)$. By Example 7.22 and Remark 7.23, we have $|\gamma_{15}(a) - \gamma(V_4)| = 2l_{45}$ and, since this holds for every admissible γ, we deduce the rigidity of e_{15} and e_{45}. Using this information one can calculate explicitly all the lengths of the edges of Γ using only the parameter x, obtaining the third class of possible minimizers.

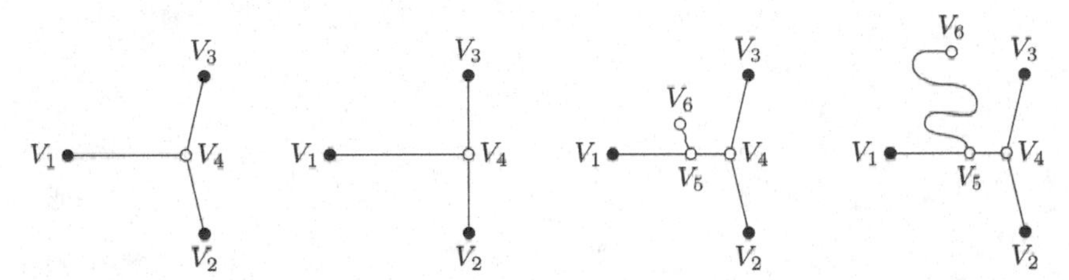

Figure 7.10. The optimal graphs for $l < 1+\sqrt{3}/2, l = 1+\sqrt{3}/2, l > 1+\sqrt{3}/2$ and $l >> 1 + \sqrt{3}/2$.

An explicit estimate of the energy shows that:

1. If $\sqrt{3} \leq l \leq 1 + \sqrt{3}/2$, we have that the solution of the problem (7.25) with $\mathcal{D} = \{D_1, D_2, D_3\}$ is of the form Γ_1 (see Figure 7.10).
2. If $l > 1 + \sqrt{3}/2$, then the solution of the problem (7.25) with $\mathcal{D} = \{D_1, D_2, D_3\}$ is of the form Γ_3.

In both cases,the parameter x is uniquely determined by the total length l and so, we have uniqueness up to rotation on $\frac{2\pi}{3}$. Moreover, in both cases the solutions are metric graphs, for which there is an embedding γ with $\gamma(V_i) = D_i$, *i.e.* they are also solutions of the problem (7.26) with $\mathcal{D} = \{D_1, D_2, D_3\}$ and $l \geq \sqrt{3}$.

List of Notations

References

[1] H.W. ALT and L.A. CAFFARELLI, *Existence and regularity for a minimum problem with free boundary*, J. Reine Angew. Math. **325** (1981), 105–144.

[2] H. W. ALT, L. CAFFARELLI and A. FRIEDMAN, *Variational problems with two phases and their free boundaries*, Trans. Amer. Math. Soc. **282** (1984), 431–461.

[3] L. AMBROSIO, V. CASELLES, S. MASNOU and J. M. MOREL, *Connected components of sets of finite perimeter and applications to image processing*, J. Eur. Math. Soc. **3** (1) (2001), 39–92.

[4] L. AMBROSIO, M. COLOMBO and S. DI MARINO, *Sobolev spaces in metric measure spaces: reflexivity and lower semicontinuity of slope*, Preprint available at http://cvgmt.sns.it/paper/2055/

[5] L. AMBROSIO, N. FUSCO and D. PALLARA, "Function of bounded Variation and Free Discontinuity Problems", Oxford University Press, 2000.

[6] L. AMBROSIO and P. TILLI, "Topics on Analysis in Metric Spaces", Oxford Lecture Series in Mathematics and its Applications, Oxford University Press, Oxford, 2004.

[7] P. R. S. Antunes and P. Freitas, *Numerical optimisation of low eigenvalues of the Dirichlet and Neumann Laplacians*, J. Optim. Theory Appl. **154** (2012), 235–257.

[8] M.S. ASHBAUGH, *Open problems on eigenvalues of the Laplacian*, In: "Analytic and Geometric Inequalities and Applications", Math. Appl. **478**, Kluwer Acad. Publ., Dordrecht, 1999, 13–28.

[9] H. ATTOUCH, "Variational convergence for functions and operators", Pitman Advanced Publishing Program, 1984.

[10] D. BAO, S. S. CHERN and Z. SHEN, "An Introduction to Riemann-Finsler Geometry", Graduate texts in mathematics, Springer-Verlag New York, 2000.

[11] J. BAXTER, G. DAL MASO and U. MOSCO, *Stopping times and* Γ*-convergence*, Trans. Amer. Math. Soc. **303** (1) (1987), 1–38.

[12] B. BOGOSEL and B. VELICHKOV, *Multiphase optimization problems for eigenvalues: qualitative properties and numerical results*, in preparation.

[13] B. BOURDIN, D. BUCUR and E. OUDET, *Optimal partitions for eigenvalues*, SIAM J. Sci. Comput. **31** (6) (2009), 4100–4114.

[14] L. BRASCO, *On torsional rigidity and principal frequencies: an invitation to the Kohler-Jobin rearrangement technique*, ESAIM COCV **20** (2) (2014), 315–338.

[15] L. BRASCO, G. DE PHILIPPIS and B. VELICHKOV, *Faber-Krahn inequalities in sharp quantitative form*, preprint available at http://cvgmt.sns.it/paper/2161/

[16] H. BREZIS, "Analyse Fonctionnelle: Théorie et applications", Collection Mathématiques appliquées pour la maîtrise, MASSON Paris Milan Barcelone Bonn, 1992.

[17] T. BRIANÇON, M. HAYOUNI and M. PIERRE, *Lipschitz continuity of state functions in some optimal shaping*, Calc. Var. Partial Differential Equations **23** (1) (2005), 13–32.

[18] T. BRIANÇON and J. LAMBOLEY, *Regularity of the optimal shape for the first eigenvalue of the Laplacian with volume and inclusion constraints*, Ann. Inst. H. Poincaré Anal. Non Linéaire **26** (4) (2009), 1149–1163.

[19] D. BUCUR, *Uniform concentration-compactness for Sobolev spaces on variable domains*, Journal of Differential Equations **162** (2000), 427–450.

[20] D. BUCUR, *Minimization of the k-th eigenvalue of the Dirichlet Laplacian*, Arch. Rational Mech. Anal. **206** (3) (2012), 1073–1083.

[21] D. BUCUR and G. BUTTAZZO, "Variational Methods in Shape Optimization Problems", Progress in Nonlinear Differential Equations, Vol. 65, Birkhäuser Verlag, Basel, 2005.

[22] D. BUCUR and G. BUTTAZZO, *On the characterization of the compact embedding of Sobolev spaces*, Calc. Var. PDE **44** (3-4) (2012), 455–475.

[23] D. BUCUR, G. BUTTAZZO and A. HENROT, *Existence results for some optimal partition problems*, Monographs and Studies in Mathematics **8** (1998), 571–579.

[24] D. BUCUR, G. BUTTAZZO and A. HENROT, *Minimization of $\lambda_2(\Omega)$ with a perimeter constraint*, Indiana University Mathematics Journal **58** (6) (2009), 2709–2728.

[25] D. BUCUR, G. BUTTAZZO and B. VELICHKOV, *Spectral optimization problems with internal constraint*, Ann. I. H. Poincaré **30** (3) (2013), 477–495.

[26] D. BUCUR, G. BUTTAZZO and B. VELICHKOV, *Spectral optimization problems for potentials and measures*, SIAM J. Math. Anal. **46** (4) (2014), 2956–2986

[27] D. BUCUR and A. HENROT, *Minimization of the third eigenvalue of the Dirichlet Laplacian*, Proc. Roy. Soc. London Ser. A **456** (2000), 985–996.

[28] D. BUCUR, D. MAZZOLENI, A. PRATELLI and B. VELICHKOV, *Lipschitz regularity of the Eigenfunctions on optimal domains*, Arch. Rational Mech. Anal. (2014), to appear, DOI: 10.1007/s00205-014-0801-6

[29] D. BUCUR and B. VELICHKOV, *Multiphase shape optimization problems*, SIAM J. Control Optim. **52** (6) (2014), 3556-3591.

[30] G. BUTTAZZO, *Spectral optimization problems*, Rev. Mat. Complut. **24** (2) (2011), 277–322.

[31] G. BUTTAZZO, "Semicontinuity, relaxation and integral representation in the calculus of variations", Pitman Research Notes in Mathematics, Vol. 207, Longman, Harlow, 1989.

[32] G. BUTTAZZO and G. DAL MASO, *Shape optimization for Dirichlet problems: relaxed formulation and optimality conditions*, Appl. Math. Optim. **23** (1991), 17–49.

[33] G. BUTTAZZO and G. DAL MASO, *An exlstence result for a class of shape optimization problems*, Arch. Rational Mech. Anal. **122** (1993), 183–195.

[34] G. BUTTAZZO, A. GEROLIN, B. RUFFINI and B. VELICHKOV, *Optimal potentials for Schrödinger operators*, JEP **1** (2014), 71–100.

[35] G. BUTTAZZO, B. RUFFINI and B. VELICHKOV, *Spectral optimization problems for metric graphs*, ESAIM: COCV **20** (1) (2014), 1–22.

[36] G. BUTTAZZO, N. VARCHON and H. SOUBAIRI, *Optimal measures for elliptic problems*, Ann. Mat. Pura Appl. **185** (2) (2006), 207–221.

[37] G. BUTTAZZO and B. VELICHKOV, *Shape optimization problems on metric measure spaces*, J. Funct. Anal. **264** (1) (2013), 1–33.

[38] G. BUTTAZZO and B. VELICHKOV, *Some new problems in spectral optimization*, Banach Center Publications **101** (2014), 19–35.

[39] L. CAFFARELLI and X. CABRÉ, "Fully Nonlinear Elliptic Equations", Amer. Math. Soc., Colloquium publications, Vol. 43, 1995.

[40] L. CAFFARELLI and A. CORDOBA, *An elementary regularity theory of minimal surfaces*, Differential Integral Equations **6** (1993), 1–13.

[41] L. CAFFARELLI, D. JERISON and C. KENIG, *Some new monotonicity theorems with applications to free boundary problems*,The Annals of Mathematics **155** (2) (2002), 369–404.

[42] L. CAFFERELLI and F. H. LIN, *An optimal partition problem for eigenvalues*, J. Sci. Comput. **31** (2007), 5–14.

[43] L.A. CAFFARELLI and S. SALSA, *A geometric approach to free boundary problems*, Graduate Studies in Mathematics **68**, AMS (2005)

[44] J. CHEEGER, *Differentiability of Lipschitz functions on metric measure spaces*, Geom. Funct. Anal. **9** (3) (1999), 428–517.

[45] M. CHIPOT and G. DAL MASO, *Relaxed shape optimization: the case of nonnegative data for the Dirichlet problem*, Adv. Math. Sci. Appl. **1** (1992), 47–81.

[46] D. CIORANESCU and F. MURAT, *Un terme etrange venu dailleurs*, Nonlinear partial differential equations and their applications. Collège de France Seminar **2** (Paris,1979/1980), pp. 98–138, 389–390; Res. Notes in Math. **60**, Pitman, Boston, Mass. London (1982).

[47] M. CONTI, S. TERRACINI and G. VERZINI, *An optimal partition problem related to nonlinear eigenvalues*, J. Funct. Anal. **198** (2003), 160–196.

[48] M. CONTI, S. TERRACINI and G. VERZINI, *A variational problem for the spatial segregation of reaction-diffusion systems*, Indiana Univ. Math. J. **54** (3) (2005), 779–815.

[49] M. CONTI, S. TERRACINI and G. VERZINI, *On a class of optimal partition problems related to the Fucik spectrum and to the monotonicity formula*, Calc. Var. **22** (2005), 45–72.

[50] G. DAL MASO, "An Introduction to Γ-convergence", Birkhäuser, Boston, 1993.

[51] G. DAL MASO and A. GARRONI, *New results on the asymptotic behaviour of Dirichlet problems in perforated domains*, Math. Models Methods Appl. Sci. **3** (1994), 373–407.

[52] G. DAL MASO and U. MOSCO, *Wiener criteria and energy decay for relaxed Dirichlet problems*, Arch.Rational Mech. Anal. **95** (1986), 345–387.

[53] G. DAL MASO and U. MOSCO, *Wiener's criterion and Γ-convergence*, Appl. Math. Optim. **15** (1987), 15–63.

[54] G. DAL MASO and F. MURAT, *Asymptotic behavior and correctors for Dirichlet problems in perforated domains with homogeneous monotone operators*, Ann. Scuola Norm. Sup. Pisa **24** (1997), 239–290.

[55] G. DA PRATO and J. ZABCZYK, "Second Order Partial Differential Equations in Hilbert Spaces", Cambridge University Press, 2002.

[56] E. DAVIES, "Heat Kernels and Spectral Theory", Cambridge University Press, 1989.

[57] E. DAVIES, "Spectral Theory and Differential Operators", Cambridge University Press, 1995.

[58] G. DE PHILIPPIS, J. LAMBOLEY, M. PIERRE and B. VELICHKOV, *TBA*, in preparation.

[59] G. DE PHILIPPIS and B. VELICHKOV, *Existence and regularity of minimizers for some spectral optimization problems with perimeter constraint*, Appl. Math. Optim. **69** (2) (2014), 199–231.

[60] K. J. ENGEL and R. NAGEL, "One-parameter Semigroups for Linear Evolution Equations", Springer, 2000.

[61] L. C. EVANS, "Partial Differential Equations", AMS Press, 2010.

[62] L. EVANS and R. GARIEPY, "Measure Theory and Fine Properties of Functions", Studies in Advanced mathematics, Crc Press, 1991.

[63] S. FRIEDLAND and W. K. HAYMAN, *Eigenvalue inequalities for the Dirichlet problem on spheres and the growth of subharmonic functions*, Comm. Math. Helv. **51** (1979), 133–161.

[64] L. FRIEDLANDER, *Extremal properties of eigenvalues for a metric graph*, Ann. Inst. Fourier **55** (1) (2005), 199–211.

[65] V. FERONE and B. KAWOHL, *Remarks on a Finsler-Laplacian*, Proceedings of the AMS **137** (1) (2007), 247–253.

[66] D. GILBARG and N. S. TRUDINGER, "Elliptic Partial Differential Equations of Second Order", Reprint of the 1998 edition, Classics in Mathematics, Springer-Verlag, Berlin, 2001.

[67] E. GIUSTI, "Minimal Surfaces and Functions of Bounded Variation", Monographs in Mathematics **80**, Birkhäuser, Boston-Basel-Stuttgart, 1984.

[68] P. HAJLASZ and P.KOSKELA, *Sobolev met Poincaré*, Memoirs of the AMS **145** (688) (2001).

[69] B. HELFFER, T. HOFFMANN-OSTENHOF and S. TERRACINI, *Nodal domains and spectral minimal partitions*, Ann. I. H. Poincaré **26** (2009), 101–138.

[70] A. HENROT, "Extremum Problems for Eigenvalues of Elliptic Operators", Frontiers in Mathematics, Birkhäuser Verlag, Basel, 2006.

[71] A. HENROT, *Minimization problems for eigenvalues of the Laplacian*, J. Evol. Equ. **3** (3) (2003), 443–461.

[72] A. HENROT and M. PIERRE, "Variation et optimisation de formes: une analyse géométrique", Springer-Berlag, Berlin, 2005.

[73] L. HÖRMANDER, *Hypoelliptic second-order differential equations*, Acta Math. **119** (1967), 147–171.

[74] H. JIANG, C. LARSEN and L. SILVESTRE, *Full regularity of a free boundary problem with two phases*, Calc. Var. Partial Differential Equations **42** (2011), 301–321.

[75] J. JOST, "Partial Differential Equations", Springer-Verlag New York, 2002.

[76] P. KUCHMENT, *Quantum graphs: an introduction and a brief survey*, In: "Analysis on Graphs and its Applications", AMS Proc. Symp. Pure. Math. **77** (2008), 291–312.

[77] N. LANDAIS, "Problemes de Regularite en Optimisation de Forme", These de doctorat de L'Ecole Normale Superieure de Cachan, 2007.

[78] E.H. LIEB and M. LOSS, "Analysis", Graduate Studies in Mathematics, Vol. 14, American Mathematical Society, Providence, Rhode Island, 1997.

[79] P.L. LIONS, *The concentration-compactness principle in the calculus of variations. The locally compact case, part 1*, Ann. Inst. H. Poincaré Anal. Non Linéaire **1** (2) (1984), 109–145.

[80] F. MAGGI, "Sets of Finite Perimeter and geometric Variational problems: an Introduction to Geometric Measure Theory", Cambridge Studies in Advanced Mathematics **135**, Cambridge University Press, 2012.

[81] D. MAZZOLENI and A. PRATELLI, *Existence of minimizers for spectral problems*, J. Math. Pures Appl. **100** (3) (2013), 433–453.

[82] F. MAZZONE, *A single phase variational problem involving the area of level surfaces*, Comm. Partial Differential Equations **28** (2003), 991–1004.

[83] A. NAGEL, E. STEIN and S. WAINGER, *Balls and metrics defined by vector fields I: Basic properties*, Acta Math. **55** (1985), 103–147.

[84] B. OSTING and C.-Y. KAO, *Minimal convex combinations of three sequential Laplace-Dirichlet eigenvalues*, preprint (2012).

[85] E. OUDET, *Numerical minimization of eigenmodes of a membrane with respect to the domain*, ESAIM Control Optim. Calc. Var. **10** (2004), 315–330.

[86] F. PACARD and P. SICBALDI, *Extremal domains for the first eigenvalue of the Laplace-Beltrami operator*, Annalles de l'Institut Fourier **59** (2) (2009), 515–542.

[87] H. J. SUSSMANN, *Orbits of families of vector fields and integrability of distributions*, Transactions of the AMS **180** (1973), 171–188.

[88] L. SIMON, "Lectures on geometric measure theory", Proceedings of the Centre for Mathematical Analysis, Australian National University, 3. Australian National University, Centre for Mathematical Analysis, Canberra, 1983.

[89] G. TALENTI, *Elliptic equations and rearrangements*, Ann. Scuola Normale Superiore di Pisa **3** (4) (1976), 697–718.

[90] I. TAMANINI, *Boundaries of Caccioppoli sets with Hölder-continuous normal vector*, J. Reine Angew. Math. **334** (1982), 27–39.

[91] I. TAMANINI, "Regularity results for almost minimal hyperurfaces in $\mathbb{R}^n$", Quaderni del Dipartimento di Matematica dell' Università di Lecce, 1984.

[92] B. VELICHKOV, *Lipschitz regularity for quasi-minimizers of the Dirichlet Integral*, in preparation.

[93] B. VELICHKOV, *Note on the monotonicity formula of Caffarelli-Jerison-Kenig*, Rend. Lincei Mat. Appl. **25** (2014), 165–189.

THESES

This series gathers a selection of outstanding Ph.D. theses defended at the Scuola Normale Superiore since 1992.

Published volumes

1. F. COSTANTINO, *Shadows and Branched Shadows of* 3 *and* 4*-Manifolds*, 2005. ISBN 88-7642-154-8
2. S. FRANCAVIGLIA, *Hyperbolicity Equations for Cusped* 3*-Manifolds and Volume-Rigidity of Representations*, 2005. ISBN 88-7642-167-x
3. E. SINIBALDI, *Implicit Preconditioned Numerical Schemes for the Simulation of Three-Dimensional Barotropic Flows*, 2007. ISBN 978-88-7642-310-9
4. F. SANTAMBROGIO, *Variational Problems in Transport Theory with Mass Concentration*, 2007. ISBN 978-88-7642-312-3
5. M. R. BAKHTIARI, *Quantum Gases in Quasi-One-Dimensional Arrays*, 2007. ISBN 978-88-7642-319-2
6. T. SERVI, *On the First-Order Theory of Real Exponentiation*, 2008. ISBN 978-88-7642-325-3
7. D. VITTONE, *Submanifolds in Carnot Groups*, 2008. ISBN 978-88-7642-327-7
8. A. FIGALLI, *Optimal Transportation and Action-Minimizing Measures*, 2008. ISBN 978-88-7642-330-7
9. A. SARACCO, *Extension Problems in Complex and CR-Geometry*, 2008. ISBN 978-88-7642-338-3
10. L. MANCA, *Kolmogorov Operators in Spaces of Continuous Functions and Equations for Measures*, 2008. ISBN 978-88-7642-336-9

11. M. LELLI, *Solution Structure and Solution Dynamics in Chiral Ytterbium(III) Complexes*, 2009. ISBN 978-88-7642-349-9
12. G. CRIPPA, *The Flow Associated to Weakly Differentiable Vector Fields*, 2009. ISBN 978-88-7642-340-6
13. F. CALLEGARO, *Cohomology of Finite and Affine Type Artin Groups over Abelian Representations*, 2009. ISBN 978-88-7642-345-1
14. G. DELLA SALA, *Geometric Properties of Non-compact CR Manifolds*, 2009. ISBN 978-88-7642-348-2
15. P. BOITO, *Structured Matrix Based Methods for Approximate Polynomial GCD*, 2011. ISBN: 978-88-7642-380-2; e-ISBN: 978-88-7642-381-9
16. F. POLONI, *Algorithms for Quadratic Matrix and Vector Equations*, 2011. ISBN: 978-88-7642-383-3; e-ISBN: 978-88-7642-384-0
17. G. DE PHILIPPIS, *Regularity of Optimal Transport Maps and Applications*, 2013. ISBN: 978-88-7642-456-4; e-ISBN: 978-88-7642-458-8
18. G. PETRUCCIANI, *The Search for the Higgs Boson at CMS*, 2013. ISBN: 978-88-7642-481-6; e-ISBN: 978-88-7642-482-3
19. B. VELICHKOV, *Existence and Regularity Results for Some Shape Optimization Problems*, 2015. ISBN: 978-88-7642-526-4; e-ISBN: 978-88-7642-527-1

Volumes published earlier

H. Y. FUJITA, *Equations de Navier-Stokes stochastiques non homogènes et applications*, 1992.

G. GAMBERINI, *The minimal supersymmetric standard model and its phenomenological implications*, 1993. ISBN 978-88-7642-274-4

C. DE FABRITIIS, *Actions of Holomorphic Maps on Spaces of Holomorphic Functions*, 1994. ISBN 978-88-7642-275-1

C. PETRONIO, *Standard Spines and 3-Manifolds*, 1995. ISBN 978-88-7642-256-0

I. DAMIANI, *Untwisted Affine Quantum Algebras: the Highest Coefficient of* $\det H_\eta$ *and the Center at Odd Roots of 1*, 1996. ISBN 978-88-7642-285-0

M. MANETTI, *Degenerations of Algebraic Surfaces and Applications to Moduli Problems*, 1996. ISBN 978-88-7642-277-5

F. CEI, *Search for Neutrinos from Stellar Gravitational Collapse with the MACRO Experiment at Gran Sasso*, 1996. ISBN 978-88-7642-284-3

A. SHLAPUNOV, *Green's Integrals and Their Applications to Elliptic Systems*, 1996. ISBN 978-88-7642-270-6

R. TAURASO, *Periodic Points for Expanding Maps and for Their Extensions*, 1996. ISBN 978-88-7642-271-3

Y. BOZZI, *A study on the activity-dependent expression of neurotrophic factors in the rat visual system*, 1997. ISBN 978-88-7642-272-0

M. L. CHIOFALO, *Screening effects in bipolaron theory and high-temperature superconductivity*, 1997. ISBN 978-88-7642-279-9

D. M. CARLUCCI, *On Spin Glass Theory Beyond Mean Field*, 1998. ISBN 978-88-7642-276-8

G. LENZI, *The MU-calculus and the Hierarchy Problem*, 1998. ISBN 978-88-7642-283-6

R. SCOGNAMILLO, *Principal G-bundles and abelian varieties: the Hitchin system*, 1998. ISBN 978-88-7642-281-2

G. ASCOLI, *Biochemical and spectroscopic characterization of CP20, a protein involved in synaptic plasticity mechanism*, 1998. ISBN 978-88-7642-273-7

F. PISTOLESI, *Evolution from BCS Superconductivity to Bose-Einstein Condensation and Infrared Behavior of the Bosonic Limit*, 1998. ISBN 978-88-7642-282-9

L. PILO, *Chern-Simons Field Theory and Invariants of 3-Manifolds*, 1999. ISBN 978-88-7642-278-2

P. ASCHIERI, *On the Geometry of Inhomogeneous Quantum Groups*, 1999. ISBN 978-88-7642-261-4

S. CONTI, *Ground state properties and excitation spectrum of correlated electron systems*, 1999. ISBN 978-88-7642-269-0

G. GAIFFI, *De Concini-Procesi models of arrangements and symmetric group actions*, 1999. ISBN 978-88-7642-289-8

N. DONATO, *Search for neutrino oscillations in a long baseline experiment at the Chooz nuclear reactors*, 1999. ISBN 978-88-7642-288-1

R. CHIRIVÌ, *LS algebras and Schubert varieties*, 2003. ISBN 978-88-7642-287-4

V. MAGNANI, *Elements of Geometric Measure Theory on Sub-Riemannian Groups*, 2003. ISBN 88-7642-152-1

F. M. ROSSI, *A Study on Nerve Growth Factor (NGF) Receptor Expression in the Rat Visual Cortex: Possible Sites and Mechanisms of NGF Action in Cortical Plasticity*, 2004. ISBN 978-88-7642-280-5

G. PINTACUDA, *NMR and NIR-CD of Lanthanide Complexes*, 2004. ISBN 88-7642-143-2

Fotocomposizione "CompoMat" Loc. Braccone, 02040 Configni (RI) Italy
Finito di stampare nel mese di febbraio 2015
dalla CSR srl, Via di Pietralata, 157, 00158 Roma

Printed in the United States
By Bookmasters